Advances in Signal Processing: Reviews

Book Series, Volume 2

Sergey Y. Yurish
Editor

Advances in Signal Processing: Reviews

Book Series, Volume 2

 International Frequency Sensor Association Publishing

Sergey Y. Yurish
Editor

Advances in Signal Processing: Reviews
Book Series, Volume 2

Published by International Frequency Sensor Association (IFSA) Publishing, S. L., 2021
E-mail (for print book orders and customer service enquires): ifsa.books@sensorsportal.com

Visit our Home Page on http://www.sensorsportal.com

ISBN: 978-84-09-28830-4
e-ISBN: 978-84-09-28831-1
BN-20210316-XX
BIC: UYS

Acknowledgments

As Editor I would like to express my undying gratitude to all authors, editorial staff, reviewers and others who actively participated in this book. We want also to express our gratitude to all their families, friends and colleagues for their help and understanding.

Contents

Contributors

Majd Abazid
SAMOVAR, Télécom SudParis, Institut Polytechnique de Paris, 9 rue Charles Fourier 91011 EVRY Cedex, France

Dorel Aiordăchioaie
'Dunarea de Jos' University of Galati, Romania

Hadj Batatia
Computer Science Research Institute of Toulouse (IRIT), National Polytechnic Institute of Toulouse (INP), University of Toulouse, Toulouse, Cedex 7 B.P. 7122-31071, France

MACS School, Heriot-Watt University, Dubai Campus, Dubai Knowledge Park, Blocks 5 & 14, Dubai 38103, UAE.
E-mail: h.batatia@hw.ac.uk

Jerome Boudy
SAMOVAR, Télécom SudParis, Institut Polytechnique de Paris, 9 rue Charles Fourier 91011 EVRY Cedex, France

Qian Chang
School of Ophthalmology and Optometry, Eye Hospital, Wenzhou Medical University, Wenzhou 325027, China

Pascal Chargé
IETR Laboratory, UMR 6164, University of Nantes, Nantes 44300, France, Sino-French Research Center in Information and Communication, Nantes 44300, France

Hao Chen
School of Ophthalmology and Optometry, Eye Hospital, Wenzhou Medical University, Wenzhou 325027, China

Anisia Culea-Florescu
'Dunarea de Jos' University of Galati, Romania

Carlos D'Giano
Epilepsy and Telemetry Integral Center, Foundation for the Fight against Pediatric Neurological Disease (FLENI), Montañeses 2325, Buenos Aires C1428AQK, Argentina

Kylliann De Santiago
SAMOVAR, Télécom SudParis, Institut Polytechnique de Paris, 9 rue Charles Fourier 91011 EVRY Cedex, France

Yuehua Ding

National Engineering Technology Research Center for Mobile Ultrasonic Detection, School of Electronic and Information Engineering, South China University of Technology, Guangzhou, 510641, China
E-mail: E-mail:eeyhding@scut.edu.cn

Bernadette Dorizzi

SAMOVAR, Télécom SudParis, Institut Polytechnique de Paris, 9 rue Charles Fourier 91011 EVRY Cedex, France

Victor Golikov

UNACAR, Ciudad del Carmen, Mexico

Rhythm Grover

Department of Mathematics and Statistics, Indian Institute of Technology Kanpur, Pin 208016, India
E-mail: rhythm@iitk.ac.in

Nesma Houmani

SAMOVAR, Télécom SudParis, Institut Polytechnique de Paris, 9 rue Charles Fourier 91011 EVRY Cedex, France

Kiyoka Kinugawa-Bourron

AP-HP, DHU FAST, GH Pitié-Salpêtrière-Charles Foix, Functional Exploration Unit of older patients, 94205, Ivry-sur-Seine, France
Sorbonne Université, CNRS, UMR 8256, Biological Adaptation and Aging, 75005, Paris, France

Andreas König

Institute of Integrated Sensor Systems, TU Kaiserslautern, Kaiserslautern, Germany

Debasis Kundu

Department of Mathematics and Statistics, Indian Institute of Technology Kanpur, Pin 208016, India
E-mail: kundu@iitk.ac.in

Jie Li

National Engineering Technology Research Center for Mobile Ultrasonic Detection, School of Electronic and Information Engineering, South China University of Technology, Guangzhou, 510641, China

Nanxi Li

School of Information Technology in Education, South China Normal University, Guangzhou 510631, China

Eric Lindahl

Volvo Group Truck Operations (GTO), 904 34 Umeå, Sweden

Jean Mariani

AP-HP, DHU FAST, GH Pitié-Salpêtrière-Charles Foix, Functional Exploration Unit of older patients, 94205, Ivry-sur-Seine, France

Sorbonne Université, CNRS, UMR 8256, Biological Adaptation and Aging, 75005, Paris, France

Swagata Nandi
Division of Theoretical Statistics and Mathematics, Indian Statistical Institute, Delhi Center, 7 SJS Sansanwal Marg, New Delhi, Pin 110016, India,
E-mail: nandi@isid.ac.in

Mehdi Nasri
Department of Electrical Engineering, Khomeinishahr branch, Islamic Azad University, Isfahan, Iran
E-mail: nasri_me@iaukhsh.ac.ir

Theodor D. Popescu
'Dunarea de Jos' University of Galati, Romania

Natalya Pya Arnqvist
Department of Mathematics and Mathematical Statistics, Umeå University, 901 87 Umeå, Sweden

Antonio Quintero-Rincón
Epilepsy and Telemetry Integral Center, Foundation for the Fight against Pediatric Neurological Disease (FLENI), Montañeses 2325, Buenos Aires C1428AQK, Argentina

Department of Electronics, Catholic University of Argentina (UCA), Av. Alicia Moreau de Justo 1300, Buenos Aires C1107AAZ, Argentina.
E-mail: antonioquintero@uca.edu.ar

Oleg Samovarov
UNACAR, Ciudad del Carmen, Mexico

Rafael Sanchez Lara
UNACAR, Ciudad del Carmen, Mexico

Kajetana Marta Snopek
Warsaw University of Technology, Faculty of Electronics and Information Technology, Nowowiejska 15/19, Warsaw, Poland
E-mail: kajetana.snopek@pw.edu.pl

Vyacheslav Tuzlukov
Department of Technical Maintenance of Aviation and Radio Electronic Equipment, Belarusian State Aviation Academy, 77, Uborevicha Str., 220096 Minsk, Belarus

Mahtab Vaezi
Department of Biomedical Engineering, Khomeinishahr branch, Islamic Azad University, Isfahn, Iran
mahtab.vaezi@iaukhsh.ac.ir

Lei Wang
School of Ophthalmology and Optometry, Eye Hospital, Wenzhou Medical University, Wenzhou 325027, China

Yide Wang,
IETR Laboratory, UMR 6164, University of Nantes, Nantes 44300, France, Sino-French Research Center in Information and Communication, Nantes 44300, France

Jun Yu
Department of Mathematics and Mathematical Statistics, Umeå University, 901 87 Umeå, Sweden

Evgeniy Zhilyakov
UNACAR, Ciudad del Carmen, Mexico

Preface

After successful publication of the first volume of '*Advances in Signal Processing Reviews*' Book Series in 2018 and positive feedback from our authors and readers, we have decided to publish the 2nd volume of this Book Series in 2021.

This volume contains 12 chapters written by 36 authors from 12 countries: Argentina, Belarus, China, France, Germany, India, Iran, Mexico, Poland, Romania, Sweden and UAE. But it is not a simple set of reviews. Each of chapter contains the extended state-of-the art followed by new, obtained by authors results, unpublished before.

Chapter 1 describes the theory of complex and hypercomplex multidimensional analytic signals. The signal-domain and frequency-domain definitions of complex and hypercomplex analytic signals have been presented, as well as some mutual relations between complex and hypercomplex spectra and polar components.

Chapter 2 investigates the performance of the quaternary the direct-sequence spread-spectrum multiple-access wireless communication system based on the generalized approach to signal processing in noise with complex signature sequences in the presence of the flat Rayleigh fading.

Chapter 3 introduces the recent work in the exploitation of communication signals' properties including the non-circularity, the cyclostationarity, the conjugate symmetry in array signal processing.

Chapter 4 overviews of different 1-D and 2-D chirp and related models, which have received a considerable amount of attention in recent years in the statistical signal processing literature. The main aim is to introduce different chirp and some other models, which are being extensively used in practice.

Chapter 5 describes a new statistical algorithms for 3D signal processing for detection of floating objects on an agitated sea surface and effective and adaptive detector of multi-pixel objects on its basis. The proposed detector offers two principal benefits. The first is that the proposed detector is able to detect no-contrast objects on the sea surface. The second advantage is that the proposed detector is slightly sensitive to background covariance matrix change and size of the secondary dataset.

Chapter 6 describes the nature of sleep and its various stages, and then, discusses the importance of automatic review of sleep stages, and due to this importance, various databases and methods for automatic classification of sleep stages are reviewed in this chapter.

Chapter 7 presents a statistical learning approach for defect detection on painted cab body surfaces, covering image acquisition, feature extraction and defect detection and classification. The main objective in developing this new approach is, in addition to providing accurate and reliable classification, to furnish uncertainty estimation in the classification results.

Chapter 8 presents some solutions to the problem of feature selection for change detection purposes, with application to change detection of (incipient) faults in bearings. The features are extracted from time-frequency images or designed with auxiliary transforms. The proposed and implemented solutions have potential on other processes, where measured or estimated signals are available, e.g. in medicine.

Chapter 9 summarizes a number of information entropy, which is widely applied in image segmentation and registration methods.

Chapter 10 describes the electroencephalography signal analysis with a statistical entropy-based measure for Alzheimer's disease detection.

Chapter 11 demonstrates the potential of artefacts detection approach in electro-encephalography, using the Hampel filter to correct different types of artefacts. Also, a complete state-of-the-art is introduced along with a recommended bibliography to research these topics.

Chapter 12 reports the in-hive multi-gas-sensor extension (BeE-Nose) to the IndusBee 4.0 system for hive air quality monitoring and varroa infestation level estimation.

Every chapter in '*Advances in Signal Processing Reviews*' Vol. 2 is independent and self-contained. The book can be useful for post-graduate students, researchers, engineers and scientist working signal processing area.

Dr. Sergey Y. Yurish

Editor
IFSA Publishing   *Barcelona, Spain*

Chapter 1
Complex and Hypercomplex Multidimensional Analytic Signals – The Theory and Chosen Properties

Kajetana Marta Snopek

1.1. Introduction

Within the last thirty years, the theory of complex and hypercomplex signals has been intensively developing and finding new interesting applications in various fields. Here we will deal with signals defined on the basis on the Cayley-Dickson algebras. The history of complex numbers began in 1748 when L. Euler (1707-1783) published his very known formula $\exp(i\varphi) = \cos\varphi + i\sin\varphi$. Only 150 years later, A. E. Kennelly (1861-1939) and C. P. Steinmetz (1865-1923) introduced a complex harmonic signal into the theory of electrical circuits and D. Gabor (1900-1979) defined his 1-D analytic signal [1]. This concept found numerous applications: in single-sideband modulation theory [2-4], in envelope detection [5], in estimation of instantaneous parameters of nonstationary signals [1, 3, 6] and in the Wigner-Ville distribution [7].

The definition of the multidimensional *analytic signal* compatible with the Gabor's approach was proposed by S. L. Hahn in 1992 [8]. In the same year, T. A. Ell defined the quaternion Fourier transformation [9] that enabled new interesting applications in analysis of linear time-invariant partial-differential systems [10]. In 1999, T. Bülow introduced the quaternion analytic signal [11] that has been applied later in analysis of 2-D seismic signals [12]. Further research in the quaternion field resulted in applications in color image watermarking and filtering [13-15]. The unified theory of complex and hypercomplex analytic signals was presented in 2011 [16]. The concept of the Octonion Fourier Transform was proposed and further developed and discussed in [17-21].

The topics presented in this Chapter are, to a large extent, the result of the author's personal and cooperative theoretical investigations in the domain of multidimensional

Kajetana Marta Snopek
Warsaw University of Technology, Faculty of Electronics and Information Technology, Warsaw, Poland

complex and hypercomplex analytic signals performed in the last twenty years. Having in mind present and future practical applications, the scope is focused on complex and hypercomplex analytic signals based on the Cayley-Dickson algebras of complex numbers, quaternions and octonions. It comprises a bridge between the complex and hypercomplex approach. Both are presented in parallel, showing the differences in formalisms and their consequences, as well as the equivalence of some presented formulas.

The organization of the chapter is as follows. In Section 1.2, we define n-D analytic signals in the complex domain. Their spectra are limited to a single-orthant of the n-D frequency space. This notion is introduced and visualized for $n = 2$ and 3. Then, the single-orthant operator as the extension of the 1-D step function is presented. Two approaches are presented to define complex analytic signals. First, we can express them as an n-fold convolution of an n-D real signal with the n-D complex delta distribution. Then using the convolution-to-multiplication property of the n-D Fourier transform, we present the alternative approach via the frequency domain. In both cases, we will put our attention on 2-D and 3-D analytic signals. The Section 1.3 is devoted to analytic signals defined in the hypercomplex domain. This part starts with the short introduction to the algebras of quaternions and octonions, i.e., Cayley-Dickson algebras of order 4 and 8 respectively. Then, we define hypercomplex analytic signals in the signal-domain as the n-fold convolution of an n-D real signal with the n-D hypercomplex delta distribution. Similarly as in the complex case, we apply the convolution-to-multiplication theorem valid for the hypercomplex Fourier transform and define hypercomplex analytic signals in the frequency domain. The formulas relating 2-D and 3-D Fourier transforms (FT) with Quaternion and Octonion FTs are recalled. The last Section 1.4 is devoted to the polar representation of complex and hypercomplex analytic signals. The whole chapter is illustrated with examples of chosen 2-D signals.

1.2. Analytic Signals in the Complex Domain

The n-D complex signal is defined as

$$\psi_c(x) = \mathrm{Re}\{\psi_c(x)\} + \mathrm{Im}\{\psi_c(x)\}\cdot e_1, \tag{1.1}$$

where e_1 is the imaginary unit (usually denoted by mathematicians with i or j) and $x = (x_1, x_2, \ldots, x_n)$. $\mathrm{Re}\{\psi_c(x)\}$ is the *real part* of $\psi_c(x)$ and $\mathrm{Im}\{\psi_c(x)\}$ - its *imaginary* part. The conjugate of (1.1) is

$$\psi_c^*(x) = \mathrm{Re}\{\psi_c(x)\} - \mathrm{Im}\{\psi_c(x)\}\cdot e_1, \tag{1.2}$$

and its norm

$$\|\psi_c(x)\| = \sqrt{\left(\mathrm{Re}\{\psi_c(x)\}\right)^2 + \left(\mathrm{Im}\{\psi_c(x)\}\right)^2} \tag{1.3}$$

All operations on complex signals are similar to operations on complex numbers. A special subclass of complex-valued signals are *n*-D *complex signals with single-orthant spectra* defined by Hahn in [8] as boundary distributions of multidimensional analytic functions, as shown in [22]. Their spectra a limited to a single orthant of the *n*-D frequency domain. Let us now explain the notion of an orthant.

1.2.1. The Notion of an *n-D Orthant*

Let us start from the simplest case of the 1-D frequency domain. It is evident that it is divided into two *half-axes* (two *orthants*): $f > 0$ and $f < 0$, labelled with 1 and 2 respectively (Fig. 1.1).

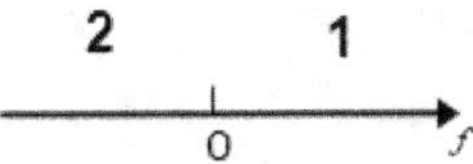

Fig. 1.1. Labelling of orthants in the 1-D frequency domain.

In the 2-D frequency plane (Fig. 1.2), we distinguish four *quadrants* (in other words: four *orthants*) labelled with 1, 2, 3 and 4 respectively. Note that their numbering differs from that commonly used by mathematicians. We apply the convention introduced by Hahn in [8] and used in all his further works concerning multidimensional analytic signals with single-orthant spectra. The advantage of such a labelling is that all higher dimensional orthants in the half-space $f_1 > 0$ are always marked with odd numbers. It can be noticed that the numeration of quadrants corresponds to the binary notation (see Table 1.1) written in reversed order. It is evident that the 2-D frequency domain is composed of two half-planes $f_1 > 0$ and $f_1 < 0$ each formed of two quadrants: (1,3) and (2,4) respectively.

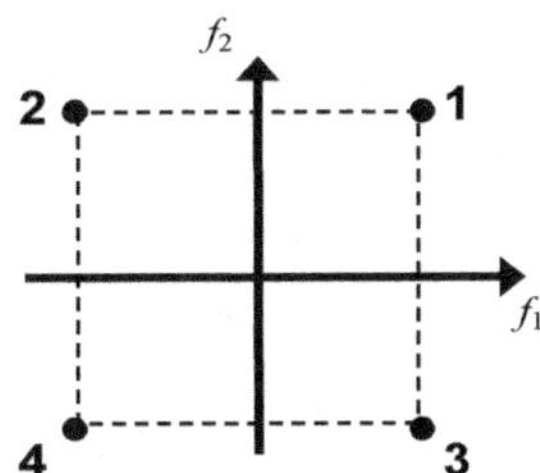

Fig. 1.2. Labelling of orthants in the 2-D frequency domain.

At last, the 3-D frequency space can be divided into eight *octants* (in other words: eight *orthants*) shown in Fig. 1.3 and in Table 1.2. We see that octants 1, 3, 5 and 7 (odd-indexed) are included in the half-space $f_1 > 0$. We notice also that following pairs of octants: (1,5) and (3,7) have the common plane $f_3 = 0$, so their sums can be considered as two *space-quadrants* in the half-space $f_1 > 0$. Analogously, in the half-space $f_1 < 0$, there are also two space-quadrants formed from octants 2, 4, 6 and 8. Thus, the 3-D frequency space can be decomposed into four space-quadrants and two half-spaces: $f_1 > 0$, $f_1 < 0$. In this Chapter, we will deal only with the space-quadrants (1,5) and (3,7) in the half-space $f_1 > 0$. Concluding, the n-D orthant is a strictly determined part of the n-D space: a half-axis in 1-D, a quadrant in 2-D, an octant in 3-D and so on.

Table 1.1. Numeration of orthants in 2-D.

No.	Binary numeration of orthants	Reversed order	Signs of f_i $0 \rightarrow$ "+" $1\rightarrow$ "–"	
			f_1	f_2
1	00	00	+	+
2	01	10	–	+
3	10	01	+	–
4	11	11	–	–

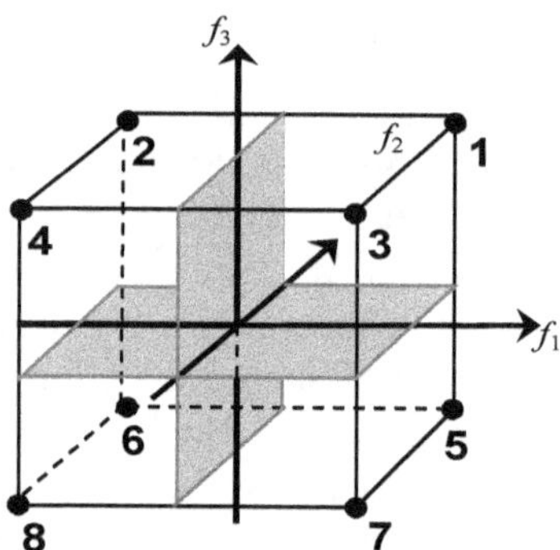

Fig. 1.3. Labelling of orthants in the 3-D frequency space.

Table 1.2. Numeration of orthants in 3-D.

No.	Binary numeration of orthants	Reversed order	Signs of f_i $0 \rightarrow$ "+" $1\rightarrow$ "–"		
			f_1	f_2	f_3
1	000	000	+	+	+
2	001	100	–	+	+
3	010	010	+	–	+
4	011	110	–	–	+
5	100	001	+	+	–
6	101	101	–	+	–
7	110	011	+	–	–
8	111	111	–	–	–

1.2.2. The Notion of a Single-orthant Operator

Here in this chapter, we will deal with signals with single-orthant spectra. A single-orthant spectrum can be written as a product $U(f) \cdot \mathbf{1}(f)$, $f = (f_1, f_2, ..., f_n)$ where $U(f)$ is the

n-D spectrum of the n-D real signal $u(\boldsymbol{x})$ and the operator $\mathbf{1}(\boldsymbol{f})$ is the n-D extension of the 1-D step function known from the system theory. It will be called the *single-orthant operator* defined as

$$\mathbf{1}(\boldsymbol{f}) = \frac{1}{2^n} \prod_{i=1}^{n} \left(1 + \operatorname{sgn} f_i\right), \tag{1.4}$$

where the sign of $\operatorname{sgn} f_i$ depends on the sign of f_i in a given orthant of the n-D frequency domain (see Tables 1.1 and 1.2). In 1-D for $f > 0$, we have

$$\mathbf{1}(f) = \frac{1}{2}\left(1 + \operatorname{sgn} f\right) \tag{1.5}$$

In 2-D for $f_1 > 0$, we define two single-quadrant operators given below

$$\mathbf{1}(f_1, f_2) = \frac{1}{4}\left(1 + \operatorname{sgn} f_1\right)\left(1 + \operatorname{sgn} f_2\right), \text{ for quadrant No. } \mathbf{1}, \tag{1.6}$$

$$\mathbf{1}(f_1, -f_2) = \frac{1}{4}\left(1 + \operatorname{sgn} f_1\right)\left(1 - \operatorname{sgn} f_2\right), \text{ for quadrant No. } 3 \tag{1.7}$$

Finally, basing on the signs of f_i in octants 1, 3, 5 and 7 from Table 1.2, we define four following single-octant operators in 3-D

$$\mathbf{1}(f_1, f_2, f_3) = \frac{1}{8}\left(1 + \operatorname{sgn} f_1\right)\left(1 + \operatorname{sgn} f_2\right)\left(1 + \operatorname{sgn} f_3\right), \text{ for octant No. } \mathbf{1}, \tag{1.8}$$

$$\mathbf{1}(f_1, -f_2, f_3) = \frac{1}{8}\left(1 + \operatorname{sgn} f_1\right)\left(1 - \operatorname{sgn} f_2\right)\left(1 + \operatorname{sgn} f_3\right), \text{ for octant No. } 3, \tag{1.9}$$

$$\mathbf{1}(f_1, f_2, -f_3) = \frac{1}{8}\left(1 + \operatorname{sgn} f_1\right)\left(1 + \operatorname{sgn} f_2\right)\left(1 - \operatorname{sgn} f_3\right), \text{ for octant No. } \mathbf{5}, \tag{1.10}$$

$$\mathbf{1}(f_1, -f_2, -f_3) = \frac{1}{8}\left(1 + \operatorname{sgn} f_1\right)\left(1 - \operatorname{sgn} f_2\right)\left(1 - \operatorname{sgn} f_3\right), \text{ for octant No. } 7 \tag{1.11}$$

We will apply these operators in the Section 1.2.4 of this chapter.

1.2.3. The Signal-domain Definition of the n-D Complex Analytic Signal

In this section, we will define the n-D analytic complex signals in the $\boldsymbol{x}$ – domain by means of the *complex delta distribution* introduced by Hahn in 1992 [8] and further developed in [23]. In order to unify the notation of n-D analytic signals, let us define the n-D identity and Hilbert operators of order 1, 2 and 3. The identity operator $\mathbf{I}$ is defined as

$$\mathbf{I}\{u(\boldsymbol{x})\} = u(\boldsymbol{x}) \underbrace{* \ldots *}_{n-\text{fold}} \delta(\boldsymbol{x}) = u(\boldsymbol{x}), \tag{1.12}$$

where $\underset{n-\text{fold}}{*...*}$ denotes the n-fold convolution in the x-domain and $\delta(x)$ is the n-D Dirac delta distribution. We will also use the Hilbert operator $\mathbf{H}$ defined as

$$\mathbf{H}\{u(x)\} = u(x)\underset{n-\text{fold}}{*...*}\frac{1}{\pi^n\prod_{k=1}^{n}x_k} = v(x),\qquad(1.13)$$

as well as the 1st–order partial Hilbert operator

$$\mathbf{H}_k\{u(x)\} = u(x)*\frac{1}{\pi x_k} = v_k(x),\qquad(1.14)$$

the 2nd–order partial Hilbert operator

$$\mathbf{H}_{ij}\{u(x)\} = u(x)**\frac{1}{\pi^2 x_i x_j} = v_{ij}(x),\ \ i<j,\qquad(1.15)$$

and finally, the 3rd– order partial Hilbert operator

$$\mathbf{H}_{ijk}\{u(x)\} = u(x)***\frac{1}{\pi^3 x_i x_j x_k} = v_{ijk}(x),\ \ i<j<k,\qquad(1.16)$$

Using the operators (1.12)-(1.13), we can write the 1-D complex analytic signal (1.1) as

$$\psi(t) = \mathbf{I}\{u(t)\} + \mathbf{H}\{u(t)\}e_1,\qquad(1.17)$$

where we replaced x_1 with t and omitted the subscript c. We have

$$\psi(t) = u(t)*\delta(t) + \left(u(t)*\frac{1}{\pi t}\right)e_1 = u(t)*\left[\delta(t)+\frac{1}{\pi t}e_1\right],\qquad(1.18)$$

that is equal to

$$\psi(t) = u(t) + v(t)e_1,\qquad(1.19)$$

i.e., a very known 1-D Gabor's analytic signal [1]. Its imaginary part $v(t)$ is the Hilbert transform of the real signal $u(t)$. The real signal $u(t)$ can be calculated from the equation

$$u(t) = \frac{\psi(t)+\psi^*(t)}{2},\qquad(1.20)$$

where $\psi^*(t)$ is the conjugate of (1.19). From (1.18), we see that the 1-D complex analytic signal is also a convolution of a real signal $u(t)$ with the *1-D complex delta distribution* [24, 25] denoted as $\Psi_c^\delta(t)$, that is

$$\Psi_{c}^{\delta}(t) = \delta(t) + \frac{1}{\pi t}e_{1} \tag{1.21}$$

Let us note that the 1-D complex delta distribution is also an analytic signal (in the Gabor's sense). Its imaginary part $1/\pi t$ is the 1-D Hilbert transform of the real part $\delta(t)$. We can express the 1-D analytic signal in the equivalent form

$$\psi(t) = u(t) * \Psi_{c}^{\delta}(t) \tag{1.22}$$

This result can easily be extended for $n > 1$ using the *n-D complex delta distribution* defined in [8]. The detailed study of properties of the complex delta distribution for $n = 1$ and $n = 2$ can also be found in [3] and [23]. The *signal-domain definition* of the *n*-D complex analytic signal is then

$$\psi(x) = u(x) \underset{n-\text{fold}}{* \ldots *} \Psi_{c}^{\delta}(x), \tag{1.23}$$

where

$$\Psi_{c}^{\delta}(x) = \prod_{k=1}^{n}\left[\delta(x_{k}) \pm \frac{1}{\pi x_{k}}e_{1}\right] \tag{1.24}$$

is the *n-D complex delta distribution*. If all signs are positive, then (1.24) defines the complex delta distribution with the Fourier spectrum in the 1st orthant of the *n*-D frequency space. An appropriate change of a sign of e_1 in multiplication factors of (1.24) corresponds to the spectral support in other orthants (see e.g. Tables 1.1 and 1.2). Let us study in detail the cases $n = 2$ and 3.

1.2.3.1. The Signal-domain Definition of the 2-D Complex Analytic Signal

In 2-D, we can derive four forms of 2-D complex delta distributions defining four 2-D complex analytic signals with spectra in corresponding quadrants of the 2-D frequency-plane. Applying the separability property of the *n*-D delta distribution: $\delta(x_{i})\delta(x_{j}) = \delta(x_{i}, x_{j})$, $i \neq j$, we can write for the half-plane $f_1 > 0$

$$\Psi_{c}^{\delta}(x) = \delta(x_{1}, x_{2}) - \frac{1}{\pi^{2}x_{1}x_{2}} + \left(\frac{\delta(x_{2})}{\pi x_{1}} + \frac{\delta(x_{1})}{\pi x_{2}}\right)e_{1}, \text{ for quadrant No. 1,} \tag{1.25}$$

$$\Psi_{c}^{\delta}(x) = \delta(x_{1}, x_{2}) + \frac{1}{\pi^{2}x_{1}x_{2}} + \left(\frac{\delta(x_{2})}{\pi x_{1}} - \frac{\delta(x_{1})}{\pi x_{2}}\right)e_{1}, \text{ for quadrant No. 3} \tag{1.26}$$

Inserting (1.25) into (1.23) we obtain the signal-domain definition of the 2-D analytic signal with spectrum limited to the 1st quadrant of the frequency plane (the superscript indicates the number of the quadrant):

$$\psi_c^1(x_1,x_2)=u(x_1,x_2)**\left[\delta(x_1,x_2)-\frac{1}{\pi^2 x_1 x_2}+\left(\frac{\delta(x_2)}{\pi x_1}+\frac{\delta(x_1)}{\pi x_2}\right)e_1\right]=$$
$$=u(x_1,x_2)-u(x_1,x_2)**\frac{1}{\pi^2 x_1 x_2}+\left[u(x_1,x_2)*\frac{1}{\pi x_1}+u(x_1,x_2)*\frac{1}{\pi x_2}\right]e_1 \qquad (1.27)$$

In the above formula (1.27), we recognize the 2-D total (v) and partial (v_1 and v_2) Hilbert transforms of $u(x_1, x_2)$. Finally, we have

$$\psi_c^1(x_1,x_2)=u-v+(v_1+v_2)e_1, \qquad (1.28)$$

where u, v, v_1 and v_2 are all functions of (x_1,x_2). Using operators given by (1.12)-(1.14), we can express (1.28) in an equivalent form

$$\psi_c^1(x_1,x_2)=\mathbf{I}\{u\}-\mathbf{H}\{u\}+\left(\mathbf{H_1}\{u\}+\mathbf{H_2}\{u\}\right)e_1 \qquad (1.29)$$

Let us repeat the same procedure for the quadrant No. 3. The complex analytic signal ψ_3 with the spectrum support in the 3rd quadrant is defined as

$$\psi_c^3(x_1,x_2)=u(x_1,x_2)**\left[\delta(x_1,x_2)+\frac{1}{\pi^2 x_1 x_2}+\left(\frac{\delta(x_2)}{\pi x_1}-\frac{\delta(x_1)}{\pi x_2}\right)e_1\right]=$$
$$=u(x_1,x_2)+u(x_1,x_2)**\frac{1}{\pi^2 x_1 x_2}+\left[u(x_1,x_2)*\frac{1}{\pi x_1}-u(x_1,x_2)*\frac{1}{\pi x_2}\right]e_1, \qquad (1.30)$$

that is equal to

$$\psi_c^3(x_1,x_2)=u+v+(v_1-v_2)e_1 \qquad (1.31)$$

It is very important to note that the following conjugate relations are fulfilled: $\psi_c^2=\left(\psi_c^3\right)^*$ and $\psi_c^4=\left(\psi_c^1\right)^*$ where ψ_c^2 and ψ_c^4 denote 2-D analytic signals with spectra limited to quadrants 2 and 4 respectively. So, the 2-D real signal $u(x_1,x_2)$ can be represented by two analytic signals ψ_c^1 and ψ_c^3 with spectra in quadrants 1 and 3 of the half-plane $f_1>0$. It can be noticed that if we calculate a sum of signals (1.28) and (1.31) (divided by 2) we obtain

$$\frac{\psi_c^1 + \psi_c^3}{2} = u + v_1 e_1 \tag{1.32}$$

Then a 2-D real signal $u(x_1,x_2)$ can be treated as the real part of a weighted sum of two analytic signals ψ_c^1 and ψ_c^3 with spectrum limited to the half-plane $f_1 > 0$:

$$u(x_1,x_2) = \mathrm{Re}\left\{\frac{\psi_c^1 + \psi_c^3}{2}\right\} \tag{1.33}$$

We can conclude that a full information about the 2-D real signal $u(x_1,x_2)$ is included in the half-plane $f_1 > 0$. Moreover, the signal $u(x_1,x_2)$ is given by the formula:

$$u(x_1,x_2) = \frac{\psi_c^1 + \psi_c^4 + \psi_c^3 + \psi_c^2}{4} = \frac{\psi_c^1 + \left(\psi_c^1\right)^* + \psi_c^3 + \left(\psi_c^3\right)^*}{4} \tag{1.34}$$

Let us illustrate this part of the theory with the example of the 2-D Cauchy separable signal.

Example 1

The 2-D Cauchy separable signal is given by

$$u(x_1,x_2) = \frac{ab}{\left(a^2 + (x_1 - c)^2\right)\left(b^2 + (x_2 - d)^2\right)} \tag{1.35}$$

Its total and partial Hilbert transforms are expressed as

$$v(x_1,x_2) = (x_1 - c)(x_2 - d)\Big/\left(a^2 + (x_1 - c)^2\right)\left(b^2 + (x_2 - d)^2\right), \tag{1.36}$$

$$v_1(x_1,x_2) = b(x_1 - c)\Big/\left(a^2 + (x_1 - c)^2\right)\left(b^2 + (x_2 - d)^2\right), \tag{1.37}$$

$$v_2(x_1,x_2) = a(x_2 - d)\Big/\left(a^2 + (x_1 - c)^2\right)\left(b^2 + (x_2 - d)^2\right) \tag{1.38}$$

The Figs. 1.4(a)-(d) show contour plots of u, v, v_1 and v_2 given by (1.35)-(1.38) for $a = 1$, $b = 2$, $c = d = 0$. We observe the differences in orientation of the partial Hilbert transforms on the (x_1,x_2)–plane and odd parity of the total Hilbert transform. Next Figs. 1.5(a)-(d) display contour lines of real (top) and imaginary (bottom) parts of 2-D analytic signals ψ_c^1 and ψ_c^3 given by (1.28) and (1.31). As we can notice, they are mirror images of each other. It is a consequence of separability of the Cauchy signal.

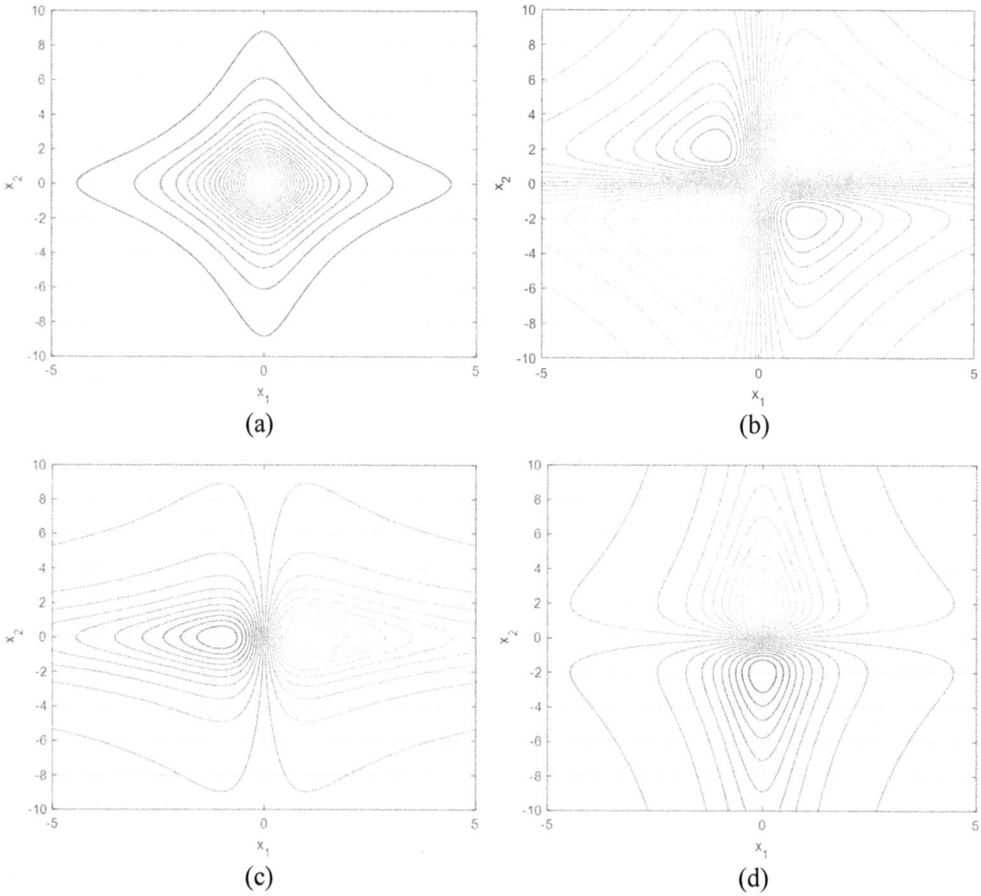

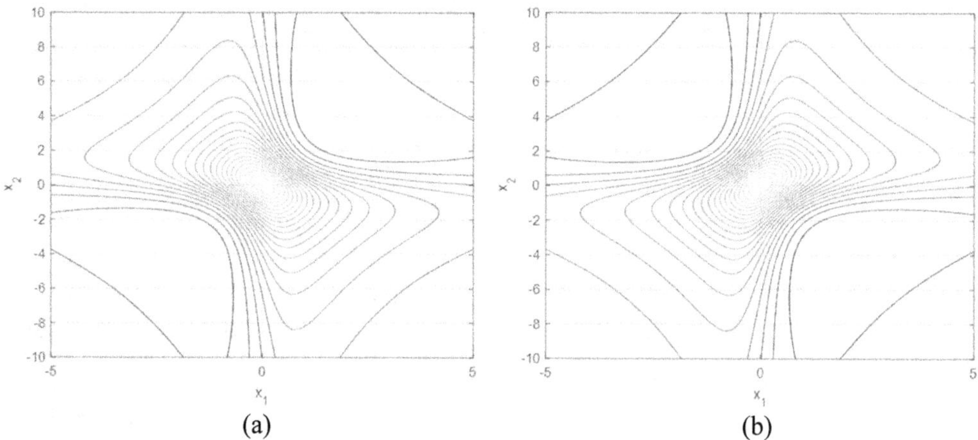

Fig. 1.4. Contour plots of (a) $u(x_1, x_2)$, (b) $v(x_1, x_2)$, (c) $v_1(x_1, x_2)$, (d) $v_2(x_1, x_2)$ of the 2-D separable Cauchy signal (1.35) for $a = 1$, $b = 2$, $c = d = 0$.

Fig. 1.5 (a, b). Contour plots of (a) u-v, (b) u+v of the 2-D separable Cauchy signal (1.35) for $a = 1$, $b = 2$.

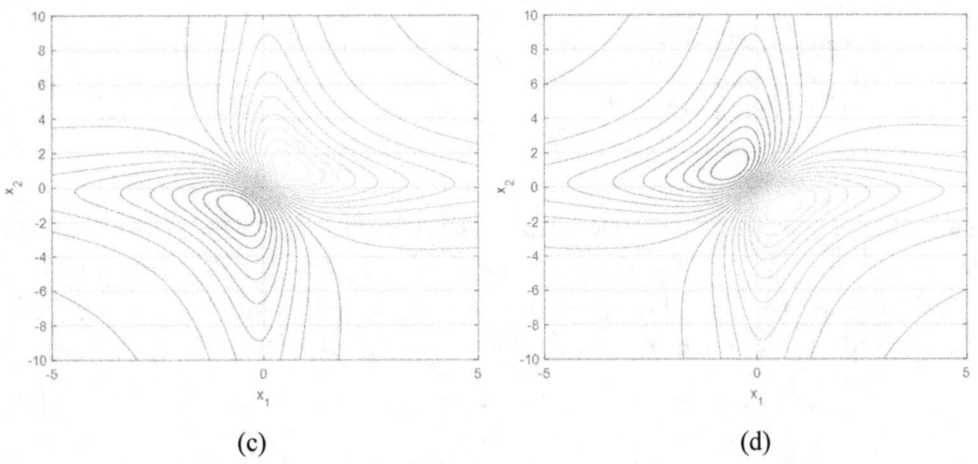

Fig. 1.5 (c, d). Contour plots of (c) v_1+v_2, (d) v_1+v_2 of the 2-D separable Cauchy signal (1.35) for $c = d = 0$.

1.2.3.2. The Signal-domain Definition of the 3-D Complex Analytic Signal

As the 3-D space is divided into eight octants, we can define eight 3-D analytic signals with spectral support in the corresponding octants. In the half-space $f_1 > 0$ (see Fig. 1.2) we have 4 octants numbered with 1, 3, 5 and 7. The 3-D complex delta distributions (1.24) are given by

$$\Psi_c^\delta(x) = \delta(x) - \frac{\delta(x_3)}{\pi^2 x_1 x_2} - \frac{\delta(x_2)}{\pi^2 x_1 x_3} - \frac{\delta(x_1)}{\pi^2 x_2 x_3} + $$
$$+ \left(\frac{\delta(x_2,x_3)}{\pi x_1} + \frac{\delta(x_1,x_3)}{\pi x_2} + \frac{\delta(x_1,x_2)}{\pi x_3} - \frac{1}{\pi^3 x_1 x_2 x_3} \right) e_1$$

for octant No. **1**, (1.39)

$$\Psi_c^\delta(x) = \delta(x) + \frac{\delta(x_3)}{\pi^2 x_1 x_2} - \frac{\delta(x_2)}{\pi^2 x_1 x_3} + \frac{\delta(x_1)}{\pi^2 x_2 x_3} + $$
$$+ \left(\frac{\delta(x_2,x_3)}{\pi x_1} - \frac{\delta(x_1,x_3)}{\pi x_2} + \frac{\delta(x_1,x_2)}{\pi x_3} + \frac{1}{\pi^3 x_1 x_2 x_3} \right) e_1$$

for octant No. **3**, (1.40)

$$\Psi_c^\delta(x) = \delta(x) - \frac{\delta(x_3)}{\pi^2 x_1 x_2} + \frac{\delta(x_2)}{\pi^2 x_1 x_3} + \frac{\delta(x_1)}{\pi^2 x_2 x_3} + $$
$$+ \left(\frac{\delta(x_2,x_3)}{\pi x_1} + \frac{\delta(x_1,x_3)}{\pi x_2} - \frac{\delta(x_1,x_2)}{\pi x_3} + \frac{1}{\pi^3 x_1 x_2 x_3} \right) e_1$$

for octant No. **5**, (1.41)

$$\Psi_c^8(x) = \delta(x) + \frac{\delta(x_3)}{\pi^2 x_1 x_2} + \frac{\delta(x_2)}{\pi^2 x_1 x_3} - \frac{\delta(x_1)}{\pi^2 x_2 x_3} +$$
$$+ \left(\frac{\delta(x_2,x_3)}{\pi x_1} - \frac{\delta(x_1,x_3)}{\pi x_2} - \frac{\delta(x_1,x_2)}{\pi x_3} - \frac{1}{\pi^3 x_1 x_2 x_3} \right) e_1 \qquad \text{for octant No. 7} \quad (1.42)$$

Similarly as previously for $n = 2$, we calculate the convolution (1.23) of the 3-D real signal $u(x)$, $x = (x_1, x_2, x_3)$, with the corresponding 3-D complex delta distribution given by (1.39)-(1.42). We get four 3-D analytic signals with spectral support in octants 1, 3, 5 and 7 respectively (superscripts indicate the corresponding octant):

$$\psi_c^1(x_1,x_2,x_3) = u - v_{12} - v_{13} - v_{23} + (v_1 + v_2 + v_3 - v) e_1, \qquad (1.43)$$

$$\psi_c^3(x_1,x_2,x_3) = u + v_{12} - v_{13} + v_{23} + (v_1 - v_2 + v_3 + v) e_1, \qquad (1.44)$$

$$\psi_c^5(x_1,x_2,x_3) = u - v_{12} + v_{13} + v_{23} + (v_1 + v_2 - v_3 + v) e_1, \qquad (1.45)$$

$$\psi_c^7(x_1,x_2,x_3) = u + v_{12} + v_{13} - v_{23} + (v_1 - v_2 - v_3 - v) e_1, \qquad (1.46)$$

where u, v, v_i, v_{ij} are functions of $x = (x_1, x_2, x_3)$. Equivalently, we can write Eqs. (1.43)-(1.46) as follows

$$\psi_c^1(x) = \mathbf{I}\{u\} - \mathbf{H}_{12}\{u\} - \mathbf{H}_{13}\{u\} - \mathbf{H}_{23}\{u\} +$$
$$+ \left(\mathbf{H}_1\{u\} + \mathbf{H}_2\{u\} + \mathbf{H}_3\{u\} - \mathbf{H}\{u\} \right) e_1, \qquad (1.47)$$

$$\psi_c^3(x) = \mathbf{I}\{u\} + \mathbf{H}_{12}\{u\} - \mathbf{H}_{13}\{u\} + \mathbf{H}_{23}\{u\} +$$
$$+ \left(\mathbf{H}_1\{u\} - \mathbf{H}_2\{u\} + \mathbf{H}_3\{u\} + \mathbf{H}\{u\} \right) e_1, \qquad (1.48)$$

$$\psi_c^5(x) = \mathbf{I}\{u\} - \mathbf{H}_{12}\{u\} + \mathbf{H}_{13}\{u\} + \mathbf{H}_{23}\{u\} +$$
$$+ \left(\mathbf{H}_1\{u\} + \mathbf{H}_2\{u\} - \mathbf{H}_3\{u\} + \mathbf{H}\{u\} \right) e_1, \qquad (1.49)$$

$$\psi_c^7(x) = \mathbf{I}\{u\} + \mathbf{H}_{12}\{u\} + \mathbf{H}_{13}\{u\} - \mathbf{H}_{23}\{u\} +$$
$$+ \left(\mathbf{H}_1\{u\} - \mathbf{H}_2\{u\} - \mathbf{H}_3\{u\} - \mathbf{H}\{u\} \right) e_1 \qquad (1.50)$$

We observe that there are four conjugate pairs: $\psi_c^2 = \left(\psi_c^7\right)^*$, $\psi_c^4 = \left(\psi_c^5\right)^*$, $\psi_c^6 = \left(\psi_c^3\right)^*$ and $\psi_c^8 = \left(\psi_c^1\right)^*$, where superscripts 2, 4, 6 and 8 denote the corresponding octant in the $f_1 < 0$ half-space (see Fig. 1.2). The 3-D real signal can also be expressed as

$$u(x_1,x_2,x_3) = \mathrm{Re}\left\{ \frac{\psi_c^1 + \psi_c^3 + \psi_c^5 + \psi_c^7}{4} \right\}, \qquad (1.51)$$

or as the weighted sum

$$u\left(x_1,x_2,x_3\right) = \frac{\psi_c^1 + \psi_c^8 + \psi_c^3 + \psi_c^6 + \psi_c^5 + \psi_c^4 + \psi_c^7 + \psi_c^2}{8} =$$

$$= \frac{\psi_c^1 + \left(\psi_c^1\right)^* + \psi_c^3 + \left(\psi_c^3\right)^* + \psi_c^5 + \left(\psi_c^5\right)^* + \psi_c^7 + \left(\psi_c^7\right)^*}{8} \tag{1.52}$$

We can remark that a full information about the 3-D real signal $u(x_1,x_2,x_3)$ is included in the half-space $f_1 > 0$.

1.2.4. The Frequency-domain Definition of the *n*-D Complex Analytic Signal

In the precedent section in the Eq. (1.23), we defined the *n*-D analytic signal with a single-orthant spectrum as an *n*-D real signal convolved with the *n*-D complex delta distribution. The convolution-to-multiplication property of the Fourier transformation implies that the spectrum of $\psi(x)$, denoted with $\Gamma(f), f = (f_1,f_2,...,f_n)$ is equal to the product

$$\Gamma(f) = U(f) \cdot F\{\Psi_\delta(x)\}, \tag{1.53}$$

where the Fourier transform $F\{\Psi_\delta(x)\}$ is given by

$$F\{\Psi_c^\delta(x)\} = F\left\{\prod_{k=1}^{n}\left[\delta(x_k) \pm \frac{1}{\pi x_k}e_1\right]\right\} \tag{1.54}$$

The separability of the complex delta distribution permits to write (1.54) as a product of Fourier transforms

$$F\{\Psi_c^\delta(x)\} = \prod_{k=1}^{n}F\left\{\delta(x_k) \pm \frac{1}{\pi x_k}e_1\right\} = \prod_{k=1}^{n}(1 \pm \operatorname{sgn} f_k) \tag{1.55}$$

Then, the definition (1.53) takes the form

$$\Gamma(f) = U(f) \cdot \prod_{k=1}^{n}(1 \pm \operatorname{sgn} f_k), \tag{1.56}$$

equivalent to

$$\Gamma(f) = 2^n \cdot U(f) \cdot 1(f) \tag{1.57}$$

So, the *n*-D analytic signal can be derived as the *n*-dimensional inverse Fourier transform of (1.57). The formula

$$\psi_c(x) = 2^n \cdot F^{-1}\{U(f) \cdot 1(f)\} \tag{1.58}$$

can be treated as the *frequency-domain definition* of the *n*-D analytic signal, where the multiplicator 2^n plays the role of a normalization factor and can be omitted. In 1-D, we have

$$\psi_c(t) = F^{-1}\left\{U(f)\cdot(1+\operatorname{sgn} f)\right\}, \qquad (1.59)$$

or, since $1+\operatorname{sgn} f = 2\cdot\mathbf{1}(f)$, we get

$$\psi_c(t) = 2\cdot F^{-1}\left\{U(f)\cdot\mathbf{1}(f)\right\} \qquad (1.60)$$

We surely recognized the very known definition of the 1-D Gabor's analytic signal. The formula (1.60) is often applied in practice to get the 1-D analytic signal $\psi_c(t)$ corresponding to a given real signal $u(t)$. Let us now analyze the case of $n = 2$.

1.2.4.1. The Frequency-domain Definition of the 2-D Complex Analytic Signal

Using the formula (1.56), we can write the 2-D analytic signals with spectra in quadrants No. 1 and 3 in the form

$$\psi_c^1(x_1,x_2) = F^{-1}\left\{U(f_1,f_2)\cdot(1+\operatorname{sgn} f_1)(1+\operatorname{sgn} f_2)\right\}, \qquad (1.61)$$

$$\psi_c^3(x_1,x_2) = F^{-1}\left\{U(f_1,f_2)\cdot(1+\operatorname{sgn} f_1)(1-\operatorname{sgn} f_2)\right\} \qquad (1.62)$$

Using single-quadrant operators defined by Eqs. (1.6) and (1.7), we get

$$\psi_c^1(x_1,x_2) = 4\cdot F^{-1}\left\{U(f_1,f_2)\cdot\mathbf{1}(f_1,f_2)\right\}, \qquad (1.63)$$

$$\psi_c^3(x_1,x_2) = 4\cdot F^{-1}\left\{U(f_1,f_2)\cdot\mathbf{1}(f_1,-f_2)\right\} \qquad (1.64)$$

The formulas (1.63) and (1.64) can easily be applied in practice. Given a real signal $u(x_1,x_2)$, we calculate its spectrum $U(f_1,f_2)$ (using e.g. Fast Fourier Transform algorithm), then take its part belonging to the corresponding frequency quadrant and calculate its 2-D inverse Fourier transform (using inverse FFT). Such a procedure is less numerically complicated than applying directly Eqs. (1.28) and (1.31) that would involve calculation of the total and partial Hilbert transforms via convolution integrals.

1.2.4.2. The Frequency-domain Definition of the 3-D Complex Analytic Signal

In Section 1.2.3.2 we defined eight 3-D analytic signals in the signal-domain. Let us now apply the formula (1.56) to obtain the frequency-domain definitions of four analytic signals with spectra limited to single octants 1, 3, 5 and 7. We respectively have

$$\psi_c^1(x_1,x_2,x_3) = F^{-1}\left\{U(f_1,f_2,f_3)\cdot(1+\operatorname{sgn} f_1)(1+\operatorname{sgn} f_2)(1+\operatorname{sgn} f_3)\right\}, \quad (1.65)$$

$$\psi_c^3(x_1,x_2,x_3) = F^{-1}\left\{U(f_1,f_2,f_3)\cdot(1+\operatorname{sgn} f_1)(1-\operatorname{sgn} f_2)(1+\operatorname{sgn} f_3)\right\}, \quad (1.66)$$

$$\psi_c^5(x_1,x_2,x_3) = F^{-1}\left\{U(f_1,f_2,f_3)\cdot(1+\operatorname{sgn} f_1)(1+\operatorname{sgn} f_2)(1-\operatorname{sgn} f_3)\right\}, \quad (1.67)$$

$$\psi_{\mathrm{c}}^{7}\left(x_1,x_2,x_3\right)=\mathrm{F}^{-1}\left\{U\left(f_1,f_2,f_3\right)\cdot\left(1+\operatorname{sgn}f_1\right)\left(1-\operatorname{sgn}f_2\right)\left(1-\operatorname{sgn}f_3\right)\right\} \qquad (1.68)$$

Applying the definitions of single-octant operators (1.8)-(1.11), we get equivalent forms of (1.65)-(1.68):

$$\psi_{\mathrm{c}}^{1}\left(x_1,x_2,x_3\right)=8\cdot\mathrm{F}^{-1}\left\{U\left(f_1,f_2,f_3\right)\cdot\mathbf{1}\left(f_1,f_2,f_3\right)\right\}, \qquad (1.69)$$

$$\psi_{\mathrm{c}}^{3}\left(x_1,x_2,x_3\right)=8\cdot\mathrm{F}^{-1}\left\{U\left(f_1,f_2,f_3\right)\cdot\mathbf{1}\left(f_1,-f_2,f_3\right)\right\}, \qquad (1.70)$$

$$\psi_{\mathrm{c}}^{5}\left(x_1,x_2,x_3\right)=8\cdot\mathrm{F}^{-1}\left\{U\left(f_1,f_2,f_3\right)\cdot\mathbf{1}\left(f_1,f_2,-f_3\right)\right\}, \qquad (1.71)$$

$$\psi_{\mathrm{c}}^{7}\left(x_1,x_2,x_3\right)=8\cdot\mathrm{F}^{-1}\left\{U\left(f_1,f_2,f_3\right)\cdot\mathbf{1}\left(f_1,-f_2,-f_3\right)\right\} \qquad (1.72)$$

Similarly as mentioned above, after Eq. (1.64), the formulas (1.69)-(1.72) are easy to use in practice.

1.3. Analytic Signals in the Hypercomplex Domain

A *hypercomplex* signal is a generalization of a hypercomplex number belonging to a chosen algebra. In this chapter, we put our attention on two Cayley-Dickson algebras: quaternions and octonions and we will present definitions of the quaternion and octonion analytic signals with spectra limited to a single orthant of the n-D ($n = 2, 3$) frequency space. Let us present a short review of properties of quaternions and octonions.

1.3.1. Cayley-Dickson Algebras of Quaternions and Octonions

Quaternions and octonions can be defined using the Cayley-Dickson construction. According to it, any complex number $z \in \mathbb{C}$ is an ordered pair of real numbers r_0 and r_1:

$$z=\left(r_0,r_1\right)=r_0+r_1\cdot e_1, \qquad (1.73)$$

where r_0 is called the *real part* and r_1 – the *imaginary part* (compare with Eq. (1.1) defining the n-D complex signal). Then, any quaternion $q \in \mathbb{H}$ is defined as a pair of complex numbers $z_0, z_1 \in \mathbb{C}$:

$$q=\left(z_0,z_1\right)=\left(r_{00},r_{01},r_{10},r_{11}\right), \qquad (1.74)$$

where $z_0 = r_{00}+r_{01}\cdot e_1$ and $z_1 = r_{10}+r_{11}\cdot e_1$. In consequence, any quaternion is a quadruple of $r_{00}, r_{01}, r_{10}, r_{11} \in \mathbb{R}$. Thus introducing the second imaginary unit e_2, we can rewrite (1.74) in a form

$$\begin{aligned} q &= z_0 + z_1\cdot e_2 = \\ &= \left(r_{00}+r_{01}\cdot e_1\right)+\left(r_{10}+r_{11}\cdot e_1\right)\cdot e_2 = r_{00}+r_{01}\cdot e_1+r_{10}\cdot e_2+r_{11}\cdot e_3, \end{aligned} \qquad (1.75)$$

where the imaginary unit e_3 is equal the product $e_1 \cdot e_2$. The rules of multiplication in the algebra of quaternions are illustrated in Fig. 1.6(a). The arrows indicate the order of multiplication. Let us remark that the multiplication of quaternions is not commutative, since, e.g. $e_1 \cdot e_2 = -e_2 \cdot e_1$. However, it is still associative, i.e., for any three quaternions p, q, r we have $(p \cdot q) \cdot r = p \cdot (q \cdot r)$. The algebra of quaternions is usually denoted with $\mathbb{H}$ as a honor to its inventor, William Rowan Hamilton (1805-1865), who in 1835 in the *"Theory of Conjugate Functions and Algebraic Couples"* constructed the algebra of complex numbers and interpreted them as ordered pairs of real numbers. The order of the algebra of quaternions is $2^2 = 4$.

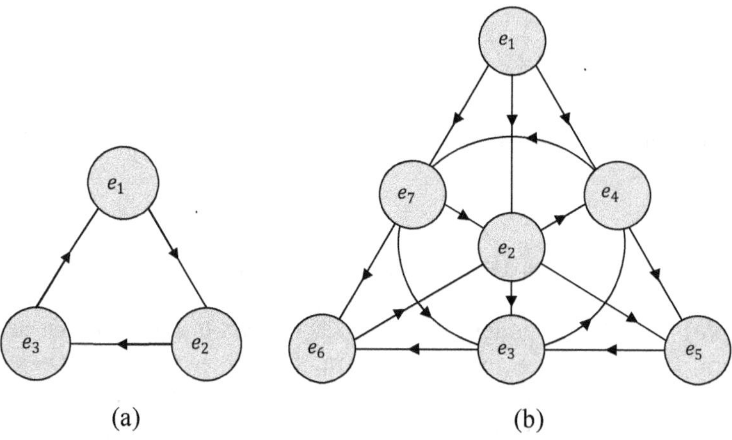

(a)

(b)

Fig. 1.6. Multiplication rules in the algebras of: (a) Quaternions; (b) Octonions.

Following the same procedure, we define an octonion $o \in \mathbb{O}$ as a pair of quaternions or a quadruple of complex numbers or finally, a 8-tuple of real numbers:

$$o = \left(q_0, q_1\right) = \left(z_{00}, z_{01}, z_{10}, z_{11}\right) = \left(r_{000}, r_{001}, r_{010}, r_{011}, r_{100}, r_{101}, r_{110}, r_{111}\right) \qquad (1.76)$$

If we now introduce the next imaginary unit e_4, we get another form of the octonion (1.76) defined as a complex sum of two quaternions, i.e.,

$$o = q_0 + q_1 \cdot e_4 \qquad (1.77)$$

Let us now write the quaternions q_0 and q_1 as complex sums (1.74) of two different complex numbers: $q_0 = z_{00} + z_{01} \cdot e_2$, $q_1 = z_{10} + z_{11} \cdot e_2$. We have

$$o = \left(z_{00} + z_{01} \cdot e_2\right) + \left(z_{10} + z_{11} \cdot e_2\right) \cdot e_4, \qquad (1.78)$$

Since $z_{00} = r_{000} + r_{001} \cdot e_1$, $z_{01} = r_{010} + r_{011} \cdot e_1$, $z_{10} = r_{100} + r_{101} \cdot e_1$, $z_{11} = r_{110} + r_{111} \cdot e_1$, we can express (1.78) as the hypercomplex sum

$$o = r_{000} + r_{001} \cdot e_1 + \left(r_{010} + r_{011} \cdot e_1\right) \cdot e_2 + \left[r_{100} + r_{101} \cdot e_1 + \left(r_{110} + r_{111} \cdot e_1\right) \cdot e_2\right] \cdot e_4 \qquad (1.79)$$

Applying the multiplication rules of the algebra $\mathbb{O}$ presented in Fig. 1.6(b) in the form of a diagram called the *Fano scheme*, that is $e_1 \cdot e_2 = e_3$, $e_1 \cdot e_4 = e_5$, $e_2 \cdot e_4 = e_6$ and $e_3 \cdot e_4 = e_7$, we get the general form of the octonion number

$$o = r_{000} + r_{001} \cdot e_1 + r_{100} \cdot e_2 + r_{101} \cdot e_3 + r_{010} \cdot e_4 + r_{011} \cdot e_5 + r_{110} \cdot e_6 + r_{111} \cdot e_7 \tag{1.80}$$

The order of the algebra of octonions is $2^3 = 8$.

1.3.2. General Form of the n-dimensional Hypercomplex Signal

The general form of a *n*-dimensional hypercomplex signal is

$$\psi_{\mathrm{h}}(x) = r_0(x) + \sum_{i=1}^{m} r_i(x)e_i , \tag{1.81}$$

where $r_0(x)$ is its real part and the imaginary part is formed of m components, all being *n*-D functions. The conjugate signal to (1.81) is

$$\psi_{\mathrm{h}}^{*}(x) = r_0(x) - \sum_{i=1}^{m} r_i(x)e_i , \tag{1.82}$$

and the norm of (1.81)

$$\left\| \psi_{\mathrm{h}}(x) \right\| = \sqrt{r_0^2(x) + \sum_{i=1}^{m} r_i^2(x)} \tag{1.83}$$

The maximal number m in (1.81)-(1.83) is related to the order 2^k ($k = 0, 1, 2, 3, \ldots$) of a given Cayley-Dickson algebra and equals $2^k - 1$. The maximum total number of components is equal 2^k. For $k = 2$, we get the general form of a *n*-D *quaternion signal* with 4 components:

$$\psi_{\mathrm{q}}(x) = r_0(x) + r_1(x)e_1 + r_2(x)e_2 + r_3(x)e_3 , \tag{1.84}$$

and for $k = 3$, a *n*-D *octonion signal* with 8 components:

$$\psi_{\mathrm{o}}(x) = r_0(x) + \sum_{i=1}^{7} r_i(x)e_i \tag{1.85}$$

It should be noticed that the order of the algebra (equal 4 for quaternions, 8 for octonions) can be different from the dimension of a signal space. For example in the following sections, we will consider 2-dimensional quaternion analytic signals and 3-dimensional octonion analytic signals.

1.3.3. The Signal-domain Definition of the *n*-D Cayley-Dickson Analytic Signal

The definition (1.22) of the *n*-D complex delta distribution inspired us to define the *n*-D *hypercomplex delta distribution* Ψ_δ^h in the Cayley-Dickson algebra [26-28] in the form

$$\Psi_\delta^h(x) = \prod_{i=1}^{n}\left[\delta(x_i) \pm \frac{1}{\pi x_i}e_{2^{i-1}}\right] =$$
$$= \left(\delta(x_1) \pm \frac{1}{\pi x_1}e_1\right)\left(\delta(x_2) \pm \frac{1}{\pi x_2}e_2\right)\left(\delta(x_3) \pm \frac{1}{\pi x_3}e_4\right)\ldots,$$

(1.86)

where the sign $\pm$ in a given factor of (1.86) is strictly connected with the form of a single-orthant operator introduced in Section 1.2.2. Please note the subscript 2^{i-1} of all imaginary units in (1.86).

In analogy to (1.23), the *n-D Cayley-Dickson analytic signal* is defined as the *n*-fold convolution of a real signal $u(x),\, x = (x_1, x_2, \ldots, x_n)$ with the *n*-D hypercomplex Cayley-Dickson delta distribution given by (1.86). We have

$$\psi_h(x) = u(x) * \ldots *_{\substack{n-\text{fold}}} \Psi_\delta^h(x)$$

(1.87)

Let us present the signal-domain definitions *signal* of $\psi_h(x)$ for $n = 2$ and 3. In the first case, we will tell about the 2-D *quaternion analytic signal*, while in the 3-D case – about the 3-D *octonion analytic signal*.

1.3.3.1. The Signal-domain Definition of the 2-D Quaternion Analytic Signal

The 2-D *quaternion analytic signal* $\psi_1^q(x_1, x_2)$ (with the 1st quadrant spectrum) is given by a two-fold convolution with the corresponding Cayley-Dickson hypercomplex delta distribution $\Psi_\delta^h(x)$:

$$\psi_q^1(x_1, x_2) = u(x_1, x_2) ** \left(\delta(x_1) + \frac{1}{\pi x_1}e_1\right)\left(\delta(x_2) + \frac{1}{\pi x_2}e_2\right)$$

(1.88)

Applying the multiplication rules in the algebra of quaternions and separability of the 2-D delta distribution, the Eq. (1.88) gets the form

$$\psi_q^1(x_1, x_2) = u(x_1, x_2) ** \left(\delta(x_1, x_2) + \frac{\delta(x_2)}{\pi x_1}e_1 + \frac{\delta(x_1)}{\pi x_2}e_2 + \frac{1}{\pi^2 x_1 x_2}e_3\right),$$

(1.89)

i.e., it has the form of a quaternion signal (compare with Eq. (1.84)):

$$\psi_q^1(x_1, x_2) = u + v_1 e_1 + v_2 e_2 + v e_3.$$

(1.90)

with all components of the sum being functions of (x_1, x_2). The signal (1.90) is known as the *quaternion analytic signal* [2, 29]. Using operators (1.12)-(1.14), we can express (1.90) in the following equivalent way as

$$\psi_q^1(x_1, x_2) = \mathbf{I}\{u\} + \mathbf{H_1}\{u\}e_1 + \mathbf{H_2}\{u\}e_2 + \mathbf{H}\{u\}e_3 \tag{1.91}$$

Analogously as above, we define next three 2-D quaternion analytic signals with spectra in Quadrants No. 2, 3 and 4 respectively. We get

$$\psi_q^2(x_1, x_2) = u(x_1, x_2) * * \left(\delta(x_1, x_2) - \frac{\delta(x_2)}{\pi x_1}e_1 + \frac{\delta(x_1)}{\pi x_2}e_2 - \frac{1}{\pi^2 x_1 x_2}e_3 \right) = \tag{1.92}$$
$$= u - v_1 e_1 + v_2 e_2 - v e_3 = \mathbf{I}\{u\} + \mathbf{H_1}\{u\}e_1 + \mathbf{H_2}\{u\}e_2 + \mathbf{H}\{u\}e_3,$$

$$\psi_q^3(x_1, x_2) = u(x_1, x_2) * * \left(\delta(x_1, x_2) + \frac{\delta(x_2)}{\pi x_1}e_1 - \frac{\delta(x_1)}{\pi x_2}e_2 - \frac{1}{\pi^2 x_1 x_2}e_3 \right) = \tag{1.93}$$
$$= u + v_1 e_1 - v_2 e_2 - v e_3 = \mathbf{I}\{u\} + \mathbf{H_1}\{u\}e_1 - \mathbf{H_2}\{u\}e_2 - \mathbf{H}\{u\}e_3,$$

$$\psi_q^4(x_1, x_2) = u(x_1, x_2) * * \left(\delta(x_1, x_2) - \frac{\delta(x_2)}{\pi x_1}e_1 - \frac{\delta(x_1)}{\pi x_2}e_2 + \frac{1}{\pi^2 x_1 x_2}e_3 \right) = \tag{1.94}$$
$$= u - v_1 e_1 - v_2 e_2 + v e_3 = \mathbf{I}\{u\} - \mathbf{H_1}\{u\}e_1 - \mathbf{H_2}\{u\}e_2 + \mathbf{H}\{u\}e_3,$$

All signals (1.90), (1.92)-(1.94) have form of quaternion signals, but there are no conjugate pairs (as in the complex case). So, the 2-D real signal $u(x_1, x_2)$ is represented in the hypercomplex domain by four different quaternion analytic signals. However, we have

$$u(x_1, x_2) = \mathrm{Re}\{\psi_q^i(x_1, x_2)\}, \quad i = 1, 2, 3, 4, \tag{1.95}$$

So, the 2-D real signal can simply be derived from its single quaternion representation. Please note that all imaginary parts of (1.91)-(1.94) are the same functions as in the 2-D complex case described in Section 1.2.3.1.

1.3.3.2. The Signal-domain Definition of the 3-D Octonion Analytic Signal

Similarly as in the 2-D case and using the multiplication rules in the algebra of octonions (see Fig. 1.6(b)), we develop the form of the 3-D octonion analytic signal (with the 1st octant spectrum) as a three-fold convolution of the 3-D real signal $u(x)$, $x = (x_1, x_2, x_3)$ with the corresponding 3-D hypercomplex delta distribution given by (1.86). We have

$$\psi_0^1(x) = u(x) *** \left[\delta(x) + \frac{\delta(x_2,x_3)}{\pi x_1} e_1 + \frac{\delta(x_1,x_3)}{\pi x_2} e_2 + \frac{\delta(x_3)}{\pi^2 x_1 x_2} e_3 + \frac{\delta(x_1,x_2)}{\pi x_3} e_4 + \right.$$
$$\left. + \frac{\delta(x_2)}{\pi^2 x_1 x_3} e_5 + \frac{\delta(x_1)}{\pi^2 x_2 x_3} e_6 + \frac{1}{\pi^3 x_1 x_2 x_3} e_7 \right], \tag{1.96}$$

that is equal to

$$\psi_0^1(x) = u + v_1 e_1 + v_2 e_2 + v_{12} e_3 + v_3 e_4 + v_{13} e_5 + v_{23} e_6 + v e_7$$
$$= \mathbf{I}\{u\} + \mathbf{H}_1\{u\} e_1 + \mathbf{H}_2\{u\} e_2 + \mathbf{H}_{12}\{u\} e_3 + \mathbf{H}_3\{u\} e_4 + \mathbf{H}_{13}\{u\} e_5 + \tag{1.97}$$
$$+ \mathbf{H}_{23}\{u\} e_6 + \mathbf{H}\{u\} e_7,$$

with all components as functions of (x_1, x_2, x_3). The signal (1.97) has been called by the authors of [16], [26-28] the *octonion analytic signal*. It is easily to develop next seven different octonion analytic signals representing the 3-D real signal $u(x_1, x_2, x_3)$ in the hypercomplex domain. We have

$$\psi_0^2(x) = u - v_1 e_1 - v_2 e_2 + v_{12} e_3 - v_3 e_4 + v_{13} e_5 + v_{23} e_6 - v e_7 =$$
$$= \mathbf{I}\{u\} - \mathbf{H}_1\{u\} e_1 - \mathbf{H}_2\{u\} e_2 + \mathbf{H}_{12}\{u\} e_3 - \mathbf{H}_3\{u\} e_4 + \mathbf{H}_{13}\{u\} e_5 + \tag{1.98}$$
$$+ \mathbf{H}_{23}\{u\} e_6 - \mathbf{H}\{u\} e_7,$$

$$\psi_0^3(x) = u + v_1 e_1 - v_2 e_2 - v_{12} e_3 + v_3 e_4 + v_{13} e_5 - v_{23} e_6 - v e_7 =$$
$$= \mathbf{I}\{u\} + \mathbf{H}_1\{u\} e_1 - \mathbf{H}_2\{u\} e_2 - \mathbf{H}_{12}\{u\} e_3 + \mathbf{H}_3\{u\} e_4 + \mathbf{H}_{13}\{u\} e_5 - \tag{1.99}$$
$$- \mathbf{H}_{23}\{u\} e_6 - \mathbf{H}\{u\} e_7,$$

$$\psi_0^4(x) = u - v_1 e_1 - v_2 e_2 + v_{12} e_3 - v_3 e_4 - v_{13} e_5 - v_{23} e_6 + v e_7 =$$
$$= \mathbf{I}\{u\} - \mathbf{H}_1\{u\} e_1 - \mathbf{H}_2\{u\} e_2 + \mathbf{H}_{12}\{u\} e_3 - \mathbf{H}_3\{u\} e_4 - \mathbf{H}_{13}\{u\} e_5 - \tag{1.100}$$
$$- \mathbf{H}_{23}\{u\} e_6 + \mathbf{H}\{u\} e_7,$$

$$\psi_0^5(x) = u + v_1 e_1 + v_2 e_2 + v_{12} e_3 - v_3 e_4 - v_{13} e_5 - v_{23} e_6 - v e_7 =$$
$$= \mathbf{I}\{u\} + \mathbf{H}_1\{u\} e_1 + \mathbf{H}_2\{u\} e_2 + \mathbf{H}_{12}\{u\} e_3 - \mathbf{H}_3\{u\} e_4 - \mathbf{H}_{13}\{u\} e_5 - \tag{1.101}$$
$$- \mathbf{H}_{23}\{u\} e_6 - \mathbf{H}\{u\} e_7,$$

$$\psi_0^6(x) = u - v_1 e_1 + v_2 e_2 - v_{12} e_3 - v_3 e_4 + v_{13} e_5 - v_{23} e_6 + v e_7 =$$
$$= \mathbf{I}\{u\} - \mathbf{H}_1\{u\} e_1 + \mathbf{H}_2\{u\} e_2 - \mathbf{H}_{12}\{u\} e_3 - \mathbf{H}_3\{u\} e_4 + \mathbf{H}_{13}\{u\} e_5 - \tag{1.102}$$
$$- \mathbf{H}_{23}\{u\} e_6 + \mathbf{H}\{u\} e_7,$$

$$\psi_o^7(\boldsymbol{x}) = u + v_1 e_1 - v_2 e_2 - v_{12} e_3 - v_3 e_4 - v_{13} e_5 + v_{23} e_6 + v e_7 =$$

$$= \mathbf{I}\{u\} + \mathbf{H}_1\{u\} e_1 - \mathbf{H}_2\{u\} e_2 - \mathbf{H}_{12}\{u\} e_3 - \mathbf{H}_3\{u\} e_4 - \mathbf{H}_{13}\{u\} e_5 - \quad (1.103)$$

$$+ \mathbf{H}_{23}\{u\} e_6 + \mathbf{H}\{u\} e_7,$$

$$\psi_o^8(\boldsymbol{x}) = u - v_1 e_1 - v_2 e_2 + v_{12} e_3 - v_3 e_4 + v_{13} e_5 + v_{23} e_6 - v e_7 =$$

$$= \mathbf{I}\{u\} - \mathbf{H}_1\{u\} e_1 - \mathbf{H}_2\{u\} e_2 + \mathbf{H}_{12}\{u\} e_3 - \mathbf{H}_3\{u\} e_4 + \mathbf{H}_{13}\{u\} e_5 + \quad (1.104)$$

$$+ \mathbf{H}_{23}\{u\} e_6 - \mathbf{H}\{u\} e_7,$$

So, the 3-D real signal can be derived from its single octonion representation as $u(x_1, x_2, x_3) = \mathrm{Re}\{\psi_o^i(x_1, x_2, x_3)\}$, $i = 1, ..., 8$.

1.3.4. The Frequency-domain Definition of the *n*-D Cayley-Dickson Analytic Signal

The *n*-D hypercomplex analytic signals are defined in a similar way as the *n*-D complex analytic signals (1.58), i.e., with the use of the *n*-D single-orthant operator $\mathbf{1}(f)$. It is known that the convolution-to-multiplication theorem applies also for the *n*-D hypercomplex Fourier transform [30] and as a result, it is possible to define the *n*-D hypercomplex analytic signals starting from their single-orthant hypercomplex spectra. The *n-D Cayley-Dickson analytic signal* is defined as the inverse *n*-D Cayley-Dickson Fourier transform

$$\psi_h(\boldsymbol{x}) = \mathrm{F}_{CD}^{-1}\{\mathbf{1}(f) \cdot U_h(f)\}, \quad (1.105)$$

where $U_h(f)$ is the hypercomplex (Cayley-Dickson) Fourier spectrum of the *n*-D real signal $u(\boldsymbol{x})$ defined in [26] as the right-side hypercomplex Fourier transform

$$U_h(f) = \int_{\mathbb{R}^n} u(\boldsymbol{x}) \prod_{i=1}^{n} \exp(-e_k 2\pi f_i x_i) d^n \boldsymbol{x}, \quad (1.106)$$

where the subscript $k = 2^{i-1}$. The inverse hypercomplex Fourier transform of (1.106) is

$$\mathrm{F}_{CD}^{-1}\{U_h(f)\} = \int_{\mathbb{R}^n} U_h(f) \prod_{i=1}^{n} \exp(e_k 2\pi f_i x_i) d^n f \quad (1.107)$$

Let us consider the case of a 2-D quaternion analytic signal and a 3-D octonion analytic signal.

1.3.4.1. The Frequency-domain Definition of the 2-D Quaternion Analytic Signal

The 2-D quaternion analytic signal is defined as the inverse right-side Quaternion Fourier Transform (QFT) of a single-quadrant quaternion spectrum of a 2-D real signal $u(x_1, x_2)$. The Eq. (1.105) has the form

$$\psi_q^1(x_1,x_2) = \mathrm{F}_{\mathrm{CD}}^{-1}\left\{\mathbf{1}(f_1,f_2)\cdot U_q(f_1,f_2)\right\}, \tag{1.108}$$

and we get the frequency-domain definition of the 2-D quaternion analytic signal with the spectrum in the quadrant No. 1 of the 2-D frequency-plane. The appropriate change of the single-quadrant operator, as described in Section 1.2, permits to define next three quaternion analytic signals with spectra in quadrants No. 2, 3 and 4

$$\psi_q^2(x_1,x_2) = \mathrm{F}_{\mathrm{CD}}^{-1}\left\{\mathbf{1}(-f_1,f_2)\cdot U_q(f_1,f_2)\right\}, \tag{1.109}$$

$$\psi_q^3(x_1,x_2) = \mathrm{F}_{\mathrm{CD}}^{-1}\left\{\mathbf{1}(-f_1,-f_2)\cdot U_q(f_1,f_2)\right\}, \tag{1.110}$$

$$\psi_q^4(x_1,x_2) = \mathrm{F}_{\mathrm{CD}}^{-1}\left\{\mathbf{1}(f_1,-f_2)\cdot U_q(f_1,f_2)\right\} \tag{1.111}$$

The Quaternion Fourier transform $U_q(f_1,f_2)$ of a 2-D real signal $u(x_1,x_2)$ is defined as follows

$$U_q(f_1,f_2) = \iint\limits_{\mathbb{R}^2} u(x_1,x_2)\, e^{-e_1 2\pi f_1 x_1}\, e^{-e_2 2\pi f_2 x_2}\, dx_1 dx_2 \tag{1.112}$$

It can be easily proved that the kernel of (1.112) can be expressed as a quaternion function:

$$e^{-e_1 2\pi f_1 x_1} e^{-e_2 2\pi f_2 x_2} = (c_1 - s_1 e_1)(c_2 - s_2 e_2) = c_1 c_2 - s_1 c_2 e_1 - c_1 s_2 e_2 + s_1 s_2 e_3, \tag{1.113}$$

with $c_i = \cos(2\pi f_i x_i)$, $s_i = \sin(2\pi f_i x_i)$, $i = 1,2$. The QFT given by (1.112) is related to the 2-D complex Fourier transform of u by a very elegant formula developed by Pei *et al.* in [31] in the form

$$U_q(f_1,f_2) = U(f_1,f_2)\frac{1-e_3}{2} + U(f_1,-f_2)\frac{1+e_3}{2} \tag{1.114}$$

This formula proves the equivalence of complex and hypercomplex approaches to the theory of 2-D analytic signals.

1.3.4.2. The Frequency-domain Definition of the 3-D Octonion Analytic Signal

The 3-D octonion analytic signal is defined as the inverse right-side Octonion Fourier Transform (OFT) of a single-octant octonion spectrum of a 3-D real signal $u(x_1,x_{2,}x_3)$. In this case, the Eq. (1.105) gets the form

$$\psi_o^1(x_1,x_2,x_3) = \mathrm{F}_{\mathrm{CD}}^{-1}\left\{\mathbf{1}(f_1,f_2,f_3)\cdot U_o(f_1,f_2,f_3)\right\} \tag{1.115}$$

If we want to derive the 3-D octonion analytic signals with spectra in Octants No. 2, 3, ...,8, we only need to use a corresponding single-octant operator $\mathbf{1}(f)$ changing signs of f_i as indicated in Table 1.2 (Section 1.2.1). The Octonion Fourier transform $U_o(f_1, f_2, f_3)$ of a 3-D real signal $u(x_1, x_2, x_3)$ defined in [16] has the following form

$$U_o(f_1, f_2, f_3) = \iiint_{\mathbb{R}^3} u(x_1, x_2, x_3) e^{-e_1 2\pi f_1 x_1} e^{-e_2 2\pi f_2 x_2} e^{-e_4 2\pi f_3 x_3} dx_1 dx_2 dx_3 \qquad (1.116)$$

Using the multiplication rules of the algebra of octonions, we can develop the kernel of (1.116) in a form of the octonion function

$$e^{-e_1 2\pi f_1 x_1} e^{-e_2 2\pi f_2 x_2} e^{-e_4 2\pi f_3 x_3} = (c_1 - s_1 e_1)(c_2 - s_2 e_2)(c_3 - s_3 e_4) =$$
$$= c_1 c_2 c_3 - s_1 c_2 c_3 e_1 - c_1 s_2 c_3 e_2 + s_1 s_2 c_3 e_3 - c_1 c_2 s_3 e_4 + s_1 c_2 s_3 e_5 + c_1 s_2 s_3 e_6 - s_1 s_2 s_3 e_7 \qquad (1.117)$$

Let us mention that the definition (1.116) is not unique and there are many other triples of imaginary units (here (e_1, e_2, e_4)) that would define the kernel of the transformation in a form of a full 3-D octonion function. The extended study of chosen properties of the OFT is included in [18-20] (symmetry property, Parseval-Plancherel and Wiener-Khintchine Theorems).

The OFT defined in (1.116) is related to the 3-D complex FT by the formula presented with the proof in [32] as follows

$$\begin{aligned} U_o(f_1, f_2, f_3) = &\tfrac{1}{4}\big[U(f_1, f_2, f_3) + U(f_1, -f_2, f_3)\big](1 - e_5) + \\ &+ \tfrac{1}{4}\big[U(f_1, f_2, -f_3) + U(f_1, -f_2, -f_3)\big](1 + e_5) + \\ &+ \tfrac{1}{4}e_3\big[U(-f_1, f_2, f_3) - U(-f_1, -f_2, f_3)\big](1 + e_5) + \\ &+ \tfrac{1}{4}e_3\big[U(-f_1, f_2, -f_3) - U(-f_1, -f_2, -f_3)\big](1 - e_5) \end{aligned} \qquad (1.118)$$

Let us emphasize that the order of imaginary units in (1.118) can not be changed since the multiplication in the algebra of octonions is not commutative. We always use the rule "from left to right".

1.4. Polar Representation of Complex and Hypercomplex Analytic Signals

This section is devoted to the polar representation of n-D complex and hypercomplex analytic signals. This problem has completely been solved only for 2-D analytic signals [3, 8, 29, 33]. However, it seems that the decomposition of higher dimensional hypercomplex analytic signals into amplitude- and phase functions is more complicated. The first approach to this problem has been proposed by Hahn and Snopek in [16]. However, according to the latest results published by Błaszczyk in [34], their hypothesis about the decomposition of 3-D octonion analytic signals into one amplitude- and seven phase functions is not validated.

1.4.1. Polar Representation of 2-D Complex Analytic Signals

Let us go back to Eqs. (1.28) and (1.31) defining the 2-D analytic signals with spectra in Quadrants No. 1 and 3 and express them in a polar form

$$\psi_c^i(x_1,x_2) = \left|\psi_c^i(x_1,x_2)\right| e^{e_1 \arg \psi_c^i(x_1,x_2)} = A_c^i(x_1,x_2) e^{e_1 \Phi_c^i(x_1,x_2)}, \qquad (1.119)$$

where $A_c^i(x_1,x_2) = \left|\psi_c^i(x_1,x_2)\right|$ are called *local amplitudes* and arguments of $\psi_c^i(x_1,x_2)$, denoted with $\Phi_c^i(x_1,x_2)$, are *local phases* of a signal. We have

$$A_c^i(x_1,x_2) = \sqrt{\left(\mathrm{Re}\{\psi_c^i(x_1,x_2)\}\right)^2 + \left(\mathrm{Im}\{\psi_c^i(x_1,x_2)\}\right)^2}, \qquad (1.120)$$

and

$$\tan \Phi_c^i(x_1,x_2) = \frac{\mathrm{Im}\{\psi_c^i(x_1,x_2)\}}{\mathrm{Re}\{\psi_c^i(x_1,x_2)\}} \qquad (1.121)$$

So, the local amplitude and the local phase of the 2-D analytic signal with spectrum in the Quadrant No. 1 of the frequency plane (1.28) respectively are

$$A_c^1(x_1,x_2) = \sqrt{(u-v)^2 + (v_1+v_2)^2} = \sqrt{u^2 + v^2 + v_1^2 + v_2^2 + 2(-uv + v_1 v_2)}, \quad (1.122)$$

$$\tan \Phi_c^1(x_1,x_2) = \frac{v_1 + v_2}{u - v} \qquad (1.123)$$

For the 2-D analytic signal with spectrum in the Quadrant No. 3, we get

$$A_c^3(x_1,x_2) = \sqrt{(u+v)^2 + (v_1-v_2)^2} = \sqrt{u^2 + v^2 + v_1^2 + v_2^2 + 2(uv - v_1 v_2)}, \quad (1.124)$$

$$\tan \Phi_c^3(x_1,x_2) = \frac{v_1 - v_2}{u + v} \qquad (1.125)$$

As has already been mentioned, the 2-D real signal $u(x_1,x_2)$ is represented by two different analytic signals ψ_c^1 and ψ_c^3, each having its local amplitude and its local phase, so in a general case, we have two different amplitudes and two different phase functions. Let us consider now the case of *separable* 2-D real signals $u(x_1,x_2) = f_1(x_1) f_2(x_2)$. Its full Hilbert transform is a product $v(x_1,x_2) = g_1(x_1) g_2(x_2)$ where g_i is the Hilbert transform of f_i. The partial Hilbert transforms are $v_1(x_1,x_2) = g_1(x_1) f_2(x_2)$ and $v_2(x_1,x_2) = f_1(x_1) g_2(x_2)$. It can easily be shown that all local amplitudes $A_c^1(x_1,x_2) = A_c^3(x_1,x_2)$, since $uv - v_1 v_2 = f_1 f_2 g_1 g_2 - g_1 f_2 f_1 g_2 = 0$. They are given by

$$A_c^i(x_1,x_2) = \sqrt{u^2 + v^2 + v_1^2 + v_2^2}, \quad i = 1, 2, 3, 4 \qquad (1.126)$$

If we introduce the phase angles:

$$\tan\varphi_1(x_1) = g_1/f_1, \quad \tan\varphi_2(x_2) = g_2/f_2, \qquad (1.127)$$

the Eq. (1.123) can be written as

$$\tan\Phi_c^1(x_1,x_2) = \frac{\dfrac{g_1}{f_1} + \dfrac{g_2}{f_2}}{1 - \dfrac{g_1}{f_1}\dfrac{g_2}{f_2}} = \tan\big(\varphi_1(x_1) + \varphi_2(x_2)\big) \qquad (1.128)$$

So, we get

$$\Phi_c^1(x_1,x_2) = \varphi_1(x_1) + \varphi_2(x_2), \qquad (1.129)$$

$$\Phi_c^3(x_1,x_2) = \varphi_1(x_1) - \varphi_2(x_2) \qquad (1.130)$$

It is possible to reconstruct the 2-D real signal and its total and partial Hilbert transforms from the polar representation derived above. The *reconstruction formulas* presented in [8] are as follows

$$u(x_1,x_2) = \frac{A_c^1(x_1,x_2)\cos\Phi_c^1(x_1,x_2) + A_c^3(x_1,x_2)\cos\Phi_c^3(x_1,x_2)}{2}, \qquad (1.131)$$

$$v(x_1,x_2) = \frac{-A_c^1(x_1,x_2)\cos\Phi_c^1(x_1,x_2) + A_c^3(x_1,x_2)\cos\Phi_c^3(x_1,x_2)}{2}, \qquad (1.132)$$

$$v_1(x_1,x_2) = \frac{A_c^1(x_1,x_2)\sin\Phi_c^1(x_1,x_2) + A_c^3(x_1,x_2)\sin\Phi_c^3(x_1,x_2)}{2}, \qquad (1.133)$$

$$v_2(x_1,x_2) = \frac{A_c^1(x_1,x_2)\sin\Phi_c^1(x_1,x_2) - A_c^3(x_1,x_2)\sin\Phi_c^3(x_1,x_2)}{2} \qquad (1.134)$$

Let us illustrate the above formulas with the example of the 2-D Cauchy separable signal defined by (1.35) in Section 1.2.3.2 of this chapter. The Fig. 1.7 shows local amplitude and two local phase functions of this signal calculated using (1.126) and (1.129), (1.130). We observe a different orientation of two local phase functions on the (x_1,x_2)–plane and their odd parity. The local amplitude is an even function w.r.t. x_1 and x_2.

1.4.2. Polar Representation of 2-D Quaternion Analytic Signals

The polar form of the 2-D quaternion analytic signal (1.90) was defined by Bülow in [2] as

$$\psi_q^1(x_1,x_2) = A_q\exp\big(e_1\Phi_1^q\big)\exp\big(e_3\Phi_3^q\big)\exp\big(e_2\Phi_2^q\big), \qquad (1.135)$$

where A_q, called the *amplitude of the quaternion analytic signal*, is given by the square root:

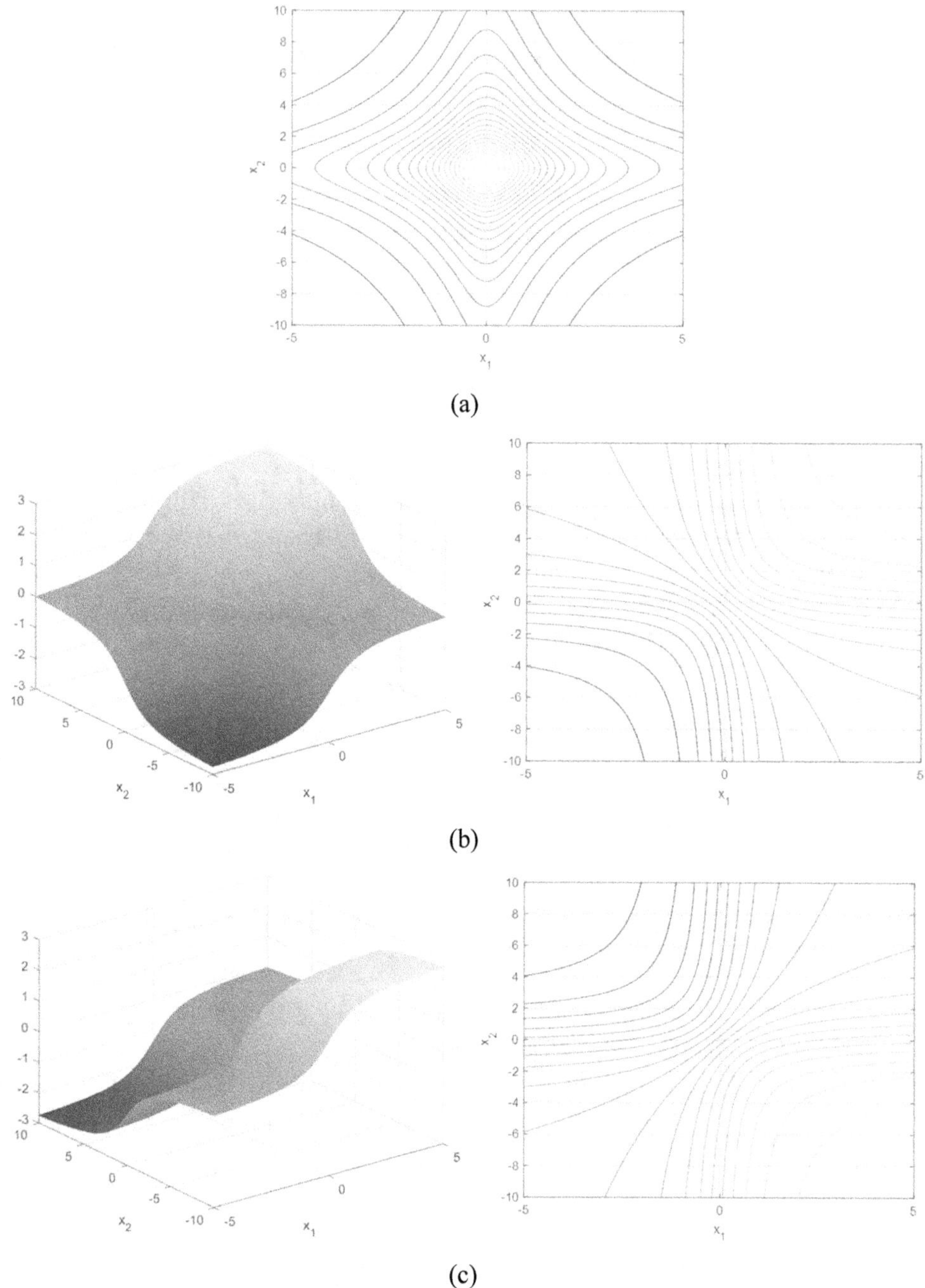

(a)

(b)

(c)

Fig. 1.7. Polar representation of the 2-D Cauchy separable signal: (a) The local amplitude $A_c^1 = A_c^3$; (b) the local phase function Φ_c^1 – mesh and contour plots; (c) the local phase function Φ_c^3 – mesh and contour plots.

$$A_{\mathrm{q}}(x_1,x_2)=\sqrt{u^2+v_1^2+v_2^2+v^2}\,, \tag{1.136}$$

and three phase functions are Euler's angles defined by

$$\tan 2\Phi_1^{\mathrm{q}}(x_1,x_2)=\frac{2(uv_1+vv_2)}{u^2-v_1^2+v_2^2-v^2}\,, \tag{1.137}$$

$$\tan 2\Phi_2^{\mathrm{q}}(x_1,x_2)=\frac{2(uv_2+vv_1)}{u^2+v_1^2-v_2^2-v^2}\,, \tag{1.138}$$

$$\sin 2\Phi_3^{\mathrm{q}}(x_1,x_2)=\frac{2(v_1v_2-uv)}{A_{\mathrm{q}}^2} \tag{1.139}$$

Let us note that $\left(\Phi_1^{\mathrm{q}},\Phi_2^{\mathrm{q}},\Phi_3^{\mathrm{q}}\right)\in\left[-\pi,\pi\right)\times\left[-\pi/2,\pi/2\right)\times\left[-\pi/4,\pi/4\right]$ [2]. The details concerning calculation of phase angles (1.137)-(1.139) are to be found in [11].

In [35], the relation between the amplitude of the 2-D quaternion analytic signal (1.136) and two amplitudes of 2-D complex analytic signals with spectra in the 1st and 3rd quadrant of the frequency plane was derived in a form:

$$A_{\mathrm{q}}=\sqrt{\frac{\left(A_{\mathrm{c}}^1(x_1,x_2)\right)^2+\left(A_{\mathrm{c}}^3(x_1,x_2)\right)^2}{2}} \tag{1.140}$$

Moreover, we have

$$\sin 2\Phi_3^{\mathrm{q}}=\frac{\left(A_{\mathrm{c}}^1(x_1,x_2)\right)^2-\left(A_{\mathrm{c}}^3(x_1,x_2)\right)^2}{\left(A_{\mathrm{c}}^1(x_1,x_2)\right)^2+\left(A_{\mathrm{c}}^3(x_1,x_2)\right)^2} \tag{1.141}$$

Let us note that the angle $\Phi_3^{\mathrm{q}}=\pm\pi/4$ (singular value) only if $A_{\mathrm{c}}^1=0$ or $A_{\mathrm{c}}^3=0$, which is almost impossible for real image functions. For other Euler's angles we have

$$\Phi_1^{\mathrm{q}}(x_1,x_2)=\frac{\Phi_{\mathrm{c}}^1+\Phi_{\mathrm{c}}^3}{2}\,, \tag{1.142}$$

$$\Phi_2^{\mathrm{q}}(x_1,x_2)=\frac{\Phi_{\mathrm{c}}^1-\Phi_{\mathrm{c}}^3}{2} \tag{1.143}$$

The formulas (1.140)-(1.143) show that both representations (quaternion and complex) are equivalent and the choice of the method is a matter of convention. In the case of 2-D separable signals, we get the following equalities:

$$A_{\mathrm{q}}=A_{\mathrm{c}}^1=A_{\mathrm{c}}^3, \tag{1.144}$$

$$\Phi_1^{\mathrm{q}}(x_1,x_2)=\varphi_1(x_1),\ \ \Phi_2^{\mathrm{q}}(x_1,x_2)=\varphi_2(x_2),\ \ \Phi_3^{\mathrm{q}}(x_1,x_2)=0 \tag{1.145}$$

The Fig. 1.8 displays all polar components of the quaternion analytic signal in the case of the 2-D separable Cauchy signal from the preceding Example 2. The quaternion amplitude given by (1.136) is exactly the same as illustrated in Fig. 1.7(a). The quaternion phase functions Φ_1^q and Φ_2^q differ in orientation. The third phase Φ_3^q is zero.

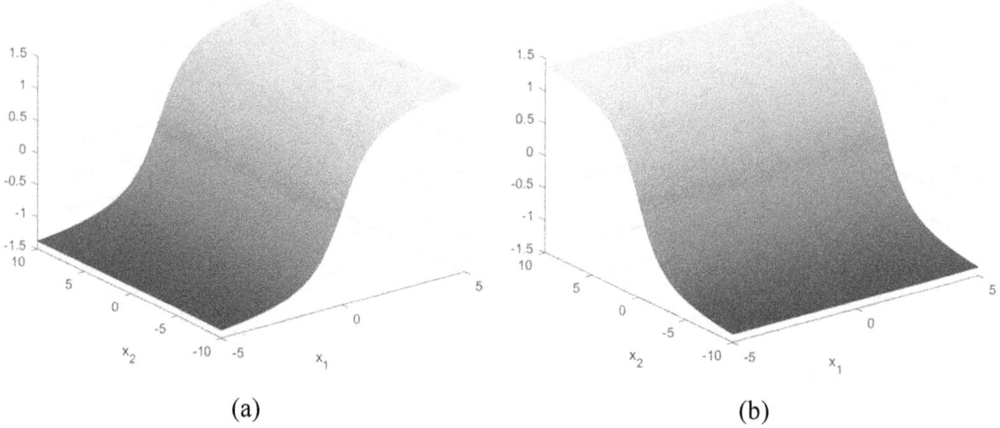

(a)   (b)

Fig. 1.8. Polar quaternion representation of the 2-D Cauchy separable signal: (a) The mesh plot of the phase function Φ_1^q, (b) The mesh plot of the phase function Φ_2^q.

It is possible to reconstruct the 2-D real signal $u(x_1,x_2)$ from its quaternion polar representation [2] by calculating the real part of (1.135):

$$u(x_1,x_2) = A_q \left[\cos\Phi_1^q \cos\Phi_2^q \cos\Phi_3^q + \sin\Phi_1^q \sin\Phi_2^q \sin\Phi_3^q \right] \tag{1.146}$$

Moreover, the other reconstruction formulas of total and partial 2-D Hilbert transforms are as follows

$$v(x_1,x_2) = A_q \left[\cos\Phi_1^q \cos\Phi_2^q \sin\Phi_3^q + \sin\Phi_1^q \sin\Phi_2^q \cos\Phi_3^q \right], \tag{1.147}$$

$$v_1(x_1,x_2) = A_q \left[\sin\Phi_1^q \cos\Phi_2^q \cos\Phi_3^q - \cos\Phi_1^q \sin\Phi_2^q \sin\Phi_3^q \right], \tag{1.148}$$

$$v_2(x_1,x_2) = A_q \left[\cos\Phi_1^q \sin\Phi_2^q \cos\Phi_3^q - \sin\Phi_1^q \cos\Phi_2^q \sin\Phi_3^q \right], \tag{1.149}$$

1.4.3. Polar Representation of 3-D Complex Analytic Signals

As has already been mentioned in Section 1.2.3.2, the 3-D real signal $u(x_1,x_2,x_3)$ is represented by four 3-D complex analytic signals $\psi_c^i(x_1,x_2,x_3)$, $i = 1, 3, 5, 7$ with spectra in corresponding octants of the half-space $f_1 > 0$. Their polar representation [8] is

$$\psi_c^i\left(x_1,x_2,x_3\right) = A_c^i\left(x_1,x_2,x_3\right)e^{e_1\Phi_c^i\left(x_1,x_2,x_3\right)}, \; i = 1,3,5,7\,, \qquad (1.150)$$

(compare with Eq. (1.119)). The functions $A_c^i\left(x_1,x_2,x_3\right)$ are *local amplitudes* and $\Phi_c^i\left(x_1,x_2,x_3\right)$ are *local phases*. They are given by

$$A_c^i\left(x_1,x_2,x_3\right) = \sqrt{\left(\mathrm{Re}\{\psi_c^i\}\right)^2 + \left(\mathrm{Im}\{\psi_c^i\}\right)^2}\,, \qquad (1.151)$$

$$\tan\Phi_c^i\left(x_1,x_2,x_3\right) = \frac{\mathrm{Im}\{\psi_c^i\}}{\mathrm{Re}\{\psi_c^i\}} \qquad (1.152)$$

So, the 3-D real signal $u(x_1,x_2,x_3)$ is defined by four different amplitudes and four different phase functions. Having in mind the conjugate relations between eight 3-D analytic signals: $\psi_c^2 = \left(\psi_c^7\right)^*$, $\psi_c^4 = \left(\psi_c^5\right)^*$, $\psi_c^6 = \left(\psi_c^3\right)^*$ and $\psi_c^8 = \left(\psi_c^1\right)^*$, we only present definitions of local amplitudes and local phase functions for signals with spectra in Octants No. 1, 3, 5 and 7. We have

$$A_c^1\left(x_1,x_2,x_3\right) = \sqrt{\left(u - v_{12} - v_{13} - v_{23}\right)^2 + \left(v_1 + v_2 + v_3 - v\right)^2}\,, \qquad (1.153)$$

$$\tan\Phi_c^1\left(x_1,x_2,x_3\right) = \frac{v_1 + v_2 + v_3 - v}{u - v_{12} - v_{13} - v_{23}}\,, \qquad (1.154)$$

$$A_c^3\left(x_1,x_2,x_3\right) = \sqrt{\left(u + v_{12} - v_{13} + v_{23}\right)^2 + \left(v_1 - v_2 + v_3 + v\right)^2}\,, \qquad (1.155)$$

$$\tan\Phi_c^3\left(x_1,x_2,x_3\right) = \frac{v_1 - v_2 + v_3 + v}{u + v_{12} - v_{13} + v_{23}}\,, \qquad (1.156)$$

$$A_c^5\left(x_1,x_2,x_3\right) = \sqrt{\left(u - v_{12} + v_{13} + v_{23}\right)^2 + \left(v_1 + v_2 - v_3 + v\right)^2}\,, \qquad (1.157)$$

$$\tan\Phi_c^5\left(x_1,x_2,x_3\right) = \frac{v_1 + v_2 - v_3 + v}{u - v_{12} + v_{13} + v_{23}}\,, \qquad (1.158)$$

$$A_c^7\left(x_1,x_2,x_3\right)=\sqrt{\left(u+v_{12}+v_{13}-v_{23}\right)^2+\left(v_1-v_2-v_3-v\right)^2}\,,\tag{1.159}$$

$$\tan\Phi_c^7\left(x_1,x_2,x_3\right)=\frac{v_1-v_2-v_3-v}{u+v_{12}+v_{13}-v_{23}}\tag{1.160}$$

Let us consider now the case of a *separable* 3-D real signal $u\left(x_1,x_2,x_3\right)=f_1\left(x_1\right)f_2\left(x_2\right)f_3\left(x_3\right)$. Its full Hilbert transform is a product $v\left(x_1,x_2,x_3\right)=g_1\left(x_1\right)g_2\left(x_2\right)g_3\left(x_3\right)$ of corresponding Hilbert transforms of f_i. The partial Hilbert transforms are:

$$v_1\left(x_1,x_2,x_3\right)=g_1\left(x_1\right)f_2\left(x_2\right)f_3\left(x_3\right),\; v_2\left(x_1,x_2,x_3\right)=f_1\left(x_1\right)g_2\left(x_2\right)f_3\left(x_3\right),$$

$$v_3\left(x_1,x_2,x_3\right)=f_1\left(x_1\right)f_2\left(x_2\right)g_3\left(x_3\right),\; v_{12}\left(x_1,x_2,x_3\right)=g_1\left(x_1\right)g_2\left(x_2\right)f_3\left(x_3\right)\tag{1.161}$$

$$v_{13}\left(x_1,x_2,x_3\right)=g_1\left(x_1\right)f_2\left(x_2\right)g_3\left(x_3\right),\; v_{23}\left(x_1,x_2,x_3\right)=f_1\left(x_1\right)g_2\left(x_2\right)g_3\left(x_3\right)$$

In the separable case, all local amplitudes A_c^i are equal and given by

$$A_c^i\left(x_1,x_2,x_3\right)=\sqrt{u^2+v_1^2+v_2^2+v_{12}^2+v_3^2+v_{13}^2+v_{23}^2+v^2}\tag{1.162}$$

If we introduce the angles: $\tan\varphi_1\left(x_1\right)=g_1/f_1$, $\tan\varphi_2\left(x_2\right)=g_2/f_2$, $\tan\varphi_3\left(x_3\right)=g_3/f_3$, then the local phase $\Phi_c^1\left(x_1,x_2,x_3\right)$ is

$$\Phi_c^1\left(x_1,x_2,x_3\right)=\varphi_1\left(x_1\right)+\varphi_2\left(x_2\right)+\varphi_3\left(x_3\right)\tag{1.163}$$

Similarly, we can show that

$$\Phi_c^3\left(x_1,x_2,x_3\right)=\varphi_1\left(x_1\right)-\varphi_2\left(x_2\right)+\varphi_3\left(x_3\right),\tag{1.164}$$

$$\Phi_c^5\left(x_1,x_2,x_3\right)=\varphi_1\left(x_1\right)+\varphi_2\left(x_2\right)-\varphi_3\left(x_3\right),\tag{1.165}$$

$$\Phi_c^7\left(x_1,x_2,x_3\right)=\varphi_1\left(x_1\right)-\varphi_2\left(x_2\right)-\varphi_3\left(x_3\right)\tag{1.166}$$

1.4.4. Polar Representation of 3-D Octonion Analytic Signals

In [34], the polar representation of the octonion was proposed and numerically validated in a form

$$o=q\cdot e^{e_4\Phi_4^o}e^{e_5\Phi_5^o}e^{e_6\Phi_6^o}e^{e_7\Phi_7^o}\,,\tag{1.167}$$

where q is the unit quaternion defined as

$$q = e^{e_1 \Phi_1^q} e^{e_3 \Phi_3^q} e^{e_2 \Phi_2^q}, \qquad (1.168)$$

with phases Φ_i^q are the same Euler angles as in (1.135). This proposition opens the way to the polar representation of 3-D octonion analytic signals – the problem solved only partially. We are certain that it is defined by a single amplitude A_0 equal the absolute value of (1.97)-(1.104):

$$A_0\left(x_1, x_2, x_3\right) = \sqrt{u^2 + v_1^2 + v_2^2 + v_{12}^2 + v_3^2 + v_{13}^2 + v_{23}^2 + v^2}, \qquad (1.169)$$

and by seven phase functions $\Phi_i^q\left(x_1, x_2, x_3\right)$, $i = 1, ..., 7$. We do not know the formulas defining them.

1.5. Conclusions

In this chapter, we presented the theory of complex and hypercomplex multidimensional analytic signals. The "analyticity" was understood in the Gabor's sense, what means that the spectrum of the n-D analytic signal is limited to a single orthant of the n-D frequency space. We showed two equivalent approaches: complex (proposed by Hahn in [8]) and hypercomplex, based on Cayley-Dickson algebras of quaternions and octonions. The signal-domain and frequency-domain definitions of complex and hypercomplex analytic signals have been presented, as well as some mutual relations between complex and hypercomplex spectra and polar components. One question remains still open: how to define the polar components of the 3-D octonion analytic signal?

References

[1]. D. Gabor, Theory of communications, Part III, *J. Inst. E. E.*, Vol. 93, November 1946, pp. 429-457.

[2]. T. Bülow, M. Felsberg, G. Sommer, Non-commutative hypercomplex Fourier Transforms of multidimensional signals, in Geometric Computing with Clifford Algebra (G. Sommer, Ed.), *Springer-Verlag*, Berlin, 2001, pp. 187-207.

[3]. S. L. Hahn, Hilbert Transforms in Signal Processing, *Artech House, Inc.*, 1996.

[4]. A. Mertins, Signal Analysis, *John Wiley & Sons, Ltd.*, 1999.

[5]. M. Feldman, Hilbert Transform Applications in Mechanical Vibration, *John Wiley & Sons, Ltd.*, 2011.

[6]. S. L. Hahn, On the uniqueness of the definition of the amplitude and phase of the analytic signal, *Signal Processing*, Vol. 83, 2003, pp. 1815-1820.

[7]. J. Ville, Théorie et Applications de la Notion de Signal Analytique, *Câbles et Transmission*, Vol. 2A, 1948, pp. 61-74.

[8]. S. L. Hahn, Multidimensional complex signals with single-orthant spectra, *Proceedings of IEEE*, Vol. 80, Issue 8, August 1992, pp. 1287-1300.

[9]. T. A. Ell, Hypercomplex spectral transformations, PhD Thesis, *Univ. of Minnesota*, Minneapolis, 1992.

[10]. T. A. Ell, Quaternion-Fourier transforms for analysis of 2-dimensional linear time-invariant partial-differential systems, in *Proceedings of the 32nd IEEE Conference on Decision and Control (CDC'93)*, San Antonio, TX, USA, 15-17 December 1993, Vols. 1-4, pp. 1830-1841.

[11]. T. Bülow, Hypercomplex spectral signal representation for the processing and analysis of images, in Bericht Nr. 99-3, Institut für Informatik und Praktische Mathematik, *Christian-Albrechts-Universität Kiel*, August 1999.

[12]. N. Le Bihan, J. Mars, New 2D attributes based on complex and hypercomplex analytic signal, in *Proceedings of the 71st Meeting of Society of Exploration Geophysicists (SEG'01)*, San Antonio, September 2001.

[13]. P. Bas, N. Le Bihan, J.-M. Chassery, Color image watermarking using quaternion Fourier transform, in *Proceedings of the International Conference on Acoustics, Speech, and Signal Processing (ICASSP'03)*, Hong Kong, 2003, pp. III-521.

[14]. P. Denis, P. Carre, C. Fernandez-Maloigne, Spatial and spectral quaternionic approaches for colour images, *Computer Vision and Image Understanding*, Vol. 107, 2007, pp. 74-87.

[15]. T. A. Ell, S. J. Sangwine, Hypercomplex Fourier transforms of color images, *IEEE Trans. Image Processing*, Vol. 16, Issue 1, January 2007, pp. 22-35.

[16]. S. L. Hahn, K. M., Snopek, The unified theory of n-dimensional complex and hypercomplex analytic signals, *Bull. Polish Ac. Sci., Tech. Sci.*, Vol. 59, Issue 2, 2011, pp. 167-181.

[17]. S. L. Hahn, K. M. Snopek, Complex and Hypercomplex Multidimensional Analytic Signals: Theory and Applications, *Artech House*, Boston/London, 2016.

[18]. Ł. Błaszczyk, K. M. Snopek, Octonion Fourier transform of real-valued functions of three variables – Selected properties and examples, *Signal Processing*, Vol. 136, 2017, pp. 29-37.

[19]. Ł. Błaszczyk, K. M. Snopek, Erratum to "Octonion Fourier transform of real-valued functions of three variables – Selected properties and examples", *Signal Processing*, Vol. 142, 2018, pp. 149-151.

[20]. Ł. Błaszczyk, Octonion spectrum of 3D octonion-valued signals – Properties and possible applications, in *Proceedings of the 26th European Signal Processing Conference (EUSIPCO'18)*, 2018, pp. 509-513.

[21]. Ł. Błaszczyk, Hypercomplex Fourier transforms in the analysis of multidimensional linear time-invariant systems, in *Progress in Industrial Mathematics at ECMI 2018* (I. Faragó, F. Izsák, P. L. Simon, Eds.), 2019, pp. 575-581.

[22]. S. L. Hahn, Complex signals with single-orthant spectra as boundary distributions of multidimensional analytic functions, *Bull. Polish Academy of Sciences, Technical Sciences*, Vol. 51, Issue 2, 2003, pp. 155-161.

[23]. S. L. Hahn, The N-dimensional complex delta distribution, *IEEE Trans. Signal Processing*, Vol. 44, Issue 7, July 1996, pp. 1833-1837.

[24]. H. Bremermann, Distributions, Complex Variables and Fourier Transforms, *Addison-Wesley*, MA, 1965.

[25]. L. Schwartz, Méthodes Mathématiqies pour les Sciences Physiques, *Hermann*, Paris, France, 1965.

[26]. K. M. Snopek, New hypercomplex analytic signals and Fourier transforms in Cayley-Dickson algebras, *Electr. Tel. Quarterly*, Vol. 55, Issue 3, 2009, pp. 403-415.

[27]. K. M. Snopek, The n-D analytic signals and Fourier spectra in complex and hypercomplex domains, in *Proceedings of the 34th Int. Conference on Telecommunications and Signal Processing (TSP'11)*, Budapest, August 18-20, 2011, pp. 423-427.

[28]. K. M. Snopek, The study of properties of n-D analytic signals in complex and hypercomplex domains, Radioengineering, Vol. 21, Issue 2, April 2012, pp. 29-36.

[29]. S. L. Hahn, K. M. Snopek, Comparison of properties of analytic, quaternionic and monogenic 2-D signals, *WSEAS Transactions on Computers*, Vol. 3, Issue 3, July 2004, pp. 602-611.

[30]. M. Felsberg, T. Bülow, G. Sommer, Commutative hypercomplex Fourier transforms of multidimensional signals, in Geometric Computing with Clifford Algebras (G. Sommer, Ed.), *Springer-Verlag*, Berlin, Heidelberg, 2001, pp. 209-229.

[31]. S.-C. Pei, J.-J. Ding, J.-H. Chang, Efficient implementation of quaternion Fourier transform, convolution, and correlation by 2-D complex FFT, *IEEE Trans. Sig. Proc.*, Vol. 49, Issue 11, November 2001, pp. 2783-2797.

[32]. K. M. Snopek, Studies on complex and hypercomplex multidimensional analytic signals, in Prace Naukowe Elektronika, *Oficyna Wydawnicza Politechniki Warszawskiej*, Warszawa, 2013.

[33]. S. J. Sangwine, N. Le Bihan, Quaternion polar representation with a complex modulus and complex argument inspired by the Cayley-Dickson form, *Adv. Appl. Clifford Alg.*, Vol. 20, 2010, pp. 111-120.

[34]. Ł. Błaszczyk, Theoretical and numerical considerations on the polar (exponential) form of octonions and elements of higher-order Cayley-Dickson algebras, arXiv:1909.04519, *arXiv: General Mathematics*, 2019.

[35]. S. L. Hahn, The relationships of analytic and quaternionic representation of 2-D signals, in *Proceedings of the X National Symposium of Radio Science*, Poznań, 14-15 March 2002, pp. 328-332.

Chapter 2
Signal Processing by Generalized Receiver in Wireless Communications Systems over Fading Channels

Vyacheslav Tuzlukov

2.1. Generalized Detector: Quaternary DS-SSMA Wireless Communications Systems

2.1.1. Problem Statement

In last several decades, the direct sequence spread spectrum multiple access (DS-SSMA) technique has attracted a lot of attention as a transmission method providing spectrum efficiency, high system capacity, multipath propagation, interference robustness, and improved quality of service [1-3]. In DS-SSMA wireless communication systems, the characteristics of the spreading codes provide a crucial effect on the performance of the whole communication system [4, 5]. How much interference from other users is received at a receiver is determined by the signature sequences. Additionally, the signature sequences influence on the extraction capability of desired signal from noise like spectrum. The orthogonal spreading sequences are characterized by the zero cross-correlation for zero delay. This fact has attracted a great attention in research and applications.

The polyphase complex sequences such as Frank-Zadoff-Chu sequences in [2, 6], and [7], have the excellent correlation properties and can be 3 dB better than the binary real Gold sequences in the maximum periodic correlation parameters. Also, the larger sets of complex sequences are available. Owing to the fact that the complex sequences exhibit the better autocorrelation and improved cross correlation properties than binary real sequences, the application of the complex sequences in DS-SSMA wireless communication systems has a great research interest [8-10]. However, these complex sequences are nonorthogonal. They can be categorized as complex valued pseudorandom

Vyacheslav Tuzlukov
Department of Technical Maintenance of Aviation and Radio Electronic Equipment, Belarusian State Aviation Academy, Minsk, Belarus

spreading sequences. A set of the orthogonal 4-phase complex sequences that are derived from the unified complex Hadamard Transform (UCHT) matrix [11] was introduced in [10].

Therein, the correlation properties of the UCHT complex sequences were obtained, and simulation results were given by applying the UCHT sequences to the binary phase shift keying (BPSK) DS-SSMA wireless communication systems over the additive white Gaussian noise (AWGN) channels. DS-SSMA wireless communication systems and their performance evaluation techniques have been discussed in [1, 4] and [12]. In [8] and [9] the performance bounds for DS-SSMA wireless communication systems with complex signature sequences were investigated for the BPSK data signalling and for M-ary PSK data signalling. The modulators and receivers discussed in these papers use the real processing. For instance, in the IS-95 CDMA mobile cellular system [1], both forward and reverse links use the forms of quadrature phase shift keying (QPSK) spread spectrum modulation in which the same baseband binary signalling data stream modulates both in phase (I) and quadrature (Q) binary real spreading sequences, and the receiver may employ separate I and Q real processing of the spread spectrum signal with binary real spreading sequences.

We consider a new receiver based on the generalized approach to signal processing in noise [13-18] employing the complex processing instead of real. Hence, for the quaternary DS-SSMA wireless communication systems, the modulators and generalized receivers are complex and not performed in separate in phase and quadrature branches. All research results in the literature mentioned above are established over AWGN channels. It is well known that the fading and multiple access interference (MAI) are two major sources of performance degradation in DS CDMA wireless communication systems. Therefore, analysis of the DS-SSMA wireless communication systems performance on fading channels is of considerable theoretical and practical interest.

Many papers deal with the error probability performance evaluations for binary DS-SSMA wireless communication systems operating in fading channels with real binary spreading sequences [19-23]. In [19] the signal-to-noise ratio (SNR) is studied at the output of the correlation receiver for Rician fading channels. The performances DS-SSMA wireless communication systems over Rayleigh fading channels are investigated in [20] for deterministic binary sequences and in [22] for random binary sequences. The influence of Rician factor of κ-μ short-term fading, κ-μ short-term fading severity parameter, Gamma long-term fading severity parameter and Gamma long-term fading correlation coefficient on level crossing rate is studied and discussed in [23] under consideration of the macro diversity with selection combining receiver and two micro diversity maximum ratio combining receivers. However, there is no the result for the error probability performance analysis of quaternary DS-SSMA wireless communication systems that operate in Rayleigh fading channels and employ the complex spreading sequence, as well as complex processing at the transmitters and receivers.

We carry out the comparative analysis between performances of the DS-SSMA wireless communication systems based on the correlation and generalized receivers for this case. The main goal of this section is to investigate the performance of the quaternary

DS-SSMA wireless communication system based on the generalized approach to signal processing in noise with complex signature sequences in the presence of the flat Rayleigh fading employing the generalized receiver and compare the performance of the quaternary DS-SSMA wireless communication system employing the correlation receiver. The complex spreading sequences include the orthogonal quaternary UCHT complex sequences [11]. In this section the bit error rate (*BER*) performance evaluation is first developed for the synchronous complex quaternary DS-SSMA wireless communication systems employing the generalized receiver. Then the *BER* is evaluated for the asynchronous complex quaternary DS-SSMA wireless communication systems employing the generalized receiver based on both the characteristic function approach and Gaussian approximation method.

2.1.2. Basic Definitions

In this section, we will define the unified complex Hadamard transform (UCHT) matrix that generates orthogonal complex sequences by its rows. The UCHT matrix U_n of order $N = 2^n$ is the square matrix and can be constructed by the following form [9, 10]

$$U_n = U_1 \otimes U_{n-1} = \underbrace{U_1 \otimes \cdots \otimes U_1}_{n \ \text{times}}, \tag{2.1}$$

where $\otimes$ denotes the Kronecker product of matrices; U_1 is defined as

$$U_1 = \begin{bmatrix} \mu_1 & \mu_1\mu_3 \\ \mu_2 & -\mu_2\mu_3 \end{bmatrix}, \tag{2.2}$$

with $\mu_1, \mu_2, \mu_3 \in \{1, -1, j, -j\}$ and $j = \sqrt{-1}$, μ_1. Note that U_1 satisfies to the condition

$$U_1 U_1^* = U_1^* U_1 = 2I_2, \tag{2.3}$$

and

$$|\det U_1|^2 = 2^2, \tag{2.4}$$

where U_1^* represents the complex conjugate transpose of the matrix U_1. Hence, U_1 is the complex orthogonal Hadamard matrix and the UCHT matrix is complex orthogonal matrix, too [24]. Moreover, it is obvious that U_1 is a unified form of 64 different matrices with elements in $\{\pm 1, \pm j\}$, and among them, there are eight matrices with all four different element values. UCHT matrices contain the Walsh-Hadamard transform (WHT) matrix as a special case with $\mu_1 = 1, \mu_2 = 1, \mu_3 = 1$ in the matrix U_1.

In addition, since the elements of the matrix U_1 are confined to four values $\{\pm 1, \pm j\}$, UCHT represents the mapping of four valued integers into a unit circle of complex plane. For example, UCHT can be used as four valued complex transform that maps four integers $(0, 1, 2, 3)$ into $(1, j, -1, -j)$ [10]. From the above discussion, 64 sets of orthogonal sequences can be generated depending on the various combinations of μ_1, μ_2 and μ_3. Depending on whether μ_3 is imaginary or real, these UCHT matrices can be categorized into two groups [11], i.e., UCHT with half spectrum property (HSP-UCHT) and UCHT without half spectrum property (NHSP-UCHT). Each group has 32 sets of UCHT matrices.

Let each row of one UCHT matrix be a complex sequence. Then the sequence $a^{(i)}$ represents the ith row of the corresponding UCHT matrix, and the mth element of $a^{(i)}$ takes the form $a_m^{(i)} = u(i,m)$, where $u(i,m)$ is the element of the UCHT matrix. By repeating the UCHT sequences, $N = 2^n$ orthogonal UCHT sequences with the periods of N are produced. For any two periodic sequences $a^{(k)}$ and $a^{(i)}$ with period N, the aperiodic cross-correlation function is defined in [25] as

$$R_{r,i}(l) = \begin{cases} \displaystyle\sum_{m=0}^{N-1-l} a_m^{(k)} a_{m+l}^{(i)^\bullet} , & 0 \le l \le N-1 \\[2mm] \displaystyle\sum_{m=0}^{N-1+l} a_{m-l}^{(k)} a_m^{(i)^\bullet} , & 1-N \le l < 0 \\[2mm] 0 , & |l| \ge N \end{cases} \tag{2.5}$$

If $k = i$, the cross-correlation function is the autocorrelation function $R_i(l) = R_{i,i}(l)$.

Introduce some notation for the complex valued random variables. The cumulative distribution function of the complex variable Z is specified by giving the joint distribution of its real and imaginary parts, i.e.

$$C_Z(z) = C_{\mathcal{R}\{Z\},\mathcal{G}\{Z\}}(x,y) = P(\mathcal{R}\{Z\} \le x, \mathcal{G}\{Z\} \le y), \tag{2.6}$$

where $P(\cdot)$ denotes the probability, $\mathcal{R}\{Z\}$ and $\mathcal{G}\{Z\}$ represent the real and imaginary parts of Z respectively, and the pair of real numbers (x,y) can be identified with a complex number $z = x + jy$, associating with (x,y) and the complex number $z = x + jy$, the probability density function of Z is defined as the joint probability density function of $(\mathcal{R}\{Z\},\mathcal{G}\{Z\})$, i.e.

$$f_Z(z) = f_{\mathcal{R}\{Z\},\mathcal{G}\{Z\}}(x,y). \tag{2.7}$$

A complex random variable is independent if

$$f_Z(z) = f_{\mathcal{R}\{Z\}}(x) f_{\mathcal{G}\{Z\}}(y). \tag{2.8}$$

The distribution of a complex random vector

$$\mathbf{Z} = (Z_1, Z_2, \ldots, Z_n) \tag{2.9}$$

is specified by the joint distribution of real random vector $(\mathcal{R}\{Z_1\},\ldots,\mathcal{R}\{Z_n\},\mathcal{G}\{Z_1\},\ldots,\mathcal{G}\{Z_n\})$.

$$\mathbf{C_Z}(\mathbf{Z}) = P(\mathcal{R}\{Z_1\} \leq \mathcal{R}\{z_1\},\ldots,\mathcal{R}\{Z_n\} \leq \mathcal{R}\{z_n\}, \; \mathcal{G}\{Z_1\} \leq \mathcal{G}\{z_1\},\ldots,\mathcal{G}\{Z_n\} \leq \mathcal{G}\{z_n\}) \, , \quad (2.10)$$

where $\mathbf{z} = (z_1, z_2,\ldots,z_n)$ denotes the complex valued vector with $z_i = x_i + jy_i, i = 1, 2,\ldots, n$. The probability density function of $\mathbf{Z}$ is defined in terms of the joint probability density function of $2n$ real random variables $\mathcal{R}\{Z_i\}$ and $\mathcal{G}\{Z_i\}$, that is

$$f_{\mathbf{Z}}(\mathbf{z}) = f_{\mathcal{R}\{Z_1\},\ldots,\mathcal{R}\{Z_n\},\mathcal{G}\{Z_1\},\ldots,\mathcal{G}\{Z_n\}}(x_1,\ldots,x_n, y_1,\ldots,y_n) \qquad (2.11)$$

Thus, the complex random variables $\{Z_i\}$ are independent if

$$f_{\mathbf{Z}}(\mathbf{z}) = \prod_{i=1}^{n} f_{Z_i}(z_i) \qquad (2.12)$$

2.1.3. System Model

In this section, a quaternary DS-SSMA wireless communication system with complex signature sequences over Rayleigh fading channel is described. As shown in Fig. 2.1, there are K simultaneous users that transmit data asynchronously. The channel inputs and outputs are complex waveforms. Both the modulator and the receiver employ a complex signal processing.

2.1.3.1. Transmitter Model

The ith user's data signal can be expressed as

$$b_i(t) = \sum_{l=-\infty}^{\infty} b_l^{(i)} p_T(t - lT), \qquad (2.13)$$

where the function $p_T(t)$ is the rectangular pulse of duration T, $b_l^{(i)} \in \{(1/\sqrt{2})(\pm 1, \pm j)\}$ denotes the lth complex quaternary data value of the ith user, and it is assumed to take values with equal probability and be independently complex uniform, i.e., $\mathcal{R}\{b_l^{(i)}\}$ and $\mathcal{G}\{b_l^{(i)}\}$ are independent and uniform with the probability

$$\begin{cases} P\{\mathcal{R}\{b_l^{(i)}\} = 1/\sqrt{2}\} = P\{\mathcal{R}\{b_l^{(i)}\} = -1/\sqrt{2}\} = 0.5 \, , \\ P\{\mathcal{G}\{b_l^{(i)}\} = 1/\sqrt{2}\} = P\{\mathcal{G}\{b_l^{(i)}\} = -1/\sqrt{2}\} = 0.5 \end{cases} \qquad (2.14)$$

The complex spreading signal for the ith user can be expressed as

$$a_i(t) = \sum_{m=-\infty}^{\infty} a_m^{(i)} \psi(t - mT_c) \,, \tag{2.15}$$

where $a_m^{(i)} \in \{\pm 1, \pm j\}$ denotes the mth complex chip value of the ith user, which is quadriphase or 4-phase complex sequence element, and the complex spreading sequence $a^{(i)} = \{a_m^{(i)}\}$ has the period N. The function $\psi(t)$ is the arbitrary chip waveform that is time limited to $[0, T_c)$, including the rectangular pulse, and T_c is the chip duration. It is assumed that there is one period of the spreading sequence per data symbol, so $T = NT_c$. Thus, the ith transmitted signal is described by

$$s_i(t) = \sqrt{2P_i}\, b_i(t) a_i(t) \exp\{j(\omega_c t + \theta_i)\} \,, \tag{2.16}$$

where $\sqrt{2P_i}\, \exp\{j(\omega_c t + \theta_i)\}$ is the complex carrier signal, as shown in Fig. 2.1; P_i represents the ith transmitted signal power; ω_c is the common complex carrier frequency; θ_i is the carrier phase of the ith carrier, which is an independently uniform random variable within the limits of the interval $[0, 2\pi)$. Power control is assumed to be perfect, and the transmitted signal power P_i is assumed to be known.

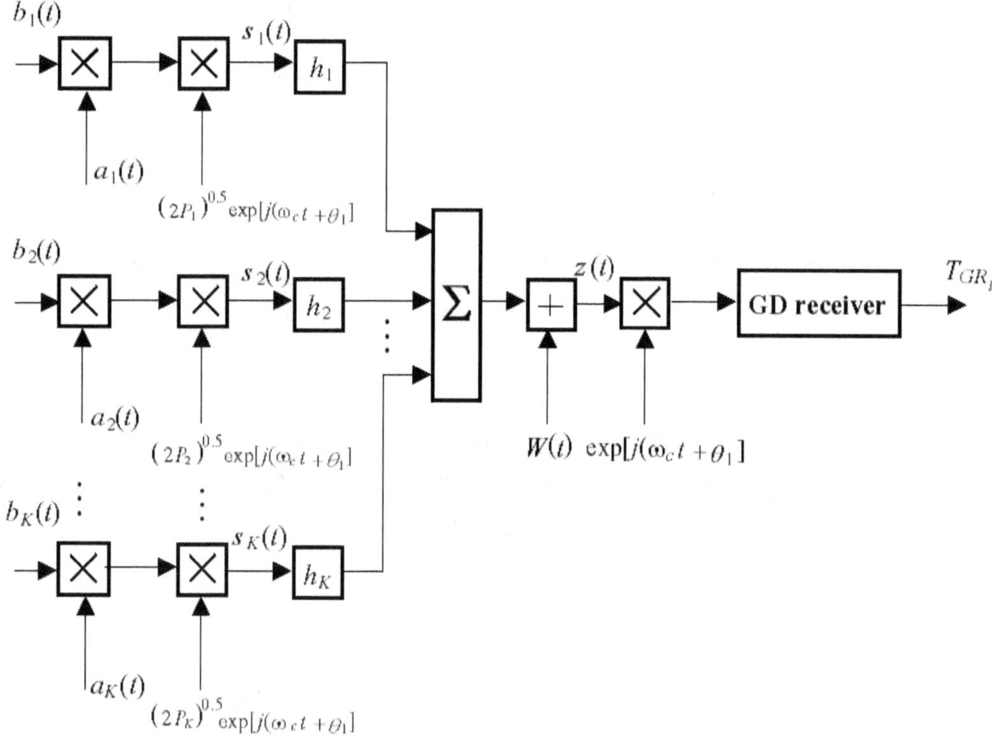

Fig. 2.1. DS-SSMA system with complex modulator and complex generalized receiver.

2.1.3.2. Rayleigh Fading Channel Model

In Fig. 2.1, each signal $s_i(t)$ is transmitted over a frequency nonselective fading channel, where the user signals and the interfering signals all experience mutually independent Rayleigh fading. The fading is also assumed to be slow such that the coherent detection is feasible. The channel impulse response for the ith transmitted signal takes the form

$$h_i(t) = A_i \exp\{j\beta_i\}\delta(t - \tau_i),\qquad(2.17)$$

where the fading random variables A_i, $1 \leq i \leq K$, are independent, Rayleigh distributed and account for the fading channel attenuation of all signals. The fading random variables are assumed to have the same probability density function (pdf) and the pdf of A_i is

$$f_{A_i}(\alpha) = \begin{cases} \alpha \, \exp\{-0.5\alpha^2\}, & \alpha \geq 0 \\ 0, & \alpha < 0 \end{cases},\qquad(2.18)$$

with $E[A_i^2] = 2$, $E[\cdot]$ denotes the mathematical expectation. In (2.17), β_i, $1 \leq i \leq K$, are the phases introduced by the fading channel and are assumed to be uniform within the limits of the interval $[0, 2\pi)$; $\delta(t)$ is the Dirac impulse function, and τ_i, $1 \leq i \leq K$, are the time delays which are assumed to be uniform within the limits of the interval $[0, T)$ and independent.

The transmitted signal is further assumed to experience an additive background thermal band pass noise process

$$w(t) = r(t)\exp\{j\omega_c t\},\qquad(2.19)$$

where

$$r(t) = x(t) + jy(t),\qquad(2.20)$$

$x(t)$ and $y(t)$ are the independent zero mean baseband Gaussian noise processes, each having power spectral density given by $S(f) = 0.25\mathcal{N}_0$ for $|f| \leq 0.5B$ and $S(f) = 0$ for $|f| > 0.5B$, where $\mathcal{N}_0$ is the power spectral density of the white Gaussian noise. We assume that the bandwidth B of the noise is much greater than the bandwidth of the baseband signals $b_i(t)$ and $a_i(t)$. The average power of the band pass noise process is

$$E[n^2(t)] = 0.5\mathcal{N}_0 B.\qquad(2.21)$$

The received signal at the input of the generalized receiver takes the following form

$$z(t) = \sum_{i=1}^{K} \sqrt{2P_i} A_i b_i(t - \tau_i) a_i(t - \tau_i)\exp\{j(\omega_c t + \phi_i)\} + w(t),\qquad(2.22)$$

where

$$\phi_i = \beta_i + \theta_i - \omega_c \tau_i.\qquad(2.23)$$

All delays are defined by modulo T and all carrier phase angles are defined by modulo 2π. This allows one to restrict attention to $0 \leq \tau_i < T$, $0 \leq \phi_i < 2\pi$, $\forall i$. It is easy to show that τ_i and ϕ_i can be modelled as mutually independent uniform random variables for very large ω_c. Hence, in this chapter, it is further assumed that all A_i, ϕ_i, τ_i are mutually independent random variables.

2.1.3.3. Generalized Receiver

To coherently demodulate the desired user signal in an asynchronous system the conventional generalized receiver (see Fig. 2.2) is employed by DS-SSMA system. As we mentioned before, the generalized receiver is constructed in accordance with the generalized approach to signal processing in noise [13-15]. The generalized approach to signal processing introduces an additional noise source that does not carry any information about the parameters of desired transmitted signal with the purpose to improve the signal processing system performance. This additional noise can be considered as the reference noise without any information about the parameters of the signal to be detected.

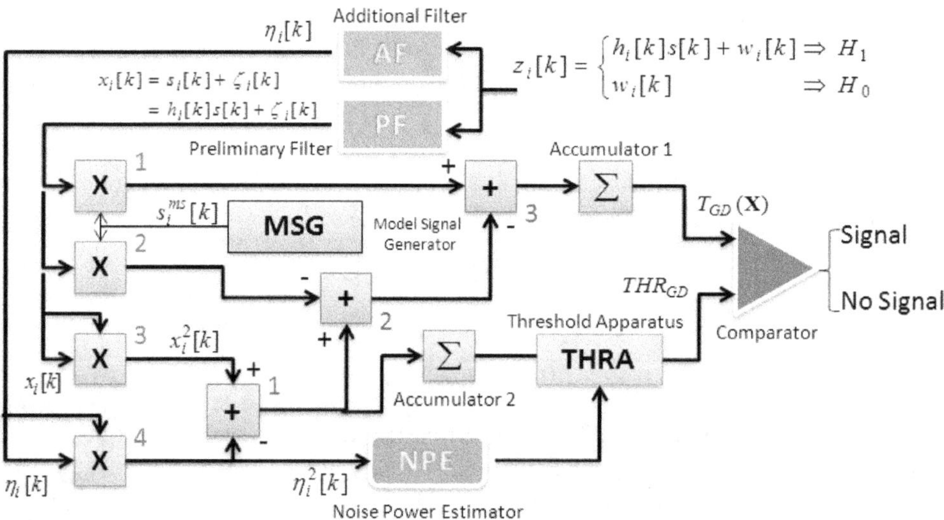

Fig. 2.2. Generalized receiver structure.

The jointly sufficient statistics of the mean and variance of the likelihood function is obtained under the generalized approach to signal processing in noise employment, while the classical and modern signal processing theories can deliver only a sufficient statistics of the mean or variance of the likelihood function. Thus, the generalized approach to signal processing in noise implementation allows us to obtain more information about the desired transmitted signal incoming at the generalized receiver input. Owing to this fact, the detectors constructed based on the generalized approach to signal processing in noise technology are able to improve the signal detection performance in comparison with other conventional detectors.

The generalized receiver (GR) consists of three channels (see Fig. 2.2): the GR correlation detector channel (GR CD) – the preliminary filter (PF), the multipliers 1 and 2, the model signal generator (MSG); the GR energy detector channel (GR ED) – the PF, the additional filter (AF), the multipliers 3 and 4, the summator 1; and the GR compensation channel (GR CC) – the summators 2 and 3, the accumulator 1. The threshold apparatus (THRA) device defines the GR threshold. As we can see from Fig. 2.2, there are two band pass filters, i.e. the linear systems, at the GR input, namely, the PF and AF. We assume for simplicity that these two filters or linear systems have the same amplitude frequency characteristics or impulse responses. The AF central frequency is detuned relative to the PF central frequency.

There is a need to note that the PF bandwidth is matched with the transmitted signal bandwidth. If the detuning value between the PF and AF central frequencies is more than 4 or 5 times the transmitted signal bandwidth to be detected, i.e. $4 \div 5 \Delta f_s$, where Δf_s is the transmitted signal bandwidth, we can believe that the processes at the PF and AF outputs are uncorrelated because the coefficient of correlation between them is negligible (not more than 0.05). This fact was confirmed experimentally in [26] and [27] independently.

Thus, the transmitted signal plus noise can be appeared at the GR PF output and the noise only is appeared at the GR AF output. The stochastic processes at the AF and PF outputs present the input stochastic samples from two independent frequency time regions. If the discrete time noise $w_i[k]$ at the PF and AF inputs is Gaussian, the discrete time noise $\zeta_i[k]$ at the PF output is Gaussian and the reference discrete time noise $\eta_i[k]$ at the AF output is Gaussian, too, owing to the fact that the PF and AF are the linear systems and we believe that these linear systems do not change the statistical parameters of the input process. Thus, the AF can be considered as a generator of the reference noise with a priori information a "no" transmitted signal (the reference noise sample) [15, Chapter 5].

The noise at the PF and AF outputs can be presented as

$$\begin{cases} \zeta_i[k] = \displaystyle\sum_{m=-\infty}^{\infty} g_{PF}[m] w_i[k-m] \ , \\[2mm] \eta_i[k] = \displaystyle\sum_{m=-\infty}^{\infty} g_{AF}[m] w_i[k-m] \ , \end{cases} \tag{2.24}$$

where $g_{PF}[m]$ and $g_{AF}[m]$ are the impulse responses of the PF and AF, respectively.

In a general, under practical implementation of any detector in communication system with sensor array, the bandwidth of the spectrum to be sensed is defined. Thus, the AF bandwidth and central frequency can be assigned, too (the AF bandwidth can not be used by the transmitted signal because it is out of its spectrum). The case when there are interfering signals within the AF bandwidth, the action of this interference on the GR detection performance, and the case of non-ideal condition when the noise at the PF and AF outputs is not the same by statistical parameters are discussed in [28] and [29].

Under the hypothesis $\mathcal{H}_1$ ("a yes" transmitted signal), the GR CD generates the signal component $s_i^m[k]s_i[k]$ caused by interaction between the model signal $s_i^m[k]$, the MSG output, and the incoming signal $s_i[k]$, and the noise component $s_i^m[k]\zeta_i[k]$ caused by interaction between the model signal $s_i^m[k]$ and the noise $\zeta_i[k]$ at the PF output. GR ED generates the transmitted signal energy $s_i^2[k]$ and the random component $s_i[k]\zeta_i[k]$ caused by interaction between the transmitted signal $s_i[k]$ and the noise $\zeta_i[k]$ at the PF output. The main purpose of the GR CC is to cancel completely in the statistical sense the GR CD noise component $s_i^m[k]\zeta_i[k]$ and the GR ED random component $s_i[k]\zeta_i[k]$ based on the same nature of the noise $\zeta_i[k]$. The relation between the transmitted signal to be detected $s_i[k]$ and the model signal $s_i^m[k]$ is defined as:

$$s_i^m[k] = \rho \, s_i[k], \tag{2.25}$$

where ρ is the coefficient of proportionality.

The main functioning condition under the GR employment in any signal processing system including the communication one with radar sensors is the equality between the parameters of the model signal $s_i^m[k]$ and the incoming signal $s_i[k]$, for example, by amplitude. Under this condition it is possible to cancel completely in the statistical sense the noise component $s_i^m[k]\zeta_i[k]$ of the GR CD and the random component $s_i[k]\zeta_i[k]$ of the GR ED. Satisfying the GR main functioning condition given by (2.25), $s_i^m[k] = s_i[k]$ and $\rho = 1$, we are able to detect the transmitted signal with the high probability of detection at the low *SNR* and define the transmitted signal parameters with high accuracy.

Practical realization of this condition (2.25) at $\rho \rightarrow 1$ requires increasing in the complexity of GR structure and, consequently, leads us to increasing in computation cost. For example, there is a need to employ the amplitude tracking system or to use the offline data samples processing. Under the hypothesis $\mathcal{H}_0$ ("a no" transmitted signal), satisfying the main GR functioning condition (2.25) at $\rho \rightarrow 1$ we obtain only the background noise $\eta_i^2[k] - \zeta_i^2[k]$ at the GR output owing to $s_i^m[k] = s_i[k]$.

Under practical implementation, the real structure of GR depends on specificity of signal processing systems and their applications, for example, the radar sensor systems, adaptive communication systems, cognitive radio systems, satellite communication systems, and mobile communication systems and so on. In the present chapter, the GR circuitry (Fig. 2.2) is demonstrated with the purpose to explain the main functioning principles. Because of this, the GR flowchart presented in the chapter should be considered under this viewpoint. Satisfying the GR main functioning condition (2.25) at $\rho \rightarrow 1$, the ideal case, for communication systems with radar sensor applications we are able to detect the transmitted signal with very high probability of detection and define accurately its parameters.

In the present chapter, we discuss the GR implementation in wireless communication systems using the radar sensor array. Since the presented GR test statistics is defined by the signal energy and noise power, the equality between the parameters of the model signal $s_i^m[k]$ and transmitted signal to be detected $s_i[k]$, in particular by amplitude, is required that leads us to high circuitry complexity in practice. For example, there is a need to employ the amplitude tracking system or offline data sample processing. Detailed discussion about the main GR functioning principles if there is no a priori information and there is an uncertainty about the parameters of transmitted signal, i.e., the transmitted signal parameters are random, can be found in [13] and [14, Chapter 6, pp. 611-621 and Chapter 7, pp. 631-695].

The complete matching between the model signal $s_i^m[k]$ and the incoming signal $s_i[k]$, for example by amplitude, is a very hard problem in practice because the incoming signal $s_i[k]$ depends on both the fading and the transmitted signal parameters and it is impractical to estimate the fading gain at the low *SNR*. This matching is possible in the ideal case only. The GD detection performance will be deteriorated under mismatching in parameters between the model signal $s_i^m[k]$ and the transmitted signal $s_i[k]$ and the impact of this problem is discussed in [18] and [30], where a complete analyze about the violation of the main GR functioning requirements is presented. The GR decision statistics requires an estimation of the noise variance σ_η^2 using the reference noise $\eta_i[k]$ at the AF output.

Under the hypothesis $\mathcal{H}_1$, the signal at the PF output, see Fig. 2.2, can be defined as

$$x_i[k] = s_i[k] + \zeta_i[k], \qquad (2.26)$$

where $\zeta_i[k]$ is the noise at the PF output and

$$s_i[k] = h_i[k]s[k], \qquad (2.27)$$

where $h_i[k]$ are the channel coefficients. Under the hypothesis $\mathcal{H}_0$ and for all i and k, the process $x_i[k] = \zeta_i[k]$ at the PF output is subjected to the complex Gaussian distribution and can be considered as the independent and identically distributed (i.i.d.) process.

In the ideal case, we can think that the signal at the AF output is the reference noise $\eta_i[k]$ with the same statistical parameters as the noise $\zeta_i[k]$. In practice, there is a difference between the statistical parameters of the noise $\eta_i[k]$ and $\zeta_i[k]$. How this difference impacts on the GR detection performance is discussed in detail in [14, Chapter 7, pp. 631-695] and in [18] and [30].

The decision statistics at the GR output presented in [13] and [14, Chapter 3] is extended for the case of antenna array when an adoption of multiple antennas and antenna arrays is effective to mitigate the negative attenuation and fading effects. The GR decision statistics can be presented in the following form:

$$T_{GR}(\mathbf{X}) = \sum_{k=0}^{N-1}\sum_{i=1}^{M} 2x_i[k]s_i^m[k] - \sum_{k=0}^{N-1}\sum_{i=1}^{M} x_i^2[k] + \sum_{k=0}^{N-1}\sum_{i=1}^{M} \eta_i^2[k] \underset{\mathcal{H}_0}{\overset{\mathcal{H}_1}{\gtrless}} THR_{GR} \,, \quad (2.28)$$

where

$$\mathbf{X} = \left[\mathbf{x}(0),...,\mathbf{x}(N-1)\right] \qquad (2.29)$$

is the vector of the random process at the PF output and THR_{GR} is the GR detection threshold.

Under the hypotheses $\mathcal{H}_1$ and $\mathcal{H}_0$ when, for example, the amplitude of the transmitted signal is equal to the amplitude of the model signal, $s_i^m[k] = s_i[k]$, the GR decision statistics $T_{GR}(\mathbf{X})$ takes the following form in the statistical sense, respectively:

$$\begin{cases} \mathcal{H}_1 : T_{GR}(\mathbf{X}) = \sum_{k=0}^{N-1}\sum_{i=1}^{M} s_i^2[k] + \sum_{k=0}^{N-1}\sum_{i=1}^{M} \eta_i^2[k] - \sum_{k=0}^{N-1}\sum_{i=1}^{M} \varsigma_i^2[k] \,, \\[2mm] \mathcal{H}_0 : T_{GR}(\mathbf{X}) = \sum_{k=0}^{N-1}\sum_{i=1}^{M} \eta_i^2[k] - \sum_{k=0}^{N-1}\sum_{i=1}^{M} \varsigma_i^2[k] \end{cases} \qquad (2.30)$$

In (2.30) the term $\sum_{k=0}^{N-1}\sum_{i=1}^{M} s_i^2[k] = E_s$ corresponds to the average transmitted signal energy, and the term $\sum_{k=0}^{N-1}\sum_{i=1}^{M} \eta_i^2[k] - \sum_{k=0}^{N-1}\sum_{i=1}^{M} \varsigma_i^2[k]$ is the background noise at the GR output. The GR output background noise is a difference between the noise power at the PF and AF outputs. Practical implementation of the GR decision statistics requires the estimation of the noise variance σ_η^2 using the reference noise $\eta_i[k]$ at the AF output.

2.1.3.4. Generalized Receiver in DS-SSMA System

Fig. 2.1 represents a conventional generalized receiver employed by DS-SSMA wireless communication system, a block diagram of which is demonstrated in Fig. 2.2, to coherently demodulate the desired user signal in an asynchronous system. It should be noted that each user signal undergoes a Rayleigh flat fading channel, thus, an equalizer is not included in the receiver. By symmetry, mathematical derivation of the average error probability is the same for all users. We assume, without loss of generality, the ith user is the target one and $\tau_i = \phi_i = 0$. Hence, all delays and carrier phase shifts are measured relative to those of the ith signal. The decision statistic is the output of a complex generalized receiver that employs the complex carrier $\exp\{-j\omega_c t\}$ and the complex spreading signal $a_i^*(t)$.

The output of the complex generalized receiver matched to the ith signal is the random variable given by

$$T_{GR_i} = \int_0^T [2z(t)a_i^*(t) - z^2(t) + \eta^2(t)]\exp\{-j\omega_c t\}dt \ . \tag{2.31}$$

Taking into consideration the general ideal condition of functioning of the generalized receiver $\rho \to 1$ and $s_i^{ms}[k] = s_i[k]$, we can rewrite (2.31) in the following form:

$$T_{GR_i} = I_i(b,\tau,\phi) + \sqrt{2P_i}\,A_i b_0^{(i)}T + \varsigma_i(t), \tag{2.32}$$

where $\varsigma_i(t)$ is the background noise defined as

$$\varsigma_i(t) = \int_0^T [\eta_i^2(t) - \zeta_i^2(t)]dt, \tag{2.33}$$

and $I_i(b,\tau,\phi)$ is the total multiple access interference (MAI) defined as

$$I_i(\overline{b},\overline{\tau},\overline{\phi}) = \sum_{i=1,i\neq k}^{K} \sqrt{2P_i}\,A_i \exp\{j\phi_i\}[b_{-1}^{(i)} R_{i,k}(\tau_i) + b_0^{(i)} \hat{R}_{i,k}(\tau_i)], \tag{2.34}$$

with

$$\begin{cases} b^{(i)} = (b_{-1}^{(i)}, b_0^{(i)}), \\ \overline{b} = (b^{(1)}, b^{(2)}, \ldots, b^{(i-1)}, b^{(i+1)}, \ldots, b^{(K)}), \\ \overline{\tau} = (\tau_1, \tau_2, \ldots, \tau_{i-1}, \tau_{i+1}, \ldots \tau_K), \\ \overline{\phi} = (\phi_1, \phi_2, \ldots, \phi_{i-1}, \phi_{i+1}, \ldots \phi_K), \end{cases} \tag{2.35}$$

$$\begin{cases} R_{i,k}(\tau_i) = \int_0^{\tau} a_i(t-\tau)a_k^*(t)dt, \\ \hat{R}_{i,k}(\tau_i) = \int_{\tau}^{T} a_i(t-\tau)a_k^*(t)dt, \end{cases} \tag{2.36}$$

for $0 \leq \tau \leq T$. Along the lines of [4], it is easy to show that for $0 \leq \tau \leq T$, these two cross-correlation functions can be written in the following form:

$$\begin{cases} R_{i,k}(\tau_i) = C_{i,k}(l-N)\hat{R}_{\psi}(\tau - lT_c) + C_{i,k}(l+1-N)R_{\psi}(\tau - lT_c), \\ \hat{R}_{i,k}(\tau_i) = C_{i,k}(l)\hat{R}_{\psi}(\tau - lT_c) + C_{i,k}(l+1)R_{\psi}(\tau - lT_c), \end{cases} \tag{2.37}$$

where $l = \lfloor \tau/T_c \rfloor$ is the integer part of τ/T_c; $R_{\psi}(s)$ and $\hat{R}_{\psi}(s)$ are the continuous time partial autocorrelation functions of the chip waveform defined as

$$\begin{cases} R_{\psi}(s) = \int_{0}^{s} \psi(t)\psi(t+T_c-s)dt, \\[2ex] \hat{R}_{\psi}(s) = \int_{s}^{T_c} \psi(t)\psi(t-s)dt, \end{cases} \qquad (2.38)$$

for $0 \le s \le T_c$ and zero otherwise. Therefore, the multiple access interference (2.34) can be presented in the following form:

$$\begin{aligned} I_i(b,\tau,\phi) &= \sum_{i=1,i\ne k}^{K} \sqrt{2P_i}\,A_i \exp\{j\phi_i\}b_0^{(i)}\{[b_{-1}^{(i)}(b_0^{(i)})^*C_{i,k}(l_i-N) + \\ &\quad + C_{i,k}(l_i)]\hat{R}_{\psi}(\tau_i-l_iT_c) + [b_{-1}^{(i)}(b_0^{(i)})^*C_{i,k}(l_i+1-N) + \\ &\quad + C_{i,k}(l_i+1)]R_{\psi}(\tau_i-l_iT_c)\} = \\ &= \sum_{i=1,i\ne k}^{K} \sqrt{2P_i}\,A_i \exp\{j\phi_i\}b_0^{(i)}[\theta_{i,k}^{(n)}(l_i)\hat{R}_{\psi}(\tau_i-l_iT_c) + \\ &\quad + \theta_{i,k}^{(n)}(l_i+1)R_{\psi}(\tau_i-l_iT_c)], \end{aligned} \qquad (2.39)$$

where $l = \lfloor \tau/T_c \rfloor$, $n \in \{0,1,2,3\}$ satisfies $\exp\{jn\pi/2\} = b_{-1}^{(i)}(b_0^{(i)})^*$ and $\theta_{i,k}^{(n)}(l)$ is the complex quaternary correlation function defined in [31] as

$$\theta_{i,k}^{(n)}(l) = C_{i,k}(l) + C_{i,k}(l-N)\exp\{j\pi n/2\}, \qquad (2.40)$$

with $0 \le l \le N-1$. Obviously, $\theta_{i,k}^{(0)}(l)$ is the complex quaternary periodic cross-correlation function and $\theta_{i,k}^{(n)}(l)$, $n=1,2,3$ is the quaternary odd cross correlation function.

2.1.4. System Performance

In this section, the exact formulas for the average bit error rate (*BER*) for the synchronous system with quaternary signal and complex sequences over Rayleigh fading channel is derived first. Thereafter, the average *BER* for quaternary asynchronous system with quaternary signal and complex sequences over Rayleigh fading channel is evaluated using the characteristic function approach. The Gaussian approximate average *BER* for the asynchronous system under consideration is obtained with the MAI being modelled as the complex Gaussian random variable. A symbol error occurs in the complex processing system if the decision statistic $T_{GDi}(\mathbf{X})$ is not in the same quadrant of the complex plane as the data symbol from the desired signal. Since $b_i^{(i)} \in \{(1/\sqrt{2})(\pm1\pm j)\}$ is the quaternary phase data, the method discussed in [32] can be adopted with some modifications to include the interference affect. Thus, simple computations show that for Gray coded quaternary signal transmission system with complex sequences in the flat Rayleigh fading, the average *BER* of the ith Rayleigh faded user is given by

$$BER = 0.5[E(P_{i\mathcal{R}}) + E(P_{i\mathcal{G}})], \tag{2.41}$$

where $P_{i\mathcal{R}}$ is the average *BER* that $\mathcal{R}\{T_{GR_i}\}$ is not of the same polarity as $\mathcal{R}\{b_0^{(i)}\} = \pm 1/\sqrt{2}$ and $P_{i\mathcal{G}}$ is the average probability that $\mathcal{G}\{T_{GR_i}\}$ is not of the same polarity as $\mathcal{G}\{b_0^{(i)}\} = \pm 1/\sqrt{2}$. The expectation operator E is taken over the random vectors $\mathbf{b}, \boldsymbol{\tau}, \phi$ and random variable A_i, which are assumed to be mutually independent.

2.1.4.1. Synchronous *BER* Analysis

In this subsection, we consider the *BER* calculation for a synchronous system, i.e., $\tau_1 = \tau_2 = \ldots = \tau_K = 0$ in (2.22). The exact *BER* for the synchronous system with complex spreading sequences in Rayleigh fading is derived here. The presented derivation both serves completeness and clarifies the derivation and understanding of the asynchronous results in the next subsection. Let us describe the *BER* property of the synchronous system under consideration. We consider the synchronous DS-SSMA system with complex signature sequences and complex generalized receiver over flat Rayleigh fading channel presented in Fig. 2.1. Then the average synchronous *BER* for a Rayleigh fading user is

$$\overline{BER} = 0.5[\overline{BER_{i\mathcal{R}}} + \overline{BER_{i\mathcal{G}}}] = 0.5(1 - \Delta_s), \tag{2.42}$$

where

$$\Delta_s = \frac{1}{\sqrt{1 + \sum_{i=1, i \neq k}^{K} \frac{2P_i}{N^2 P_k} C_{i,k}(0)C_{i,k}^*(0) + \frac{\mathcal{N}_0^2}{16E_i^2}}} \tag{2.43}$$

and $E_i = P_i T$ is the energy per data symbol.

Let us show it. The output of the complex generalized receiver matched to the ith signal is given by (2.31) and (2.32), where $\varsigma_i(t) = \varsigma_{i\mathcal{R}}(t) + \varsigma_{i\mathcal{G}}(t)$ is as (2.33) with $\varsigma_{i\mathcal{R}}(t)$ and $\varsigma_{i\mathcal{G}}(t)$ being independent random variable distributed according to the modified second kind Bessel function of an imaginary argument or, as it also called, McDonald function each having the mean $\sigma_{\eta_{i\mathcal{R}}}^2 = \sigma_{\varsigma_{i\mathcal{G}}}^2 = \mathcal{N}_0 T/4$ and variance $(1/16)\mathcal{N}_0^2 T^2$ in the simplest ideal case when $\sigma_\eta^2 = \sigma_\varsigma^2$, i.e., the power of the noise at the PF and AF of the generalized receiver are equal between each other. In the statistical sense we can assume that $\varsigma_i(t)$ can be considered as the random process with asymptotic Gaussian distribution with the mean $\sigma_{\eta_{i\mathcal{R}}}^2 = \sigma_{\varsigma_{i\mathcal{G}}}^2 = \mathcal{N}_0 T/4$ and variance $(1/16)\mathcal{N}_0^2 T^2$.

In view of (2.37) and (2.38)

$$I_i(b,\tau,\phi) = \sum_{i=1,i\neq k}^{K} \sqrt{2P_i}\, A_i \exp\{j\phi_i\} b_0^{(i)} C_{i,k}(0) T_c = \sum_{i=1,i\neq k}^{K} \sqrt{2P_i}\, D_i b_0^{(i)} C_{i,k}(0) \frac{T}{N}, \quad (2.44)$$

where $D_i = A_i \exp\{j\phi_i\} = D_{i_1} + jD_{i_2}$ is the zero mean complex Gaussian variable with the unit variance, i.e., D_{i_1} and D_{i_2} are the independent zero mean real Gaussian variables with unit variance. This is true since A_i is the Rayleigh distributed; ϕ_i is the uniformly distributed within the limits of the interval $[0,2\pi)$ and A_i, ϕ_i are independent; $b_0^{(i)}$ takes the values within the limits of the interval $\{(1/\sqrt{2})(\pm 1 \pm j)\}$ with independent complex uniform, and is independent of D_i, so $D_i' = D_i b_0^{(i)}$ is the independent zero mean complex Gaussian variable with the unit variance. Note that $\{A_i,\phi_i,b_0^{(i)}\}_{i=1}^{K}$ are mutually independent complex random variables, hence, I_i is the complex Gaussian random variable with the real and imaginary parts being zero mean real independent Gaussian random variables and each having variance

$$\sigma_{I,\mathfrak{R}}^2 = \sigma_{I,\mathfrak{I}}^2 = \sum_{i=1,i\neq k}^{K} \frac{2P_i\,T^2}{N^2} |C_{i,k}(0)|^2 = \sum_{i=1,i\neq k}^{K} \frac{2P_i\,T^2}{N^2} C_{i,k}(0) C_{i,k}^*(0) \,. \quad (2.45)$$

Define

$$I_i' = I_i(b,\tau,\phi) + \varsigma_i(t) = I_{i_1}' + jI_{i_2}', \quad (2.46)$$

with I_{i_1}' and I_{i_2}' being the real and imaginary parts of I_i'. Then it follows from (2.44) and (2.45) that I_{i_1}' and I_{i_2}' are the independent asymptotic zero mean Gaussian random variables with the variances $\sigma_{I,\mathfrak{R}}^2 + \sigma_{\varsigma,\mathfrak{R}}^2$ and $\sigma_{I,\mathfrak{I}}^2 + \sigma_{\varsigma,\mathfrak{I}}^2$, respectively, i.e., I_i' is the complex Gaussian random variable. By symmetry and using the independence of I_i and ς_i, given A_i, the average conditional *BER*s for the real and imaginary parts of the ith user are

$$\overline{BER}_{i\mathfrak{R}|A_i} = \overline{BER}_{i\mathfrak{I}|A_i} = Q\left\{ \frac{\sqrt{P_i}\,A_i T}{\sqrt{\displaystyle\sum_{i=1,i\neq k}^{K} \frac{2P_i T^2}{N^2} C_{i,k}(0) C_{i,k}^*(0) + \dfrac{\mathcal{N}_0^2 T^2}{16}}} \right\}, \quad (2.47)$$

where

$$Q(x) = \frac{1}{2\pi} \int_{x}^{\infty} \exp\{-0.5t^2\}dt \,. \quad (2.48)$$

Averaging over A_i with respect to the Rayleigh distribution in (2.18) and using the integral identity

$$\int_{0}^{\infty} x \exp\{-0.5x^2\} Q(x/\sigma)dx = \frac{1}{2} - \frac{1}{2\sqrt{\sigma^2+1}}, \tag{2.49}$$

the average synchronous *BER* for a Rayleigh faded user is obtained in (2.42). Note that UCHT sequences are orthogonal, i.e., $C_{i,k}(0) = 0, i \neq k$; Δ_s in (2.42) for quaternary synchronous systems with complex UCHT sequences reduces to

$$\Delta_s = 1 + \left(\sqrt{\frac{\mathcal{N}_0^2}{16E_i^2}} \right)^{-1} = 1 + \frac{4E_i}{\mathcal{N}_0} . \tag{2.50}$$

This shows that the output of the ith generalized receiver is not affected at all by the signals of other users, and it is only affected by the background noise and the Rayleigh fading.

2.1.4.2. Asynchronous *BER* Analysis

In this subsection, the average *BER* calculation for asynchronous DS-SSMA system with complex signature sequences is investigated. The channel is assumed to be the Rayleigh fading and complex processing is adopted for the modulator and generalized receiver.

2.1.4.2.1. Characteristic Function Approach

Define

$$d_{i,k}^{(n)} = \theta_{i,k}^{(n)}[(l_i)\hat{R}_\psi(\tau_i - l_i T_c) + (l_i + 1)R_\psi(\tau_i - l_i T_c), \tag{2.51}$$

where $n = 0,1,2,3$ and

$$I_{i,k} = \sqrt{2P_i}D_i b_0^{(i)} d_{i,k}^{(n)} . \tag{2.52}$$

Then, (2.39) can be written in the following form:

$$I_i(b,\tau,\phi) = \sum_{i=1,i\neq k}^{K} I_{i,k} = \sum_{i=1,i\neq k}^{K} \sqrt{2P_i}D_i b_0^{(n)} d_{i,k}^{(n)} . \tag{2.53}$$

Since

$$D_i = A_i \exp\{j\phi_i\} = D_{i_1} + jD_{i_2} \tag{2.54}$$

is the independent zero mean complex Gaussian random variable with the unit variance, then, given $b^{(i)} = (b_0^{(i)}, b_{-1}^{(i)})$ and τ_i,

$$I_{i,k} = I_{i,k,\mathcal{R}} + jI_{i,k,\mathcal{G}} \tag{2.55}$$

is the independent zero mean complex Gaussian random variable with the variance

$$\sigma^2_{i,k,\mathcal{R}} = \sigma^2_{i,k,\mathcal{I}} = 2P_i \mid d^{(n)}_{i,k} \mid^2 . \tag{2.56}$$

This implies that the conditional probability density function for $I_{i,k}$ takes the following form:

$$f_{I_{i,k} \mid b^{(i)}, \tau_i}(x,y) = \frac{1}{4P_i \pi \mid d^{(n)}_{i,k} \mid^2} \exp\left\{ -\frac{x^2 + y^2}{4P_i \mid d^{(n)}_{i,k} \mid^2} \right\}. \tag{2.57}$$

Note that

$$b^{(i)} = (b^{(i)}_0, b^{(i)}_{-1}) \in \{(1/\sqrt{2})(\pm 1 \pm j)\}^2, \tag{2.58}$$

averaging over $b^{(i)}$, we obtain

$$f_{I_{i,k} \mid b^{(i)}, \tau_i}(x,y) = \frac{1}{4}\sum_{n=0}^{3} \frac{1}{4P_i \pi \mid d^{(n)}_{i,k} \mid^2} \exp\left\{ -\frac{x^2 + y^2}{4P_i \mid d^{(n)}_{i,k} \mid^2} \right\}. \tag{2.59}$$

Since $\mid d^{(n)}_{i,k} \mid^2$, $n = 0,1,2,3$ appears in the denominators of the exponential function arguments, thus it is difficult to average over τ_i. In order to solve this problem, we employ the characteristic function approach. The characteristic function of $I_{i,k}$, given τ_i, takes the following form:

$$\Theta_{I_{i,k} \mid \tau_i}(\omega_1,\omega_2) = \frac{1}{4}\sum_{n=0}^{3} \exp\{-(\omega_1^2 + \omega_2^2)P_i \mid d^{(n)}_{i,k} \mid^2\}. \tag{2.60}$$

Averaging over τ_i, the characteristic function of $I_{i,k}$ takes the form:

$$\begin{aligned}
\Theta_{I_{i,k} \mid \tau_i}(\omega_1,\omega_2) &= \frac{1}{4}\sum_{n=0}^{3} \int_0^T \exp\{-(\omega_1^2 + \omega_2^2)P_i \mid d^{(n)}_{i,k} \mid^2\}d\tau_i = \\
&= \frac{1}{4}\sum_{n=0}^{3}\sum_{l_i=0}^{N-1} \int_{l_i}^{(l_i+1)T_c} \exp\{-(\omega_1^2 + \omega_2^2)P_i \mid d^{(n)}_{i,k} \mid^2\}d\tau_i
\end{aligned} \tag{2.61}$$

If the chip waveform $\psi(t)$ is the rectangular pulse of duration T_c, i.e., $\psi(t) = 1$ for $0 \leq t < T_c$ and $\psi(t) = 0$ otherwise. Then for this waveform, $R_\psi(\tau) = \tau$ and $\hat{R}_\psi(\tau) = T_c - \tau$. Let

$$u_i = \frac{\tau_i - l_i T_c}{T_c}, \tag{2.62}$$

so $u_i \in [0,1)$. Then (2.61) reduces to

$$\Theta_{I_{i,k}|\tau_i}(\omega_1,\omega_2) = \frac{1}{4N}\sum_{n=0}^{3}\sum_{l_i=0}^{N-1}\int_0^1 \exp\{-(\omega_1^2+\omega_2^2)P_iT_c^2 \mid \theta_{i,k}^{(n)}(l_i)(1-u_i)+\theta_{i,k}^{(n)}(l_i+1)u_i \mid^2\}du_i =$$

$$\cong \frac{1}{4N}\sum_{n=0}^{3}\sum_{l_i=0}^{N-1}\int_0^1 \exp\{-(\omega_1^2+\omega_2^2)P_iT_c^2 \mathscr{K}_{i,k}^{(n)}\}du_i,$$

(2.63)

where

$$\mathscr{K}_{i,k}^{(n)} = \mid \theta_{i,k}^{(n)}(l_i)(1-u_i)+\theta_{i,k}^{(n)}(l_i+1)u_i \mid^2.$$

(2.64)

In order to calculate $\Theta_{I_{i,k}|\tau_i}(\omega_1,\omega_2)$, let us show that

$$\mathscr{K}_{i,k}^{(n)} = \begin{cases} A^2 + B^2 = \mid \theta_{i,k}^{(n)}(l_i)\mid^2, & \text{if } \theta_{i,k}^{(n)}(l_i)=\theta_{i,k}^{(n)}(l_i+1) \\ G^2(u_i-F)^2+H & \text{otherwise} \end{cases},$$

(2.65)

where

$$\begin{cases} A = \mathscr{R}\{\theta_{i,k}^{(n)}(l_i)\}, & B = \mathscr{G}\{\theta_{i,k}^{(n)}(l_i)\}, \\ C = \mathscr{R}\{\theta_{i,k}^{(n)}(l_i)\}, & J = \mathscr{G}\{\theta_{i,k}^{(n)}(l_i)\}, \end{cases}$$

(2.66)

and in the case of $\theta_{i,k}^{(n)}(l_i) \neq \theta_{i,k}^{(n)}(l_i+1)$, the parameters are defined as

$$\begin{cases} E = \sqrt{(C-A)^2+(J-B)^2} = \mid \theta_{i,k}^{(n)}(l_i+1)-\theta_{i,k}^{(n)}(l_i)\mid, \\ F = \dfrac{A^2+B^2-AC-BJ}{(C-A)^2+(D-B)^2}, \\ H = \dfrac{(AJ-BC)^2}{(C-A)^2+(J-B)^2} \end{cases}$$

(2.67)

In this case

$$J_{i,k}^{(n)} = \mid [(C-A)+j(J-D)]u_i + A + jB\mid^2 = [(C-A)u_i + A]^2 + [(J-B)u_i + B]^2 =$$

$$= [(C-A)^2+(J-B)^2]u_i^2 + 2[(C-A)A+(J-B)B]u_i + A^2 + B^2 =$$

$$= \begin{cases} A^2+B^2, & \mid \theta_{i,k}^{(n)}(l_i+1)=\theta_{i,k}^{(n)}(l_i)\mid \\ E^2(u_i-F)^2+H, & \text{otherwise} \end{cases}$$

(2.68)

Consider the next statement. Let

$$\phi_1^{(n)}(l_i,\omega_1,\omega_2) \cong \int_0^1 \exp\{-(\omega_1^2+\omega_2^2)P_iT_c^2[E^2(u_i-F)^2+H]du_i.$$

(2.69)

Then

$$\phi_1^{(n)}(l_i,\omega_1,\omega_2) = \frac{\sqrt{2\pi}\,\mathcal{M}_1(i)}{T_c E\sqrt{2P_i(\omega_1^2+\omega_2^2)}}\exp\{-(\omega_1^2+\omega_2^2)P_iT_c^2 H\}, \qquad (2.70)$$

where

$$\mathcal{M}_1(i) = Q\left[T_c E(F-1)\sqrt{2P_i(\omega_1^2+\omega_2^2)}\right] - Q\left[T_c EF)\sqrt{2P_i(\omega_1^2+\omega_2^2)}\right]. \qquad (2.71)$$

Based on definition [32] we obtain

$$\phi_1^{(n)}(l_i,\omega_1,\omega_2) = \exp\{-(\omega_1^2+\omega_2^2)P_iT_c^2 H\}\int_0^1 \exp\{-(\omega_1^2+\omega_2^2)P_iT_c^2[E^2(u_i-F)^2]\}du_i =$$

$$= \frac{1}{T_c E\sqrt{2P_i((\omega_1^2+\omega_2^2)}}\exp\{-(\omega_1^2+\omega_2^2)P_iT_c^2 H\}\int_{T_c E(F-1)\sqrt{2P_i(\omega_1^2+\omega_2^2)}}^{T_c EF\sqrt{2P_i(\omega_1^2+\omega_2^2)}} \exp(-0.5t^2)dt. \qquad (2.72)$$

Thus, (2.70) is obtained. Now, the characteristic function of $I_{i,k}$ can be presented in the following form:

$$\Theta_{I_{i,k}}(\omega_1,\omega_2) = \frac{1}{4T}\sum_{n=0}^{3}\sum_{l_i=0}^{N-1}\phi^{(n)}(l_i,\omega_1,\omega_2), \qquad (2.73)$$

where

$$\phi^{(n)}(l_i,\omega_1,\omega_2) = \begin{cases} \exp\{-(\omega_1^2+\omega_2^2)P_iT_c^2\,|\,\theta_{i,k}^{(n)}(l_i)\,|^2\}, & |\,\theta_{i,k}^{(n)}(l_i+1) = \theta_{i,k}^{(n)}(l_i)\,| \\ \phi_1^{(n)}(l_i,\omega_1,\omega_2), & \text{otherwise} \end{cases}, \qquad (2.74)$$

where $\phi_1^{(n)}(l_i,\omega_1,\omega_2)$ is given by (2.70). Note that the above characteristic function is given as an explicit closed form expression involving only the exponential function and Q-functions. It can be readily programmed for direct evaluation. Thus, it is easy to show that

$$\phi_1^{(n)}(l_i,\omega_1,\omega_2) \to \exp\{-(\omega_1^2+\omega_2^2)P_iT_c^2\,|\,\theta_{i,k}^{(n)}(l_i)\,|^2\}, \qquad (2.75)$$

as

$$\theta_{i,k}^{(n)}(l_i) - \theta_{i,k}^{(n)}(l_i+1)\,|\to 0. \qquad (2.76)$$

Since the complex random variables $I_{i,k}$, $1\le i\le K$, $i\ne k$ from different interferers are independent, the characteristic function for the total multiple access interference term I_i is given by

$$\Theta_{I_{i,k}}(\omega_1,\omega_2) = \prod_{i=1,i\ne k}^{K}\Theta_{I_i}(\omega_1,\omega_2). \qquad (2.77)$$

Let $\xi_i = I_i + \varsigma_i$, where ς_i is the complex zero mean Gaussian random variable in the statistical sense representing the background noise [14]. Since the other user interference and background noise are independent, the characteristic function of ξ_i takes the following form:

$$\Theta_{\xi_i}(\omega_1,\omega_2) = \Theta_{I_i}(\omega_1,\omega_2)\Theta_{\varsigma_i}(\omega_1,\omega_2),$$ (2.78)

where $\Theta_{\varsigma_i}(\omega_1,\omega_2)$ is the characteristic function of the background noise ς_i. Then the marginal characteristic functions of ξ_i are given by

$$\begin{cases} \Theta_{\xi_i\mathfrak{R}}(\omega) = \Theta_{I_i}(\omega,0)\Theta_{\varsigma_i}(\omega,0) = \Theta_{I_i\mathfrak{R}}(\omega)\Theta_{\varsigma_i\mathfrak{R}}(\omega), \\ \Theta_{\xi_i\mathfrak{I}}(\omega) = \Theta_{I_i}(0,\omega)\Theta_{\varsigma_i}(0,\omega) = \Theta_{I_i\mathfrak{I}}(\omega)\Theta_{\varsigma_i\mathfrak{I}}(\omega) \end{cases}.$$ (2.79)

It follows from (2.73)-(2.79) that the marginal characteristic functions of $\xi_i = I_i + \varsigma_i$ satisfy the following equality $\Theta_{\xi_i\mathfrak{R}}(\omega) = \Theta_{\xi_i\mathfrak{I}}(\omega)$. Therefore, by symmetry, the conditional *BER* for the target user can be given as

$$\overline{BER}_{A_i} = 0.5(\overline{BER}_{i\mathfrak{R}|A_i} + \overline{BER}_{i\mathfrak{I}|A_i}) = \overline{BER}_{i\mathfrak{R}|A_i} = BER\{\mathfrak{R}(\xi_i) < -\sqrt{P_i}A_iT\} =$$

$$= \frac{1}{2} - \frac{1}{\pi}\int_0^\infty \frac{\sin(\sqrt{P_i}A_iT\omega)}{\omega}\Theta_{\xi_i\mathfrak{R}}(\omega)d\omega =$$ (2.80)

$$= Q\left(\frac{\sqrt{P_i}A_iT}{\sigma_{\varsigma_i\mathfrak{R}}}\right) + \frac{1}{\pi}\int_0^\infty \frac{\sin(\sqrt{P_i}A_iT\omega)}{\omega}[1 - \Theta_{I_i\mathfrak{R}}(\omega)]\Theta_{\varsigma_i\mathfrak{R}}(\omega)d\omega$$

Using the integral identity

$$\int_0^\infty \sin(kx)x\exp(-0.5x^2)dx = \sqrt{0.5\pi}k\exp(-0.5k^2),$$ (2.81)

and averaging over A_i, we obtain the average *BER* for a Rayleigh faded user given by

$$BER = \frac{1}{2}\left\{1 - \left(\sqrt{1 + \frac{\mathcal{N}_0^2}{16E_i^2}}\right)^{-1}\right\} + \sqrt{\frac{P_i}{2\pi}}T\int_0^\infty \exp(-0.5P_iT^2\omega^2)[1 - \Theta_{I_i\mathfrak{R}}(\omega)]\Theta_{\varsigma_i\mathfrak{R}}(\omega)d\omega.$$ (2.82)

Consider the DS-SSMA quaternary asynchronous system with complex sequences and complex receivers over Rayleigh fading channels shown in Fig. 2.1. If the chip waveform is a rectangular pulse, then the average *BER* of Rayleigh faded user can be given by (2.82). Note that the above mentioned *BER* in (2.82) is expressed as the Q-functions and single integral. Numerical integration techniques similar to [33] that use Simpson's rule and series expansion method can be applied to obtain the *BER* with high accuracy.

2.1.4.2.2. Gaussian Approximation

The high accuracy *BER* for asynchronous system under consideration can be obtained from the previous discussion. However, it would generally require excessive computations. Hence, here we shall compute the average *BER* for asynchronous complex DS-SSMA system in the flat Rayleigh fading by modelling multiple access interference in (2.34) as a complex Gaussian process. Obviously, the mean of the multiple access interference reduce to zero and the variance is given by

$$
\begin{aligned}
\sigma_{I_k \mathcal{R}}^2 = \sigma_{I_k \mathcal{G}}^2 &= \sum_{i=1, i \neq k}^{K} 2 P_i E[|R_{i,k}(\tau_i)|^2 + |\hat{R}_{i,k}(\tau_i)|^2] = \\
&= \sum_{i=1, i \neq k}^{K} \frac{2 P_i}{T} \int_0^T [|R_{i,k}(\tau_i)|^2 + |\hat{R}_{i,k}(\tau_i)|^2] d\tau = \\
&= \sum_{i=1, i \neq k}^{K} \sum_{k=0}^{N-1} \frac{2 P_i}{T} \int_{lT_c}^{(l+1)T_c} [|R_{i,k}(\tau_i)|^2 + |\hat{R}_{i,k}(\tau_i)|^2] d\tau,
\end{aligned}
\tag{2.83}
$$

where the mathematical expectation has been computed with respect to mutually independent complex random variables $\{A_i, \phi_i, b_{-1}^{(i)}, b_0^{(i)}\}|_{i=1}^{K}$ and the random variables given by (2.54) being zero mean complex Gaussian random variable with the unit variance. It is trivial to show that

$$
E[\mathcal{R}(I_k)\mathcal{G}(I_k)] = E[\mathcal{R}(I_k)]E[\mathcal{G}(I_k)] = 0,
\tag{2.84}
$$

with that, the real and imaginary parts of random variable I_k are uncorrelated and orthogonal. Therefore, if I_k is assumed to be complex Gaussian random variable, then it is independent. In the case of a rectangular wave shape function $\psi(t) = 1, \; 0 \leq t \leq T_c$, then $R_\psi(s) = s$ and $\hat{R}_\psi(s) = T_c - s$. Hence, following the treatment in [4] and substituting (2.37) in (2.83), we obtain

$$
\sigma_{I_k \mathcal{R}}^2 = \sigma_{I_k \mathcal{G}}^2 = \sum_{i=1, i \neq k}^{K} \frac{2 P_i T^2}{3 N^3} r_{i,k},
\tag{2.85}
$$

where

$$
\begin{aligned}
r_{i,k} &= \sum_{l=0}^{N-1} [C_{i,k}(l-N) C_{i,k}^*(l-N) + C_{i,k}(l+1-N) \times \\
&\times C_{i,k}^*(l+1-N) + C_{i,k}(l) C_{i,k}^*(l) + \\
&+ C_{i,k}(l+1) C_{i,k}^*(l+1) \mathcal{R}\{C_{i,k}(l-N) C_{i,k}^*(l+1-N)\} + \mathcal{R}\{C_{i,k}(l) C_{i,k}^*(l+1)\} = \\
&= 2\mu_{i,k}(0) + \mathcal{R}\{\mu_{i,k}(1)\},
\end{aligned}
\tag{2.86}
$$

with

$$\mu_{i,k}(n) = \sum_{l=1-N}^{N-1} C_{i,k}(l)C_{i,k}^*(l+n).$$
(2.87)

Similar to analysis for the synchronous case, given A_k, the average conditional *BER*s for the real and imaginary parts of the ith user in the asynchronous case are

$$BER_{k\mathcal{R}|A_k} = BER_{k\mathcal{G}|A_k} = Q\left\{\frac{\sqrt{P_k}\,A_k T}{\sqrt{\sum_{i=1,i\neq k}^{K}\frac{2P_i T^2}{3N^3}r_{i,k} + \frac{\mathcal{N}_0^{\circ 2}}{16}}}\right\}.$$
(2.88)

Averaging over A_k with respect to the Rayleigh distribution in (2.18), the average asynchronous *BER* in the case of the Rayleigh fading channel takes the following form:

$$\overline{BER} = 0.5(\overline{BER_{i\mathcal{R}}} + \overline{BER_{i\mathcal{G}}}) = 0.5(1 - \Delta_a),$$
(2.89)

where

$$\Delta_a = \frac{1}{\sqrt{1 + \sum_{i=1,i\neq k}^{K}\frac{2P_i}{3N^3 P_k}r_{i,k} + \frac{\mathcal{N}_0^{\circ 2}}{16E_k^2}}}.$$
(2.90)

Based on the Gaussian approximation of the multiple access interference, the average *BER* evaluation is approximately obtained in (2.89) for the quaternary asynchronous DS-SSMA system with complex signature sequences, as well as the complex transmitters and receivers in Rayleigh fading as depicted in Fig. 2.1.

From [25]

$$\sum_{l=1-N}^{N-1} C_{i,k}(l)C_{i,k}^*(l+n) = \sum_{l=1-N}^{N-1} C_i(l)C_{k,k}^*(l+n).$$
(2.91)

Therefore

$$r_{i,k} = 2N^2 + 4\sum_{l=1}^{N-1}C_i(l)C_{k,k}^*(l) + \mathcal{R}\left\{\sum_{l=1-N}^{N-1}C_i(l)C_k^*(l+1)\right\}.$$
(2.92)

This shows that the *BER* performance for asynchronous system over Rayleigh fading channel can be obtained using the periodic autocorrelation functions.

2.1.5. Numerical Results

In this section, we report the numerical evaluation of the system performance for Gold sequences, near optimum four phase sequences of family $\mathcal{A}$ [34] and UCHT sequences.

As was discussed previously, there are 64 sets of UCHT matrices, and each set can generate 2^n complex sequences with the period duration of $N = 2^n$. It has been investigated in [10] that not all the UCHT sequences generated from one UCHT matrix can be simultaneously used as the spreading sequences, as this will lead to the large periodic and periodic cross correlation values.

In order to reduce the multiple access interference effectively, the subset of the complex sequences with smaller sum of the periodic cross-correlation values is selected for simulation, but a consideration of trade-off between the autocorrelation and cross correlation properties for spreading sequences suggests that the subset with the smallest sum must be avoided. High-speed UCHT sequences are also recommended to be more suitable for the asynchronous DS-SSMA communication systems than not high speed UCHT sequences [9].

Hence, in this section, we will choose $\mu_1 = j, \mu_2 = 1, \mu_3 = j$ and $N = 2^6 = 64$ as one set of not high speed UCHT sequences which have relatively good autocorrelation and cross -correlation functions among the UCHT sequences [10]. On the other hand, the Gold sequences are produced by a preferred pair of m sequences u and v, where u is obtained by using primitive polynomials

$$f_1 = x^6 + x + 1, \tag{2.93}$$

and v is obtained by using

$$f_2 = x^6 + x^5 + x^2 + x + 1. \tag{2.94}$$

Thus, the period of the Gold sequences is $N = 63$, and it has 65 different sequences belonging to the set

$$\mathscr{P}(u,v) = \{u, v, u \oplus v, u \oplus \mathscr{T}v, u \oplus \mathscr{T}^2 v, \ldots, u \oplus \mathscr{T}^{N-1} v\}, \tag{2.95}$$

where $\mathscr{T}$ is the shift operator. The 4-phase family $\mathscr{A}$ sequences are produced by using the primitive polynomials

$$f_3 = x^6 + 2x^3 + 3x + 1. \tag{2.96}$$

So, the period of 4-phase sequences is $N = 63$ with 65 different sequences. In the simulations, the equal transmission power, for simplicity, is assumed for each user and

$$P_1 = P_2 = \cdots = P_K = 1. \tag{2.97}$$

The *BER* performance of quaternary synchronous DS-SSMA system over Rayleigh fading channels is shown in Figs. 2.3 and 2.4, where the active users are $K = 8, K = 16$ and $K = 30$, respectively. It can be seen that the performance of the UCHT sequences is the best among the three sequences, and the Gold sequences also perform the better than the 4-phase family $\mathscr{A}$ sequences at the synchronous conditions. The reason is that the UCHT sequences are orthogonal, i.e., $C_{i,k}(0)=0, i \neq k$. As to the Gold sequences, $C_{i,k}(0)-1$, if $i \neq k$ and the v in the set of Gold sequences $\mathscr{P}(u,v)$ is not chosen as spreading sequence. For the 4-phase family $\mathcal{A}$ sequences, $C_{i,k}(0)=0, i \neq k$, could be $-1, 7, -9, -1 \pm 8j$ and each occurs many times. Hence, in view of (2.42), the *BER* performance in Figs. 2.3 and 2.4 are reasonable. Furthermore, for the UCHT sequences, the curves of *BER*s are the same for all different number of users, such as $K=8,16,30$. However, for the Gold sequences and 4-phase family $\mathcal{A}$ sequences, the curves of *BER*s are different for $K=8,16,30$.

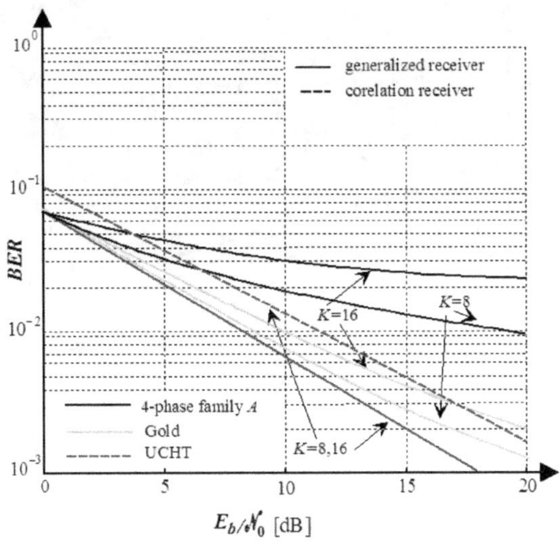

Fig. 2.3. *BER* performance for synchronous DS-SSMA with active users $K=8,16$.

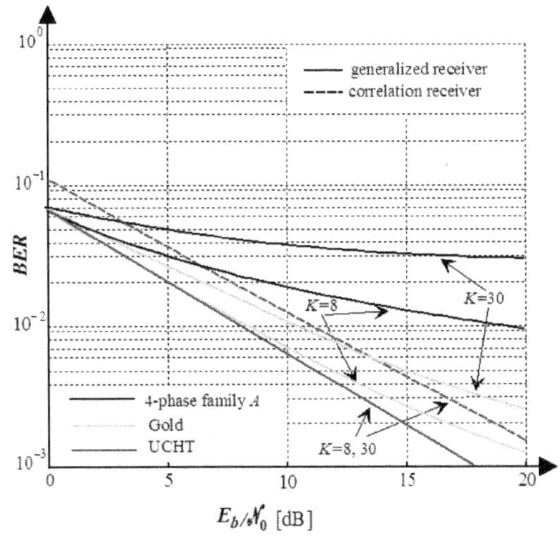

Fig. 2.4. *BER* performance for synchronous DS-SSMA with active users $K=8,30$.

Finally, Figs. 2.5-2.7 show the accurate *BER* computed using the characteristic function approach as well as the *BER* obtained by the Gaussian approximation method described in the previous section for quaternary asynchronous DS-SSMA systems with the Rayleigh fading channels, where the active users are K=8,16,30, respectively. In the asynchronous case, the cross-correlation function of the Gold sequences $C_{i,k}(l), i \neq k, |l|, 63$, could be $-1, -17, 15$ and each value occurs many times. The distributions of the periodic cross-correlation function values are different for the Gold and 4-phase family $\mathcal{A}$ sequences, but they are similar. Thus, the Gold and 4-phase family $\mathcal{A}$ sequences belong to pseudorandom spreading sequences and have similar correlation properties.

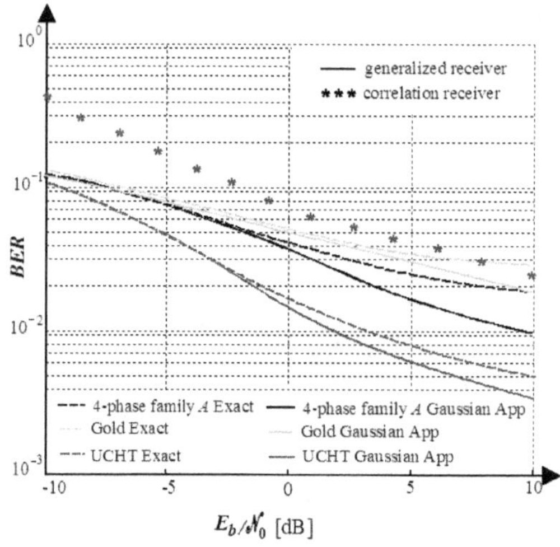

Fig. 2.5. *BER* performance for asynchronous DS-SSMA with active users $K = 8$. Correlation receiver is represented for UCHT and Gaussian approximation.

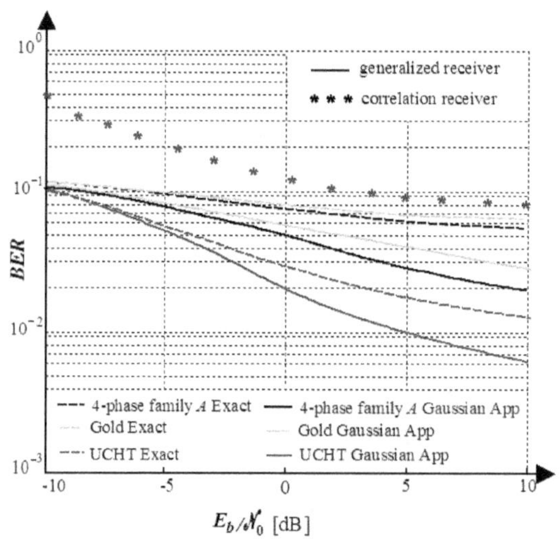

Fig. 2.6. *BER* performance for asynchronous DS-SSMA with active users $K = 16$.

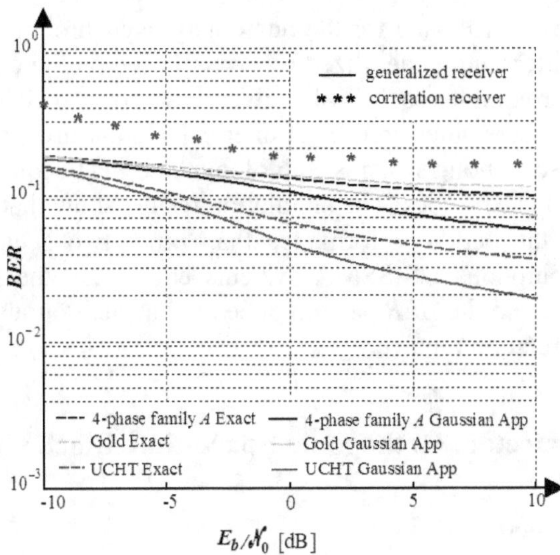

Fig. 2.7. *BER* performance for asynchronous DS-SSMA with active users $K = 30$.

It has also been shown that the UCHT complex sequences offer the better cross-correlation function properties compared with the Gold sequences [10]. In view of (2.82) and (2.89), the *BER* performance is mainly determined by the cross correlation function properties of the signature sequences. Figs. 2.5-2.7 have verified this. That is to say, the asynchronous system under consideration with UCHT sequences outperforms that with the Gold and 4-phase family $\mathscr{A}$ sequences, and the *BER* performance using the Gold sequences is also close to that using the 4-phase family $\mathscr{A}$ sequences.

We also derive from our numerical results presented in Figs. 2.5-2.7 that the *BER* estimates based on the complex Gaussian approximation consistently overestimate the accurate *BER* computed via the characteristic function method for the asynchronous quaternary DS-SSMA system in the Rayleigh fading channels with complex processing at the transmitter and receiver. Comparative analysis between the asynchronous DS-SSMA systems constructed on the basis of the generalized receiver and correlation one is represented in Figs. 2.5-2.7. We can see superiority in the *BER* performance under employment of the generalized receiver in DS-SSMA communication systems with active users in comparison with implementation of the correlation one in these systems.

2.1.6. Conclusions

In this section we investigated the performance of quaternary DS-SSMA wireless communications, which are constructed on the basis of the generalized approach to signal processing in noise, with complex signature sequences, as well as complex transmitters and receivers over the Rayleigh fading channels. The average *BER* is first derived for the quaternary synchronous systems. Due to the orthogonal property of UCHT sequences, the probability of bit errors of the synchronous DS-SSMA systems with UCHT sequences is lower in comparison with the synchronous systems with other nonorthogonal sequences.

The average *BER* is also evaluated for the quaternary asynchronous DS-SSMA systems with complex transmitters and generalized receivers based on the characteristic function approach, and the approximate result is also given based on the Gaussian approximation method of multiple access interference. Numerical results are presented to illustrate performance comparison among systems employing the UCHT sequences, 4-phase family $\mathcal{A}$ sequences and Gold sequences. The simulation results show that the UCHT complex sequences can yield the better performance than other two sequences. Comparative analysis of the asynchronous DS-SSMA systems employing the generalized receiver demonstrates superiority in the *BER* performance over asynchronous DS-SSMA systems implementing the correlation receiver.

2.2. Generalized Detector: Orthogonal Space-time Block Coding

2.2.1. Problem Statement

The growing demand for high rate date services through wireless channels experienced in recent years motivates the design of multiple antenna wireless communication systems to transmit increase data rates without substantial bandwidth expansion. In particular, the antenna diversity can be used to improve the performance of wireless communication systems such as the code division multiple access (CDMA). First presented in [35] for two transmit antennas and generalized in [36] and [37] for an arbitrary number of transmit antennas, the orthogonal space-time block coding (STBC) is the remarkable technique which can provide a full diversity gain with very low computational complexity.

However, a loss in capacity, characterized by the code rate and the number of receive antennas, is shown in [38] and [39] for an arbitrary channel. Following the analysis in [37-39], a characterization based on the equivalent scalar AWGN channel multiplied by a coefficient, which is a function of the Frobenius norm of the channel matrix with multiple antennas, was given in [4] assuming a full code rate. The Shannon and outage capacity for the scalar AWGN channel were also given.

The Shannon capacity

$$C = W \log_2 (1 + SNR),\qquad(2.93)$$

where *SNR* is the signal-to-noise ratio and W is the channel bandwidth, predicts the channel capacity C for an AWGN channel with continuous valued inputs and continuous valued outputs. However, a channel employing the STBC with the pulse amplitude modulation (PAM), phase shift keying (PSK) or quadrature amplitude modulation (QAM) has the discrete valued inputs and continuous valued outputs, which imposes an additional constraint on the capacity calculation.

In this section, we generalize the effective channel representation in [38] for all rate orthogonal STBCs, including the rate 0.5 STBCs $\mathbf{G}_N$, $N > 2$ and the rate 0.75 STBCs $\mathbf{H}_3$ and $\mathbf{H}_4$, given in [36]. A new capacity calculation taking into account the constraint of discrete valued inputs is presented here, as well as the capacity loss incurred by employing STBC.

In [40], an analysis of the bit error probability (*BEP*) of $\mathbf{G}_2$ for q-ary PSK was presented for the Rayleigh fading channel using the probability density function (pdf) of the received signal phase from [41]. However, the *BEP* for q-PSK ($q > 4$) is very complex using this approach. In [42], a union bound on the symbol error probability (*SEP*) for the STBC was presented. A general form for the exact pair wise error probability of STBCs was obtained in [43] based on the moment generating function of the Gaussian tail function. In [44], the STBC was applied to a DS-CDMA wireless communication downlink channel, and a novel decoding algorithm was presented. In [45], the exact expression for the pair wise error probability in a flat Rayleigh fading channel was derived in terms of the symbol distance between two message vectors for quaternary phase shift keying (QPSK), 16-QAM, 64-QAM and 256-QAM. In [46], a unified approach for calculating the error rates of linearly modulated signals over generalized fading channels was presented. However, the results of both [45] and [46] are given in an open form that has to be evaluated via numerical integration.

In this section, we analyze the error probability of STBCs for PAM, PSK, QAM modulated signals as a function of an *SNR* perspective based upon the equivalent scalar channel induced by the STBC. Using the *SNR* pdf, the closed form *SEPs* are given for various combinations of modulation and fading channels. Furthermore, these results are extended to a multiuser DS-CDMA wireless communication system with STBC employing the generalized receiver. Expressions for the capacity and error probability of the DS-CDMA wireless communication system with STBC are derived and analyzed.

2.2.2. General Channel Model

2.2.2.1. Channel Model

The channel model is the same as in [36-38]. Consider the wireless communication system with N transmitting and M receiving antennas. The channel is assumed to be quasistatic with flat fading, which means that the channel parameters are constant within the limits of one frame period, but are varied independently between frames. Furthermore, the perfect channel state information is assumed available at the receiver input, but the channel parameters are unknown at the transmitter.

Let T represents the number of time slots used to transmit S symbols. Hence, a general form for the transmission matrix of STBC is

$$\mathbf{G} = \begin{pmatrix} g_{11} & g_{21} & \cdots & g_{N1} \\ g_{12} & g_{22} & \cdots & g_{N2} \\ \cdots & \cdots & \cdots & \cdots \\ g_{1T} & g_{2T} & \cdots & g_{NT} \end{pmatrix}, \tag{2.94}$$

where g_{ij} represent a linear combination of the signal constellation components and their conjugates, and are transmitted simultaneously by the ith transmit antenna in the jth time slot for $i = 1 \ldots N$ and $j = 1 \ldots T$. Since there are S symbols transmitted over T time slots, the code rate of the STBC is given by

$$R = S/T. \tag{2.95}$$

It is shown in [36], based on the theory of orthogonal designs, that full rate STBCs exist for any number of transmitting antennas using an arbitrary real constellation such as PAM. For an arbitrary complex constellation such as PSK, QAM, the half rate STBCs exist for any number of transmit antennas, while full rate STBCs exist only for two transmitting antennas. As specific cases for two, three, and four transmit antennas, the rate 1, 0.5, and 0.75 STBCs are given in [36], and are denoted as $\mathbf{G}_2$, $\mathbf{G}_3$, $\mathbf{G}_4$, $\mathbf{H}_3$, and $\mathbf{H}_4$, respectively. At the particular time nT, the received signal corresponding to the nth input block spanning T time slots is

$$\mathbf{Y}_{nT} = \mathbf{H}\mathbf{G}^T + \mathbf{W}_{nT}, \tag{2.96}$$

where $\mathbf{Y}_{nT}$ is the $M \times T$ matrix; $\mathbf{H}$ is the $M \times N$ fading channel coefficient matrix with independent identically distributed (i.i.d.) entries modelled as the circular complex Gaussian random variables; $\mathbf{G}^T$ is the transpose of $\mathbf{G}$ with the size $N \times T$, and $\mathbf{W}_{nT}$ is the $M \times N$ receiver noise matrix with i.i.d. entries modelled as the circular complex Gaussian random variables with the zero mean and variance $\sigma_W = \Delta f \times \mathcal{N}_0/2$ in each dimension, where Δf is the bandwidth of linear system at the receiver input and $\mathcal{N}_0/2$ is the AWGN power spectral density.

2.2.2.2. Effective Scaled AWGN Channel

In [38], the equivalent scaled AWGN channel induced by the STBC for the complex constellations was given as

$$\mathbf{y}_{nT} = \| \mathbf{H} \|_F^2 \, \mathbf{x}_{nT} + \mathbf{w}_{nT}, \tag{2.97}$$

where $\mathbf{y}_{nT}$ is the $S \times 1$ complex matrix after STBC decoding from the received matrix $\mathbf{Y}_{nT}$; $\mathbf{x}_{nT}$ is the input $S \times 1$ complex input matrix with each entry having the energy E_s/N, E_s is the maximum total transmitted energy on the N transmit antennas per symbol time; and $\mathbf{w}_{nT}$ is the complex Gaussian noise with the zero mean and variance $\| \mathbf{H} \|_F^2 \, 0.5 \times \mathcal{N}_0$ in each real dimension;

$$\| \mathbf{H} \|_F^2 = \sum_{i=1}^{N} \sum_{j=1}^{M} \| h_{ij} \|^2 \tag{2.98}$$

is the squared Frobenius norm of $\mathbf{H}$; h_{ij} is the channel gain from the ith transmit antenna to the jth receive antenna. Taking into account the code rate, the equivalent AWGN scaled channel with a STBC is

$$\mathbf{y}_{nT} = (1/R)\|\mathbf{H}\|_F^2\,\mathbf{x}_{nT} + \mathbf{w}_{nT}, \tag{2.99}$$

where $\mathbf{w}_{nT}$ is the complex Gaussian noise with the zero mean and variance

$$Var = (1/2R)\|\mathbf{H}\|_F^2\,\mathcal{N}_0, \tag{2.100}$$

in each real dimension. Therefore, the effective instantaneous *SNR* denoted as γ_s at the receiver input is

$$\gamma_s = \frac{E_s}{NR\mathcal{N}_0}\|\mathbf{H}\|_F^2. \tag{2.101}$$

Let

$$h = \frac{1}{R}\|\mathbf{H}\|_F^2 = \sum_{i=1}^{N}\sum_{j=1}^{M}\frac{\|h_{ij}\|^2}{R}, \tag{2.102}$$

then the STBC channel model of (2.7) can be simplified to

$$\mathbf{y}_{nT} = h\mathbf{x}_{nT} + \mathbf{w}_{nT}, \tag{2.103}$$

and γ_s can be written as

$$\gamma_s = \frac{E_s}{N\mathcal{N}_0}h. \tag{2.104}$$

2.2.2.3. Distribution of Channel Coefficients and *SNR*

2.2.2.3.1. Rayleigh Fading

With Rayleigh fading, the coefficients h_{ij} can be modelled as the complex Gaussian variables with the zero mean and variance σ_w^2 in each dimension. The pdf of the coefficient h is then a central chi-square distribution with $2MN$ degrees of freedom

$$f_{Rayleigh}(h) = \frac{R^{MN}}{(2\sigma^2)^{MN}\Gamma(MN)}h^{MN-1}\exp\left\{-\frac{hR}{2\sigma_w^2}\right\},\ h \geq 0. \tag{2.105}$$

Consequently, the instantaneous *SNR* per symbol γ_s is also chi-square distributed. Using a change of variables, the pdf of γ_s is given by

$$f_{Rayleigh}(h) = \frac{R^{MN}}{\overline{\gamma}_{ch}^{MN}\Gamma(MN)}\gamma_s^{MN-1}\exp\left\{-\frac{\gamma_s}{\overline{\gamma}_{ch}}\right\},\ \gamma_s \geq 0 , \qquad (2.106)$$

where $\overline{\gamma}_{ch}$ is the average *SNR* per channel, which is assumed to be identical for all channels, i.e.

$$\overline{\gamma}_{ch} = \frac{E_s}{NR\mathcal{N}_0}E[\|\,h_{ij}\,\|^2] = \frac{2\sigma_w^2 E_s}{NR\mathcal{N}_0}, \qquad (2.107)$$

where $E[\cdot]$ is the mathematical expectation. It can easily be shown that the instantaneous *SNR* per bit γ_b has the same pdf except that

$$\overline{\gamma}_{ch} = \frac{E_s}{NR\mathcal{N}_0 \log_2 \mathcal{M}}, \qquad (2.108)$$

for $\mathcal{M}$-ary signal constellation.

2.2.2.3.2. Rician Fading

For Rician fading, the coefficients h_{ij} can be modelled as the complex Gaussian variables with the means μ_I and μ_Q for the real and imaginary parts, respectively, and the variance σ_w^2 in each dimension. In this case, $\|\mathbf{H}\|_F^2$ has a non-central chi-square distribution with $2MN$ degrees of freedom. The pdf of h is given by

$$f_{Rician}(h) = \frac{R}{2\sigma_w^2}\left[\frac{Rh}{s^2}\right]^{\frac{MN-1}{2}}\exp\left\{-\frac{s^2 + Rh}{2\sigma_w^2}\right\}I_{MN-1}\left(\frac{\sqrt{Rh}s}{\sigma_w^2}\right),\ h\geq 0, \qquad (2.109)$$

where

$$s^2 = MN(\mu_I^2 + \mu_Q^2) \qquad (2.110)$$

is the non-centrality parameter, $I_\alpha(x)$ is the α-th order modified Bessel function of the first kind, which may be represented by the infinite series

$$I_m(x) = \sum_{n=0}^{\infty}\frac{(0.5x)^{n+2k}}{k!\Gamma(n+k+1)}. \qquad (2.111)$$

Introducing the Rician parameter

$$\beta = \frac{\mu_I^2 + \mu_Q^2}{2\sigma_w^2}, \qquad (2.112)$$

the pdf in (2.105) can be written in the following form

$$f_{Rician}(h) = \sum_{i=0}^{\infty} \frac{(MN\beta)^i \exp\{-MN\beta\} R^{MN+i}}{\Gamma(i+1)\Gamma(MN+i)(2\sigma_w^2)^{MN+i}} \times$$
$$\times h^{MN+i-1} \exp\{-Rh/2\sigma_w^2\}, \ h \geq 0 \tag{2.113}$$

Using a change of variables, the pdf of the instantaneous *SNR* takes the following form:

$$f_{Rician}(h) = \sum_{i=0}^{\infty} \frac{(MN\beta)^i \exp\{-MN\beta\} \gamma_s^{MN+i-1} \exp\{-\gamma_s/\overline{\gamma}_c\}}{\Gamma(i+1)\Gamma(MN+i)\overline{\gamma}_c^{MN+i}}, \ \gamma_s \geq 0, \tag{2.114}$$

where $\overline{\gamma}_c$ is the average *SNR* per channel, which is assumed to be identical for all channels, as given in (2.107).

2.2.2.3.3. Nakagami-*m* Fading

For Nakagami-*m* fading with integer *m*, $\| h_{ij} \|$ is the amplitude of the channel coefficient h_{ij} and has the Nakagami-*m* distribution with the zero mean and variance σ_N^2 in each dimension. The random variable

$$y = R^{-1} \| h_{ij} \|^2, \tag{2.115}$$

then has the pdf

$$f(y) = \frac{R^m}{(2\sigma^2)^m \Gamma(m)} y^{m-1} \exp\{-Ry/2\sigma_w^2\}, \tag{2.116}$$

where $\sigma_w^2 = \sigma_N^2/m$. Observing that the pdf for Nakagami-*m* fading has the same form as the pdf for the Rayleigh fading but with $2m$ degree in (2.105), a single Nakagami-*m* fading channel is equivalent to an *m* diversity system for a Rayleigh fading channel. It is then straightforward to show that the results for STBCs over Nakagami-*m* fading channels can be obtained by considering the Rayleigh fading channels with the channel diversity order increased from *MN* to *mMN*. Consequently, the pdf of the instantaneous *SNR* per symbol γ_s can be obtained directly from (2.102) as

$$f_{Nakagami-m}(\gamma_s) = \frac{1}{\overline{\gamma}_c^{mMN} \Gamma(mMN)} \gamma_s^{mMN-1} \exp\{-\gamma_s/\overline{\gamma}_c\} \ , \tag{2.117}$$

where $\overline{\gamma}_c$ is the average *SNR* per channel, which is assumed to be identical for all channels, as in (2.107).

2.2.3. Capacity Analysis of STBC

2.2.3.1. Shannon Capacity over Fading Channels

The capacity of a multiple antenna wireless communication system over a fading channel with continuous valued inputs and continuous valued outputs is given in [47] in the following form:

$$C = E\left[\log_2 \det\left(\mathbf{I} + \frac{E_s}{N\mathcal{N}_0}\mathbf{HH}^{T*}\right)\right], \ [(b/s)/\text{Hz}], \tag{2.118}$$

where $E[\cdot]$ is the mathematical expectation; $\mathbf{I}$ is the identity matrix with M dimension; $\det(\mathbf{x})$ denotes the determinant of the matrix $\mathbf{x}$, and the superscript $T*$ denotes the matrix transpose and conjugate. The capacity of the equivalent STBC channel in (2.7) with the continuous valued inputs and continuous valued outputs for complex signals is given in [38] and [39] as

$$\overline{C} = E\left[R\log_2 \det\left(1 + \frac{E_s}{RN\mathcal{N}_0}\|\mathbf{H}\|_F^2\right)\right] = E\left[R\log_2\left(1 + \gamma_s\right)\right], \ [(b/s)/\text{Hz}]. \tag{2.119}$$

Given the pdf of γ_s, the capacity of the equivalent STBC channel can be obtained based on

$$\overline{C} = R\int_0^\infty \log_2(1+\gamma_s)f(\gamma_s)d\gamma_s, \ [(b/s)/\text{Hz}]. \tag{2.120}$$

2.2.3.2. Capacity of $\mathcal{M}$-ary Signal Constellations over Fading Channels

Both (2.118) and (2.119) were obtained assuming continuous valued inputs. Here, we consider modulation channels with discrete valued multilevel phase inputs and continuous valued outputs. Assuming maximum likelihood (ML) soft decoding with perfect channel state information at the receiver input, it is well known [48-50] that the capacity C_{STBC}^* of the STBC channel (2.99) can be obtained by averaging the corresponding conditional capacity $\tilde{C}^*(\mathbf{H})$ with respect to the joint pdf of the channel matrix $\mathbf{H}$. By doing so, the following expression for the capacity C_{STBC}^* of the fading channel is obtained

$$C_{STBC}^* = E[\tilde{C}^*(\mathbf{H})] = \int \tilde{C}^*(\mathbf{H})f(\mathbf{H})d\mathbf{H}, \ [(b/s)/\text{Hz}], \tag{2.121}$$

with

$$\tilde{C}^*(\mathbf{H}) = R\left\{\log_2 q - \frac{1}{q\pi R^{-1}\|\mathbf{H}\|_F^2 \mathcal{N}_0} \sum_{j=1}^{q} \int_{y\in C} \exp\left\{-\frac{\|y - R^{-1}\|\mathbf{H}\|_F^2 \alpha_j\|^2}{R^{-1}\|\mathbf{H}\|_F^2 \mathcal{N}_0}\right\} \times \right.$$

$$\left. \times \log_2\left[\sum_{s=1}^{q}\exp\left\{\frac{\|y - R^{-1}\|\mathbf{H}\|_F^2 \alpha_j\|^2 - \|y - R^{-1}\|\mathbf{H}\|_F^2 \alpha_s\|^2}{R^{-1}\|\mathbf{H}\|_F^2 \mathcal{N}_0}\right\}\right] dy\right\}, \tag{2.122}$$

where $\alpha_j, j = 1,\ldots,q$ is a real signal in the q-ary PAM constellation or a complex in the q-ary PSK and QAM constellation, and $f(\mathbf{H})$ is the joint pdf of the $M \times N$ random elements of the channel matrix $\mathbf{H}$ for the fading channel.

Applying the channel model (2.103), (2.121) can be simplified to the one-dimensional integral containing the pdf of h

$$C^*_{STBC,R} = E[\tilde{C}^*(h)] = \int \tilde{C}^*(h)f(h)dh, \ [(b/s)/\text{Hz}], \tag{2.123}$$

where

$$\tilde{C}^*(h) = R\left\{\log_2 q - \frac{1}{q\pi h\mathcal{N}_0}\sum_{j=1}^{q}\int_{y\in C}\exp\left\{-\frac{\|\mathbf{y} - h\alpha_j\|^2}{h\mathcal{N}_0}\right\} \times \right.$$

$$\left. \times \log_2\left[\sum_{s=1}^{q}\exp\left\{\frac{\|\mathbf{y} - h\alpha_j\|^2 - \|\mathbf{y} - h\alpha_s\|^2}{h\mathcal{N}_0}\right\}\right]dy\right\} \tag{2.124}$$

Note that (2.123) applies to both real signal constellations such as PAM, and complex signal constellations such as PSK and QAM.

2.2.3.3. Capacity Comparison

It is shown in [38] that the difference between (2.118) and (2.119) is the capacity loss incurred by using a STBC in the multiple-input multiple-output (MIMO) fading channel with continued valued inputs. The capacity of MIMO fading channel with PAM, PSK, QAM is given in [49] as

$$C^*_{M,N} = \int \tilde{C}^*_{M,N}(\mathbf{H})f(\mathbf{H})d\mathbf{H}, \ [(b/s)/\text{Hz}] \ , \tag{2.125}$$

where

$$\tilde{C}^*_{M,N}(\mathbf{H}) = N\log_2 q - q^{-N}(\pi\mathcal{N}_0)^{-M} \times$$

$$\times \sum_{\mathbf{x}\in(Ax)^N}\int_{\mathbf{y}\in C}\exp\left\{-\frac{\|\mathbf{y} - \mathbf{Hx}\|^2}{\mathcal{N}_0}\right\}\log_2\left[\sum_{\mathbf{x}'\in(Ax)^N}\exp\left\{-\frac{\|\mathbf{y} - \mathbf{Hx}\|^2 - |\mathbf{y} - \mathbf{Hx}'\|^2}{\mathcal{N}_0}\right\}\right]dy \tag{2.126}$$

Here, N and M are the number of transmit and receive antennas, respectively;

$$Ax \equiv \{\alpha_1, \ldots, \alpha_q\} \qquad (2.127)$$

is the q-ary complex signal constellation; $(Ax)^N$ is the N-fold Cartesian product of Ax with itself; the coded vector

$$\mathbf{x} \equiv [x_1, \ldots, x_N] \in (Ax)^N \qquad (2.128)$$

is the q^N-variate random variable with outcomes taking values from the expanded signal constellation $(Ax)^N$; and

$$\mathbf{y} \equiv [y_1, \ldots, y_N]^T \qquad (2.129)$$

is the M-dimensional output vector of the receive antennas. It can easily be shown that the second terms in (2.124) and (2.126) vanish as the *SNR* increases. This fact implies the capacity of a MIMO fading channel approaches $N \log_2 q$ bit/channel use while the capacity with STBC approaches only $R \log_2 q$ bit/channel use for large *SNR*. While the capacity loss of $(N - R) \log_2 q$ bit/channel incurred by using STBC is fairly significant, it will be shown below that the *SNR* threshold for reliable data transmission is reduced because of the STBC diversity gain.

2.2.4. Analysis of Error Probability over Fading Channels

2.2.4.1. Error Probability with Rayleigh Fading

Let $P_q^{error}(\gamma_s)$ is the error probability of q-ary signal constellation with STBC in the AWGN channel. The error probability with Rayleigh fading can be obtained by averaging the $P_q^{error}(\gamma_s)$ over the pdf of γ_s

$$P_{STBC,q}^{error}(\gamma_s) = \int_0^\infty P_q^{error}(\gamma_s) f_{Rayleigh}(\gamma_s) d\gamma_s . \qquad (2.130)$$

Note, $P_q^{error}(\gamma_s)$ can be *SEP* or *BEP*, respectively.

2.2.4.1.1. Error Probability for PAM

Since the full rate STBCs exist for any number of transmit antennas using a real PAM constellation [36], $R = 1$ for q-ary PAM. The average *SEP* for PAM over the AWGN channel is [41]

$$P_q^{error}(\gamma_s) = 2(1 - q^{-1})Q\left[\sqrt{6(q^2 - 1)^{-1}\gamma_s}\right], \qquad (2.131)$$

where $Q(\cdot)$ is the Gaussian tail function. Substituting (2.106) and (2.131) into (2.130), the average *SEP* for PAM takes the form

$$P_{STBC,PAM,q}^{error}(\gamma_s) = \int_0^\infty 2(1-q^{-1})Q\left[\sqrt{6(q^2-1)^{-1}\gamma_s}\right] \times$$
$$\times \frac{\gamma_s^{MN-1}}{\overline{\gamma}_c^{MN}\Gamma(MN)}\exp\{-\gamma_s/\overline{\gamma}_c\}d\gamma_s \tag{2.132}$$

To evaluate the integral in (2.132), the following integral function can be employed:

$$g(L) = \int_0^\infty Q(\sqrt{ax})x^{L-1}\exp\{-x/u\}dx =$$
$$= 0.5u^L\Gamma(L)\left[1-\sum_{k=0}^{L-1}\mu\left(\frac{1-\mu^2}{4}\right)^k\binom{2k}{k}\right], \tag{2.133}$$

where

$$\mu = \sqrt{\frac{au}{2+au}}. \tag{2.134}$$

The proof is given in Appendix 1. The closed form of the *SEP* for PAM is then given by

$$P_{STBC,PAM,q}^{error}(\gamma_s) = (1-q^{-1})\left[1-\sum_{k=0}^{MN-1}\mu\left(\frac{1-\mu^2}{4}\right)^k\binom{2k}{k}\right], \tag{2.135}$$

where

$$\mu = \sqrt{\frac{3\overline{\gamma}_c}{q^2-1+3\overline{\gamma}_c}}. \tag{2.136}$$

2.2.4.1.2. Error Probability for PSK

Based on the equivalent scalar AWGN channel model presented before, the error probability for q-ary PSK is equivalent to the analysis in [41] and [51] for adaptive reception of multiphase signals in Rayleigh fading but with *MN* branch diversity. This approach was also employed in [40]. Following the same steps as in [51], the *SEP* is given by

$$P_{STBC,q}^{error} = \frac{(-1)^{MN-1}(1-\mu^2)^{MN}}{\pi\,\Gamma(MN)}\left\{\frac{\partial^{MN-1}}{\partial s^{MN-1}}\left[\frac{1}{s-\mu^2}\left(\frac{\pi}{M}(M-1)-\right.\right.\right.$$
$$\left.\left.\left.-\frac{\mu\sin(\pi/M)}{\sqrt{s-\mu^2\cos^2(\pi/M)}}\times\cot^{-1}\left(-\frac{\mu\cos(\pi/M)}{\sqrt{s-\mu^2\cos^2(\pi/M)}}\right)\right)\right]\right\}\Bigg|_{s=1}, \tag{2.137}$$

where

$$\mu = \sqrt{\frac{\overline{\gamma}_c}{1+\overline{\gamma}_c}}, \tag{2.138}$$

and the notation

$$\left. \frac{\partial^{MN-1}}{\partial s^{MN-1}} f(s,\mu) \right|_{s=1}$$

Denotes the $(MN-1)$-th partial derivative of the function $f(s,\mu)$ evaluated at $s=1$. Note that the coherent detection with perfect channel state information at the receiver input is assumed in (2.137). Following approach presented in [41, 51], performing the differentiation indicated in (2.137) and evaluating the resulting function at $s=1$ for $q=2$ and 4, we obtain the following closed form of *BEP*s for binary phase shift keying (BPSK) and QPSK

$$P_{STBC,2}^{error} = 0.5\left[1 - \mu \sum_{k=0}^{MN-1} \binom{2k}{k}\left(\frac{1-\mu^2}{4}\right)^k\right], \tag{2.139}$$

$$P_{STBC,4}^{error} = 0.5\left[1 - \frac{\mu}{\sqrt{2-\mu^2}} \sum_{k=0}^{MN-1} \binom{2k}{k}\left(\frac{1+\mu^2}{4-2\mu^2}\right)^k\right], \tag{2.140}$$

respectively. Note that the Gray coding was assumed in the *BEP* calculation for QPSK. The same procedure can be applied to calculate the *SEP* for q-ary PSK at $q=8,16,32$, however, the expressions are not as simple as in (2.139) and (2.140). In the remainder of this section we employ (2.130) in order to derive a simpler expression for the error probability.

It is well known [41] that the *BEP* for BPSK and QPSK over the AWGN channel are given as

$$P_{BPSK}^{error}(\gamma_s) = Q(\sqrt{2\gamma_s}), \tag{2.141}$$

$$P_{QPSK}^{error}(\gamma_s) = Q(\sqrt{\gamma_s}), \tag{2.142}$$

respectively. As shown in [52], the exact *SEP* of *M*-ary PSK for the AWGN channel can be presented in the following form

$$P_{PSK,AWGN,M}^{error}(\gamma_s) = 2Q\left[\sqrt{2\gamma_s}\,\sin(\pi/q)\right] -$$
$$- \frac{1}{\pi} \int_{\pi/2-\pi/q}^{\pi/2} \exp\left\{-\gamma_s \frac{\sin^2(\pi/q)}{\cos^2\theta}\,d\theta\right\}. \tag{2.143}$$

For large *SNR* and large values of q, the *SEP* of q-ary PSK in the AWGN channel can be approximated as

$$P^{error}_{PSK,AWGN,q}(\gamma_s) \approx 2Q\left\{\sqrt{2\gamma_s}\,\sin(\pi/q)\right\}, \tag{2.144}$$

and the equivalent *BEP* is given by

$$P^{error}_{PSK,AWGN,q}(\gamma_s) \approx \frac{SEP_q}{\log q}, \tag{2.145}$$

where the Gray coding is assumed. This approximation is good for large values of q, however, for $q=2$ there is a difference in the factor of 2 with the exact probability given in (2.139).

By substituting for $P^{error}_{PSK,AWGN,q}(\gamma_s)$ in (2.144) and using (2.133), (2.130) can be written in the following form

$$P^{error}_{STBC,PSK,q}(\gamma_s) \approx \int_0^\infty 2Q\left[\sqrt{2\gamma_s}\,\sin(\pi/q)\right]\frac{\gamma_s^{MN-1}}{\overline{\gamma}_c^{MN}\Gamma(MN)}\exp\{-\gamma_s/\overline{\gamma}_c\}d\gamma_s =$$

$$= 1 - \sum_{k=0}^{MN-1}\mu\left(\frac{1-\mu^2}{4}\right)^k\binom{2k}{k}, \tag{2.146}$$

where

$$\mu = \sqrt{\frac{\sin^2(\pi\gamma_c/q)}{1+\sin^2(\pi\gamma_c/q)}}. \tag{2.147}$$

Therefore, the *BEP* can be approximated as

$$P^{error}_{STBC,PSK,q}(\gamma_s) \approx \frac{1}{\log q}\left[1 - \sum_{k=0}^{MN-1}\mu\left(\frac{1-\mu^2}{4}\right)^k\binom{2k}{k}\right]. \tag{2.148}$$

By substituting for $P^{error}_{BPSK}(\gamma_s)$ or $P^{error}_{QPSK}(\gamma_s)$ in (2.141) and (2.142), and using (2.133), the exact *BEP* for BPSK and QPSK can be derived from (2.130) as

$$P^{error}_{STBC,PSK,q}(\gamma_s) = \frac{1}{2}\left[1 - \sum_{k=0}^{MN-1}\mu\left(\frac{1-\mu^2}{4}\right)^k\binom{2k}{k}\right], \tag{2.149}$$

where

$$\mu = \sqrt{\frac{\overline{\gamma}_c}{1+\overline{\gamma}_c}}, \tag{2.150}$$

for BPSK, and

$$\mu = \sqrt{\frac{\overline{\gamma}_c}{2+\overline{\gamma}_c}}, \tag{2.151}$$

for QPSK, respectively. It can easily be shown that (2.149) is equivalent to (2.139) and (2.140). Using (2.148), approximations for the *BEP*s of BPSK and QPSK can be obtained as

$$P_{STBC,BPSK,2}^{error} \approx 1 - \sum_{k=0}^{MN-1} \mu \left(\frac{1-\mu^2}{4}\right)^k \binom{2k}{k}, \tag{2.152}$$

with μ given by (2.150) and

$$P_{STBC,QPSK,4}^{error} \approx \frac{1}{2}\left[1 - \sum_{k=0}^{MN-1} \mu \left(\frac{1-\mu^2}{4}\right)^k \binom{2k}{k}\right], \tag{2.153}$$

with μ given by (2.151), respectively. As expected, this approximation is unsuitable for BPSK, but equation (2.153) brings us the exact *BEP* for QPSK. It is shown later that this approximation is very accurate for the $q > 4$ and large *SNR*.

2.2.4.1.3. Error Probability for QAM

Rectangular QAM signal constellations are frequently employed because they are equivalent to two PAM signals on quadrature carriers. For q-ary, $q = 2^k$ (k is even), rectangular QAM, the *SEP* is given in [41] as

$$P_q^{error} = 1 - [1 - P_{\sqrt{q}}^{error}(\gamma_s)]^2, \tag{2.154}$$

where

$$P_{\sqrt{q}}^{error}(\gamma_s) = 2(1-q^{-0.5})Q\left(\sqrt{\frac{3}{q-1}\gamma_s}\right). \tag{2.155}$$

By substituting for P_q^{error} in (2.155), (2.130) can be written as

$$P^{error}_{STBC,PAM,\sqrt{q}}(\gamma_s) = \int_0^\infty 2(1-q^{-0.5})Q\left(\sqrt{\frac{3}{q-1}\gamma_s}\right)\frac{\gamma_s^{MN-1}\exp\{\gamma_s/\overline{\gamma}_c\}}{\overline{\gamma}_c^{MN}\Gamma(MN)}d\gamma_s. \qquad (2.156)$$

Using (2.133), the closed form *SEP* for the rectangular q-ary QAM is then given by

$$P^{error}_{STBC,QAM,q}(\gamma_s) = 1-[1-P^{error}_{STBC,PAM,\sqrt{q}}(\gamma_s)]^2, \qquad (2.157)$$

where

$$P^{error}_{STBC,PAM,\sqrt{q}}(\gamma_s) = (1-q^{-0.5})\left[1-\sum_{k=0}^{MN-1}\mu\left(\frac{1-\mu^2}{4}\right)^k\binom{2k}{k}\right], \qquad (2.158)$$

and

$$\mu = \sqrt{\frac{3\overline{\gamma}_c}{2q-2+3\overline{\gamma}_c}}. \qquad (2.159)$$

2.2.4.2. Error Probability over Rician Fading

Substituting $f_{Rician}(\gamma_s)$ given by (2.114) in (2.130) instead of $f_{Rayleigh}(\gamma_s)$ and following the same procedure used previously, the *SEP* over Rician fading channel can be presented in the following form

$$P^{error}_{STBC,Rician,q}(\gamma_s) =$$

$$= \sum_{n=0}^\infty \frac{(MN\beta)^n\exp\{-MN\beta\}\lambda}{\Gamma(n+1)}\left[1-\sum_{k=0}^{MN+n-1}\mu\left(\frac{1-\mu^2}{4}\right)^k\binom{2k}{k}\right], \qquad (2.160)$$

where

$$\mu = \sqrt{\frac{3\overline{\gamma}_c}{q^2-1+3\overline{\gamma}_c}} \quad \text{and} \quad \lambda = 1-q^{-1}, \qquad (2.161)$$

for q-ary PAM,

$$\mu = \sqrt{\frac{\sin^2(\pi\gamma_c/q)}{1+\sin^2(\pi\gamma_c/q)}}, \qquad (2.162)$$

with $\lambda = 0.5$ at $q=2$ and $\lambda = 1$ at $q>2$ for q-ary PSK, and

$$\mu = \sqrt{\frac{3\overline{\gamma}_c}{2q - 2 + 3\overline{\gamma}_c}} \quad \text{and} \quad \lambda = 1 - q^{-0.5}, \tag{2.163}$$

for $\sqrt{q}$ -ary PAM in q-ary rectangular QAM. The *SEP* of q-ary rectangular QAM can then be calculated using (2.157). Note that each term in the second part of (2.160) is a monotonically decreasing function of i, and is strictly smaller than 1 for all i. The truncation of the first L terms will introduce an error of at most

$$Error = \frac{\lambda\mu(1 - \mu^{2MN+L}}{1 - \sum_{k=0}^{L} \frac{(MNK)^k \exp\{-MNK\}}{\Gamma(k+1)}}, \tag{2.164}$$

in the probability of error. The proof is given in Appendix 2.

2.2.4.3. Error Probability with Nakagami-*m* Fading

The probability of error over Nakagami-*m, m* is the integer, can be obtained from the results presented in Section 2.2.4.1 by increasing the diversity order from *MN* to *mMN*. The *SEP* for STBC over a Nakagami-*m* fading channel takes the following form

$$P_{STBC,Nakagami,q}^{error}(\gamma_s) = \lambda\left[1 - \sum_{k=0}^{MN+n-1} \mu\left(\frac{1-\mu^2}{4}\right)^k \binom{2k}{k}\right], \tag{2.165}$$

where μ and λ for q-ary PAM, q-ary PAM, and $\sqrt{q}$ -ary PAM in q-ary rectangular QAM are given by (2.161)-(2.163), respectively. The *SEP* of q-ary rectangular QAM can be calculated by (2.157).

2.2.5. Extension to STBC DS-CDMA

There is a great interest in the application of STBCs to practical wireless communications systems constructed based on the generalized approach to signal processing in noise and employing the generalized receiver (see Fig. 2.2).

2.2.5.1. System Model

To facilitate the analysis, we generalize the CDMA multiple access interference model from [53] and [54] to accommodate multiple antennas. The system model is presented in Fig. 2.8. In the transmitter, S information symbols for K users are encoded by the respective STBC encoders, and then spread by each user's pseudo noise code, modulated and transmitted from N transmit antennas over the symbol duration T, simultaneously. At the receivers, each user has M receive antennas, and the filtered signals are first dispread, and then sent to STBC decoders. The symbol decisions are made based on the M STBC decoder outputs. The signal at the jth receive input antenna is given by

$$z_j(t) = \sum_{n=1}^{N}\sum_{k=1}^{K}\sqrt{2P_k}\,h_{knj}(t)a_k(t-\tau_{knj})b_{nk}(t-\tau_{knj})\cos[\omega_c(t-\tau_{knj})]+w(t),\quad (2.166)$$

where $h_{knj}(t)$ is the channel coefficient from the nth transmit antenna to the jth receive antenna for the kth user; $P_k = E_s/T_s$ is the symbol power of the kth user; E_s is the symbol energy; T_s is the symbol duration; $a_k(t)$ is the PN spreading chip sequence with chip duration T_c; $w(t)$ is the zero mean AWGN with the power spectral density $\mathcal{N}_0/2$ in each real dimension; τ_{knj} is the time delay from the nth transmit antenna to the jth receive antenna; ω_c is the carrier frequency; $b_{nk}(t)$ is the encoded signal transmitted from the nth antenna of user k. Binary modulation is assumed in this model.

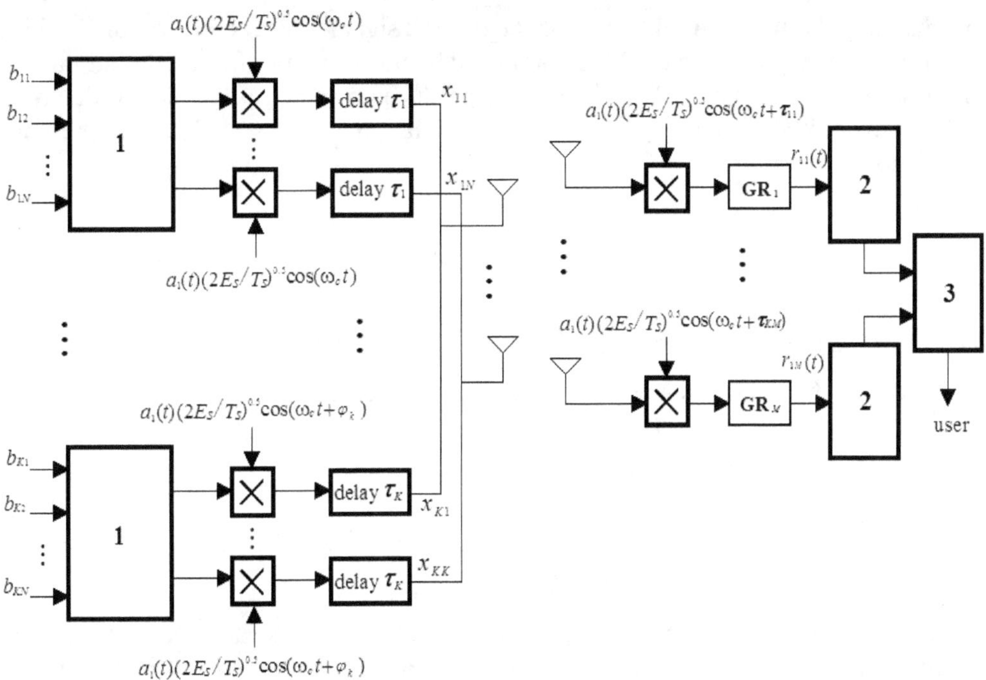

Fig. 2.8. DS-CDMA system model with STBC and multiple antennas employing the generalized receiver: 1 – STBC encoder; 2 – STBC decoder; 3 – decision device; GR - generalized receiver.

2.2.5.2. Channel Model Analysis

To facilitate the analysis, we assume the first user is the desired one, without loss of generality. The dispread signal input to the jth STBC decoder for the ith received symbol takes the form

$$\hat{z}_{1j}(t) = \sum_{q=1}^{N} \int_{(i-1)T_s}^{iT_s} \left[\sqrt{2P_k} \sum_{n=1}^{N}\sum_{k=1}^{K} h_{knj}(t)a_k(t-\tau_{knj})b_{nk}(t-\tau_{knj})\cos[\omega_c(t-\tau_{knj})]+w(t) \right]$$

$$\times a_1(t-\tau_{1qj})\cos[\omega_c(t-\tau_{1nj})]dt$$

$$= \sqrt{\frac{P_1}{2}}T_s\sum_{n=1}^{N}h_{1nj}^{(i)}b_{n1}^{(i)} + \int_{(i-1)T_s}^{iT_s}\left[\sqrt{2P_k}\sum_{n=1}^{N}\sum_{\substack{q=1\\q\neq n}}^{N}h_{1nj}(t)a_1(t-\tau_{1nj})b_{n1}(t-\tau_{1nj})\cos[\omega_c(t-\tau_{1nj})]\right]$$

$$\times a_1(t-\tau_{1qj})\cos[\omega_c(t-\tau_{1nj})]dt \tag{2.167}$$

$$+\int_{(i-1)T_s}^{iT_s}\left[\sqrt{2P_k}\sum_{n=1}^{N}\sum_{k=2}^{K}\sum_{\substack{q=1\\q\neq n}}h_{knj}(t)a_k(t-\tau_{knj})b_{nk}(t-\tau_{knj})\cos[\omega_c(t-\tau_{knj})]\right]dt$$

$$+\sum_{q=1}^{N}\int_{(i-1)T_s}^{iT_s} w(t)a_1(t-\tau_{1qj})\cos[\omega_c(t-\tau_{1nj})]dt \quad .$$

Note that only the first term of (2.167) is the desired signal; the second term of (2.167) is the multiple access interference (MAI) produced by the same user from different transmit antennas; the third term of (2.167) is the MAI produced by other users; the fourth term of (2.167) is the AWGN. Using the Gaussian approximation in [30], (2.167) can be presented in the following form

$$\hat{z}_{1j}(t) = \sqrt{0.5P_1}T_s\sum_{n=1}^{N}h_{1nj}b_{n1} + \eta_{1j}, \tag{2.168}$$

where $\hat{z}_{1j}(t)$ is the Gaussian random variable; η_{1j} is the combination of the interference and noise. The expected mean and variance of η_{1j} are given by

$$\begin{cases} E[\eta_{1j}] = 0, \\ \text{Var}[\eta_{1j}] = N(N-1)\dfrac{T_s^2}{6G}\sum_{k=1}^{K}P_k + 4\sigma_w^4 NT_s, \end{cases} \tag{2.169}$$

respectively, where G is the processing gain of the DS-CDMA system. Note that (2.168) has the same form as (2.96). After STBC decoding is performed on $\hat{z}_{1j}(t)$ given in (2.168), the decision statistic for user 1 over T symbol durations takes the form

$$\hat{\mathbf{z}}_1(t) = R^{-1}\sqrt{0.5P_1}T_s \parallel \mathbf{H} \parallel_F^2 \mathbf{b}_{1T} + \boldsymbol{\eta}_T, \tag{2.170}$$

where $\eta_{iT}(t)$ have the zero mean and the variance is defined as

$$Var\{\eta_{iT}(t)\} = \frac{1}{R}\parallel \mathbf{H} \parallel_F^2 \left[\frac{N(N-1)T_s^2}{6G}\sum_{k=1}^{K}P_k + NT_s4\sigma_w^4 \right]. \tag{2.171}$$

Therefore, the effective instantaneous *SNR* at the generalized receiver output is given by

$$\gamma_s = \left\{ \frac{N(N-1)K}{3G} + \frac{4N\sigma_w^4}{E_s^2} \right\}^{-1} R^{-1} \parallel \mathbf{H} \parallel_F^2 . \tag{2.172}$$

Note that the perfect power control is assumed in (2.172), i.e., $P_k = P_1$.

2.2.5.3. Capacity Analysis of STBC-DS-CDMA

To facilitate the capacity analysis, we first normalize the equivalent channel by $\sqrt{0.5T_s}$, then (2.170) can be written in the same form as (2.99)

$$\mathbf{y}_{nT} = R^{-1} \parallel \mathbf{H} \parallel_F^2 \mathbf{x}_{nT} + \mathbf{\eta}_{nT} , \tag{2.173}$$

where $\mathbf{x}_{nT}$ is the $S \times 1$ complex input matrix with each entry having the symbol energy E_s; the combination of the interference and noise $\mathbf{\eta}_{nT}$ has the zero mean and variance

$$Var\{\mathbf{\eta}_{nT}\} = R^{-1} \frac{\parallel \mathbf{H} \parallel_F^2 N(N-1)K}{3G} + 4\sigma_w^4 N . \tag{2.174}$$

The capacity of this STBC DS-CDMA wireless communication system constructed based on the generalized approach to signal processing in noise [13-15] and employing a *q*-ary signal constellation can be obtained directly from (2.123).

2.2.5.4. Probability of Error for DS-CDMA with STBC

The average *SNR* per channel is determined as

$$\bar{\gamma}_c = \frac{E\{\parallel h_{ij} \parallel^2\}}{R\left\{ \dfrac{N(N-1)K}{3G} + \dfrac{4N\sigma_w^4}{E_s^2} \right\}} = \frac{2\sigma^2}{R\left\{ \dfrac{N(N-1)K}{3G} + \dfrac{4N\sigma_w^4}{E_s^2} \right\}}, \tag{2.175}$$

and this can be used with the probability of error results in Section 2.2.4 to obtain the performance in fading channels. In particular, the exact *BEP* of BPSK is given by

$$P_{STBC\_CDMA,Rayleigh,2b}^{error} = \frac{1}{2}\left[1 - \sum_{k=0}^{MN-1} \mu \left(\frac{1-\mu^2}{4} \right)^k \binom{2k}{k} \right], \tag{2.176}$$

for the Rayleigh fading channel;

$$P^{error}_{STBC\_CDMA,Rician,2b} =$$

$$= \sum_{n=0}^{\infty} \frac{(MN\beta)^n \exp\{-MN\beta\}}{2\Gamma(n+1)} \lambda \left[1 - \sum_{i=0}^{MN+n-1} \mu \left(\frac{1-\mu^2}{4} \right)^i \binom{2i}{i} \right], \qquad (2.177)$$

for the Rician fading channel;

$$P^{error}_{STBC\_CDMA,Naragami,q} = \frac{\lambda}{2} \left[1 - \sum_{k=0}^{mMN-1} \mu \left(\frac{1-\mu^2}{4} \right)^k \binom{2k}{k} \right], \qquad (2.178)$$

for the Nakagami-m fading channel, where μ is given by (2.150) and $\overline{\gamma}_c$ is defined in (2.175). Note that there are two factors in $\overline{\gamma}_c$ which determine the *BEP* of the DS-CDMA system with STBC designed based on the generalized approach to signal processing in noise [13-15], $N(N-1)K/3G$ and $4\sigma_w^4 N$. The first term corresponds to the MAI from other users and the self-interference from different transmitting antennas. The second term corresponds to the system noise, i.e., AWGN. At the high *SNR*, $\overline{\gamma}_c$ will be dominated on the MAI, i.e., the number of users limits the performance, as expected.

2.2.6. Numerical Results

In this subsection, some numerical results are presented to illustrate and verify the capacity and probability of error results obtained before. Fig. 2.9 demonstrates the capacity using SBTC $\mathbf{G}_2$ over Rician fading channel with one, two and four receive antennas for BPSK, QPSK, and 8-PSK. The Rician parameter is $\beta = 100$. This figure shows that the capacity with STBC $\mathbf{G}_2$ is not increased as the number of receiving antennas increases. However, the *SNR* threshold required to achieve capacity improves as the number of receive antennas increased. The capacity with a single antenna over Rician fading channel for PSK is included for reference. Additionally, Fig. 2.9 represents a superiority of employment of the generalized receiver in DS-CDMA wireless communication system with STBC in comparison with the correlation one.

The capacity using several STBCs over Rayleigh fading channel is presented in Fig. 2.10. As expected, it shows that $\mathbf{G}_2$ is the optimal code from a capacity perspective, and $\mathbf{H}_4$ is more efficient in comparison with $\mathbf{G}_4$. Also, a demonstration of superiority to employ the generalized receiver in comparison with the correlation one in DS-CDMA wireless communication system with STBCs is demonstrated in Fig. 2.10. Comparison between Figs. 2.9 and 2.10 demonstrates that the DS-CDMA wireless communication system with STBC codes achieves the capacity at the lower *SNR* over the Rician fading channel at $\beta = 100$ than over the Rayleigh fading channel. Fig. 2.11 demonstrates the relationship between the Rician parameter β and the capacity if STBC code $\mathbf{G}_2$ is used. Note that the capacity is insensitive to the Rician parameter when $\beta > 15$ dB. Additionally, we can see a superiority of implementation of the generalized receiver over the correlation one in DS-CDMA system with STBC codes.

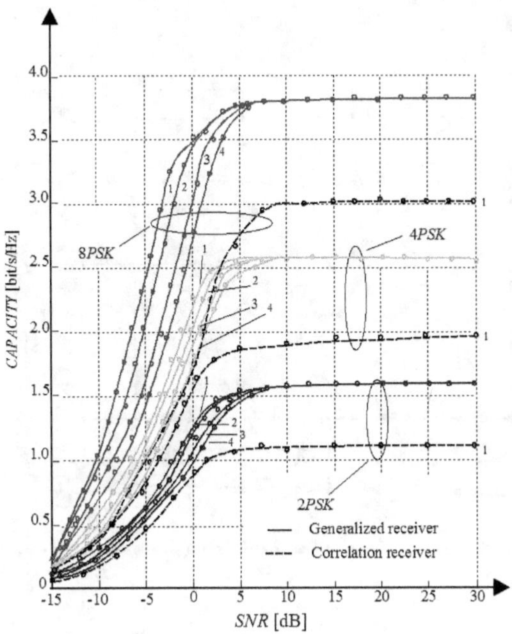

Fig. 2.9. Capacity of DS-CDMA system employing STBC $\mathbf{G}_2$ over Rician fading channel with Rician parameter $\beta = 100$: 1 – MIMO 4×2; 2 – MIMO 2×2; 3 – MIMO 1×2; 4 – MIMO 1×1.

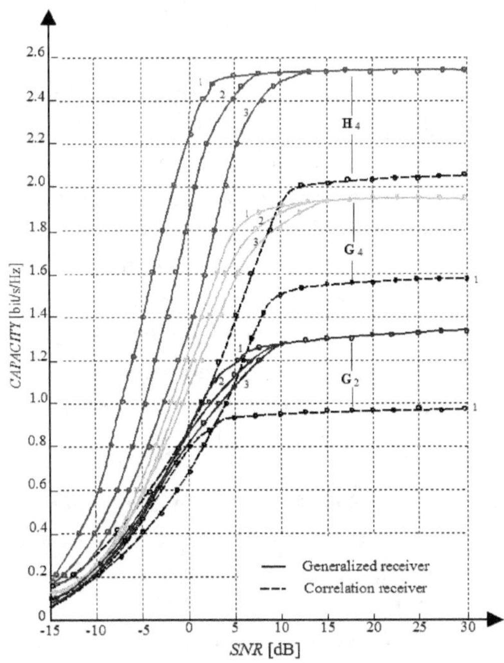

Fig. 2.10. Capacity of DS-CDMA system employing various STBCs over Rayleigh fading channel: 1 – MIMO 4×2; 2 – MIMO 2×2; 3 – MIMO 1×2.

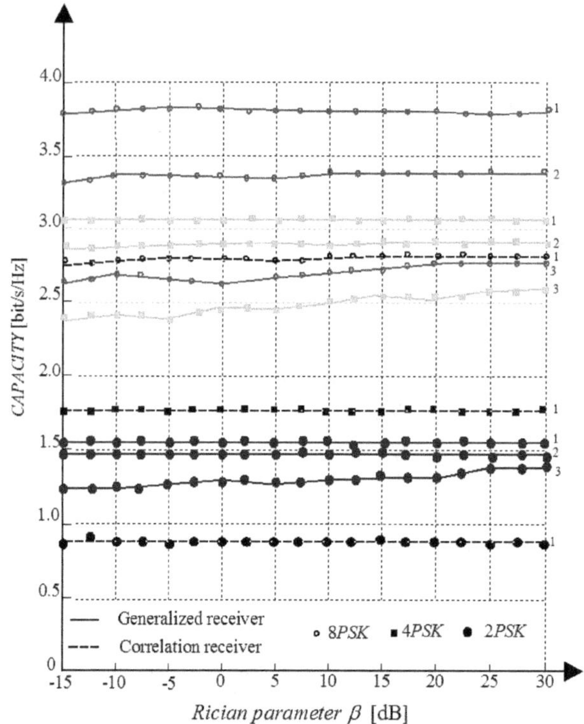

Fig. 2.11. Capacity of DS-CDMA system with STBC $\mathbf{G}_2$ versus the Rician parameter β; $SNR = 5$ dB: 1 – MIMO 4×2;2 – MIMO 2×2; 3 – MIMO 1×2.

Simulation was used to verify the exact and approximate error probabilities given in Section 2.4. In Fig. 2.12, the *BEP* of QPSK at STBCs $\mathbf{G}_2$, $\mathbf{G}_3$, $\mathbf{G}_4$, $\mathbf{H}_3$ and $\mathbf{H}_4$ with one and two receive antennas are presented, and these results are identical to those obtained using (2.140). Also, a superiority of employment of the generalized receiver in DS-CDMA wireless communication system with STBC in comparison with the correlation one is evident. Fig. 2.13 shows the comparison between the approximate and exact (via simulation) *SEP* for 8-PSK. Note that the approximation error is negligible. Fig. 2.14 demonstrates the *SEP* for 16-QAM with one and two receive antennas for different STBCs. A superiority of implementation of the generalized receiver in DS-CDMA wireless communication system with STBC over the correlation one is presented, too. These results are identical to those obtained with (2.39).

In Fig. 2.15 the capacity of DS-CDMA based on the generalized approach to signal processing in noise [13-15] with several STBCs is demonstrated for the case of BPSK modulation over the Rayleigh fading channel. We can see that with a given number of users and signal processing gain, the system may not be able to achieve the full channel capacity even with infinite *SNR* due to the dominant MAI component. In this case, increasing the number of antennas will increase the achievable system capacity. Note that the capacity increases significantly as the number of receive antennas increases for a given number of users and signal processing gain, as expected. However, it should be noted that if the system can already achieve the channel capacity for a given number of antennas, users and signal processing gain, increasing the number of antennas cannot increase the system capacity.

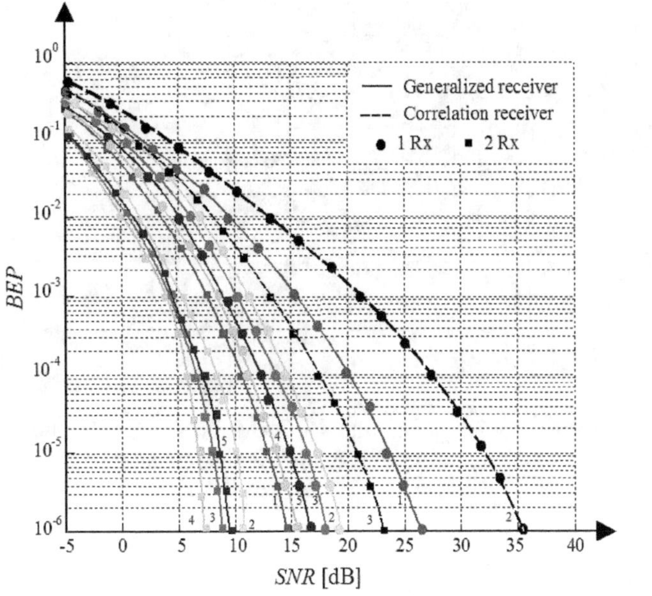

Fig. 2.12. *BEP* of DS-CDMA system with QPSK for STBC using one
and two receive antennas over Rayleigh fading channel:
$1-$ STBC $\mathbf{G}_2$; $2-$ STBC $\mathbf{G}_3$; $-$ STBC $\mathbf{H}_3$; $4-$ STBC $\mathbf{G}_4$; $5-$ STBC $\mathbf{H}_4$.

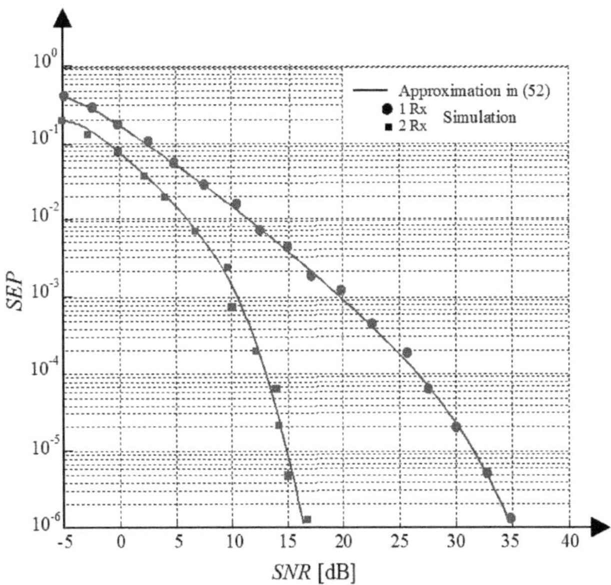

Fig. 2.13. *SEP* of DS-CDMA wireless communication system with QPSK for STBC using one
and two receive antennas over Rayleigh fading channel:
$1-$ STBC $\mathbf{G}_2$; $2-$ STBC $\mathbf{G}_3$; $3-$ STBC $\mathbf{H}_3$; $4-$ STBC $\mathbf{G}_4$; $5-$ STBC $\mathbf{H}_4$.

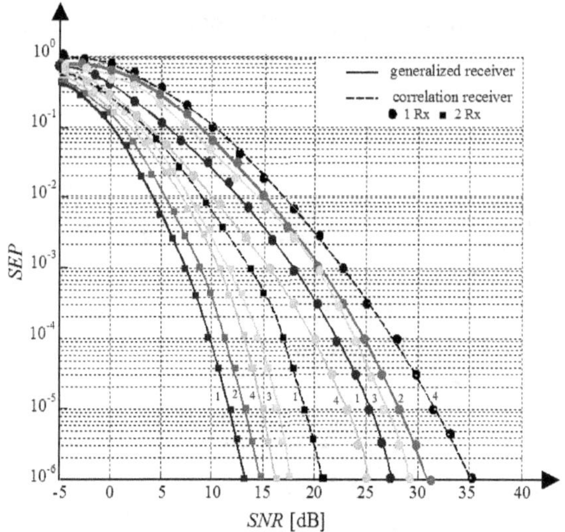

Fig. 2.14. *SEP* of DS-CDMA system with 16-QAM for STBC using one and two receive antennas over Rayleigh fading channel:
$1 - \text{STBC } \mathbf{H}_4 ; 2 - \text{STBC } \mathbf{H}_3 ; 3 - \text{STBC } \mathbf{G}_3 ; 4 - \text{STBC } \mathbf{G}_4 .$

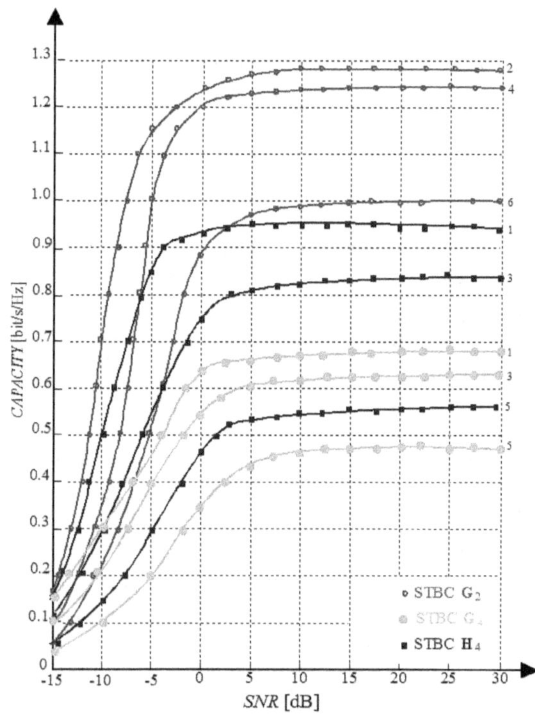

Fig. 2.15. Capacity of DS-CDMA system with BPSK, STBC over Rayleigh fading channel
$G = 32, K = 30$: 1 – MIMO 4×4; 2 – MIMO 4×2; 3 – MIMO 2×4; 4– MIMO 2×2;
5 – MIMO 1×4; 6 – MIMO 1×2.

Fig. 2.16 presents us the relationship between the system capacity and the number of users given the processing gain and *SNR* over the Rayleigh fading channel. It can be seen that the capacity decreases rapidly as the number of users is increased for one and two receive antennas. However, with four receive antennas increasing the number of users has much less effect on the capacity.

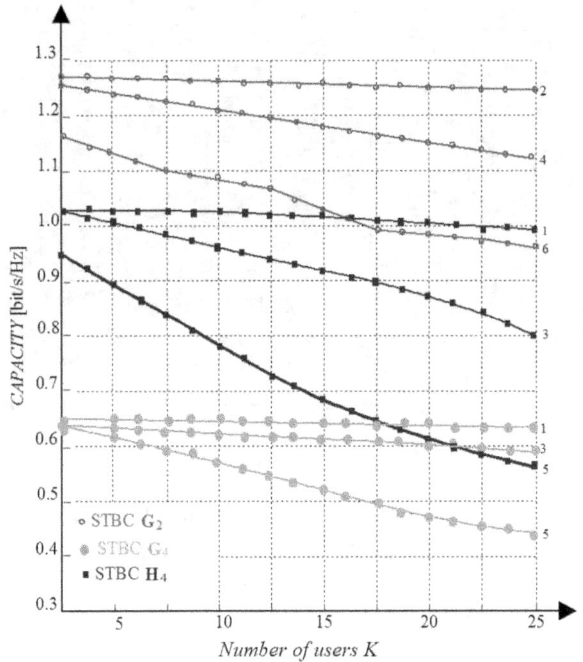

Fig. 2.16. Capacity of DS-CDMA wireless communication system with BPSK and STBC over Rayleigh fading channel $G = 32, SNR = 7$ dB: 1 – MIMO 4×4; 2 – MIMO 4×2; 3 – MIMO 2×4; 4 – MIMO 2×2; 5 – MIMO 1×4; 6 – MIMO 1×2.

The *BEP* for STBC $\mathbf{G}_2$ with DS-CDMA wireless communication system with STBC constructed on the basis of the generalized approach to signal processing in noise using BPSK modulated signals is presented in Fig. 2.17 for one, two, three, and four receive antennas. The signal processing gain is 64, the number of users is 20 and the Rayleigh fading is employed for all DS-CDMA wireless communication system with STBC figures. Fig. 2.17 presents that significant performance gain can be obtained with multiple receive antennas. For comparison the simulation results for the correlation receiver is presented, too. We can see a great superiority of implementation of the generalized receiver in DS-CDMA wireless communication system with STBC over the correlation one.

Fig. 2.18 demonstrates the relationship between the *BEP* and the number of users in the system at *SNR* equal to 7 dB with $\mathbf{G}_2$ and $\mathbf{G}_4$. As the number of users is increased, the performance degrades, but at the four receive antennas the DS-CDMS wireless communication system with STBC is capable of accommodating far more users than with one receive antenna. As it is shown in the Fig. 2.18, $\mathbf{G}_4$ provides the better performance than $\mathbf{G}_2$, however, this is obtained at the price of the capacity loss as shown in Fig. 2.16.

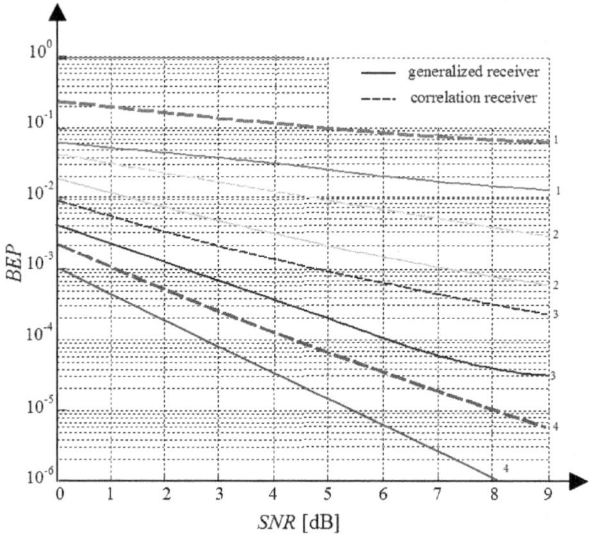

Fig. 2.17. *BEP* of DS-CDMA system with BPSK and STBC $\mathbf{G}_2$ over Rayleigh fading channel $G = 64, K = 20$: 1 – one receive antenna; 2 – two receive antennas; 3 – three receive antennas; 4 – four receive antennas.

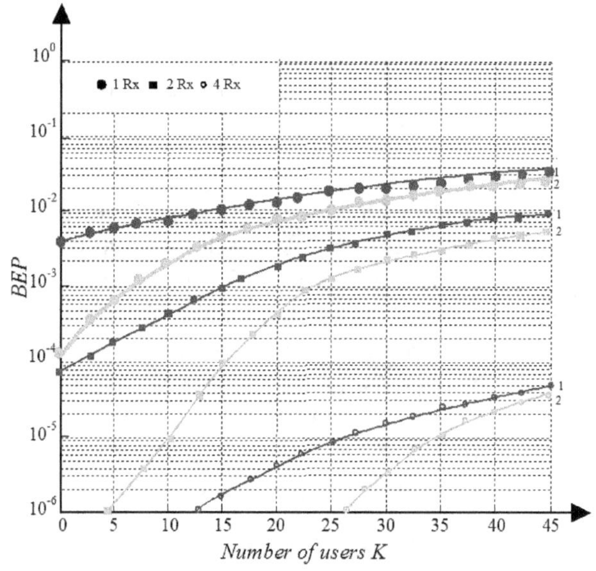

Fig. 2.18. *BEP* versus the number of users for DS-CDMA wireless communication system with BPSK, STBC $\mathbf{G}_2$ (the curve 1) and $\mathbf{G}_4$ (the curve 2) over the Rayleigh fading channel with $G = 64$, $SNR = 7$ dB.

2.2.7. Conclusions

The capacity and error probability of STBCs employed by DS-CDMA wireless communication systems with STBC constructed on the basis of the generalized approach to signal processing in noise have been studied in fading channels for q-ary signal constellations. Rayleigh, Rician, and Nakagami-m fading channels were investigated for PAM, PSK, and QAM modulations. The closed form of the error probabilities under employment of the generalized receiver based on the generalized approach to signal processing in noise were derived for various fading and modulation combinations. This analysis has been employed to determine the performance of STBC multiuser DS-CDMA wireless communication system. Comparative analysis concerning implementation of the generalized receiver and correlation one in DS-CDMA system with STBCs demonstrates a great superiority in favour of the first.

References

[1]. J. S. Lee, L. E. Miller, CDMA Systems Engineering Handbook, *Artech House*, Boston, MA., USA, 1998.

[2]. R. L. Frank, S. A. Zadoff, Phase shift pulse codes with good periodic correlation properties, *IEEE Transactions on Information Theory*, Vol. 8, Issue 5, 1962, pp. 381-382.

[3]. D. Torrier, Principles of Spread-Spectrum Communication Systems, *Springer*, Switzerland, 2018.

[4]. M. B. Pursley, Performance evaluation for phase code spread spectrum multiple access communication – Part I: System analysis, *IEEE Transactions on Communications*, Vol. 25, Issue 8, 1977, pp. 795-799.

[5]. A. J. Viterbi, CDMA: Principles of Spread Spectrum Communication, *Addison-Wesley*, Reading, MA, USA, 1995.

[6]. D. C. Chu, Polyphase codes with good periodic correlation properties, *IEEE Transactions on Information Theory*, Vol. 18, Issue 4, 1972, pp. 531-532.

[7]. D. V. Sarwate, Bounds on cross-correlation and autocorrelation of sequences, *IEEE Transactions on Information Theory*, Vol. 25, Issue 6, 1979, pp. 720-724.

[8]. A. W. Lam, F. M. Ozluturk, Performance bounds for DS-SSMA communications with complex signature sequences, *IEEE Transactions on Communications*, Vol. 40, Issue 10, 1992, pp. 1607-1614.

[9]. F. M. Ozluturk, S. Tantaratana, A. W. Lam, Performances for DS-SSMA communications with MPSK signalling and complex signature sequences, *IEEE Transactions on Communications*, Vol. 43, Issues 2-4, 1995, pp. 1127-1133.

[10]. S. Rahardja, W. Ser, Z. Lin, UCHT-based complex sequences for synchronous CDMA system, *IEEE Transactions on Communications*, Vol. 51, Issue 4, 2003, pp. 618-626.

[11]. S. Rahardja, B. J. Falkowski, Family of unified complex Hadamard transforms, *IEEE Transactions on Circuits and Systems II*, Vol. 46, Issue 8, 1999, pp. 1094-1100.

[12]. J. S. Lehnert, M. B. Pursley, Error probabilities for binary direct sequences spread spectrum communications with random signature sequences, *IEEE Transactions on Communications*, Vol. 37, Issue 8, 1989, pp. 851-858.

[13]. V. P. Tuzlukov, A new approach to signal detection theory, *Digital Signal Processing*, Vol. 8, Issue 3, 1998, pp. 166-184.

[14]. V. P. Tuzlukov, Signal Detection Theory, *Springer-Verlag*, New York, 2001.

[15]. V. P. Tuzlukov, Signal Processing Noise, *CRC Press, Taylor & Francis Group*, Boca Raton, New York, Washington D.C., USA, London, U.K., 2002.

[16]. V. P. Tuzlukov, DS-CDMA downlink systems with fading channel employing the generalized, *Digital Signal Processing*, Vol. 21, Issue 6, 2011, pp. 725-733.

[17]. V. P. Tuzlukov, Signal processing by generalized detector in DS-CDMA wireless communication systems with frequency selective channels. *Circuits, Systems, and Signal Processing*, Vol. 30, Issue 6, 2011, pp. 1197-1230.

[18]. M. Shbat, V. P. Tuzlukov, Primary signal detection algorithms for spectrum sensing at low SNR over fading channels in cognitive radio, *Digital Signal Processing*, Vol. 93. Issue 5, 2019, pp. 187- 207.

[19]. D. E. Borth, M. B. Pursley, Analysis of direct sequence spread spectrum multiple access communication over Rician fading channels, *IEEE Transactions on Communications*, Vol. 27, Issue 10, 1979, pp. 1566-1577.

[20]. C. S. Gardner, J. A. Orr, Fading effects on the performance of a spread spectrum multiple access communication system, *IEEE Transactions on Communications*, Vol. 27, Issue 1, 1979, pp. 143-149.

[21]. J. H. Cho, Y. K. Jeomg, J. S. Lehnert, Average bit error rate performance of band limited DS-SSMA communications, *IEEE Transactions on Communications*, Vol. 50, Issue 7, 2002, pp. 1150-1159.

[22]. C. Trabelsi, A. Yongacoglu, Biterrorrate performance for asynchronous DS-CDMA over multipath fading channels, *IEEE Proceedings Communications*, Vol. 142, Issue 5, 1980, pp. 307-314.

[23]. S. Minić, D. Krstić, D. Bandjur, V. Milenković, S. S. Suljović, M. Stefanović, Level crossing rate of macro diversity in the presence of gamma long-term fading, κ-μ short term fading and Rayleigh short-term fading. *WSEAS Transactions on Communications*, Vol. 6, 2017, 1.

[24]. W. D. Wallis, A. P. Street, J. S. Wallis, Combinatorics: Room Squares, Sum-Free Sets, Hadamard Matrices, *Springer-Verlag*, New York, USA, 1972.

[25]. D. V. Sarwate, M. B. Pursley, Cross-correlation properties of pseudorandom and related sequences, *Proceedings of IEEE*, Vol. 68, Issue 5, 1980, pp. 593-619.

[26]. M. Maximov, Joint correlation of fluctuative noise at outputs of frequency filters, *Radio Engineering*, Issue 9, 1956, pp. 28-38.

[27]. Y. Chernyak, Joint correlation of noise voltage at amplifier outputs with non-overlapping Responses, *Radio Physics and Electronics*, Issue 4, 1960, pp. 551-561.

[28]. M. Shbat, V. P. Tuzlukov, Evaluation of detection performance under employment of the generalized detector in radar sensor systems, *Radioengineering*, Vol. 23, Issue 1, 2014, pp. 50-65.

[29]. M. Shbat, V. P. Tuzlukov, Definition of adaptive detection threshold under employment of the generalized detector in radar sensor systems, *IET Signal Processing*, Vol. 8, Issue 6, 2014, pp. 622-632.

[30]. M. Shbat, V. P. Tuzlukov, SNR wall effect alleviation by generalized detector employment in cognitive radio networks, *Sensors,* Vol. 15, Issue 7, 2015, pp. 16105-16135.

[31]. T. G. Macdonald, M. B. Pursley, The performance of direct sequence spread spectrum with complex processing and quaternary data modulation, *IEEE Journal of Selected Areas of Communications*, Vol. 18, Issue 8, 2000, pp. 1408-1417.

[32]. M. I. Irshid, I. S. Salous, Bit error probability for coherent *M*-ary PSK systems, *IEEE Transactions on Communications*, Vol. 39, Issue 3, 1991, pp. 349-352.

[33]. E. A. Geraniotis, M. B. Pursley, Error probability for direct sequence spread spectrum multiple access communications – Part II: Approximations, *IEEE Transactions on Communications,* Vol. 39, Issue 3, 1991, pp. 349-352.

[34]. S. Boztas, R. Hammons, P. V. Kumar, 4-phase sequences with near optimum correlation Properties, *IEEE Transactions on Communications,* Vol. 39, Issue 3, 1991, pp. 349-352.

[35]. S. M. Alamouti, A simple transmit diversity technique for wireless communications, *IEEE Journal of Selection Areas on Communications,* Vol. 16, Issue 8, 1998, pp. 1452-1458.

[36]. V. Tarokh, H. Jafarkhani, A. R. Claderbank, Space-time block codes from orthogonal designs, *IEEE Transactions on Information Theory,* Vol. 45, Issue 5, 1999, pp. 1456-1467.

[37]. V. Tarokh, Space-time block coding for wireless communications: Performance results, *IEEE Journal of Selection Areas on Communications,* Vol. 16, Issue 8, 1999, pp. 1452-1458.

[38]. S. Sandhu, A. Paulraj, Space-time block codes: A capacity perspective, *IEEE Communications Letters,* Vol. 4, Issue 12, 2000, pp. 384-386.

[39]. D. W. Yue, Q. Wang, Capacity of orthogonal space-time block codes in MISO fading channels with co-channel interferences and noise, *Science in China. Series F: Information Sciences,* Vol. 52, 2009, pp. 1697-1703.

[40]. L. Dai, S. Sfar, K. B. Lefaief, Optimal antenna selection based on capacity maximization for MIMO systems in correlated channels, *IEEE Transactions on Communications,* Vol. 54, Issue 3, 2006, pp. 563-573.

[41]. J. G. Proakis, M. Salehi, Digital Communications, 5th Edition, *McGraw-Hill,* New York, USA, 2007.

[42]. X. Li, T. Luo, G. Yue, C. Yin, A squaring method to simplify the decoding of orthogonal space-time block codes, *IEEE Transactions on Communications,* Vol. 49, Issue 10, 2001, pp. 1700-1703.

[43]. G. Taricco, E. Biglieri, Exact pair wise error probability of space-time codes, *IEEE Transactions on Information Theory,* Vol. 48, Issue 2, 2002, pp. 510-514.

[44]. R. Kshetrimayum, Advanced topics in MIMO wireless communications, in Fundamentals of MIMO Wireless Communications, *Cambridge University Press,* Cambridge, U.K., 2017, pp. 270-308.

[45]. C. Martins, M. L. Brandao, E. Brandani da Silva, New space-time block codes from spectral norm, *PLoS ONE,* Vol. 14, Issue 9, 2019, e02227083.

[46]. M.-S. Alouini, A. J. Goldsmith, A unified approach for calculating error rates of linearly modulated signals over generalized fading channels, *IEEE Transactions on Communications,* Vol. 47, Issue 9, 1999, pp. 1324-1334.

[47]. G. J. Foschini, M. J. Gans, On limits of wireless communications in a fading environment when using multiple antennas, *Wireless Personal Communications,* Vol. 6, 1998, pp. 311-335.

[48]. G. Ungerboeck, Channel coding with multiple phase signals, *IEEE Transactions on Information Theory,* Vol. 28, Issue 1, 1982, pp. 55-67.

[49]. E. Baccarelli, A. Fasano, Some simple bounds on the symmetric capacity, outage probability for QAM wireless channels with Rice and Nakagami fadings, *IEEE Journal of Selected Areas on Communications,* Vol. 18, Issue 3, 2000, pp. 361-368.

[50]. E. Baccarelli, Evaluation of the reliable data rates supported by multiple antenna coded wireless links for QAM transmissions, *IEEE Journal of Selected Areas on Communications,* Vol. 19, Issue 2, 2001, pp. 295-304.

[51]. G. J. Proakis, Probabilities of error for adaptive reception of M-phase signals, *IEEE Transactions on Communication Technology,* Vol. 16, Issue 1, 1968, pp. 71-81.

[52]. M. Shayesteh, A. Aghamohammadi, On the error probability of linearly modulated signals on frequency flat Rician, Rayleigh, and AWGN channels, *IEEE Transactions on Communications,* Vol. 43, Issue 2, 1995, pp. 1454-1466.

[53]. S. X. Wei, An alternative derivation for the signal-to-noise ratio of a SSMA system, *IEEE Transactions on Communications,* Vol. 42, Issue 11, 1994, pp. 2224-2226.

[54]. T. S. Rappaport, Wireless Communications: Principles and Practice, *Prentice-Hall,* Englewood Cliffs, N.J., USA, 1996.

Appendix 1.

Using integration by parts, the proof of (2.133) is as follows:

$$g(L) = \int_0^\infty Q(\sqrt{ax})x^{L-1}\exp\{-x/u\}dx = -\int_0^\infty Q(\sqrt{ax})x^{L-1}ud\{\exp[-x/u]\} =$$

$$= -Q(\sqrt{ax})x^{L-1}u\exp\{-x/u\} + \int_0^\infty \exp\{-x/u\}u\left[Q(\sqrt{ax})x^{L-1}\right]' dx =$$

$$= u(L-1)\int_0^\infty \exp\{-x/u\}Q(\sqrt{ax})x^{L-2}dx - u\sqrt{a/8\pi}\int_0^\infty x^{L-3/2}\exp\{-x/u\}\exp\{-ax/2\}dx =$$

$$= u(L-1)g(L-1) - \frac{1}{2}u^L\mu\left(\frac{1-\mu^2}{4}\right)^{L-1}\frac{(2L-2)!}{(L-1)!}$$

(2.179)

where

$$\mu = \sqrt{\frac{au}{2+au}}$$

(2.180)

Repeating the process, we have

$$g(L) = u^{L-1}\Gamma(L)g(1) - \frac{1}{2}u^L\Gamma(L)\sum_{k=2}^{L}\mu\left(\frac{1-\mu^2}{4}\right)^{k-1}\binom{2k-2}{k-1}$$

(2.181)

Now, using the fact that

$$g(1) = \int_0^\infty Q(\sqrt{ax})\exp\{-x/u\}dx = \frac{1}{2}u(1-\mu),$$

(2.182)

we obtain (2.133)

$$g(L) = \frac{1}{2}u^L\Gamma(L)\left[1 - \sum_{k=0}^{L-1}\mu\left(\frac{1-\mu^2}{4}\right)^{k}\binom{2k}{k}\right]$$

(2.183)

Appendix 2.

The truncation of (2.160) to the first L terms will introduce the error

$$Error = \lambda\sum_{i=L}^{\infty}\mu\left(\frac{1-\mu^2}{4}\right)^{MN+i}\binom{2(MN+i)}{MN+i}\left[1 - \sum_{n=0}^{i}\frac{(MNK)^n\exp\{-MNK\}}{\Gamma(n+1)}\right]$$

(2.184)

Given the fact that

$$4^{-MN-i}\binom{2(MN+i)}{MN+i} < 1,\qquad(2.185)$$

for all i, we have

$$\lambda\sum_{i=L}^{\infty}\mu\left(\frac{1-\mu^2}{4}\right)^{MN+i}\binom{2(MN+i)}{MN+i}\left[1-\sum_{n=0}^{i}\frac{(MNK)^n\exp\{-MNK\}}{\Gamma(n+1)}\right] <$$
$$< \lambda\mu\sum_{i=L}^{\infty}(1-\mu^2)^{MN+i}\left[1-\sum_{n=0}^{L}\frac{(MNK)^n\exp\{-MNK\}}{\Gamma(n+1)}\right]\qquad(2.186)$$

The right-hand side of (2.179) can be simplified to

$$\lambda\mu\sum_{i=L}^{\infty}(1-\mu^2)^{MN+i}\left[1-\sum_{n=0}^{L}\frac{(MNK)^n\exp\{-MNK\}}{\Gamma(n+1)}\right] =$$
$$= \lambda\mu\left[1-\sum_{n=0}^{L}\frac{(MNK)^n\exp\{-MNK\}}{\Gamma(n+1)}\right](1-\mu^2)^{MN+L}\sum_{i=0}^{\infty}(1-\mu^2)^i\qquad(2.187)$$

Given that

$$\sum_{i=0}^{\infty}q^i = \frac{1}{1-q}\ ,\qquad -1\le q\le 1,\qquad(2.188)$$

(2.179) can be written as

$$\lambda\sum_{i=L}^{\infty}\mu\left(\frac{1-\mu^2}{4}\right)^{MN+i}\binom{2(MN+i)}{MN+i}\left[1-\sum_{n=0}^{i}\frac{(MNK)^n\exp\{-MNK\}}{\Gamma(n+1)}\right] <$$
$$< \lambda\frac{(1-\mu^2)^{MN+L}}{\mu}\times\left[1-\sum_{n=0}^{L}\frac{(MNK)^n\exp\{-MNK\}}{\Gamma(n+1)}\right]\qquad(2.189)$$

Thus, truncation of (2.160) to the first N terms will introduce an error

$$Error = \lambda\left[1-\sum_{n=0}^{L}\frac{(MNK)^n\exp\{-MNK\}}{\Gamma(n+1)}\right](1-\mu^2)^{MN+L},\qquad(2.190)$$

in the error probability.

Chapter 3
Exploitation of the Properties of Communication Signals: From Non-circular Signal to Circular Signal

Yuehua Ding, Yide Wang, Pascal Chargé, Jie Li and Nanxi Li

3.1. Introduction

In wireless communications, the signals are usually assumed to be circular. This assumption is true in many cases, especially in the digital communication systems adopting two-dimensional digital linear modulation for high data rate. At the same time, there are also considerable cases where the non circular signals are used. One typical example is military communications, where the reliability of communication is extremely important, one dimensional modulation is used to ensure the reliability in harsh communication environment. Even in civil wireless communications, rectilinear signals, or non circular signals, such as Amplitude Modulated (AM), Amplitude Shift Keying (ASK) or Binary Phase Shift Keying (BPSK) signals are used. In some unwanted cases, the non circular signals can be produced by unperfect hardware even if the circular signals are chosen as a transmission strategy. In addition to circularity and non circularity, almost all the man-made modulation signals exhibit a cyclostationarity (or periodic correlation) property, corresponding to the underlying periodicity arising from carrier frequencies or baud rates. Also linear modulated signals have conjugate symmetric property, namely every constellation point has a corresponding complex conjugate point in the constellation.

These properties can serve as a priori informations, and their exploitation may produce significant performance gain with respect to the conventional processing, in DoA estimation [1, 2], beamforming [3], signal detection [4, 5] etc. This chapter introduces the recent work in the exploitation of signals' properties.

Yuehua Ding

National Engineering Technology Research Center for Mobile Ultrasonic Detection, School of Electronic and Information Engineering, South China University of Technology, Guangzhou, China

3.2. Exploitation of Non-circularity

3.2.1. Non-circularity

Non circularity can be introduced from its opposite aspect, circularity. A random signal $\mathbf{x}$ is said to be second order circular if it satisfies the following conditions:

1. $E[\mathbf{x}(t)\mathbf{x}^T(t+\tau)] = \mathbf{0}$ for any couple (t, τ);

2. $E[\mathbf{x}(t)] = 0.$

All the second order statistic characteristics of a circular signal $\mathbf{x}$ are contained in its covariance matrix $E[\mathbf{x}(t)\mathbf{x}^H(t+\tau)]$. Although circularity is a common hypothesis in narrowband system, there are still numerous non circular signals. These situations occur in rectilinear signals, for example: AM, ASK, or BPSK signals in code-division-multiple-access (CDMA) system. Minimum shift keying (MSK), Gaussian Minimum Shift Keying (GMSK), offset quadrature amplitude modulated (OQAM) signals are also in these situations [6]. A signal $\mathbf{x}$ is second order non circular if $E[\mathbf{x}(t)\mathbf{x}^T(t+\tau)] \neq \mathbf{0}$ for at least one couple (t, τ) [7, 8].

3.2.2. DoA Estimation by Exploiting Non-circularity

3.2.2.1. Problem Statement

A uniform linear array (ULA) of M antennas is considered. The distance between two adjacent antennas is denoted by Δ. Suppose K not correlated electromagnetic waves are impinging on the array from angular directions $\theta_k, k = 1, \cdots, K$. The snapshots number of the received signals is denoted by N. The incoming signals are assumed to be plane waves and narrowband. The received signal at the antenna array can be written as:

$$\mathbf{y}(t) = \sum_{k=1}^{K} s_k(t)\mathbf{a}(\theta_k) + \mathbf{n}(t), \tag{3.1}$$

where $\mathbf{n}(t)$ represents the complex circular spatially and temporally white zero-mean Gaussian noise. Let $s_k(t)$ denote the kth signal waveform arriving at the first antenna, and $\mathbf{a}(\theta_k)$ be the corresponding steering vector given by

$$\mathbf{a}(\theta_k) = [1, e^{j2\pi \frac{\Delta \sin \theta_k}{\lambda}}, \cdots, e^{j2\pi(M-1)\frac{\Delta \sin \theta_k}{\lambda}}]^T,$$

where λ is the wavelength of the incoming signal. The data model (3.1) can be written in the following matrix form:

$$\mathbf{y}(t) = \mathbf{A}\mathbf{s}(t) + \mathbf{n}(t), \tag{3.2}$$

with $\mathbf{A} = [\mathbf{a}(\theta_1), \mathbf{a}(\theta_2), \cdots, \mathbf{a}(\theta_K)]$, $\mathbf{s}(t) = [s_1(t), s_2(t), \cdots, s_K(t)]^T$.

The signals are assumed to be independent of the noise. Accounting for the path loss and channel phase shift from the sources to the receiver and assuming they are relatively stable during the observation time, the data model can be re-expressed in sampling instant nT_s, with T_s the sampling period, as follows:

$$\mathbf{y}(n) = \mathbf{A}\mathbf{\Psi}\mathbf{x}(n) + \mathbf{n}(n), n = 1, 2, ..., N, \tag{3.3}$$

where $\mathbf{x}(n)$ contains the temporal signals transmitted by K sources, $\mathbf{\Psi} = diag(h_1 e^{j\psi_1}, h_2 e^{j\psi_2}, \cdots, h_K e^{j\psi_K})$ with h_k and ψ_k representing the path loss and channel phase rotation for the kth element of $\mathbf{x}(n)$, respectively. To facilitate the discussion, $\mathbf{x}(n)$ is supposed to be real signal, and it can be extended to general non circular signals [2].

3.2.2.2. Extended Data Model

The extended data model is given in the following form:

$$\begin{bmatrix} \mathbf{y}(n) \\ \mathbf{y}^*(n) \end{bmatrix} = \begin{bmatrix} \mathbf{A}\mathbf{\Psi} \\ \mathbf{A}^*\mathbf{\Psi}^* \end{bmatrix} \mathbf{x}(n) + \begin{bmatrix} \mathbf{n}(n) \\ \mathbf{n}^*(n) \end{bmatrix}, \tag{3.4}$$

(3.4) can be re-written in a more compact form as follows:

$$\tilde{\mathbf{y}}(n) = \sum_{k=1}^{K} \tilde{\mathbf{a}}(\theta_k) x_k(n) + \tilde{\mathbf{n}}(n), \tag{3.5}$$

where

$$\tilde{\mathbf{y}}(n) = \begin{bmatrix} \mathbf{y}(n) \\ \mathbf{y}^*(n) \end{bmatrix}, \tag{3.6}$$

$$\tilde{\mathbf{a}}(\theta_k) = \begin{bmatrix} h_k \mathbf{a}(\theta_k) e^{j\psi_k} \\ h_k \mathbf{a}^*(\theta_k) e^{-j\psi_k} \end{bmatrix}, \tag{3.7}$$

$$\tilde{\mathbf{n}}(n) = \begin{bmatrix} \mathbf{n}(n) \\ \mathbf{n}^*(n) \end{bmatrix} \tag{3.8}$$

In matrix form, (3.4) becomes:

$$\tilde{\mathbf{y}}(n) = \tilde{\mathbf{A}}\mathbf{x}(n) + \tilde{\mathbf{n}}(n), \tag{3.9}$$

where $\tilde{\mathbf{A}} = [\tilde{\mathbf{a}}(\theta_1), \tilde{\mathbf{a}}(\theta_2), ..., \tilde{\mathbf{a}}(\theta_K)]$.

3.2.2.3. DoA Estimation

Based on the extended data model, the MUSIC method can be used to calculate the DoA of incoming signal. The covariance matrix of $\tilde{\mathbf{y}}(n)$ is given by:

$$\widetilde{\mathbf{R}} = E\left[\tilde{\mathbf{y}}(n)\tilde{\mathbf{y}}^H(n)\right] = \widetilde{\mathbf{A}}\mathbf{R}_{xx}\widetilde{\mathbf{A}}^H + \sigma_n^2 \mathbf{I}_{2M}, \tag{3.10}$$

where the signal covariance matrix is $\mathbf{R}_{xx} = E[\mathbf{x}(n)\mathbf{x}^H(n)]$.

The eigenvalue decomposition of $\widetilde{\mathbf{R}}$ is:

$$\widetilde{\mathbf{R}} = \widetilde{\mathbf{U}}\mathbf{\Lambda}\widetilde{\mathbf{U}}^H = \widetilde{\mathbf{U}}_s\mathbf{\Lambda}_s\widetilde{\mathbf{U}}_s^H + \sigma_n^2\widetilde{\mathbf{U}}_n\widetilde{\mathbf{U}}_n^H, \tag{3.11}$$

where $\mathbf{\Lambda}_s$ is a diagonal matrix with eigenvalues being the diagonal elements in descending order. $\widetilde{\mathbf{U}} = [\tilde{\mathbf{u}}_1, \tilde{\mathbf{u}}_2, ..., \tilde{\mathbf{u}}_{2M}]$ is a unitary matrix, the columns of $\widetilde{\mathbf{U}}_s = [\tilde{\mathbf{u}}_1, \tilde{\mathbf{u}}_2, ..., \tilde{\mathbf{u}}_K]$ span the signal space, and the columns of $\widetilde{\mathbf{U}}_n = [\tilde{\mathbf{u}}_{K+1}, \tilde{\mathbf{u}}_{K+2}, ..., \tilde{\mathbf{u}}_{2M}]$ span the noise space. One can observe that the maximum number of resolvable non circular sources are $2M - 1$. Theoretically, $\hat{\theta}_k$ can be estimated by solving the following equation:

$$\tilde{\mathbf{a}}^H(\theta_k)\widetilde{\mathbf{U}}_n\widetilde{\mathbf{U}}_n^H\tilde{\mathbf{a}}(\theta_k) = 0 \tag{3.12}$$

Let $z = e^{j2\pi\frac{\Delta\sin\theta}{\lambda}}$. The above equation can be transformed into the following form:

$$\tilde{\mathbf{a}}^T(z_k^{-1})\widetilde{\mathbf{U}}_n\widetilde{\mathbf{U}}_n^H\tilde{\mathbf{a}}(z_k) = 0 \tag{3.13}$$

By finding the K roots of (3.13) which are inside and closest to the unit circle, the direction angles $\hat{\theta}_1, \hat{\theta}_2, \cdots, \hat{\theta}_K$ can be estimated. Equation (3.13) depends on the unknown parameters ψ_k, which makes the root-finding problem impossible. In order to fix the problem, the noise subspace matrix $\widetilde{\mathbf{U}}_n$ is split into two equal-sized sub-matrices $\mathbf{U}_{n1}$ and $\mathbf{U}_{n2}$, namely, $\widetilde{\mathbf{U}}_n = \begin{bmatrix} \mathbf{U}_{n1} \\ \mathbf{U}_{n2} \end{bmatrix}$.

(3.13) can be rewritten as follows:

$$\mathbf{q}^H\mathbf{T}\mathbf{q} = 0, \tag{3.14}$$

where $\mathbf{q} = \begin{bmatrix} 1 \\ e^{-j2\psi_k} \end{bmatrix}$, and $\mathbf{T} = \begin{bmatrix} \mathbf{a}^T(z_k^{-1})\mathbf{U}_{n1}\mathbf{U}_{n1}^H\mathbf{a}(z_k) & \mathbf{a}^T(z_k^{-1})\mathbf{U}_{n1}\mathbf{U}_{n2}^H\mathbf{a}(z_k^{-1}) \\ \mathbf{a}^T(z_k)\mathbf{U}_{n2}\mathbf{U}_{n1}^H\mathbf{a}(z_k) & \mathbf{a}^T(z_k)\mathbf{U}_{n2}\mathbf{U}_{n2}^H\mathbf{a}(z_k^{-1}) \end{bmatrix}$.

Due to the non negativity of the quadratic form $\{\mathbf{q}^H \mathbf{T} \mathbf{q}\}$, the minimum of $\{\mathbf{q}^H \mathbf{T} \mathbf{q}\}$ depends on the smallest eigenvalue of $\mathbf{T}$, which is always non negative. Thus (3.14) can be solved by finding the roots of $\det\{\mathbf{T}\}$, that is:

$$\det\{\mathbf{T}\} = 0 \tag{3.15}$$

Obviously, (3.15) is independent of parameters ψ_k. It is proved in [9] that $\mathbf{U}_{n2}^* \mathbf{U}_{n2}^T = \mathbf{U}_{n1} \mathbf{U}_{n1}^H$, (3.14) is equivalent to the following simplified equation:

$$\mathbf{a}^T(z_k^{-1})\mathbf{U}_{n1}\mathbf{U}_{n1}^H\mathbf{a}(z_k) - \left|\mathbf{a}^T(z_k^{-1})\mathbf{U}_{n1}\mathbf{U}_{n2}^H\mathbf{a}(z_k^{-1})\right| = 0 \tag{3.16}$$

By finding the roots of (3.16), the DoA can be estimated. Alternatively, the DoA can also be estimated by searching the peaks of the following spatial pseudo-spectrum:

$$P(\theta_k) = \arg\max_{\theta_k \in [-\frac{\pi}{2},\frac{\pi}{2}]} \frac{1}{\mathbf{a}^T(z_k^{-1})\mathbf{U}_{n1}\mathbf{U}_{n1}^H\mathbf{a}(z_k) - \left|\mathbf{a}^T(z_k^{-1})\mathbf{U}_{n1}\mathbf{U}_{n2}^H\mathbf{a}(z_k^{-1})\right|} \tag{3.17}$$

One can refer to [2, 10] for more details.

3.2.3. MIMO Detection by Exploiting Non-circularity

3.2.3.1. System Model

The MIMO system model considered here is the V-BLAST system [11] with N_R antennas at the receiver and N_T antennas at the transmitter, as described by the following equation:

$$\mathbf{y} = \mathbf{H}\mathbf{x} + \mathbf{n}, \tag{3.18}$$

where $\mathbf{y}$ is the received signal, the transmitted signal $\mathbf{x}$ is BPSK/ASK modulated, $\mathbf{n}$ is the additive zero-mean complex circular white Gaussian noise, $\mathbf{H}$ represents the frequency-flat channel. In this model, the following assumptions are used:

- $E[\mathbf{x}\mathbf{x}^H] = P\mathbf{I}_{N_T}$, where P is the average transmitting power of each transmitting antenna;

- $E[\mathbf{n}\mathbf{n}^H] = \sigma^2\mathbf{I}_{N_R}$, where σ^2 is the power of noise at each receiving antenna;

- $E[\mathbf{n}\mathbf{x}^H] = \mathbf{0}$, which means that the signal and noise are independent;

- Constellation points are equally probable.

3.2.3.2. Extended ZF Detector

For real signals (for example, BPSK or ASK modulated signals), $\mathbf{x} = \mathbf{x}^*$, and thus $E[\mathbf{x}(t)\mathbf{x}^T(t+\tau)] \neq \mathbf{0}$ when $\tau = 0$. An extended model can be obtained as

$$\tilde{\mathbf{y}} = \widetilde{\mathbf{H}}\mathbf{x} + \tilde{\mathbf{n}}, \tag{3.19}$$

where $\tilde{\mathbf{y}} = [\mathbf{y}^T \ \mathbf{y}^H]^T$, $\widetilde{\mathbf{H}} = [\mathbf{H}^T \ \mathbf{H}^H]^T$, $\tilde{\mathbf{n}} = [\mathbf{n}^T \ \mathbf{n}^H]^T$. The transmitted signal $\mathbf{x}$ can be estimated through the ZF receiver.

$$\hat{\mathbf{x}}_{EZF} = \mathbf{C}_{EZF}\,\tilde{\mathbf{y}}, \tag{3.20}$$

where $\mathbf{C}_{EZF} = \widetilde{\mathbf{H}}^{\dagger}$, $(\cdot)^{\dagger}$ denotes the pseudo inverse. Since the linear space under consideration has been extended to $2N_R \times N_T$ dimensions, this method is called extended ZF (EZF) to distinguish it from the conventional ZF.

3.2.3.3. Extended MMSE

If MMSE principle is applied to the extended MIMO model, the MMSE detector can be obtained as:

$$\mathbf{C}_{EMMSE} = P\widetilde{\mathbf{H}}^{H}\mathbf{R}^{-1}, \tag{3.21}$$

where

$$\mathbf{R} = E[\tilde{\mathbf{y}}\tilde{\mathbf{y}}^{H}] = \begin{bmatrix} \mathbf{R}_y & \mathbf{R}_c \\ \mathbf{R}_c^* & \mathbf{R}_y^* \end{bmatrix}, \tag{3.22}$$

where

$$\mathbf{R}_y = E[\mathbf{y}\mathbf{y}^{H}] = P\mathbf{H}\mathbf{H}^{H} + \sigma^2 \mathbf{I}_{N_R}, \tag{3.23}$$

$$\mathbf{R}_c = E[\mathbf{y}\mathbf{y}^{T}] = P\mathbf{H}\mathbf{H}^{T} \tag{3.24}$$

The signal vector $\mathbf{x}$ can be estimated as

$$\hat{\mathbf{x}}_{EMMSE} = \mathbf{C}_{EMMSE}\,\tilde{\mathbf{y}} \tag{3.25}$$

Similarly, this method is called extended MMSE (EMMSE) to distinguish it from the conventional MMSE.

3.2.3.4. Nulling and Cancelling

System (3.19) is considered as a $2N_R \times N_T$ MIMO system. The ordered successive interference cancelation (OSIC) can be additionally applied. Especially, the performance of EZF and EMMSE is further improved by extended ZF-OSIC (EZF-OSIC) and extended MMSE-OSIC (EMMSE-OSIC) respectively, just like ZF-OSIC [11] and MMSE-OSIC [12] which improve ZF and MMSE respectively. The details and optimal order are discussed in [11, 12].

The exploitation of the non circularity can extend the conventional MIMO model to an observation space with doubled dimensions. The diversity of EZF/EMMSE is proven to be $N_R - \dfrac{N_T - 1}{2}$, and the diversity of EZF/EMMSE-OSIC is between $N_R - \dfrac{N_T - 1}{2}$ and N_R [4]. For more details, one can refer to [5].

3.2.4. MIMO-NOMA Framework by Exploiting Non-circularity

3.2.4.1. System Model

The downlink of a MIMO-NOMA communication system is considered. The number of transmitting antennas is N_T, and the number of receiving antennas at a user is N_R. It is assumed that $N_R \geq N_T$. This case is possible in the ultra-dense small cells of 5G networks [13]. In this scenario, low-cost small-cell base station with same (or less) number of antennas as user handsets is likely to be used. Cloud radio access networks (C-RANs) can also be another example, where users are served by a small number of low cost remote radio heads (RRHs) in order to reduce the fronthaul overhead [14]. The signal vector transmitted by the BS is represented as:

$$\bar{\mathbf{s}} = \begin{bmatrix} \bar{s}_1 & \bar{s}_2 & \cdots & \bar{s}_{N_T} \end{bmatrix}^T, \tag{3.26}$$

where $\bar{s}_m = \boldsymbol{\alpha}_m^H \mathbf{s}_m$ denotes the signals intended for the mth cluster.

$$\mathbf{s}_m = \begin{bmatrix} s_{m,1} & \cdots s_{m,K} \end{bmatrix}^T, \quad \boldsymbol{\alpha}_m = \begin{bmatrix} \alpha_{m,1} & \cdots \alpha_{m,K} \end{bmatrix}^T, \tag{3.27}$$

where K is the number of users in the mth cluster, $s_{m,k}$ denotes the signal intended for the kth user in the mth cluster with $k \in \{1, 2, \cdots, K\}$, $\alpha_{m,k}$ is the power allocation coefficient. Conventionally, $\alpha_{m,k}$ is a real positive number and meet $\alpha_{m,1}^2 + \alpha_{m,2}^2 + \cdots + \alpha_{m,K}^2 = 1$. The signal received by the kth user in the ith cluster is given by:

$$\mathbf{y}_{i,k} = \mathbf{H}_{i,k} \mathbf{P} \bar{\mathbf{s}} + \mathbf{n}_{i,k}, \tag{3.28}$$

where $\mathbf{H}_{i,k}$ is a $N_R \times N_T$ Rayleigh fading channel from the base station to the kth user in the ith cluster, and $\mathbf{n}_{i,k}$ is a Gaussian noise vector with dimension of $N_R \times 1$. Similar to [14], let $\mathbf{P} = \mathbf{I}_{N_T}$. The following assumptions are used:

- $E[\bar{\mathbf{s}}\,\bar{\mathbf{s}}^H] = \mathbf{I}_{N_T}$ (each transmitter has unit power);

- $E[\mathbf{n}_{i,k}\mathbf{n}_{i,k}^H] = \sigma^2 \mathbf{I}_{N_R}$;

- $s_{m,k}$ is real except in the discussion of mixed signal.

3.2.4.2. MIMO-NOMA Based on Widely Linear Processing (WLP)

Case 1: WL-MIMO-NOMA with complex power coefficients ($K = 2$)

This is the case of user pairing. Most of the existing works use real and positive power allocation coefficients. However, phase angles can be added to the power coefficients within a cluster [15]. This operation can produce a staggering angle between users in each cluster. It facilitates the receiving end to decode the signals. The complex allocation coefficients can be expressed as follows:

$$\boldsymbol{\alpha}_m = \begin{bmatrix} \alpha_{m,1}e^{-j\theta_1} & \alpha_{m,2}e^{-j\theta_2} \end{bmatrix}^T , \tag{3.29}$$

where $\alpha_{m,k}$ is the strength coefficient of power allocation, taking real positive values and $\alpha_{m,1}^2 + \alpha_{m,2}^2 = 1$; θ_1 and θ_2 are the introduced phase angles. $\mathbf{y}_{i,k}$ can be expressed in more details with $\mathbf{h}_{m,ik}$ denoting the mth column of $\mathbf{H}_{i,k}$.

$$\mathbf{y}_{i,k} = \begin{bmatrix} \mathbf{h}_{1,ik}, \mathbf{h}_{2,ik}, \cdots, \mathbf{h}_{N_T,ik} \end{bmatrix} \begin{bmatrix} \boldsymbol{\alpha}_1^H \mathbf{s}_1 \\ \boldsymbol{\alpha}_2^H \mathbf{s}_2 \\ \vdots \\ \boldsymbol{\alpha}_{N_T}^H \mathbf{s}_{N_T} \end{bmatrix} + \mathbf{n}_{i,k} \tag{3.30}$$

The above equation can be further expressed as:

$$\mathbf{y}_{i,k} = \begin{bmatrix} \mathbf{h}_{1,ik}e^{j\theta_1}, \mathbf{h}_{1,ik}e^{j\theta_2}, \cdots, \mathbf{h}_{N_T,ik}e^{j\theta_1}, \mathbf{h}_{N_T,ik}e^{j\theta_2} \end{bmatrix} \tilde{\mathbf{s}} + \mathbf{n}_{i,k} , \tag{3.31}$$

where $\tilde{\mathbf{s}} = \begin{bmatrix} \alpha_{1,1}s_{1,1}, \alpha_{1,2}s_{1,2}, \cdots, \alpha_{N_T,1}s_{N_T,1}, \alpha_{N_T,2}s_{N_T,2} \end{bmatrix}^T$. An extended model is constructed as follows by jointly euxploiting the received signal $\mathbf{y}_{i,k}$ and its conjugate version:

$$\tilde{\mathbf{y}}_{i,k} = \begin{bmatrix} \mathbf{y}_{i,k} \\ \mathbf{y}_{i,k}^* \end{bmatrix} = \widetilde{\mathbf{H}}_{i,k}\tilde{\mathbf{s}} + \tilde{\mathbf{n}}_{i,k} , \tag{3.32}$$

where

$$\widetilde{\mathbf{H}}_{i,k} = \begin{bmatrix} \mathbf{h}_{1,ik}e^{j\theta_1}, \ \mathbf{h}_{1,ik}e^{j\theta_2}, \cdots, \mathbf{h}_{N_T,ik}e^{j\theta_1}, \ \mathbf{h}_{N_T,ik}e^{j\theta_2} \\ \mathbf{h}_{1,ik}^{*}e^{-j\theta_1}, \mathbf{h}_{1,ik}^{*}e^{-j\theta_2}, \cdots, \mathbf{h}_{N_T,ik}^{*}e^{-j\theta_1}, \mathbf{h}_{N_T,ik}^{*}e^{-j\theta_2} \end{bmatrix} \tag{3.33}$$

One notes that the WL-MIMO-NOMA model is transformed into a classical MIMO model. The advantage is that both the inter-cluster interference and intra-cluster interference can be eliminated completely. Without loss of generality, the first cluster is discussed. In the first cluster, the signal detection for the two users can be performed by adopting zero-forcing (ZF) method. Denote $\mathbf{W}_{1,1} = \widetilde{\mathbf{H}}_{1,1}^{\dagger}$ and $\mathbf{W}_{1,2} = \widetilde{\mathbf{H}}_{1,2}^{\dagger}$ respectively. Let $\mathbf{w}_{1,11}^{H}$ be the first row of $\mathbf{W}_{1,1}$, and $\mathbf{w}_{2,12}^{H}$ the second row of $\mathbf{W}_{1,2}$, the two users are detected respectively as:

$$\mathbf{w}_{1,11}^{H}\widetilde{\mathbf{y}}_{1,1} = \alpha_{1,1}s_{1,1} + \mathbf{w}_{1,11}^{H}\widetilde{\mathbf{n}}_{1,1}, \tag{3.34}$$

$$\mathbf{w}_{2,12}^{H}\widetilde{\mathbf{y}}_{1,2} = \alpha_{1,2}s_{1,2} + \mathbf{w}_{2,12}^{H}\widetilde{\mathbf{n}}_{1,2}, \tag{3.35}$$

where $\|\mathbf{w}_{1,11}\|^{2}$ and $\|\mathbf{w}_{2,12}\|^{2}$ stand for the noise amplification coefficients of ZF on the first and the second users, respectively. For ZF, the user's gain is fixed to one, so the noise amplification coefficient determines the user's channel condition. Without loss of generality, we assume that[1]:

$$\|\mathbf{w}_{1,11}\|^{2} \leq \|\mathbf{w}_{2,12}\|^{2}, \tag{3.36}$$

which means that the first user's detection vector brings less noise gain, i.e. the first user's channel condition is better than the second user. According to the NOMA power allocation strategy, the power allocation coefficients should be ordered as follows:

$$\alpha_{1,1} < \alpha_{1,2} \tag{3.37}$$

Based on the signal model described above, the second user in the first cluster directly demodulates its own information with the following signal to noise ratio (SNR):

$$SNR_{1,2} = \frac{\alpha_{1,2}^{2}}{\|\mathbf{w}_{2,12}\|^{2}\dfrac{1}{\rho}}, \tag{3.38}$$

---

[1] The base station needn't know the users' channel matrices, it needs only the noise amplification coefficient for each user, which reflects the user's channel condition, this brings a much less demanding requirement than acquiring the global channel condition.

where $\rho = \dfrac{P}{\sigma^2}$. For the first user, the detection performance can be improved by SIC, namely it firstly detects the second user's signal as follows:

$$\mathbf{w}_{2,11}^{H}\tilde{\mathbf{y}}_{1,1} = \alpha_{1,2}s_{1,2} + \mathbf{w}_{2,11}^{H}\tilde{\mathbf{n}}_{1,1},$$
(3.39)

where $\mathbf{w}_{2,11}^{H}$ denotes the second row of $\mathbf{W}_{1,1}$. Following the representation style of [14], let $SNR_{1,1}^{2}$ be the SNR of the second user at the first user, $SNR_{1,1}^{2}$ is given by:

$$SNR_{1,1}^{2} = \frac{\alpha_{1,2}^{2}}{\left\| \mathbf{w}_{2,11} \right\|^{2}\dfrac{1}{\rho}}$$
(3.40)

Once the second user's message is decoded, the received signal is updated as $\overline{\mathbf{y}}_{1,1}$ by removing the second user's signal. Then the first user detects its own signal, the detection model in (3.34) is rewritten as:

$$\overline{\mathbf{v}}_{1,11}^{H}\overline{\mathbf{y}}_{1,1} = \alpha_{1,1}s_{1,1} + \overline{\mathbf{v}}_{1,11}^{H}\tilde{\mathbf{n}}_{1,1}$$
(3.41)

For simplicity, denote $\widetilde{\mathbf{H}}_{1,1} = [\tilde{\mathbf{h}}_{1,11} \quad \tilde{\mathbf{h}}_{2,11} \quad \overline{\overline{\mathbf{H}}}_{1,1}]$, where $\overline{\mathbf{v}}_{1,11}^{H}$ is given by the first row of $\overline{\mathbf{H}}_{1,1}^{\dagger}$, and $\overline{\mathbf{H}}_{1,1}$ is obtained by removing the second column from $\widetilde{\mathbf{H}}_{1,1}$, namely $\overline{\mathbf{H}}_{1,1} = [\tilde{\mathbf{h}}_{1,11} \quad \overline{\overline{\mathbf{H}}}_{1,1}]$. The SNR in the detection of the first user's signal is then given by:

$$SNR_{1,1} = \frac{\alpha_{1,1}^{2}}{\left\| \overline{\mathbf{v}}_{1,11} \right\|^{2}\dfrac{1}{\rho}}$$
(3.42)

If the fixed power allocation strategy is adopted, the SNR of the two users are determined by the noise amplification effect of ZF detection vector $\overline{\mathbf{v}}_{1,11}$ and $\mathbf{w}_{2,12}$, which can be optimized if $\theta_1 - \theta_2 = k\pi + \dfrac{\pi}{2}$ [15].

Case 2: Extension to K-user clusters ($K > 2$)

Different from 2-user cluster, widely linear processing can not completely separate $K > 2$ users in a cluster, because the available degrees of freedom for a complex-valued power coefficient are only 2. In a K-user cluster, the interference can be partially canceled by the following strategy.

1. The mth cluster is divided into two subsets, denoted by $\Theta_{m,1}$ and $\Theta_{m,2}$, respectively. Assume that $\left|\Theta_{m,1}\right| = K_1$ and $\left|\Theta_{m,2}\right| = K_2$, where $K_1 + K_2 = K$. Without loss of generality, $\Theta_{m,1}$ and $\Theta_{m,2}$ can be represented as follows:

$$\Theta_{m,1} = \{s_{m,1}, s_{m,2}, \cdots, s_{m,K_1}\}, \quad \Theta_{m,2} = \{s_{m,K_1+1}, s_{m,K_1+2}, \cdots, s_{m,K}\}$$

2. The users in subsets $\Theta_{m,1}$ and $\Theta_{m,2}$ respectively use the power coefficient vectors $\boldsymbol{\alpha}_m$, $\boldsymbol{\beta}_m$, given by

$$\boldsymbol{\alpha}_m = \left[\alpha_{m,1}e^{-j\theta_1} \, \alpha_{m,2}e^{-j\theta_1} \cdots \alpha_{m,K_1}e^{-j\theta_1}\right]^T,$$

$$\boldsymbol{\beta}_m = \left[\beta_{m,K_1+1}e^{-j\theta_2} \, \beta_{m,K_1+2}e^{-j\theta_2} \cdots \beta_{m,K}e^{-j\theta_2}\right]^T,$$

where $\alpha_{m,1}^2 + \cdots + \alpha_{m,K_1}^2 + \beta_{m,K_1+1}^2 + \cdots + \beta_{m,K}^2 = 1$. One notes that, the complex-valued power coefficients in $\Theta_{m,1}$ have the identical phase θ_1, and those in $\Theta_{m,2}$ share another phase θ_2. Recalling (3.26), the signal intended for the mth cluster is given by:

$$\overline{s}_m = \underbrace{\left(\sum_{k=1}^{K_1}\alpha_{m,k}s_{m,k}\right)e^{j\theta_1}}_{\Theta_{m,1}} + \underbrace{\left(\sum_{k=K_1+1}^{K}\beta_{m,k}s_{m,k}\right)e^{j\theta_2}}_{\Theta_{m,2}} \tag{3.43}$$

The above signal model shows that the users in $\Theta_{m,1}$ are overlapped in phase, and so are the users in $\Theta_{m,2}$. However, $\Theta_{m,1}$ and $\Theta_{m,2}$ are staggered by θ_1 and θ_2. This is quite similar to the user pairing if $\Theta_{m,1}$ and $\Theta_{m,2}$ are considered as two 'big users' in a cluster.

It is proven in [15] that the system reaches the optimal performance if $\theta_1 - \theta_2 = k\pi + \pi/2$. Therefore, if θ_1 and θ_2 are orthogonal, a general system model is obtained as follows:

$$\tilde{\mathbf{y}}_{i,k} = \widetilde{\mathbf{H}}_{i,k}\tilde{\mathbf{s}} + \tilde{\mathbf{n}}_{i,k} \, .$$

The $2N_T \times 1$ vector $\tilde{s}$ is given by:

$$\tilde{\mathbf{s}} = \begin{bmatrix} \alpha_{1,1}s_{1,1} + \cdots + \alpha_{1,K_1}s_{1,K_1} \\ \beta_{1,K_1+1}s_{1,K_1+1} + \cdots + \beta_{1,K}s_{1,K} \\ \vdots \\ \alpha_{N_T,1}s_{N_T,1} + \cdots + \alpha_{N_T,K_1}s_{N_T,K_1} \\ \beta_{N_T,K_1+1}s_{N_T,K_1+1} + \cdots + \beta_{N_T,K}s_{N_T,K} \end{bmatrix} \tag{3.44}$$

Without loss of generality, the first cluster is considered. Denote $\mathbf{W}_{1,k} = \widetilde{\mathbf{H}}_{1,k}^{\dagger}$ the ZF matrix at the kth user. Let $\mathbf{w}_{1,k}$ be the ZF detection vector of the kth user. For $k \in \{1, 2, \cdots K_1\}$, the kth user is in subset $\Theta_{1,1}$, the detection vector $\mathbf{w}_{1,k}$ is the first row of $\mathbf{W}_{1,k}$; and for $k \in \{K_1 + 1, \cdots K\}$, the kth user belongs to $\Theta_{1,2}$, and $\mathbf{w}_{1,k}$ is the second row of $\mathbf{W}_{1,k}$. So the kth user is detected as:

$$\mathbf{w}_{1,k}^{H} \widetilde{\mathbf{y}}_{1,k} = \sum_{n=1}^{K_1} \alpha_{1,n} s_{1,n} + \mathbf{w}_{1,k}^{H} \widetilde{\mathbf{n}}_{1,k}, k \in \{1, \cdots K_1\}, \tag{3.45}$$

$$\mathbf{w}_{1,k}^{H} \widetilde{\mathbf{y}}_{1,k} = \sum_{n=K_1+1}^{K} \beta_{1,n} s_{1,n} + \mathbf{w}_{1,k}^{H} \widetilde{\mathbf{n}}_{1,k}, k \in \{K_1 + 1, \cdots K\}, \tag{3.46}$$

According to the above analysis, $\left\| \mathbf{w}_{1,k} \right\|^2$ stands for the noise amplification coefficients of ZF, reflecting the user's channel condition. Without loss of generality, we assume that:

$$\left\| \mathbf{w}_{1,1} \right\|^2 \le \left\| \mathbf{w}_{1,2} \right\|^2 \le \cdots \le \left\| \mathbf{w}_{1,K_1} \right\|^2, \textit{ for } \Theta_{1,1}, \tag{3.47}$$

$$\left\| \mathbf{w}_{1,K_1+1} \right\|^2 \le \left\| \mathbf{w}_{1,K_1+2} \right\|^2 \cdots \le \left\| \mathbf{w}_{1,K} \right\|^2, \textit{ for } \Theta_{1,2} \tag{3.48}$$

Although the channel conditions are ordered within $\Theta_{1,1}$ (or $\Theta_{1,2}$), one should note that, in general, there is no specific magnitude order for any pair across the subsets, i.e. $\| \mathbf{w}_{1,k_1} \|^2$ in $\Theta_{1,1}$ and $\| \mathbf{w}_{1,K_1+k_2} \|^2$ in $\Theta_{1,2}$, with $1 \le k_1 \le K_1$, $1 \le k_2 \le K_2$. $\| \mathbf{w}_{1,k_1} \|^2$ can be bigger or smaller than $\| \mathbf{w}_{1,K_1+k_2} \|^2$. Following the principle of NOMA power allocation, the power allocation coefficients should be ordered as follows:

$$\alpha_{1,1} \le \cdots \le \alpha_{1,K_1}, \ \beta_{1,K_1+1} \le \cdots \le \beta_{1,K} \tag{3.49}$$

For the detection in subset $\Theta_{1,1}$, as shown by (3.45), the K_1^{th} user in subset $\Theta_{1,1}$ can directly decode its messages, it will be detected with the following signal-to-interference-plus-noise ratio (SINR):

$$SINR_{1,K_1} = \frac{\alpha_{1,K_1}^2}{\sum_{m=1}^{K_1-1} \alpha_{1,m}^2 + \left\| \mathbf{w}_{1,K_1} \right\|^2 \frac{1}{\rho}} \tag{3.50}$$

The kth user, $1 \le k < K_1$, needs to use SIC technique to decode the signal of the jth user, $1 + k \le j \le K_1$, and then remove all these users' signal before detecting its own. So the SINR of the jth user at the kth user is given by:

$$SINR_{1,k}^{j} = \frac{\alpha_{1,j}^{2}}{\sum\limits_{m=1}^{j-1}\alpha_{1,m}^{2} + \left\|\mathbf{w}_{1,k}\right\|^{2}\dfrac{1}{\rho}} \tag{3.51}$$

The first user in $\Theta_{1,1}$, needs to decode all the other users' messages (from the K_{1}^{th}-th user to the second user) and remove their contribution, the first user is then detected with the following SNR:

$$SNR_{1,1} = \frac{\alpha_{1,1}^{2}}{\left\|\mathbf{w}_{1,1}\right\|^{2}\dfrac{1}{\rho}} \tag{3.52}$$

Similar to the detection in subset $\Theta_{1,1}$, the Kth user in subset $\Theta_{1,2}$ is detected with the following SINR:

$$SINR_{1,K} = \frac{\beta_{1,K}^{2}}{\sum\limits_{m=K_{1}+1}^{K-1}\beta_{1,m}^{2} + \left\|\mathbf{w}_{1,K}\right\|^{2}\dfrac{1}{\rho}} \tag{3.53}$$

The kth user, $K_{1}+1 \leq k < K$, needs to perform SIC to decode the signal of the jth user, $1+k \leq j \leq K$, and then remove all these users' signals before detecting its own. So the SINR of the jth user at the kth user is:

$$SINR_{1,k}^{j} = \frac{\beta_{1,j}^{2}}{\sum\limits_{m=K_{1}+1}^{j-1}\beta_{1,m}^{2} + \left\|\mathbf{w}_{1,k}\right\|^{2}\dfrac{1}{\rho}} \tag{3.54}$$

For the $(K_{1}+1)^{th}$ user in $\Theta_{1,2}$, if other users can be detected successfully, its SNR is given by:

$$SNR_{1,K_{1}+1} = \frac{\beta_{1,K_{1}+1}^{2}}{\left\|\mathbf{w}_{1,K_{1}+1}\right\|^{2}\dfrac{1}{\rho}} \tag{3.55}$$

3.2.4.3. Coexistence of Real and Complex Circular Signals

In practice, users are likely to suffer from various channel fadings. To deal with the problem, communication system usually adopts an adaptive strategy by using real signals for users with poor channels and complex circular signals for those experiencing good channels. Therefore, the coexistence of real and complex circular signals is quite usual. The above model can be extended to the mixed case where real and complex circular signals coexist in a cluster.

Recalling (3.43), $\Theta_{m,1}$ and $\Theta_{m,2}$ are supposed to be the sets for the complex circular and real signals, respectively. An extended mixed signal model can be expressed as follows:

$$\tilde{\mathbf{y}}_{i,k} = \begin{bmatrix} \mathbf{y}_{i,k} \\ \mathbf{y}_{i.k}^* \end{bmatrix} = \begin{bmatrix} e^{j\theta}\mathbf{H}_{i,k}\mathbf{s}_c \\ e^{-j\theta}\mathbf{H}_{i.k}^*\mathbf{s}_c^* \end{bmatrix} + \begin{bmatrix} \mathbf{H}_{i,k} \\ \mathbf{H}_{i.k}^* \end{bmatrix}\mathbf{s}_n + \begin{bmatrix} \mathbf{n}_{i,k} \\ \mathbf{n}_{i.k}^* \end{bmatrix}, \tag{3.56}$$

where θ is the difference of two phase angles (θ_1 and θ_2 for $\Theta_{m,1}$ and $\Theta_{m,2}$, respectively), which can simplify the expression, $\mathbf{s}_n$ and $\mathbf{s}_c$ denote the real signals and complex circular signals, respectively. They are expressed as follows:

$$\mathbf{s}_c = \begin{bmatrix} \alpha_{1,1}s_{1,1} + \cdots + \alpha_{1,K_1}s_{1,K_1} \\ \vdots \\ \alpha_{N_T,1}s_{N_T,1} + \cdots + \alpha_{N_T,K_1}s_{N_T,K_1} \end{bmatrix}, \tag{3.57}$$

$$\mathbf{s}_n = \begin{bmatrix} \beta_{1,K_1+1}s_{1,K_1+1} + \cdots + \beta_{1,K}s_{1,K} \\ \vdots \\ \beta_{N_T,K_1+1}s_{N_T,K_1+1} + \cdots + \beta_{N_T,K}s_{N_T,K} \end{bmatrix} \tag{3.58}$$

Without loss of generality, in the first cluster, let $\mathbf{c}_{1,k}$ be the detection vector at the kth user, so the detected signal is described as:

$$\tilde{s}_{1,k} = \mathbf{c}_{1,k}^H \tilde{\mathbf{y}}_{1,k} \tag{3.59}$$

The detection vector $\mathbf{c}_{1,k}$ is obtained by minimizing the following criterion:

$$\min_{\mathbf{c}_{1,k}} \ L_{1,k} = E\left[|\tilde{s}_{1,k} - s_{1,k}|^2 \right] \tag{3.60}$$

Let $\dfrac{\partial L_{1,k}}{\partial \mathbf{c}_{1,k}} = 0$, $\mathbf{c}_{1,k}$ can be obtained:

$$\mathbf{c}_{1,k} = \left(E\left[s_{1,k}\tilde{\mathbf{y}}_{1,k}^H \right] \left(E\left[\tilde{\mathbf{y}}_{1,k}\tilde{\mathbf{y}}_{1,k}^H \right] \right)^{-1} \right)^H,$$

$$L_{1,k} = E\left[|s_{1,k}|^2 \right] - E\left[s_{1,k}\tilde{\mathbf{y}}_{1,k}^H \right] \left(E\left[\tilde{\mathbf{y}}_{1,k}\tilde{\mathbf{y}}_{1,k}^H \right] \right)^{-1} \left(E\left[s_{1,k}\tilde{\mathbf{y}}_{1,k}^H \right] \right)^H$$

The SINR of kth user is given by:

$$SINR_{1,k} = \frac{|s_{1,k}|^2}{L_{1,k}} \tag{3.61}$$

It is assumed that the channel conditions of complex circular signals are better than those of real signals, and the channel condition in each subsets is ordered as (3.47) and (3.48). The Kth user in real signals subset $\Theta_{1,2}$ can directly decode its messages. The outage probability of the Kth user (real signal) is given by:

$$P_{1,K}^o = P\left(SINR_{1,K} < \zeta_{1,K}\right) = P\left(\frac{\beta_{1,K}^2}{L_{1,K}} < \zeta_{1,K}\right) \tag{3.62}$$

The kth user, $k < K$, needs to use SIC technique to decode the signal of jth user, $1+k \leq j \leq K$, and then remove all these users' signal before detecting its own. The SIC process is similar to those in the above sections. And the outage probability experienced by the kth user is shown as follows:

$$P_{1,k}^o = 1 - P\left(SINR_{1,k}^j > \zeta_{1,j}, j \in \{k, \cdots, K\}\right) \tag{3.63}$$

One should note that, for the case of circular complex signals, it is well known that WLP is equivalent to its traditional counterpart [16].

3.3. Exploitation of Cyclostationarity

3.3.1. Cyclostationarity

A signal $s(k)$ is said to be *cyclostationary* if its cyclic conjugate or cyclic correlation functions defined respectively as:

$$r_{ss}(\alpha, \tau) = \langle s(n)s(n+\tau)e^{-j2\pi\alpha n}\rangle_\infty, \tag{3.64}$$

$$r_{ss*}(\alpha, \tau) = \langle s(n)s^*(n+\tau)e^{-j2\pi\alpha n}\rangle_\infty \tag{3.65}$$

is nonzero at cycle frequency α for some time shift τ [17, 18], where $\langle \cdot \rangle_\infty = \lim_{N\to\infty} \frac{1}{N}\sum_{n=1}^N (\cdot)$

. Most man-made signals exhibit cyclostationarity with cycle frequency equal to the twice of the carrier frequency, multiples of the baud rate, or combinations of these [18]. For a vector $\mathbf{s}(t)$, its cyclic conjugate and cyclic correlation matrices are defined respectively as:

$$\mathbf{R}_{ss}(\alpha, \tau) = \langle \mathbf{s}(n)\mathbf{s}^T(n+\tau)e^{-j2\pi\alpha n}\rangle_\infty, \tag{3.66}$$

$$\mathbf{R}_{ss*}(\alpha, \tau) = \langle \mathbf{s}(n)\mathbf{s}^H(n+\tau)e^{-j2\pi\alpha n}\rangle_\infty \tag{3.67}$$

3.3.2. DoA Estimation by Exploiting Cyclostationarity

3.3.2.1. Problem Statement

An array of M antennas is considered. K electromagnetic waves impinging on the array are from angular directions θ_k, $k=1,\ldots,K$. The incident waves are assumed to be narrowband plane waves, K_α sources emit cyclostationary signals with cycle frequency α (with $K_\alpha \le K$). $\mathbf{s}(t)$ contains only the K_α signals that exhibit cycle frequency α, and all of the remaining $K - K_\alpha$ signals (that have not cycle frequency α) and any noise are lumped into a vector $\mathbf{i}(t)$. The signal received by the array can be written as:

$$\mathbf{y}(t) = \mathbf{As}(t) + \mathbf{i}(t), \tag{3.68}$$

where $\mathbf{s}(t) = [s_1(t),\ldots,s_{K_\alpha}(t)]^T$ contains the temporal signals having cycle frequency α, $\mathbf{i}(t)$ represents interfering sources and noise. $\mathbf{A} = \left[\mathbf{a}(\theta_1),\ldots,\mathbf{a}(\theta_{K_\alpha})\right]$ contains the steering vectors of the impinging signals of interest (SOI). The received signals are sampled at N distinct times t_n, $n=1,2,\ldots,N$.

3.3.2.2. Extended Cyclostationary-exploiting Data Model

In order to exploit the cyclostationarity of the incoming signals, an extended-data vector is constructed:

$$\mathbf{y}_{CE}(t) = \begin{bmatrix} \mathbf{y}(t) \\ \mathbf{y}^*(t) \end{bmatrix} = \sum_{k=1}^{K_\alpha} \mathbf{B}(\theta_k) \begin{bmatrix} s_k(t) \\ s_k^*(t) \end{bmatrix} + \begin{bmatrix} \mathbf{i}(t) \\ \mathbf{i}^*(t) \end{bmatrix}, \tag{3.69}$$

with

$$\mathbf{B}(\theta) = \left[\mathbf{a}_1(\theta) \quad \mathbf{a}_2(\theta)\right], \tag{3.70}$$

and

$$\mathbf{a}_1(\theta) = \begin{bmatrix} \mathbf{a}(\theta) \\ \mathbf{0} \end{bmatrix} \quad , \quad \mathbf{a}_2(\theta) = \begin{bmatrix} \mathbf{0} \\ \mathbf{a}^*(\theta) \end{bmatrix} \tag{3.71}$$

For any angle θ, note that:

$$\mathbf{a}_1^H(\theta)\,\mathbf{a}_2(\theta) = 0,$$
$$\mathbf{a}_1^H(\theta)\,\mathbf{a}_1(\theta) = \mathbf{a}_2^H(\theta)\,\mathbf{a}_2(\theta) = \beta, \tag{3.72}$$

where β is a real positive constant such that $\|\mathbf{a}(\theta)\|^2 = \beta$. Without losing generality, let β be unity, and in this case the matrix $\mathbf{B}^H(\theta)\mathbf{B}(\theta)$ is equal to the identity matrix.

3.3.2.3. Extended Autocorrelation Matrix

The cyclic correlation matrix for the extended data model is calculated as:

$$\mathbf{R}_{CE}^{\alpha}(\tau) = \frac{1}{N}\sum_{n=1}^{N}\mathbf{I}_{2M}^{\alpha}(t_n)\mathbf{y}_{CE}(t_n + \tau/2)\mathbf{y}_{CE}^{H}(t_n - \tau/2), \qquad (3.73)$$

where the time dependent matrix $\mathbf{I}_{2M}^{\alpha}(t)$ is defined by:

$$\mathbf{I}_{2M}^{\alpha}(t) = \begin{bmatrix} \mathbf{I}_M e^{-j2\pi\alpha t} & \mathbf{0} \\ \mathbf{0} & \mathbf{I}_M e^{+j2\pi\alpha t} \end{bmatrix}, \qquad (3.74)$$

and $\mathbf{I}_M$ is the M-dimensional identity matrix. The extended cyclic correlation matrix can be developed as:

$$\mathbf{R}_{CE}^{\alpha}(\tau) = \begin{bmatrix} \mathbf{R}_{yy}(\alpha,\tau) & \mathbf{R}_{yy*}(\alpha,\tau) \\ \mathbf{R}_{yy*}^{*}(\alpha,\tau) & \mathbf{R}_{yy}^{*}(\alpha,\tau) \end{bmatrix}, \qquad (3.75)$$

where $\mathbf{R}_{yy}(\alpha,\tau)$ and $\mathbf{R}_{yy*}(\alpha,\tau)$ are estimated according to (3.66) and (3.67), respectively.

3.3.2.4. Extended-Cyclic-MUSIC

According to the extended data model (3.69), the extended cyclic correlation matrix (3.75) can be developed as:

$$\mathbf{R}_{CE}^{\alpha}(\tau) = \begin{bmatrix} \mathbf{A} & \mathbf{0} \\ \mathbf{0} & \mathbf{A}^{*} \end{bmatrix} \begin{bmatrix} \mathbf{R}_{ss}(\alpha,\tau) & \mathbf{R}_{ss*}(\alpha,\tau) \\ \mathbf{R}_{ss*}^{*}(\alpha,\tau) & \mathbf{R}_{ss}^{*}(\alpha,\tau) \end{bmatrix} \begin{bmatrix} \mathbf{A} & \mathbf{0} \\ \mathbf{0} & \mathbf{A}^{*} \end{bmatrix}^{H}, \qquad (3.76)$$

where $\mathbf{R}_{CE}^{\alpha}(\tau)$ is a $(2M \times 2M)$ - dimensional matrix but its rank is equal to $K_{\alpha'}$ with $K_{\alpha} \leq K_{\alpha'} \leq 2K_{\alpha}$.

In the extended data model, for any DoA θ_k, there are two signal components: a non conjugate signal component and a conjugate signal component. That is, each source can be considered as the combination of these two signal components. Moreover, these two components are associated with the steering vectors $\mathbf{a}_1(\theta_k)$ and $\mathbf{a}_2(\theta_k)$ in (3.71), respectively. These two components can be viewed as two sources with a same DoA.

In order to exploit simultaneously the contribution of the two components, the following extended normalized steering vector is designed:

$$\mathbf{b}(\theta,\mathbf{c}) = \frac{\mathbf{B}(\theta)\mathbf{c}}{\|\mathbf{B}(\theta)\mathbf{c}\|}, \tag{3.77}$$

where vector $\mathbf{c} = \begin{bmatrix} c_1 & c_2 \end{bmatrix}^T$ contains unknown coefficients, and matrix $\mathbf{B}(\theta)$ is defined by (3.70). According to the subspace based methods principle and by using this extended steering vector, the DoA of the SOI is given by the minima of the following function:

$$\overline{P}(\theta,\mathbf{c}) = \|\mathbf{U}_n^H \mathbf{b}(\theta,\mathbf{c})\|^2 \tag{3.78}$$

The values of θ and $\mathbf{c}$ are to be determined to minimize $\overline{P}(\theta,\mathbf{c})$. The spatial spectrum $P(\theta)$ is given as:

$$P(\theta) = \left[\min_{\mathbf{c}} \frac{\mathbf{c}^H \mathbf{B}^H(\theta)\mathbf{U}_n \mathbf{U}_n^H \mathbf{B}(\theta)\mathbf{c}}{\mathbf{c}^H \mathbf{B}^H(\theta)\mathbf{B}(\theta)\mathbf{c}} \right]^{-1} \tag{3.79}$$

Matrix $\mathbf{B}^H(\theta)\mathbf{B}(\theta)$ being equal to the identity matrix, according to (3.72), it can be shown that the minimum value within the brackets is given by the minimum eigenvalue of $\mathbf{P}(\theta) = \mathbf{B}^H(\theta)\mathbf{U}_n \mathbf{U}_n^H \mathbf{B}(\theta)$, and the minimizing vector $\mathbf{c}$ is the corresponding eigenvector. The (2×2)-dimensional matrix $\mathbf{P}(\theta)$ can be written as:

$$\mathbf{P}(\theta) = \mathbf{B}^H(\theta) \begin{bmatrix} \mathbf{U}_{n1}\mathbf{U}_{n1}^H & \mathbf{U}_{n1}\mathbf{U}_{n2}^H \\ \mathbf{U}_{n2}\mathbf{U}_{n1}^H & \mathbf{U}_{n2}\mathbf{U}_{n2}^H \end{bmatrix} \mathbf{B}(\theta), \tag{3.80}$$

where $\mathbf{U}_{n1}$ and $\mathbf{U}_{n2}$ are two sub-matrices of same dimension defined by:

$$\mathbf{U}_n = \begin{bmatrix} \mathbf{U}_{n1} \\ \mathbf{U}_{n2} \end{bmatrix} \tag{3.81}$$

By showing that $\mathbf{U}_{n2}^* \mathbf{U}_{n2}^T = \mathbf{U}_{n1}\mathbf{U}_{n1}^H$, hence:

$$\mathbf{P}(\theta) = \begin{bmatrix} \mathbf{a}^H(\theta)\mathbf{U}_{n1}\mathbf{U}_{n1}^H\mathbf{a}(\theta) & (\mathbf{a}^T(\theta)\mathbf{U}_{n2}\mathbf{U}_{n1}^H\mathbf{a}(\theta))^* \\ \mathbf{a}^T(\theta)\mathbf{U}_{n2}\mathbf{U}_{n1}^H\mathbf{a}(\theta) & \mathbf{a}^H(\theta)\mathbf{U}_{n1}\mathbf{U}_{n1}^H\mathbf{a}(\theta) \end{bmatrix}, \tag{3.82}$$

Then, by calculating the analytical expression of the eigenvalues of the (2×2)-dimensional matrix $\mathbf{P}(\theta)$, it can be easily shown that the spatial spectrum of the Extended-Cyclic-MUSIC method is given by:

$$P(\theta) = \frac{1}{\mathbf{a}^H(\theta)\mathbf{U}_{n1}\mathbf{U}_{n1}^H\mathbf{a}(\theta) - |\mathbf{a}^T(\theta)\mathbf{U}_{n2}\mathbf{U}_{n1}^H\mathbf{a}(\theta)|} \tag{3.83}$$

For more details, one can refer to [1].

3.3.3. Beamforming by Exploiting Cyclostationarity

3.3.3.1. Problem Statement

Consider an array of M elements. Suppose $P+1$ $(P+1<M)$ far-filed signals impinging on the array, the received signal at time n is given as:

$$\mathbf{y}(n) = \sum_{p=0}^{P} \mathbf{a}_p s_p(n) + \mathbf{n}(n), \tag{3.84}$$

where $\mathbf{a}_p$ is the steering vector of the pth source signal $s_p(n)$, $\mathbf{n}(n)$ is the noise vector. Here, $s_0(n)$ is considered as the desired signal while $s_p(n), p = 1,...,P$ are interferences. The output signal-to-interference-plus-noise ratio (SINR) of the beamformer is given as:

$$SINR = \frac{\sigma_0^2 \, | \, \mathbf{w}^H \mathbf{a}_0 \, |^2}{\mathbf{w}^H \mathbf{R}_{in} \mathbf{w}}, \tag{3.85}$$

where $\mathbf{w}$ is the beamformer weight vector, $\mathbf{R}_{in}$ is the interference plus noise covariance matrix (INCM), σ_0^2 is the power of the desired signal. To maximize the SINR (3.85), MVDR beamformer obtains the weight $\mathbf{w}$ by solving the following optimization problem [19]

$$\min_{\mathbf{w}} \mathbf{w}^H \mathbf{R}_{in} \mathbf{w} \quad s.t. \ \mathbf{w}^H \mathbf{a}_0 = 1, \tag{3.86}$$

whose solution is $\mathbf{w} = \mathbf{R}_{in}^{-1} \mathbf{a}_0 / (\mathbf{a}_0^H \mathbf{R}_{in}^{-1} \mathbf{a}_0)$. In practice, $\mathbf{R}_{in}$ is unavailable, and is replaced by the sample array covariance matrix $\hat{\mathbf{R}} = \frac{1}{N} \sum_{n=1}^{N} \mathbf{x}(n) \mathbf{x}^H(n)$ with N snapshots. The resultant beamformer $\mathbf{w} = \hat{\mathbf{R}}^{-1} \mathbf{a}_0 / (\mathbf{a}_0^H \hat{\mathbf{R}}^{-1} \mathbf{a}_0)$ is commonly referred to as the sample matrix inversion (SMI) beamformer. However, when N is not large enough, the estimated $\hat{\mathbf{R}}$ will dramatically affect the beamformer's performance [20]. More importantly, since $\hat{\mathbf{R}}$ includes the desired signal component, the SMI beamformer dose not provide sufficient robustness against steering vector mismatch for the desired signal [21].

Efforts have been made in [25, 26] to replace $\hat{\mathbf{R}}$ with a reconstructed INCM. The INCM in [26] is reconstructed as:

$$\hat{\mathbf{R}}_{in} = \int_{\bar{\Theta}} \frac{\mathbf{a}(\theta) \mathbf{a}^H(\theta)}{\mathbf{a}(\theta)^H \hat{\mathbf{R}}^{-1} \mathbf{a}(\theta)} d\theta, \tag{3.87}$$

where $\overline{\Theta}$ is the angle region excluding that of the desired signal. Therefore, the desired signal is totally removed from $\hat{\mathbf{R}}_{in}$. However, $\mathbf{a}(\theta)$ in (3.87) is based on a known array structure. In practice, if there is unknown gain and phase errors in $\mathbf{a}(\theta)$, the interference will not be well suppressed by using such $\hat{\mathbf{R}}_{in}$, and the SINR of the beamformer will degrade especially for strong interferences. [27] developed an INCM reconstruction method by exploiting the interferences' cyclostationarity.

According to (3.66), the cyclic conjugate correlation matrix is given by:

$$\mathbf{R}_{yy*}(\alpha,\tau) = \langle \mathbf{y}(n)\mathbf{y}^T(n+\tau)e^{-j2\pi\alpha n}\rangle_\infty \tag{3.88}$$

In practice, $\mathbf{R}_{yy*}(\alpha,\tau)$ is estimated by finite-number of snapshot data as $\hat{\mathbf{R}}_{yy*}(\alpha,\tau) = \langle \mathbf{y}(n)\mathbf{y}^T(n+\tau)e^{-j2\pi\alpha n}\rangle_N$. Suppose $s_p(n), p=0,...,P$ are cyclostationary signals with different cycle frequencies $\alpha_p, p=0,...,P$. Then for each α_p, $\mathbf{R}_{yy*}(\alpha_p,\tau)$ can be expressed as

$$\mathbf{R}_{yy*}(\alpha_p,\tau) = r_{s_p s_p^*}(\alpha_p,\tau)\mathbf{a}_p\mathbf{a}_p^T, \quad p=0,...,P, \tag{3.89}$$

which is a rank-one and non Hermitian matrix [22]. Therefore, the dominant left singular vector of $\mathbf{R}_{yy*}(\alpha_p,\tau)$, denoted as $\mathbf{c}_p$, is proportional to $\mathbf{a}_p$ [18, 23]. Based on this fact, the constrained cyclic adaptive beamforming (CCAB) algorithm [18] is given as:

$$\min_{\mathbf{w}} \ \mathbf{w}^H\hat{\mathbf{R}}\mathbf{w} \quad s.t. \ \mathbf{w}^H\mathbf{c}_0 = 1, \tag{3.90}$$

which exploits the cyclostationarity of the desired signal. However, CCAB is also based on the sample covariance matrix $\hat{\mathbf{R}}$, thus has similar sensitivity problem as the SMI beamformer. Therefore, [27] proposes to reconstruct the INCM further by exploiting the cyclostationarity of interferences.

3.3.3.2. INCM Reconstruction by Exploiting the Cyclostationarity of Interferences

Theoretically, if all the source signals and noises are uncorrelated with each other, the INCM can be expressed as:

$$\mathbf{R}_{in} = \sum_{p=1}^{P}\sigma_p^2\mathbf{a}_p\mathbf{a}_p^H + \sigma_n^2\mathbf{I}, \tag{3.91}$$

where σ_p^2 is the power of the pth interference and σ_n^2 is the noise power, $\mathbf{I}$ is the identity matrix. Hence, in order to reconstruct $\mathbf{R}_{in}$, the steering vector and the power of each interference signal should be estimated.

In [18, 23], it is shown that $\mathbf{c}_p$ which is the dominant left singular vector of $\mathbf{R}_{\mathbf{xx}^*}(\alpha_p,\tau)$ is proportional to $\mathbf{a}_p$. Thus, for each cyclostationary interference, $\mathbf{c}_p$ can be considered as the estimate of $\mathbf{a}_p, p=1,..,P$. Note here that τ is not unique to make a non zero $\mathbf{R}_{\mathbf{xx}^*}(\alpha_p,\tau)$. However, the optimal τ should be the one which maximizes the dominant singular value of $\mathbf{R}_{\mathbf{xx}^*}(\alpha_p,\tau)$.

After that, the Capon power estimator [24] can be used to further obtain the power of each interference as:

$$\hat{\sigma}_p^2 = \frac{1}{\mathbf{c}_p^H \hat{\mathbf{R}}^{-1} \mathbf{c}_p}, \quad p=1,...,P \tag{3.92}$$

Note that to avoid the scaling ambiguity of the power estimate, a scaling process is applied to $\mathbf{c}_p$ in (3.92) to make it have the same norm as $\mathbf{a}_p$, i.e.

$$\mathbf{c}_p \leftarrow \sqrt{M}\,\mathbf{c}_p / \|\mathbf{c}_p\|, \quad p=1,...,P \tag{3.93}$$

For the noise power σ_n^2, it can be estimated as γ_{min}, which is the minimum eigenvalue of the sample covariance matrix $\hat{\mathbf{R}}$.

To summarize, the INCM reconstruction steps are described as follows:

1. For each interference with cycle frequency α_p, obtain $\hat{\mathbf{R}}_{\mathbf{xx}^*}(\alpha_p,\tau), p=1,...,P$;

2. Perform the singular value decomposition (SVD) of $\hat{\mathbf{R}}_{\mathbf{xx}^*}(\alpha_p,\tau)$, and obtain the dominant left singular vector $\mathbf{c}_p, p=1,...,P$;

3. Scale $\mathbf{c}_p$ according to (3.93). Then obtain each interference's power as (3.92);

4. Perform the eigenvalue decomposition (EVD) of $\hat{\mathbf{R}}$, and obtain its minimum eigenvalue γ_{min};

5. Reconstruct the INCM as $\hat{\mathbf{R}}_{in} = \sum_{p=1}^{P} \hat{\sigma}_p^2 \mathbf{c}_p \mathbf{c}_p^H + \gamma_{min}\mathbf{I}$.

With the reconstructed $\hat{\mathbf{R}}_{in}$, the robustness of MVDR and CCAB can be greatly improved since the desired signal is removed from $\hat{\mathbf{R}}_{in}$. This performance improvement can be attributed to the exploitation of the cyclostationarity of interferences.

3.4. Exploitation of Conjugate Symmetry

3.4.1. Conjugate Symmetry

Most of the practically used digital linearly modulated signals have conjugate symmetric constellations, such as MPSK/MQAM signals. For an N_T-dimensional MPSK/MQAM modulated signal $\mathbf{x} \in \Omega^{N_T}$, its rotated version $\mathbf{z} = \mathbf{x}\,exp(j\theta_0)$, $\mathbf{z} \in \Omega'^{N_T}$, has the following property [4]:

$$\mathbf{z}^* = \mathbf{M}_{N_T}\mathbf{z}, \tag{3.94}$$

Ω'^{N_T} represents the rotated version of Ω^{N_T} with respect to $exp(j\theta_0)$, $\mathbf{M}_{N_T} = diag(\mathbf{l})$ is the phase rotation matrix (PRM) with $\mathbf{l} = [l_1, l_2, \cdots, l_{N_T}]^T$ and l_i is defined as the phase rotation factor between two conjugate constellation points. $l_i \in \Xi$, the value of l_i depends on the phase of z_i (the ith element of $\mathbf{z}$). $\mathbf{l}$ is called phase-rotation vector.

$\mathbf{M}_{N_T} \in \Xi^{N_T}$ is data-dependent. Ξ^{N_T} is represented as $\Xi^{N_T} = \{\mathbf{M}_{N_T}^{(1)}, \mathbf{M}_{N_T}^{(2)}, \cdots, \mathbf{M}_{N_T}^{(|\Xi^{N_T}|)}\}$. $\mathbf{M}_{N_T}$ can be viewed from a perspective of constellation partition, as shown in Fig. 3.1. The constellation set Ω'^{N_T} is divided into $|\Xi^{N_T}|$ distinct subsets $\Omega_1'^{N_T}, \Omega_2'^{N_T}, \cdots, \Omega_{|\Xi^{N_T}|}'^{N_T}$ by the PRM-s $\mathbf{M}_{N_T}^{(1)}, \mathbf{M}_{N_T}^{(2)}, \cdots, \mathbf{M}_{N_T}^{(|\Xi^{N_T}|)}$, we have the following relations:

$$\Omega'^{N_T} = \Omega_1'^{N_T} \cup \Omega_2'^{N_T} \cup \cdots \cup \Omega_{|\Xi^{N_T}|}'^{N_T}, \tag{3.95}$$

$$\Omega_i'^{N_T} \cap \Omega_k'^{N_T} = \varnothing, \quad i \neq k, \tag{3.96}$$

where $\varnothing$ represents the empty set. For any $\mathbf{z} \in \Omega_i'^{N_T}$, its PRM must be $\mathbf{M}_{N_T}^{(i)}$. The following special cases are taken to illustrate this property.

- For the extreme case of real signals, e.g. BPSK/ASK, $\mathbf{M}_{N_T} = \mathbf{I}_{N_T}$ is the sole PRM. The constellation partition is unnecessary.

- For MPSK signals, each diagonal element of $\mathbf{M}_{N_T}^{(i)}$ represents a pair of anti-polar constellation points. The size of each subset is 2^{N_T}, just like the case of BPSK signals. For MQAM signals, the sizes of subsets are not always 2^{N_T}, because one real diagonal element defines multiple constellation points. One example is the 16QAM, whose subset with $\mathbf{M}_{N_T}^{(i)} = \mathbf{I}_{N_T}$ has the 4^{N_T} vectors. To further illustrate the concept of constellation partition, examples of QPSK signals are employed as follows:

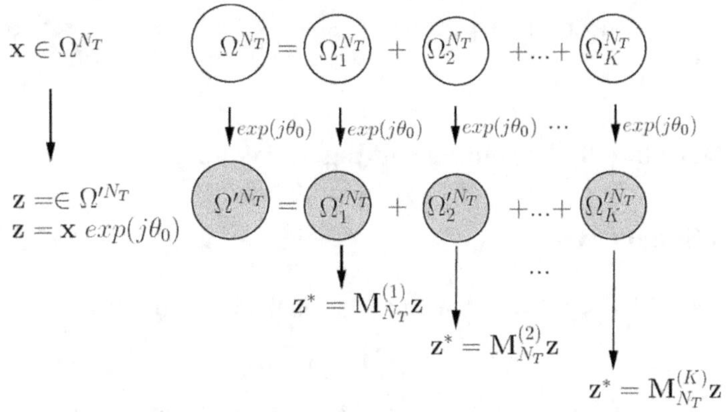

Fig. 3.1. Constellation partition ($K =| \Xi^{N_T} |$), $\Omega_i'^{N_T} \cap \Omega_k'^{N_T} = \varnothing, \Omega_i^{N_T} \cap \Omega_k^{N_T} = \varnothing, i \neq k$.

Example 1. $z \in \Omega'^{N_T}$ is QPSK modulated, $N_T = 1$.

$\Omega'^{N_T} = \{\pm 1, \pm j\}$, $\Xi^{N_T} = \{\pm 1\}$, $| \Omega'^{N_T} |= 4$, $| \Xi^{N_T} | = 2$.

$\mathbf{M}_{N_T}^{(1)} = 1$, $\Omega_1'^{N_T} = \{\pm 1\}$, $| \Omega_1'^{N_T} |= 2$.

$\mathbf{M}_{N_T}^{(2)} = -1$, $\Omega_2'^{N_T} = \{\pm j\}$, $| \Omega_2'^{N_T} |= 2$.

Example 2. $z \in \Omega'^{N_T}$ is QPSK modulated, $N_T = 2$.

$\Omega'^{N_T} = \{[\pm 1 \pm 1]^T, [\pm 1 \pm j]^T, [\pm j \pm 1]^T, [\pm j \pm j]^T\}$, $\Xi^{N_T} = \{diag([\pm 1 \pm 1]^T)\}$, $| \Omega'^{N_T} |= 16$, $| \Xi^{N_T} |= 4$.

$\mathbf{M}_{N_T}^{(1)} = diag([1 \ 1]^T)$, $\Omega_1'^{N_T} = \{[\pm 1 \pm 1]^T\}$, $| \Omega_1'^{N_T} |= 4$.

$\mathbf{M}_{N_T}^{(2)} = diag([1 \ \ -1]^T)$, $\Omega_2'^{N_T} = \{[\pm 1 \pm j]^T\}$, $| \Omega_2'^{N_T} |= 4$.

$\mathbf{M}_{N_T}^{(3)} = diag([-1 \ 1]^T)$, $\Omega_3'^{N_T} = \{[\pm j \pm 1]^T\}$, $| \Omega_3'^{N_T} |= 4$.

$\mathbf{M}_{N_T}^{(4)} = diag([-1 \ \ -1]^T)$, $\Omega_4'^{N_T} = \{[\pm j \pm j]^T\}$, $| \Omega_4'^{N_T} |= 4$.

- For a linearly modulated signal $\mathbf{x}$, the same conclusions can be obtained for its constellation Ω^{N_T}. There also exist distinct subsets $\Omega_1^{N_T}, \Omega_2^{N_T}, \cdots, \Omega_{|\Xi^{N_T}|}^{N_T}$ satisfying:

$$\Omega^{N_T} = \Omega_1^{N_T} \cup \Omega_2^{N_T} \cup \cdots \cup \Omega_{|\Xi^{N_T}|}^{N_T}, \tag{3.97}$$

$$\Omega_j^{N_T} \cap \Omega_k^{N_T} = \varnothing, \quad j \neq k \tag{3.98}$$

135

Accordingly, $\Omega_i'^{N_T}$ is the rotated version of space $\Omega_i^{N_T}$ with respect to $exp(j\theta_0)$, as shown in Fig. 3.1.

3.4.2. MIMO Detection by Exploiting Conjugate Symmetry

3.4.2.1. MIMO System Model

The previously considered MIMO system model is taken as follows:

$$\mathbf{y} = \mathbf{Hx} + \mathbf{w} \tag{3.99}$$

The only difference is that the transmitted signal $\mathbf{x}$ is MPSK/MQAM modulated, and the constellation points in MQAM (or MPSK) are equally probable.

3.4.2.2. Widely Linear Model for Detection

The MIMO system model can be transformed into (3.100) by a phase shift of θ_0:

$$\mathbf{y}\, exp(j\theta_0) = \mathbf{Hz} + \mathbf{w}\, exp(j\theta_0), \tag{3.100}$$

where $\mathbf{z} = \mathbf{x}\, exp(j\theta_0)$. By exploiting the property in (3.94), the conjugate version of (3.100) is obtained as follows:

$$\mathbf{y}^*\, exp(-j\theta_0) = \mathbf{H}^*\mathbf{M}_{N_T}\mathbf{z} + \mathbf{w}^*\, exp(-j\theta_0) \tag{3.101}$$

Combining (3.100) and (3.101), an extended model is constructed as:

$$\begin{bmatrix} \mathbf{y}\, exp(j\theta_0) \\ \mathbf{y}^*\, exp(-j\theta_0) \end{bmatrix} = \begin{bmatrix} \mathbf{H} \\ \mathbf{H}^*\mathbf{M}_{N_T} \end{bmatrix} \mathbf{z} + \begin{bmatrix} \mathbf{w}\, exp(j\theta_0) \\ \mathbf{w}^*\, exp(-j\theta_0) \end{bmatrix} \tag{3.102}$$

Intuitively, (3.102) can be seen as an extended MIMO system model with N_T transmitting antennas and $2N_R$ receiving antennas. $\mathbf{z}$ can be detected by applying the ZF or MMSE principle as follows:

$$\hat{\mathbf{z}} = \mathbf{C}\begin{bmatrix} \mathbf{y}\, exp(j\theta_0) \\ \mathbf{y}^*\, exp(-j\theta_0) \end{bmatrix}, \tag{3.103}$$

where $\mathbf{C}$ is the extended ZF (EZF) matrix or extended MMSE (EMMSE) matrix, given by:

$$\mathbf{C} = \begin{cases} \widetilde{\mathbf{H}}^\dagger & \text{for EZF receiver} \\ (\widetilde{\mathbf{H}}^H\widetilde{\mathbf{H}} + \dfrac{\sigma^2}{P}\mathbf{I}_{N_T})^{-1}\widetilde{\mathbf{H}}^H & \text{for EMMSE receiver} \end{cases}, \tag{3.104}$$

with

$$\widetilde{\mathbf{H}} = \begin{bmatrix} \mathbf{H} \\ \mathbf{H}^{*}\mathbf{M}_{N_T} \end{bmatrix} \qquad (3.105)$$

Finally, the transmitted signal $\mathbf{x}$ is recovered by

$$\hat{\mathbf{x}} = \hat{\mathbf{z}} \, exp(-j\theta_0) \qquad (3.106)$$

The computation of $\mathbf{C}$ depends on $\mathbf{M}_{N_T}$, which is unknown and data-dependent. One basic way is to calculate $\hat{\mathbf{x}}$ using (3.103) and (3.106) for each $\mathbf{M}_{N_T} \in \Xi^{N_T}$ and a set of $\{\hat{\mathbf{x}}^{(1)}, \hat{\mathbf{x}}^{(2)}, \cdots, \hat{\mathbf{x}}^{(|\Xi^{N_T}|)}\}$ is obtained accordingly. Candidate vectors $\{\mathbf{s}_1, \mathbf{s}_2, \cdots, \mathbf{s}_{|\Xi^{N_T}|}\}$ are produced by the quantization process $\mathbf{s}_i = Q(\hat{\mathbf{x}}^{(i)})$, and $\mathbf{s}_i \in \Omega_i^{N_T}$. Finally, the likelihood test is performed to determine the individual which minimizes $\|\mathbf{y} - \mathbf{H}\mathbf{s}_i\|^2$, with $i = 1, 2, \cdots, |\Xi^{N_T}|$. In this case, $\mathbf{M}_{N_T}$ and $\mathbf{x}$ are jointly estimated. The process can be described by Algorithm 1 and Fig. 3.2. Note that Algorithm 1 should perform an inverse matrix for each $\mathbf{M}_{N_T}^{(i)}$, leading to a substantial computational overhead. In the following, widely linear sphere decoder (WLSD) is introduced to avoid the repetitive inversions.

Algorithm 1. EZF/EMMSE based on ML criterion.

1:   Input: $\mathbf{H}, \mathbf{y}$

2:   Output: $\mathbf{s}$

3:   Initialization:

4:   $d^2 = inf$;   % Error distance initialization

5:   set $\Xi^{N_T} = \{\mathbf{M}_{N_T}^{(1)}, \mathbf{M}_{N_T}^{(2)}, \cdots, \mathbf{M}_{N_T}^{(|\Xi^{N_T}|)}\}$

6:

7:   **for** $i = 1{:}1{:}|\Xi^{N_T}|$ **do**

8:      $\widetilde{\mathbf{H}}^{(i)} = [\mathbf{H}^T \quad \mathbf{M}_{N_T}^{(i)} \mathbf{H}^H]^T$

9:      $\mathbf{C}^{(i)} = (\widetilde{\mathbf{H}}^{(i)})^{\dagger}$ % ZF receiver

10:      or $\mathbf{C}^{(i)} = [(\widetilde{\mathbf{H}}^{(i)})^H \widetilde{\mathbf{H}}^{(i)} + \frac{\sigma^2}{P}\mathbf{I}_{N_T}]^{-1}(\widetilde{\mathbf{H}}^{(i)})^H$

11:                 % MMSE receiver

12:      $\hat{\mathbf{z}}^{(i)} = \mathbf{C}^{(i)}[\mathbf{y}^T exp(j\theta_0) \quad \mathbf{y}^H exp(-j\theta_0)]^T$

13:      $\hat{\mathbf{x}}^{(i)} = \hat{\mathbf{z}}^{(i)} exp(-j\theta_0)$

14:      $\mathbf{s}_i = Q(\hat{\mathbf{x}}^{(i)})$ %quantization

15:      **if** $\|\mathbf{y} - \mathbf{H}\mathbf{s}_i\|^2 < d^2$ **then**

16:          $d^2 = \|\mathbf{y} - \mathbf{H}\mathbf{s}_i\|^2$ % The distance is updated

17:          $\mathbf{s} = \mathbf{s}_i$ %Optimal solution s is updated

18:      **end if**;

19:   **end for**

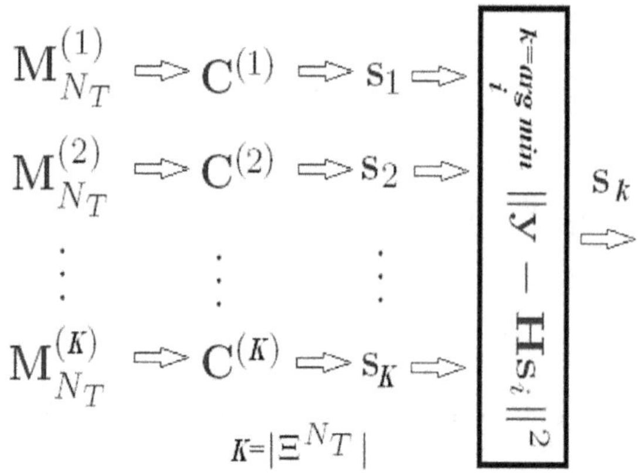

$$
\begin{array}{ccccc}
\mathbf{M}_{N_T}^{(1)} & \Rightarrow & \mathbf{C}^{(1)} & \Rightarrow & s_1 \Rightarrow \\
\mathbf{M}_{N_T}^{(2)} & \Rightarrow & \mathbf{C}^{(2)} & \Rightarrow & s_2 \Rightarrow \\
\vdots & & \vdots & & \vdots \\
\mathbf{M}_{N_T}^{(K)} & \Rightarrow & \mathbf{C}^{(K)} & \Rightarrow & s_K \Rightarrow
\end{array}
\qquad k = \underset{i}{\arg\min}\; \| \mathbf{y} - \mathbf{H}s_i \|^2 \quad \Rightarrow s_k
$$

$$K = | \Xi^{N_T} |$$

Fig. 3.2. EZF/EMMSE based on ML criterion.

3.4.2.3. Widely Linear Sphere Decoder

Motivated by sphere decoding, WLSD can jointly estimate $\mathbf{M}_{N_T}$ and $\mathbf{x}$. To this end, (3.102) can be rewritten as follows by applying QR decomposition:

$$\mathbf{y}_Q = \mathbf{H}_Q \mathbf{z} + \mathbf{w}_Q, \tag{3.107}$$

where

$$\mathbf{y}_Q = \widetilde{\mathbf{Q}}[\mathbf{y}^T exp(j\theta_0) \;\; \mathbf{y}^H exp(-j\theta_0)]^T,$$

$$\mathbf{H}_Q = \widetilde{\mathbf{Q}}[\mathbf{H}^T \;\; \mathbf{M}_{N_T} \mathbf{H}^H]^T,$$

$$\mathbf{w}_Q = \widetilde{\mathbf{Q}}[\mathbf{w}^T exp(j\theta_0) \;\; \mathbf{w}^H exp(-j\theta_0)]^T,$$

$\widetilde{\mathbf{Q}}$ is a block diagonal unitary matrix defined as:

$$\widetilde{\mathbf{Q}} = \begin{bmatrix} \mathbf{Q}^H & \mathbf{0} \\ \mathbf{0} & \mathbf{Q}^T \end{bmatrix} \tag{3.108}$$

For an EZF receiver, unitary matrix $\mathbf{Q}$ is given by QR decomposition $\mathbf{H} = \mathbf{QR}$, and $\mathbf{R}$ is an upper triangular matrix. For EMMSE detection, QR decomposition is applied to the matrix $[\mathbf{H}^T \; \frac{\sigma}{\sqrt{P}} \mathbf{I}_{N_T}]^T$, which is based on the following equivalent model [28, 29]:

$$\begin{bmatrix} \mathbf{y}\, exp(j\theta_0) \\ \mathbf{0}_{N_T \times 1} \end{bmatrix} = \begin{bmatrix} \mathbf{H} \\ \dfrac{\sigma}{\sqrt{P}} \mathbf{I}_{N_T} \end{bmatrix} \mathbf{z} + \begin{bmatrix} \mathbf{w}\, exp(j\theta_0) \\ -\dfrac{\sigma}{\sqrt{P}} \mathbf{z} \end{bmatrix}, \tag{3.109}$$

$\mathbf{R}$ can be obtained by taking the first N_T rows after the QR decomposition on model (3.109). Note that $\mathbf{H}_Q$ can be expressed by $\mathbf{H}_Q = [\mathbf{R}^T \quad \mathbf{M}_{N_T} \mathbf{R}^H]^T$. Because $\mathbf{R}$ is an upper triangular matrix, $\mathbf{H}_Q$ can be written as follows:

$$\mathbf{H}_Q = \begin{bmatrix} r_{11} & r_{12} & \cdots & r_{1N_T} \\ & r_{22} & \cdots & r_{2N_T} \\ & & \ddots & \vdots \\ & & & r_{N_T N_T} \\ r_{11}^* l_1 & r_{12}^* l_2 & \cdots & r_{1N_T}^* l_{N_T} \\ & r_{22}^* l_2 & \cdots & r_{2N_T}^* l_{N_T} \\ & & \ddots & \vdots \\ & & & r_{N_T N_T}^* l_{N_T} \end{bmatrix}, \tag{3.110}$$

where r_{pq} is the element of $\mathbf{R}$ in the pth row and q^{th} column, and the phase-rotation factor l_q is the qth diagonal element of matrix $\mathbf{M}_{N_T}$. $\mathbf{H}_Q$ can be re-arranged to a "quasi-upper triangular" matrix $\widetilde{\mathbf{R}}$ by respectively gathering its inter-conjugate pairs as follows:

$$\widetilde{\mathbf{R}} = \begin{bmatrix} \mathbf{r}_{11} & \mathbf{r}_{12} & \cdots & \mathbf{r}_{1N_T} \\ & \mathbf{r}_{22} & \cdots & \mathbf{r}_{2N_T} \\ & & \ddots & \vdots \\ & & & \mathbf{r}_{N_T N_T} \end{bmatrix}, \tag{3.111}$$

with $\mathbf{r}_{pq} = [r_{pq} \quad r_{pq}^* l_q]^T$. Accordingly, by combining the inter-conjugate pairs in $\mathbf{y}_Q$ and $\mathbf{w}_Q$, a new system model is given as follows:

$$\widetilde{\mathbf{y}} = \widetilde{\mathbf{R}} \mathbf{z} + \widetilde{\mathbf{w}} \tag{3.112}$$

Let $\mathbf{y}_Q = [y_1, y_2, \cdots, y_{N_T}, y_1^*, y_2^*, \cdots, y_{N_T}^*]^T$, then $\widetilde{\mathbf{y}}$ is given by $\widetilde{\mathbf{y}} = [\mathbf{y}_1^T, \mathbf{y}_2^T, \cdots, \mathbf{y}_{N_T}^T]^T$, with $\mathbf{y}_p = [y_p, y_p^*]^T$, and vector $\widetilde{\mathbf{w}}$ is obtained by applying the same arrangement. It is easy to verify that $E[\widetilde{\mathbf{w}}\widetilde{\mathbf{w}}^H] = \sigma^2 \mathbf{I}_{2N_R}$. To obtain the true $\mathbf{M}_{N_T}$ (or phase-rotation vector $\mathbf{l}$), a new successive searching is performed inside a sphere of radius d. This is described as:

$$\begin{cases} \mathbf{M}_{N_T} = \arg \min_{\mathbf{l} \in \Xi^{N_T}} \| \widetilde{\mathbf{y}} - \widetilde{\mathbf{R}} \mathbf{z} \|^2, \\ s.t. \quad \| \widetilde{\mathbf{y}} - \widetilde{\mathbf{R}} \mathbf{z} \|^2 \le d^2, \end{cases} \tag{3.113}$$

(3.113) can be further specified as:

$$\begin{cases} \mathbf{M}_{N_T} = \arg \min_{l \in \Xi^{N_T}} \sum_{p=1}^{N_T} \| \mathbf{y}_p - \sum_{q=p}^{N_T} \mathbf{r}_{pq} z_q \|^2, \\[2mm] s.t. \quad \sum_{p=1}^{N_T} \| \mathbf{y}_p - \sum_{q=p}^{N_T} \mathbf{r}_{pq} z_q \|^2 \le d^2, \end{cases}$$ (3.114)

To meet the requirement of (3.114), the following inequality holds for $p = N_T$:

$$\| \mathbf{y}_{N_T} - \mathbf{r}_{N_T N_T} z_{N_T} \|^2 \le d^2,$$ (3.115)

and for any $p = 1, 2, \cdots, N_T - 1$, the similar inequality is obtained as follows:

$$\| \mathbf{y}_p - \sum_{q=p}^{N_T} \mathbf{r}_{pq} z_q \|^2 \le d^2 - \sum_{p'=p+1}^{N_T} \| \mathbf{y}_{p'} - \sum_{q=p'}^{N_T} \mathbf{r}_{p'q} z_q \|^2$$ (3.116)

According to (3.115), (3.116), the search of (3.113) is performed in an iterative manner based on matrix $\widetilde{\mathbf{R}}$ from its lower right corner to its upper left corner. The diagonal elements of $\mathbf{M}_{N_T}$ are searched one by one from l_{N_T} to l_1. Accordingly, with the obtained l_i, the transmitted symbols are estimated from z_{N_T} to z_1. In the estimation of z_i, widely linear processing is employed, this is called sphere searching "widely linear sphere decoder (WLSD)". The WLSD searching for general MPSK/MQAM signals is described by the pseudo codes in Algorithm 2 and by Fig. 3.3.

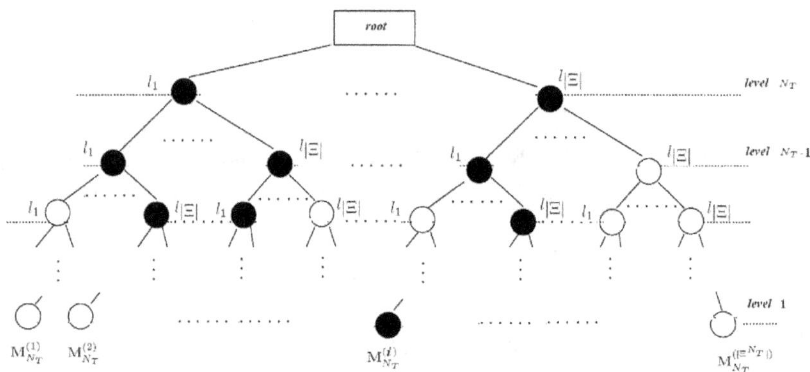

Fig. 3.3. Tree Searching of WLSD.

Algorithm 2 describes the sphere searching process inside the sphere within radius d, whose value is updated in iterations. Set Ξ is used to build matrix $\mathbf{L}$, which serves as a data base of $\mathbf{M}_{N_T}$. *depth* denotes the depth of sphere searching, it is initialized to N_T. Matrix $\mathbf{B}$ is defined to save the initial states of $\mathbf{y}_Q(1:N_T)$ at N_T levels, which is very useful when the searching is switched from one node to another in the same level or when the searching is withdrawn from current level (e.g. level k) to upper level (e.g. level $k+1$). During the searching, $\mathbf{M}_{N_T}$ and $\mathbf{z}$ hold current optimal PRM and its corresponding solution respectively. The final optimal solution is saved in $\mathbf{M}_{N_T}$ and $\mathbf{z}$ once the searching is terminated. The initial value of d can be infinity and the PRM can be initialized with the first column of matrix $\mathbf{L}$; or alternatively, the radius and PRM can be initialized by ZF-OSIC/MMSE-OSIC.

3.4.2.4. Widely Linear Signal Estimation

In "line 21" of Algorithm 2, one should estimate and quantize the signal component z_q to update the distance. One classic method is successive interference cancelling [11]. For $q = N_T$, z_{N_T} is estimated by jointly exploiting the information contained in y_{N_T} and $y_{N_T}^*$:

$$\hat{z}_{N_T} = Q(\frac{\mathbf{r}_{N_T N_T}^H \mathbf{y}_{N_T}}{\parallel \mathbf{r}_{N_T N_T} \parallel^2}) = Q(\frac{r_{N_T N_T}^* y_{N_T} + l_{N_T}^* r_{N_T N_T} y_{N_T}^*}{2 \mid r_{N_T N_T} \mid^2}), \tag{3.117}$$

and for any $q = 1, 2, \cdots, N_T - 1$:

$$\hat{z}_q = Q(\frac{\mathbf{r}_{qq}^H \mathbf{y}_{q'}}{\parallel \mathbf{r}_{qq} \parallel^2}) = Q(\frac{r_{qq}^* y_q' + l_q^* r_{qq} y_q'^*}{2 \mid r_{qq} \mid^2}), \tag{3.118}$$

where $\mathbf{y}_{q'} = [y_{q'}, y_{q'}^*]^T = \mathbf{y}_q - \sum_{p=q+1}^{N_T} \mathbf{r}_{qp} z_p$. Let $r_{N_T N_T}^* y_{N_T} = a_{N_T} + b_{N_T} j$, $r_{qq}^* y_q' = a_q + b_q j$ and $l_q = \alpha_q + \beta_q j$, a more general expression for $\hat{z}_q$ can be obtained as follows:

$$\hat{z}_q = Q(\frac{(a_q + a_q \alpha_q - b_q \beta_q) + (b_q - b_q \alpha_q - a_q \beta_q)j}{2 \mid r_{qq} \mid^2}) \tag{3.119}$$

3.4.2.5. Quantization

The general rule of quantization $Q(\cdot)$ is somewhat simple. For an MPSK/4QAM signal, there is only one pair of anti-polar constellation points for each value of l_q, the quantization can be realized according to the 'polarity' of $\hat{z}_q$. For MQAM signal (M > 4), the quantization is exactly the same as the MPSK signal except for the constellation points lying on the x-axis or y-axis. For the points lying on the axes, they are quantized in the same way as M-ary ASK (MASK) signals.

One notes that, for the extreme case of real signals, e.g. BPSK/ASK, $\mathbf{M}_{N_T} = \mathbf{I}_{N_T}$ always holds. ML is unnecessary, **Algorithm 1** (or **Algorithm 2**) becomes EZF/EMMSE (or EZF-OSIC/EMMSE-OSIC) [5]. For more details, one can refer to [5].

Algorithm 2. WLSD.

1: Input: $\mathbf{R}$, $\mathbf{y}_Q(1:N_T)$

2: %$\mathbf{R}$, $\mathbf{y}_Q(1:N_T)$ can be obtained in model (107)

3: Output: $\mathbf{M}_{N_T}$, $\mathbf{z}$

4: %$\mathbf{M}_{N_T}$ returns the estimated phase-rotation matrix

5: %$\mathbf{z}$ returns the estimated rotated signal

6: Initialization:

7: $depth=N_T$; % Depth of tree search from N_T to 1

8: $d^2 = inf$; % The initial radius of sphere is set to infinity

9: $\mathbf{z}_0 = \mathbf{0}$; % The vector holds the value of currently visited branch

10: $w = 0$; % w is defined for distance accumulation

11: $\Xi_0=[\Xi,0]$; % The row vector Ξ_0 includes the elements in set Ξ

12: %'0' is added as a stop point of search in current level

13: $\mathbf{L}=ones(N_T,1)*\Xi_0$; % A database of phase-rotation matrix $\mathbf{M}_{N_T}$

14: % An $\mathbf{M}_{N_T}$ is built by drawing an element in each row of $\mathbf{L}$

15: %$\mathbf{M}_{N_T} = diag([l_1,\cdots,l_{N_T}])$, $l_i \in \Xi$

16: $\mathbf{idx} = [1,1,\cdots,1]$; %The search starts with the first $\mathbf{M}_{N_T}$

17: %The first $\mathbf{M}_{N_T}$ includes the first element of each row

18: $\mathbf{B}(:,N_T) = \mathbf{y}_Q(1:N_T)$; %Save the state of $\mathbf{y}_Q$ in level N_T

19: **while** $depth \leq N_T$ **do**

20: **if** $\mathbf{L}(depth,\mathbf{idx}(depth)) \neq 0$ **then**

21: $\mathbf{z}_0(depth)=Q[\frac{\mathbf{r}_{depth,depth}^{H}[y_{depth},\, y_{depth}^*]^T}{\|\mathbf{r}_{depth,depth}\|^2}]$; %WLP and quantization

22: $\mathbf{B}(:,depth) = \mathbf{y}_Q(1:N_T)$; %Save the state of $\mathbf{y}_Q$ in level $depth$ before signal cancelling

23: $\mathbf{y}_Q(1:depth) = \mathbf{y}_Q(1:depth) - \mathbf{R}(1:depth,depth)\mathbf{z}_0(depth)$;

24: $w(depth) = w(depth+1) + \|\mathbf{y}_Q(depth)\|^2$; %distance accumulation

25: **if** $w(depth) < d^2$ **then** %If the searching is inside the sphere

26: **if** $depth = 1$ **then** %If the searching reaches the bottom

27: $d^2 = w(1)$; %The radius is updated.

28: $\mathbf{M}_{N_T} = diag([\mathbf{L}(1,\mathbf{idx}(1)),\cdots,\mathbf{L}(N_T,\mathbf{idx}(N_T))])$;

29: $\mathbf{z} = \mathbf{z}_0$; %Currently optimal solution is saved.

30: $\mathbf{idx}(depth) = \mathbf{idx}(depth) + 1$; %Try the next phase-rotation factor

31: $\mathbf{y}_Q(1:N_T) = \mathbf{B}(:,depth)$; %Go back to the former state

32: **else** %If the searching does not reach the bottom

33: $depth = depth - 1$; %The searching goes to lower level

34: **end if**;

35: **else** %If the searching is not inside the sphere

36: $\mathbf{idx}(depth) = \mathbf{idx}(depth) + 1$; %Try the next phase-rotation factor

37: $\mathbf{y}_Q(1:N_T) = \mathbf{B}(:,depth)$; %Go back to the former state

38: **end if**;

39: **else** %All phase-rotation factors are tried

40: $\mathbf{idx}(depth) = 1$; %Go back to the first phase-rotation factor

41: $depth = depth + 1$; %Go back to the upper level

42: **if** $depth \leq N_T$ **then** %If the searching does not reach the root

43: $\mathbf{idx}(depth) = \mathbf{idx}(depth) + 1$;

44: $\mathbf{y}_Q(1:N_T) = \mathbf{B}(:,depth)$;

45: **end if**;

46: **end if**;

47: **end while**

3.4.3. DoA Estimation by Exploiting Conjugate Symmetry

3.4.3.1. Problem Statement

The DoA finding model considered is exactly the same as in the previous discussion, except that $\mathbf{x}(n)$ can be BPSK, MPSK or QAM signal.

3.4.3.2. Extended Data Model

Without loss of generality, the discussion is firstly focused on the case where the transmitted signals are sharing a same rotation matrix $\mathbf{M}_K^{(i)}$, the corresponding received signals are included in subset Θ_i. Hence, the sample in Θ_i meets:

$$\mathbf{y}(n_i) = \mathbf{A}\mathbf{\Psi}\mathbf{x}(n_i) + \mathbf{n}(n_i) \tag{3.120}$$

The complex conjugate version of (3.120) is written as:

$$\mathbf{y}^*(n_i) = \mathbf{A}^*\mathbf{\Psi}^*\mathbf{M}_K^{(i)}\mathbf{x}(n_i) + \mathbf{n}^*(n_i) \tag{3.121}$$

Extended data model : The extended data model is given in the following form:

$$\begin{bmatrix} \mathbf{y}(n_i) \\ \mathbf{y}^*(n_i) \end{bmatrix} = \begin{bmatrix} \mathbf{A}\mathbf{\Psi} \\ \mathbf{A}^*\mathbf{\Psi}^*\mathbf{M}_K^{(i)} \end{bmatrix} \mathbf{x}(n_i) + \begin{bmatrix} \mathbf{n}(n_i) \\ \mathbf{n}^*(n_i) \end{bmatrix} \tag{3.122}$$

Let $\mathbf{M}_K^{(i)} = diag(l_1^{(i)}, l_2^{(i)}, \cdots, l_K^{(i)})$, (3.122) can be rewritten in a more compact form as follows:

$$\tilde{\mathbf{y}}(n_i) = \sum_{k=1}^{K} \tilde{\mathbf{a}}^{(i)}(\theta_k)\mathbf{x}_k(n_i) + \tilde{\mathbf{n}}(n_i), \tag{3.123}$$

where

$$\tilde{\mathbf{y}}(n_i) = \begin{bmatrix} \mathbf{y}(n_i) \\ \mathbf{y}^*(n_i) \end{bmatrix}, \tag{3.124}$$

$$\tilde{\mathbf{a}}^{(i)}(\theta_k) = \begin{bmatrix} h_k\mathbf{a}(\theta_k)e^{j\psi_k} \\ h_k\mathbf{a}^*(\theta_k)e^{-j\psi_k}l_k^{(i)} \end{bmatrix}, \tag{3.125}$$

$$\tilde{\mathbf{n}}(n_i) = \begin{bmatrix} \mathbf{n}(n_i) \\ \mathbf{n}^*(n_i) \end{bmatrix} \tag{3.126}$$

In matrix form, (3.122) becomes:

$$\tilde{\mathbf{y}}(n_i) = \tilde{\mathbf{A}}^{(i)} \mathbf{x}(n_i) + \tilde{\mathbf{n}}(n_i),$$ (3.127)

where $\tilde{\mathbf{A}}^{(i)} = [\tilde{\mathbf{a}}^{(i)}(\theta_1), \tilde{\mathbf{a}}^{(i)}(\theta_2), ..., \tilde{\mathbf{a}}^{(i)}(\theta_K)]$.

3.4.3.3. DoA Estimation

Based on the extended data model, the Root-MUSIC algorithm is applied to calculate the DoA of incoming signal. In each subset Θ_i, the covariance matrix of $\tilde{\mathbf{y}}(n_i)$ is given as:

$$\tilde{\mathbf{R}}_{\Theta_i} = E\left[\tilde{\mathbf{y}}(n_i)\tilde{\mathbf{y}}^H(n_i)\right] = \tilde{\mathbf{A}}^{(i)} \mathbf{R}_{\Phi_i} (\tilde{\mathbf{A}}^{(i)})^H + \sigma_n^2 \mathbf{I}_{2M},$$ (3.128)

where the signal covariance matrix is $\mathbf{R}_{\Phi_i} = E[\mathbf{x}(n_i)\mathbf{x}^H(n_i)]$.

The eigenvalue decomposition of $\tilde{\mathbf{R}}_{\Theta_i}$ is:

$$\tilde{\mathbf{R}}_{\Theta_i} = [\tilde{\mathbf{U}}\Lambda\tilde{\mathbf{U}}^H]^{(i)} = [\tilde{\mathbf{U}}_s\Lambda_s\tilde{\mathbf{U}}_s^H + \sigma_n^2\tilde{\mathbf{U}}_n\tilde{\mathbf{U}}_n^H]^{(i)},$$ (3.129)

$\tilde{\mathbf{U}} = [\tilde{\mathbf{u}}_1, \tilde{\mathbf{u}}_2, ..., \tilde{\mathbf{u}}_{2M}]$ is a unitary matrix, the columns of $\tilde{\mathbf{U}}_s = [\tilde{\mathbf{u}}_1, \tilde{\mathbf{u}}_2, ..., \tilde{\mathbf{u}}_K]$ span the signal space, and the columns of $\tilde{\mathbf{U}}_n = [\tilde{\mathbf{u}}_{K+1}, \tilde{\mathbf{u}}_{K+2}, ..., \tilde{\mathbf{u}}_{2M}]$ span the noise space. Theoretically, in subset Θ_i, $\hat{\theta}_k$ can be estimated by solving the following equation:

$$[\tilde{\mathbf{a}}^H(\theta_k)\tilde{\mathbf{U}}_n\tilde{\mathbf{U}}_n^H\tilde{\mathbf{a}}(\theta_k)]^{(i)} = 0$$ (3.130)

Let $z = e^{j2\pi\frac{\Delta\sin\theta}{\lambda}}$. The above equation can be transformed into the following form:

$$[\tilde{\mathbf{a}}^T(z_k^{-1})\tilde{\mathbf{U}}_n\tilde{\mathbf{U}}_n^H\tilde{\mathbf{a}}(z_k)]^{(i)} = 0$$ (3.131)

By finding the K roots of (3.131) inside and closest to the unit circle, the direction angles $\hat{\theta}_1, \hat{\theta}_2, \cdots, \hat{\theta}_K$ can be estimated. Equation (3.131) depends on unknown parameters ψ_k and $l_k^{(i)}$, which makes the root-finding problem impossible. In order to fix the problem, the noise subspace matrix $\tilde{\mathbf{U}}_n$ is split into two equal-sized sub-matrices $\mathbf{U}_{n1}$ and $\mathbf{U}_{n2}$, namely,

$$\tilde{\mathbf{U}}_n = \begin{bmatrix} \mathbf{U}_{n1} \\ \mathbf{U}_{n2} \end{bmatrix},$$

(3.131) is rewritten as follows:

$$(\mathbf{q}^{(i)})^H \mathbf{T}^{(i)} \mathbf{q}^{(i)} = 0, \tag{3.132}$$

where $\mathbf{q}^{(i)} = \begin{bmatrix} 1 \\ e^{-j2\psi_k} l_k^{(i)} \end{bmatrix}$, and $\mathbf{T}^{(i)} = \begin{bmatrix} \mathbf{a}^T(z_k^{-1})\mathbf{U}_{n1}\mathbf{U}_{n1}^H\mathbf{a}(z_k) & \mathbf{a}^T(z_k^{-1})\mathbf{U}_{n1}\mathbf{U}_{n2}^H\mathbf{a}(z_k^{-1}) \\ \mathbf{a}^T(z_k)\mathbf{U}_{n2}\mathbf{U}_{n1}^H\mathbf{a}(z_k) & \mathbf{a}^T(z_k)\mathbf{U}_{n2}\mathbf{U}_{n2}^H\mathbf{a}(z_k^{-1}) \end{bmatrix}^{(i)}.$

Due to the non negativity of the quadratic form $\{(\mathbf{q}^{(i)})^H \mathbf{T}^{(i)} \mathbf{q}^{(i)}\}$, the minimum of $\{(\mathbf{q}^{(i)})^H \mathbf{T}^{(i)} \mathbf{q}^{(i)}\}$ depends on the smallest eigenvalue of $\mathbf{T}^{(i)}$, which is always non negative. Thus (3.132) can be solved by finding the roots of $\det\{\mathbf{T}^{(i)}\}$, that is:

$$\det\{\mathbf{T}^{(i)}\} = 0, \tag{3.133}$$

Obviously, (3.133) is independent of parameters ψ_k and $l_k^{(i)}$. It can be proved that $\mathbf{U}_{n2}^*\mathbf{U}_{n2}^T = \mathbf{U}_{n1}\mathbf{U}_{n1}^H$, (3.133) is equivalent to the following simplified equation:

$$\left\{\mathbf{a}^T(z_k^{-1})\mathbf{U}_{n1}\mathbf{U}_{n1}^H\mathbf{a}(z_k) - \left|\mathbf{a}^T(z_k^{-1})\mathbf{U}_{n1}\mathbf{U}_{n2}^H\mathbf{a}(z_k^{-1})\right|\right\}^{(i)} = 0 \tag{3.134}$$

By finding the roots of (3.134), the DoA can be estimated. Alternatively, the DoA can also be estimated by searching the peaks of a new spatial pseudo-spectrum, given by:

$$P(\theta_k) = \underset{\theta_k \in [-\frac{\pi}{2}, \frac{\pi}{2}]}{\arg\ \max} \frac{1}{\left\{\mathbf{a}^T(z_k^{-1})\mathbf{U}_{n1}\mathbf{U}_{n1}^H\mathbf{a}(z_k) - \left|\mathbf{a}^T(z_k^{-1})\mathbf{U}_{n1}\mathbf{U}_{n2}^H\mathbf{a}(z_k^{-1})\right|\right\}^{(i)}} \tag{3.135}$$

The DoA estimation can be further refined by synthesizing the results over ρ subsets, e.g. one can add and average the DoA estimates.

An synthesizing method is used in [10]. Due to the non negativity of the quadratic form $[\tilde{\mathbf{a}}^H(\theta_k)\widetilde{\mathbf{U}}_n\widetilde{\mathbf{U}}_n^H\tilde{\mathbf{a}}(\theta_k)]^{(i)}$ in (3.130), the statistical average over the whole samples is used, the DoA estimation is refined by synthesizing all the polynomials as:

$$\sum_{i=1}^{\rho} \frac{N_i}{N} [\tilde{\mathbf{a}}^T(z_k^{-1})\widetilde{\mathbf{U}}_n\widetilde{\mathbf{U}}_n^H\tilde{\mathbf{a}}(z_k)]^{(i)} = 0 \tag{3.136}$$

Hence

$$\sum_{i=1}^{\rho} \frac{N_i}{N} \det\{\mathbf{T}^{(i)}\} = 0, \tag{3.137}$$

which is equivalent to the following equation:

$$\sum_{i=1}^{\rho} \frac{N_i}{N} \left\{\mathbf{a}^T(z_k^{-1})\mathbf{U}_{n1}\mathbf{U}_{n1}^H\mathbf{a}(z_k) - \left|\mathbf{a}^T(z_k^{-1})\mathbf{U}_{n1}\mathbf{U}_{n2}^H\mathbf{a}(z_k^{-1})\right|\right\}^{(i)} = 0 \tag{3.138}$$

By finding the roots of (3.138), the improved DoA estimator can be obtained.

One notes that the above discussion focuses on the received signals belonging to subset Θ_i, whose elements are sharing a same rotation matrix $\mathbf{M}_K^{(i)}$. Actually, before the above processing, the classification on the received signals is needed. Generally, in digital communications, the receiver does not know the exact value of each received sample, which is a challenge in the classification. However, reference signals, e.g. training signal and pilot signal, are known by the receiver. Before the classification, the data symbols other than reference signals are estimated firstly. For details, one can refer to [10].

3.5. Summary

This chapter gives a brief introduction to the recent work on the exploitation of signals' properties, including the non circularity, the cyclostationarity, the conjugate symmetry in array signal processing. Besides in DoA estimation, the non circularity is exploited in signal detection/demodulation in MIMO communication system; and this property is further utilized in the MIMO-NOMA downlink system to jointly eliminate or control the intra/inter-cluster interferences. The cyclostationarity is used for filtering out the interferences with cycle frequencies distinct from the desired signals to enhance the performance of DoA estimation or beamforming. The non circularity can be considered as a special case of the conjugate symmetry, which exhibits in the general linearly modulated digital signals. Therefore, the non circularity exploiting method can be extended to the exploitation of conjugate symmetry, which can be equally applied in DoA estimation and signal detection/demodulation.

Acknowledgment

This work is funded by Guangzhou Science & Technology Program (No. 201807010071), the funds of central universities (No.2018ZD09), National Natural Science Foundation of China under Grant 61971198, Guangdong Basic and Applied Basic Research Foundation under Grant 2019A1515011040, and Guangdong Natural Science Foundation Project "Hybrid I-Q multiplexing in power domain for MIMO-NOMA system based on widely linear processing".

References

[1]. P. Chargé, Y. Wang, J. Saillard, An extended cyclic MUSIC algorithm, *IEEE Transactions on Signal Processing*, Vol. 51, Issue 7, July 2003, pp. 1695-1701.

[2]. P. Chargé, Y. Wang, J. Saillard, A non-circular sources direction finding method using polynomial rooting, *Signal Processing*, Vol. 81, Issue 8, 2001, pp. 1765-1770.

[3]. P. Chargé, Y. Wang and J. Saillard, Cyclostationarity-exploiting direction finding algorithms, *IEEE Transactions on Aerospace and Electronic Systems*, Vol. 39, Issue 3, July 2003, pp. 1051-1056.

[4]. Y. Ding, N. Li, Y. Wang, S. Feng, H. Chen, Widely linear sphere decoder in MIMO systems by exploiting the conjugate symmetry of linearly modulated signals, *IEEE Transactions on Signal Processing*, Vol. 64, Issue 24, 2016, pp. 6428-6442.

[5]. Y. Ding, Y. Wang, J.F. Diouris, Efficient detection algorithms for multiple-input/multiple-output systems by exploiting the non circularity of transmitted signal source, *IET Signal Processing*, Vol. 5, Issue 2, 2011, pp. 180-186.

[6]. P. Chevalier, F. Pipon, New insights into optimal widely linear array receivers for the demodulation of BPSK, MSK, and GMSK signals corrupted by noncircular interferences-application to SAIC, *IEEE Transactions on Signal Processing*, Vol. 54, Issue 3, March 2006, pp. 870-883.

[7]. P. Chevalier, A. Blin, Widely linear MVDR beamformers for the reception of an unknown signal corrupted by noncircular interferences, *IEEE Transactions on Signal Processing*, Vol. 55, Issue 11, Nov. 2007, pp. 5323-5336.

[8]. B. Picinbono, On circularity, *IEEE Transactions on Signal Processing*, Vol. 42, Issue 12, Dec. 1994, pp. 3473-3482.

[9]. J. Galy, Antenne adaptative: du second ordre aux ordres supérieurs, applications aux signaux de télécommunications, PhD Thesis, *Université de Toulouse 3*, Toulouse, 1998.

[10]. C. Wen, Y. Ding, C. Fu, Y. Wang, A MPSK sources direction finding method by exploiting the property of signal sources, in *Proceedings of the IEEE 87th Vehicular Technology Conference (VTC-Spring'18)*, Porto, 2018, pp. 1-5.

[11]. P. W. Wolniansky, G. J. Foschini, G. D. Golden, R. A. Valenzuela, V-BLAST: An architecture for realizing very high data rates over the rich-scattering wireless channel, in *Proceedings of the URSI International Symposium on Signals, Systems, and Electronics*, Pisa, Italy, 1998, pp. 295-300.

[12]. D. Tse, P. Viswanath, Fundmentals of Wireless Communication, *Cambridge University Press*, 2007.

[13]. X. Ge, H. Cheng, M. Guizani, T. Han, 5G wireless backhaul networks: challenges and research advances, *IEEE Network*, Vol. 28, Issue 6, Nov.-Dec. 2014, pp. 6-11.

[14]. Z. Ding, F. Adachi, H. V. Poor, The application of MIMO to non-orthogonal multiple access, *IEEE Transactions on Wireless Communications*, Vol. 15, Issue 1, Jan. 2016, pp. 537-552.

[15]. D. Tong, Y. Ding, Y. Liu, Y. Wang, A MIMO-NOMA framework with complex-valued power coefficients, *IEEE Transactions on Vehicular Technology*, Vol. 68, Issue 3, March 2019, pp. 2244-2259.

[16]. B. Picinbono, P. Chevalier, Widely linear estimation with complex data, *IEEE Transactions on Signal Processing*, Vol. 43, Issue 8, Aug. 1995, pp. 2030-2033.

[17]. W. A. Gardner, A. Napolitano, L. Paura, Cyclostationarity: Half a century of research, *Signal Processing*, Vol. 86, Issue 4, 2006, pp. 639-697.

[18]. Q. Wu, K.-M. Wong, Blind adaptive beamforming for cyclostationary signals, *IEEE Transactions on Signal Processing*, Vol. 44, Issue 11, November 1996, pp. 2757-2767.

[19]. H. L. Van Trees, Detection, Estimation, and Modulation Theory, Part IV: Optimum Array Processing, *John Wiley&Sons*, New York, 2002.

[20]. B. D. Carlson, Covariance matrix estimation errors and diagonal loading in adaptive arrays, *IEEE Transactions on Aerospace and Electronic Systems*, Vol. 24, Issue 4, July 1988, pp. 397-401.

[21]. S.-A. Vorobyov, A.-B. Gershman, Z.-Q. Luo, Robust adaptive beamforming using worst-case performance optimization: a solution to the signal mismatch problem, *IEEE Transactions on Signal Processing*, Vol. 51, Issue 2, February 2003, pp. 313-324.

[22]. J. Zhang, G. Liao, J. Wang, Robust direction finding for cyclostationary signals with cycle frequency error, *Signal Processing*, Vol. 85, Issue 12, 2005, pp.2386-2393.

[23]. J.-H. Lee, C.-C. Huang, Robust cyclic adaptive beamforming using a compensation method, *Signal Processing*, Vol. 92, Issue 4, 2012, pp. 954-962.

[24]. J. Capon, High-resolution frequency-wavenumber spectrum analysis, *Proceedings of the IEEE*, Vol. 57, Issue 8, August 1969, pp. 1408-1418.

[25]. R. Mallipeddi, J. P. Lie, P. N. Suganthan, S. G. Razul, C. M. S. See, Robust adaptive beamforming based on covariance matrix reconstruction for look direction mismatch, *Progress in Electromagnet Research Letters*, Vol. 25, 2011, pp. 37-46.

[26]. Y. Gu, A. Leshem, Robust adaptive beamforming based on interference covariance matrix reconstruction and steering vector estimation, *IEEE Transactions on Signal Processing*, Vol. 60, Issue 7, July 2012, pp. 3881-3885.

[27]. J. Li, G. Wei, Y. Ding, Adaptive beamforming based on covariance matrix reconstruction by exploiting interferences' cyclostationarity, *Signal Processing*, Vol. 93, Issue 9, 2013, pp. 2543-2547.

[28]. D. Wübben, R. Böhnke, V. Kühn, K. D. Kammeyer, MMSE extension of V-BLAST based on sorted QR decomposition, in *Proceedings of the IEEE 58th Vehicular Technology Conference (VTC-Fall'03)*, Vol. 1, Orlando, FL, 2003, pp. 508-512.

[29]. Q. Wen, X. Ma, Efficient greedy LLL algorithms for lattice decoding, *IEEE Transactions on Wireless Communications*, Vol. 15, Issue 5, May 2016, pp. 3560-3572.

Chapter 4
A Review on Chirp and Some Other Related Signal Processing Models

Debasis Kundu, Rhythm Grover and Swagata Nandi

4.1. Introduction

We observe periodic phenomenon everyday in our daily lives. For example, the number of tourists visiting the famous Taj Mahal in India, the daily average temperature in Toronto, Canada, the number of passengers travelling daily between New York and London, the Electro Cardio Graph (ECG) signal of a normal human being are all periodic in nature. Sometimes the data or signals may not be exactly periodic but they may be nearly periodic. Figs. 4.1 and 4.2 are examples of an ECG plot of a normal human being and 'UUU' vowel sound of a male, respectively. Evidently, both of them exhibit periodic behavior.

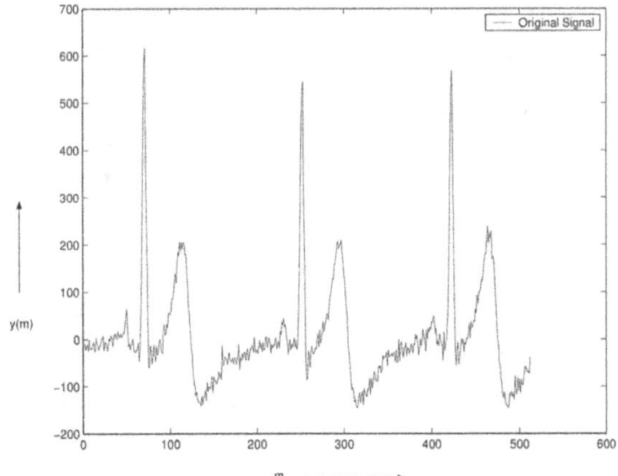

Fig. 4.1. ECG Plot of a normal human being.

Debasis Kundu
Department of Mathematics and Statistics, Indian Institute of Technology Kanpur, Pin 208016, India

149

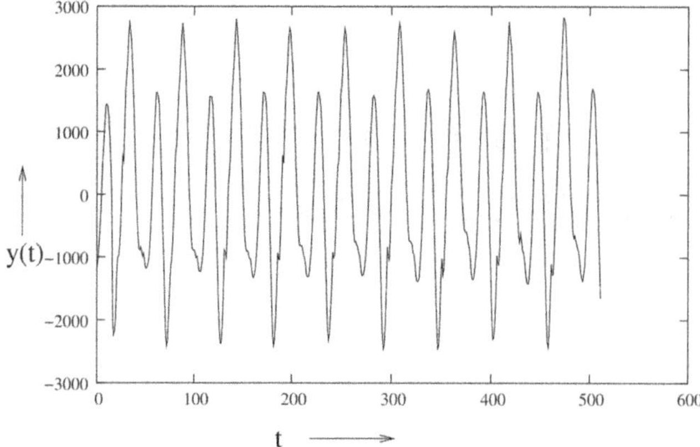

Fig. 4.2. UUU-sound of a male.

The periodic or nearly periodic phenomena may be observed even in higher dimensions also. See for example a black and white symmetric image plot in Fig. 4.3 and a colored symmetric image plot in Fig. 4.4. Both of them indicate periodic nature in higher dimensions.

Fig. 4.3. Black and white symmetric image plot.

As unforeseeable glitches occur in every experiment, the observed signals are usually corrupted with random noise. Due to random nature of the noise corrupted signal, statistics plays an important role in Signal Processing. In adopting a statistical model, we assume that the observed signal has two components- a deterministic component and a stochastic component. The deterministic component constitutes the important characteristics of the noise free signal and the stochastic component takes care of the contaminated part of the observed signal. To recover the deterministic component from the noisy signal, different statistical techniques may be used quite effectively. The main aim of this chapter is to provide different methods which are available till date to analyze periodic or nearly periodic signal in presence of additive noise. One natural question that arises why somebody wants to analyze a periodic or nearly periodic signal. One reason might be purely academic curiosity. However, some of the more practical reasons can be: a) data compression, b) to better understand the observed phenomenon, and/ or c) to make predictions about the future.

Fig. 4.4. Colored symmetric image plot.

A simple periodic function can be represented as follows:

$$y(t) = A\cos(\omega t) + B\sin(\omega t).$$

Here $0 < \omega < 2\pi$. In this case the period of the function $y(t)$, i.e. the shortest time it needs to repeat itself is $2\pi / \omega$. Any mean zero smooth periodic function can be written in general as $y(t) = \sum_{k=1}^{\infty}\{A_k \cos(k\omega t) + B_k \sin(k\omega t)\}$, and it mainly follows from the Fourier expansion of a periodic function. For any given smooth periodic function $y(t)$, the coefficients A_k's and B_k's are unique and they can be recovered from $y(t)$ as follows:

$$A_k = \frac{\omega}{\pi}\int_0^{2\pi/\omega} \cos(k\omega t)y(t)dt \quad and \quad B_k = \frac{\omega}{\pi}\int_0^{2\pi/\omega} \sin(k\omega t)y(t)dt .$$

Since an observed signal is usually corrupted with noise, it is more reasonable to write it in a general form as

$$y(t) = \sum_{k=1}^{\infty}\{A_k \cos(k\omega t) + B_k \sin(k\omega t)\} + e(t) . \tag{4.1}$$

Here $e(t)$ is the noise component of the signal with mean zero. It is difficult to estimate infinite number of parameters, hence, (4.1) is approximated as

$$y(t) = \sum_{k=1}^{p}\{A_k \cos(k\omega t) + B_k \sin(k\omega t)\} + e(t) \text{ with general form } \omega_k \text{ for } k\omega \tag{4.2}$$

Here $0 < \omega_1, \ldots, \omega_k < 2\pi$. The sum of sinusoidal model (4.2) plays an important role in Signal Processing. The problem is to extract the deterministic component of the signal

$$\mu(t) = \sum_{k=1}^{p} \{A_k \cos(\omega_k t) + B_k \sin(\omega_k t)\},$$

based on the noisy signal at the discrete time points $\{y(1), \ldots, y(N)\}$. Hence, the problem becomes the estimation of 'p' as well as A_k, B_k, ω_k, for $k = 1, \ldots, p$. Here, 'p' is the order of sinusoidal model, $A_k^2 + B_k^2$ is known as the amplitude and ω_k is the frequency, for $k = 1, \ldots, p$.

Sometimes it is more convenient to work with the associated complex version of model (4.2), and with the abuse of notation it can be written as

$$y(t) = \sum_{k=1}^{p} A_k e^{i\omega_k t} + e(t). \tag{4.3}$$

In model (4.3), $y(t)$ is complex valued signal, $e(t)$ is the complex valued error, A_k is a complex number, and $|A_k|^2$ is the amplitude and $i = \sqrt{-1}$. Model (4.3) can be obtained from model (4.2) by taking the Hilbert transformation [131]. Both the models have been used to analyze noisy periodic signals in a wide variety of applications. The two models are equivalent, and results which have developed for model (4.2) can be modified suitably for model (4.3) and vice versa. An extensive amount of work has been done on models (4.2) and (4.3). We provide a brief overview of the sinusoidal model and some of the well-established parameter estimation methods for this model in Section 4.3.

Many natural signals are not exactly periodic, however they can be nearly periodic. The sinusoidal model in such a scenario becomes unsuitable because of its harmonic structure. Several alternative models have been introduced in literature to address this concern. One of the most popular models among them is the chirp model, where the frequency is allowed to vary linearly in time, making the model more capable of capturing the aperiodicities of the signal. A chirp model can be mathematically described as follows:

$$y(t) = A\cos(\alpha t + \beta t^2) + B\sin(\alpha t + \beta t^2) + e(t). \tag{4.4}$$

Here also A and B are real numbers, and $|A|^2 + |B|^2$ is the amplitude of the signal, α and β are known as frequency and frequency rate, respectively, $y(t)$ is the observed signal and $e(t)$ is a mean zero noise. This one component chirp model has been used quite extensively in different applications in signal processing. One most important application can be found in modeling radar signals. For example, suppose a radar is illuminating a target, then the transmitted signal will undergo a phase shift induced by the distance and relative speed between the target and the receiver. Based on the assumption that the speed is continuous and differentiable, the phase shift can be approximated as

$\phi(t) = a_0 + a_1 t + a_2 t^2$. Here the parameters a_1 and a_2 are either related to speed and acceleration or range and speed depending on what the radar is intended for, and on the kind of waveform transmitted. For explicit details see Rihaczek [108] (pages 56-65). A more general form of the chirp signal can be written as follows:

$$y(t) = \sum_{k=1}^{p} \{ A_k \cos(\alpha_k t + \beta_k t^2) + B_k \sin(\alpha_k t + \beta_k t^2) \} + e(t). \tag{4.5}$$

Model (4.5) can be found in several engineering applications also, see for example Wehner [121].

In case of chirp signals also, sometimes it is more convenient to work with the associated complex model. With the abuse of notation, it can be written as follows:

$$y(t) = \sum_{k=1}^{p} A_k e^{i(\alpha_k t + \beta_k t^2)} + e(t). \tag{4.6}$$

Similar to the complex exponential model, in this case also, for model (4.6) $y(t)$s are complex valued signals, A_ks are complex numbers, and $e(t)$s are complex valued errors. Models (4.5) and (4.6) are equivalent models. Therefore, the methodologies and results which have been developed for model (4.6) can be used for model (4.5) and vice versa.

This problem is a challenging problem both from the theoretical and computational view points, since it is a highly non-linear model. Although model (4.5) or (4.6) can be seen as a non-linear regression problem, it does not satisfy some of the standard sufficient conditions needed to establish the basic properties like consistency and asymptotic normality of least squares estimators (LSEs) of maximum likelihood estimators (MLEs). Hence, it is not guaranteed whether the LSEs or the MLEs will be consistent or not. Moreover, even if it can be established that the LSEs or the MLEs will be consistent, finding them is not a trivial issue. Most of the standard algorithms do not work in this case. During the last fifteen years an extensive amount of work can be found both in Signal Processing and in Statistics literature developing different efficient estimation procedures and establishing their properties. We provide a comprehensive review of the different methods which have been developed in the last fifteen to twenty years related to one dimensional one component chirp model in Section 4.4 and one dimensional multicomponent chirp model in Section 4.5.

A more general model than the one component chirp model was proposed by Djurić and Kay [24], it is known as the polynomial phase model and it can be written as follows:

$$y(t) = A e^{i(\alpha_1 t + \alpha_2 t^2 + \dots + \alpha_p t^p)} + e(t). \tag{4.7}$$

Here also $y(t)$, A and $e(t)$ are complex valued. The corresponding real valued model can be written as

$$y(t) = A \cos(\alpha_1 t + \alpha_2 t^2 + \dots + \alpha_p t^p) + B \sin(\alpha_1 t + \alpha_2 t^2 + \dots + \alpha_p t^p) + e(t). \tag{4.8}$$

Here $y(t)$, A, B and $e(t)$ are all real valued, and (4.7) and (4.8) are equivalent models. Clearly, model (4.8) is more flexible than model (4.4), and it has also been used quite extensively in different signal processing applications. Model (4.7) or (4.8) is a highly non-linear model. In Section 4.6 we have provided various applications, theoretical developments and different estimation procedures related to this model.

Another interesting, yet absurdly scarce in literature, is a random amplitude chirp model. This model is a generalization of the random amplitude sinusoidal model put forward by Besson and Stoica [9]. The latter is defined as follows:

$$y(t) = \alpha(t)e^{i(a_0 + a_1 t)} + e(t) . \tag{4.9}$$

Here $\alpha(t)$ is assumed to be real valued Gaussian stationary process and $e(t)$ is a complex valued noise. The random amplitude sinusoidal model (4.9) has several applications in different signal processing applications, for example when a radar on-board a train emits a continuous wave towards the track, the echo received is known to have a slowly varying amplitude and it can be modeled by using (4.9), see Besson and Castanié [8]. Similarly, in underwater systems, it has been observed that the acoustic signals propagating through the ocean are perturbed by multiplicative noise, and Dwyer [30] observed that model (4.9) can be used for this purpose.

Besson, Ghogho and Swami [10] proposed the random amplitude chirp signal model to capture the linearly time-varying phase shift as well as amplitude distortion, and it can be represented as follows:

$$y(t) = \alpha(t)e^{i(a_0 + a_1 t + a_2 t^2)} + e(t) . \tag{4.10}$$

Here $\alpha(t)$ and $e(t)$ are same as defined above. Clearly, the random amplitude chirp model (4.10) is more flexible than the one component chirp model. If $\alpha(t)$ is constant, then the random amplitude chirp model becomes the one component chirp model. Recently Nandi and Kundu [94] proposed a multicomponent random amplitude chirp model and it can be described as follows:

$$y(t) = \sum_{k=1}^{p} \alpha_k(t)e^{i(\theta_{1k} t + \theta_{2k} t^2)} + e(t) . \tag{4.11}$$

Here $\{\alpha_k(t)\}$ is a sequence of independent and identically distributed (i.i.d.) real valued random variables and $e(t)$ is the complex valued noise. Again, the multicomponent random amplitude chirp model (4.11) is more flexible than the multiple chirp model (4.6), and model (4.6) can be obtained from model (4.11), when $\alpha_k(t)$s are assumed to be constant. In Section 4.7 we provide different applications, theoretical development and various estimation procedures related to one component and multicomponent random amplitude chirp models.

Recently, Grover [45] proposed a model which is a combination of the sinusoidal and the chirplet component and it can be expressed as follows:

$$y(t) = A\cos(\alpha t) + B\sin(\alpha t) + C\sin(\beta t^2) + D\cos(\beta t^2) + e(t). \qquad (4.12)$$

Here, A, B, C and D are real numbers, $|A|^2 + |B|^2$ is the amplitude corresponds to the sinusoidal component, $|C|^2 + |D|^2$ is the amplitude corresponds to the chirplet component, $|A|^2 + |B|^2 + |C|^2 + |D|^2$ is the amplitude corresponds to the signal $y(t)$, α is the sinusoidal frequency parameter, β is the chirp rate parameter and $e(t)$ is the real valued noise component. It is observed that model (4.12) exhibits same type of behavior as the one component chirp model. In her thesis, the author proposed a more general multicomponent 'chirp like' model as follows:

$$y(t) = \sum_{k=1}^{P} \left\{ A_k \cos(\alpha_k t) + B_k \sin(\alpha_k t) \right\}$$
$$+ \sum_{k=1}^{q} \left\{ C_k \sin(\beta_k t^2) + D_k \cos(\beta_k t^2) \right\} + e(t). \qquad (4.13)$$

The multicomponent chirp like model (4.13) behaves very similar to the multicomponent chirp model. See for example Fig. 4.5, where a multicomponent chirp model has been approximated by a chirp like model (4.13). They are very similar in nature as evidenced from Fig. 4.5.

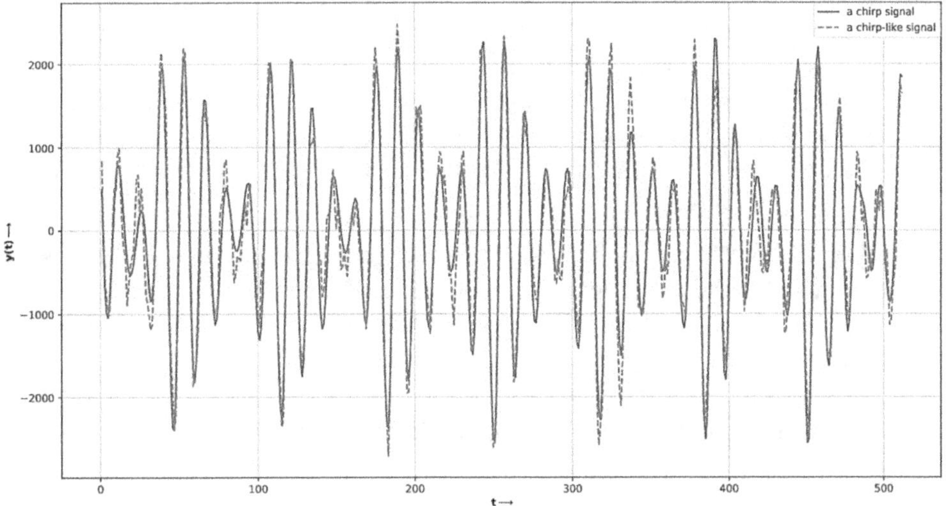

Fig. 4.5. A chirp model fitting and a chirp like model fitting to a sound vowel data.

It is observed that any signal which can be modelled using the multicomponent chirp model can be modelled using the multicomponent chirp like model also and vice versa. The main advantage about the chirp like model is that it is more convenient to use it in practice than the chirp model. In Section 4.8 we provide theoretical developments about one component and multicomponent chirp like models, different estimation procedures and the properties of these estimators have been discussed in details.

A special case of the chirp model is a Harmonic chirp model where frequencies and frequency rates instead of being arbitrary are integer multiples of a common fundamental frequency and a fundamental chirp rate. This model was put forward quite recently by Jensen et al. [59] as a generalization of the well-established fundamental frequency sinusoidal model proposed by Hannan [48]. The latter is mathematically expressed as follows:

$$y(t) = \sum_{k=1}^{p} \{A_k \cos(k\alpha t) + B_k \sin(k\alpha t)\} + e(t).$$ (4.14)

Due to the particular structure of the model, the number of non-linear parameters to be estimated reduces to only one. Model (4.14) has been used quite extensively for analyzing speech data. In short span periodic speech signal often it is observed that there is one fundamental frequency, and it has several harmonics. However, many a times it is also observed that frequency is not constant but time-varying. This had motivated the introduction of harmonic chirp model, mathematically expresses as follows:

$$y(t) = \sum_{k=1}^{p} \{A_k \cos(k\alpha t + k\beta t^2) + B_k \sin(k\alpha t + k\beta t^2)\} + e(t).$$ (4.15)

Here, α and β are the fundamental frequency and fundamental frequency rate, respectively. Due to presence of only two non-linear parameters, numerically it is easier to handle than the multiple chirp model. Therefore, if a non-stationary signal can be modelled using the harmonic chirp model, it is always preferable. Jensen et al. [59] proposed estimation procedures of the unknown parameters of model (4.15) and recently Grover [42] provided theoretical properties of different estimators. In Section 4.9 we describe different estimation procedures and their theoretical properties in details under various error assumptions.

In this chapter, we also look into the modelling of two dimensional signals. The most basic and popular model is the traditional sinusoidal model proposed by Barbieri and Barone [6] in two dimensions to model symmetric texture data. A 'p' component two dimensional sinusoidal frequency model can be written as follows:

$$y(m,n) = \sum_{k=1}^{p} \left[A_k \cos(m\lambda_k + n\mu_k) + B_k \sin(m\lambda_k + n\mu_k) \right] + e(m,n).$$ (4.16)

Here, A_k, B_k are real numbers, λ_k, μ_k are the frequencies, and $e(m,n)$ is the real valued error component. A typical texture image as shown in Fig. 4.3 can be obtained using model (4.16). This model is used in a wide range of practical applications such as antenna array processing, geophysical perception, biomedical spectral analysis etc. Since the introduction of the model, an extensive amount of work has been done for developing different estimation procedures and establishing their properties for the two dimensional sinusoidal model.

A related but more general two dimensional model is the two dimensional polynomial phase signal model suggested by Friedlander and Francos [38]. This model has found several applications in image processing and finger print imaging. Mathematically, it can be expressed as follows:

$$y(m,n) = Ae^{i(\alpha_1 m + \beta_1 n + \alpha_2 m^2 + \beta_2 n^2 + \gamma mn)} + e(m,n) . \tag{4.17}$$

Here, A is a complex number, $e(m,n)$ is the complex valued two dimensional noise and $y(m,n)$ is the two dimensional signal. The associated real model becomes

$$y(m,n) = A\cos(\alpha_1 m + \beta_1 n + \alpha_2 m^2 + \beta_2 n^2 + \gamma mn)$$
$$+ B\sin(\alpha_1 m + \beta_1 n + \alpha_2 m^2 + \beta_2 n^2 + \gamma mn) + e(m,n). \tag{4.18}$$

A typical image generated by using model (4.18) can be seen in Fig. 4.6.

The most general two dimensional model has been considered by Lahiri and Kundu [75], and it can be expressed as follows:

$$y(m,n) = A\cos\left(\sum_{p=1}^{r}\sum_{j=0}^{p}\alpha(j,p-j)m^j n^{p-j}\right) + B\sin\left(\sum_{p=1}^{r}\sum_{j=0}^{p}\alpha(j,p-j)m^j n^{p-j}\right) + e(m,n).$$

$$\tag{4.19}$$

Fig. 4.6. A simulated image generated using the two dimensional model (4.18).

Here as before A and B are real numbers, $\alpha(j,p-j)$s are distinct frequency rates of order $(j,p-j)$, respectively. Different forms of (4.19) have been considered in the literature. A significant amount of work has been done in developing efficient estimation procedures of the unknown parameters of the two dimensional polynomial phase signal. Lahiri and Kundu [75] discussed about the theoretical properties of the LSEs of the unknown parameters of the most general two dimensional polynomial phase model (4.19) under a fairly general error assumptions. In Section 4.10 we provide different theoretical developments and numerical procedures available related to this model. In all the sections we provide several open problems for future research, and finally we conclude the chapter in Section 4.11.

4.2. Notations and Preliminaries

In this section, we present some useful notations that we use throughout the chapter. We also provide some preliminaries (Table 4.1) foreshadowing the analytical development of many of the existing methods and algorithms.

Table 4.1. List of symbols and notations.

Symbol	Notation	Example
lower or uppercase letters	scalar quantity	A
lower or upper case bold typefaces	a vector or a matrix	$\boldsymbol{A}$
$(.)^{\mathrm{T}}$	transpose of a real matrix	A^{T} is the transpose of real matrix A
$(.)^{H}$	complex conjugate transpose of a complex matrix	A^{H} is the complex conjugate transpose of complex matrix A
$\overline{(.)}$	complex conjugate of a complex number	$\overline{c}$ is the complex conjugate of c
$\mathrm{diag}\{a_1,...,a_m\}$	$m \times m$ diagonal matrix with diagonal elements $a_1,...,a_m$	$\mathrm{diag}\{a_1,...,a_m\}$
$\boldsymbol{P}_A = A(A^TA)^{-1}A^T$	projection matrix on the column space of a real matrix A	$\boldsymbol{P}_A$
$\boldsymbol{P}_A = A(A^HA)^{-1}A^H$	projection matrix on the column space of a complex matrix A	$\boldsymbol{P}_A$
$\xrightarrow{a.s.}$	convergence almost surely	$X_n \xrightarrow{a.s.} X$
$\xrightarrow{d}$	convergence in distribution	$X_n \xrightarrow{d} X$
$N_m(\boldsymbol{\mu},\Sigma)$	m-variate normal distribution with mean vector $\boldsymbol{\mu}$ and dispersion matrix Σ	$X \sim N_m(\boldsymbol{\mu},\Sigma)$
$\boldsymbol{\theta}$	parameter vector that denotes a set of unknown parameters of a model	$\boldsymbol{\theta}$
$\hat{\boldsymbol{\theta}}$	an estimator of $\boldsymbol{\theta}$	$\hat{\boldsymbol{\theta}}$
$\boldsymbol{I}$	an identity matrix	$I = \begin{bmatrix} 1 & 0 \\ 0 & 1 \end{bmatrix}$, an identity matrix of order 2
$\boldsymbol{D}$	a diagonal matrix	$D = \mathrm{diag}\{N^{-1/2}, N^{-1/2}, N^{-3/2}, N^{-5/2}\}$

4.2.1. Prony's Equation

In 1795, in his seminal paper, Prony proposed a method for fitting sum of real valued exponential model to real-life data. The interested reader is referred to numerical analysis textbooks by Froberg [39] or Hilderband [50]. Later on, different variants of Prony's equations have been developed. Some basic features of Prony's method can be briefly described as follows. Suppose $\alpha_1,\ldots,\alpha_M$ are arbitrary non-zero real numbers and $\beta_1,\ldots,\beta_M$ are distinct real numbers. If

$$\mu(t) = \alpha_1 e^{\beta_1 t} + \ldots + \alpha_M e^{\beta_M t}; \quad t = 1,\ldots,N, \tag{4.20}$$

and $M < N$, then there exists $(M+1)$ constants $\{g_0,\ldots,g_M\}$, such that

$$Ag = 0, \tag{4.21}$$

where

$$A = \begin{bmatrix} \mu(1) & \cdots & \mu(M+1) \\ \vdots & \ddots & \vdots \\ \mu(N-M) & \cdots & \mu(N) \end{bmatrix}, \; g = \begin{bmatrix} g_0 \\ \vdots \\ g_M \end{bmatrix}, 0 = \begin{bmatrix} 0 \\ \vdots \\ 0 \end{bmatrix}.$$

The matrix A is of rank M, and g is the eigen vector corresponds to the zero eigenvalue. It is always possible to put restrictions on $g_0,\ldots,g_M$, such as

$$\sum_{j=0}^{M} g_j^2 = 1 \quad and \quad g_0 > 0, \tag{4.22}$$

so that g becomes unique. The set of linear equations (4.21) is known as Prony's equations. It is possible to recover $\beta_1,\ldots,\beta_M$ from $g_0,\ldots,g_M$ as follows. The following polynomial equation:

$$p(x) = g_0 + g_1 x + \ldots + g_M x^M = 0 \tag{4.23}$$

has M distinct roots, and they are $e^{\beta_1},\ldots,e^{\beta_M}$. Therefore, there is a one to one correspondence between $\{\beta_1,\ldots,\beta_M\}$ and $\{g_0,\ldots,g_M\}$, such that (4.22) hold. Further, $\{g_0,\ldots,g_M\}$ does not depend on $\{\alpha_1,\ldots,\alpha_M\}$. Once, $\beta_1,\ldots,\beta_M$ are obtained from $\mu(1),\ldots,\mu(N)$, $\alpha_1,\ldots,\alpha_M$ can be obtained from the linear equation

$$\mu = X\alpha, \tag{4.24}$$

where $\mu = (\mu(1),\ldots,\mu(N))^{\mathrm{T}}$, $\alpha = (\alpha_1,\ldots,\alpha_M)^{\mathrm{T}}$ and

$$
X = \begin{bmatrix} e^{\beta_1} & \dots & e^{\beta_M} \\ \dots & \ddots & \dots \\ e^{N\beta_1} & \dots & e^{N\beta_M} \end{bmatrix}.
$$

The matrix X is of rank M, hence, $a = (X^{\mathrm{T}} X)^{-1} X^{\mathrm{T}} \mu$.

The idea of Prony has been extended for the complex case also. Consider the following complex exponential model

$$
\mu(t) = \alpha_1 e^{i\omega_1 t} + \dots + \alpha_M e^{i\omega_M t}; \quad t = 1, \dots, N . \tag{4.25}
$$

Here, $\alpha_1, \dots, \alpha_M$ are non-zero complex numbers, $0 < \omega_1, \dots, \omega_M < 2\pi$ and they are distinct. Then, there exists $\{g_0, \dots, g_M\}$ such that (4.21) is satisfied. Moreover, the complex polynomial equation

$$
p(z) = g_0 + g_1 z + \dots + g_M z^M = 0
$$

has the roots $z_1 = e^{i\omega_1}, \dots, z_M = e^{i\omega_M}$ and $a = (\alpha_1, \dots, \alpha_M)^{\mathrm{T}} = (X^H X)^{-1} X^H \mu$. Here $X = ((X_{jk})), X_{jk} = e^{ij\omega_k}$. Note that the roots of the complex polynomial $p(z)$ satisfy

$$
|z_1| = \dots = |z_M| = 1, \ \bar{z}_k = z_k^{-1}; \quad k = 1, \dots, M . \tag{4.26}
$$

Define the new complex polynomial

$$
q(z) = z^{-M} \bar{p}(z) = \bar{g}_0 z^{-M} + \dots + \bar{g}_M . \tag{4.27}
$$

Due to (4.26), the polynomials $p(z)$ and $q(z)$ have same roots. Therefore,

$$
\frac{g_k}{g_M} = \frac{\bar{g}_{M-k}}{\bar{g}_0}; \quad k = 0, \dots, M
$$

can be obtained by comparing the coefficients of $p(z)$ and $q(z)$. Hence,

$$
b_k = \bar{b}_{M-k}; \quad k = 0, \dots, M , \tag{4.28}
$$

where $b_k = g_k \left(\dfrac{\bar{g}_0}{g_M} \right)^{-\frac{1}{2}}; \ k = 0, \dots, M.$ The condition (4.28) is known as the conjugate symmetry property and in matrix notation it can be written as

$$
b = J \bar{b} ,
$$

here $\boldsymbol{b} = (b_0,\ldots,b_M)^{\mathrm{T}}$, and $\boldsymbol{J}$ is an exchange matrix as follows:

$$\boldsymbol{J} = \begin{bmatrix} 0 & 0 & \cdots & 0 & 1 \\ 0 & 0 & \cdots & 1 & 0 \\ \vdots & \vdots & \ddots & \vdots & \vdots \\ 0 & 1 & \cdots & 0 & 0 \\ 1 & 0 & \cdots & 0 & 0 \end{bmatrix}.$$

4.2.2. Number Theoretic Results and Conjecture

The theoretical derivation of the asymptotic properties of the various estimators proposed for the chirp signal model are established using some number theoretic results and a famous conjecture by Vinogradov [115]. We present these results and conjecture here. However, before stating these, we first present some useful trigonometric identities used to establish various asymptotic results for the sinusoidal model. For $\omega \in (0, \pi)$,

$$\frac{1}{N} \sum_{t=1}^{N} \cos(\omega t) = o(1), \tag{4.29}$$

$$\frac{1}{N} \sum_{t=1}^{N} \sin(\omega t) = o(1), \tag{4.30}$$

$$\frac{1}{N} \sum_{t=1}^{N} \cos^2(\omega t) = \frac{1}{2} + o(1), \tag{4.31}$$

$$\frac{1}{N} \sum_{t=1}^{N} \sin^2(\omega t) = \frac{1}{2} + o(1), \tag{4.32}$$

$$\frac{1}{N^{k+1}} \sum_{t=1}^{N} t^k \cos^2(\omega t) = \frac{1}{2(k+1)} + o(1), \tag{4.33}$$

$$\frac{1}{N^{k+1}} \sum_{t=1}^{N} t^k \sin^2(\omega t) = \frac{1}{2(k+1)} + o(1), \tag{4.34}$$

$$\frac{1}{N^{k+1}} \sum_{t=1}^{N} t^k \cos(\omega t)\sin(\omega t) = o(1), \tag{4.35}$$

here $a_N = o(1)$ means $a_N \to 0$ as $N \to \infty$.

These results can be easily established and for reference one may look into the handbook by Mangulis [86].

The following lemma provides the number theoretic results by Vinogradov [115].

Lemma 1. *Let* $\theta_1, \theta_2 \in (0, \pi)$ *and* $k = 0, 1, \ldots$ *Then except for countable number of points the followings are true:*

$$\lim_{N \to \infty} \frac{1}{N} \sum_{t=1}^{N} \cos(\theta_1 t + \theta_2 t^2) = 0, \tag{4.36}$$

$$\lim_{N \to \infty} \frac{1}{N} \sum_{t=1}^{N} \sin(\theta_1 t + \theta_2 t^2) = 0, \tag{4.37}$$

$$\lim_{N \to \infty} \frac{1}{N^{k+1}} \sum_{t=1}^{N} t^k \cos^2(\theta_1 t + \theta_2 t^2) = \frac{1}{2(k+1)}, \tag{4.38}$$

$$\lim_{N \to \infty} \frac{1}{N^{k+1}} \sum_{t=1}^{N} t^k \sin^2(\theta_1 t + \theta_2 t^2) = \frac{1}{2(k+1)}, \tag{4.39}$$

$$\lim_{N \to \infty} \frac{1}{N^{k+1}} \sum_{t=1}^{N} t^k \cos(\theta_1 t + \theta_2 t^2) \sin(\theta_1 t + \theta_2 t^2) = 0. \tag{4.40}$$

Proof: The proof of this lemma can be seen in Lahiri, Kundu and Mitra [76] in detail.

The following conjecture has been used to establish some of the crucial results related to chirp model. It was proposed by Vinogradov [115] and it still remains an open problem. It is important to note that even though it has not been proved theoretically, extensive simulation results suggest the validity of the conjecture.

Conjecture 1. *Let* $\theta_1, \theta_2, \theta_1', \theta_2' \in (0, \pi)$ *and* $k = 0, 1, \ldots$ *Then except for countable number of points the followings are true:*

$$\lim_{N \to \infty} \frac{1}{\sqrt{N} N^k} \sum_{t=1}^{N} t^k \cos(\theta_1 t + \theta_2 t^2) \sin(\theta_1' t + \theta_2' t^2) = 0. \tag{4.41}$$

In addition if $\theta_2 \neq \theta_2'$, *then*

$$\lim_{N \to \infty} \frac{1}{\sqrt{N} N^k} \sum_{t=1}^{N} t^k \cos(\theta_1 t + \theta_2 t^2) \cos(\theta_1' t + \theta_2' t^2) = 0,$$

$$\lim_{N \to \infty} \frac{1}{\sqrt{N} N^k} \sum_{t=1}^{N} t^k \sin(\theta_1 t + \theta_2 t^2) \sin(\theta_1' t + \theta_2' t^2) = 0.$$

4.3. Sinusoidal Model

In this section we briefly provide different estimation procedures of the frequencies of a periodic signal and their properties. Due to vast amount of scientific literature dealing with solving this problem (see for example the books by Kay [60] or Stoica and Mosses

[112]), the overview is limited to the estimation methods that can be extended and generalized for the chirp model as well.

Also, of the many different kinds of sinusoidal models previously mentioned, we limit our discussion here to the basic multicomponent sinusoidal model, mathematically expressed as follows:

$$y(t) = \sum_{k=1}^{p} \{A_k^0 \cos(\omega_k^0 t) + B_k^0 \sin(\omega_k^0 t)\} + e(t); \quad 1,2,\ldots,N, \tag{4.42}$$

where A_k^0, B_k^0, ω_k^0 for $k=1,\ldots,p$ are true parameter values and they are unknown. Our problem is to estimate the unknown parameters based on the sample $\{y(1),\ldots,y(N)\}$. It is important to note that, for the estimators that we will discuss subsequently, p is assumed to be known.

Most of the results in this chapter are asymptotic in nature, so we make certain assumptions about the asymptotic framework. In particular, we present assumptions on the error component $e(t)$.

Assumption 1. *$\{e(t)\}$ is a sequence of i.i.d. normal random variables with mean zero and variance σ^2.*

Assumption 2. *$\{e(t)\}$ is a sequence of i.i.d. random variables with $E(e(t))=0$ and $V(e(t))=\sigma^2$.*

Assumption 3. *$\{e(t)\}$ is a stationary linear process with the following form:*

$$e(t) = \sum_{j=0}^{\infty} a(j) X(t-j),$$

here $\{X(t); t=1,2,\ldots\}$ are i.i.d. random variables with $E(X(t)) = 0$ and $V(X(t))=\sigma^2$, and $\sum_{j=0}^{\infty} |a(j)| < \infty$.

The explicit structure of $e(t)$ can follow either of the above assumptions and will depend on the estimation method used.

It is important to note that in practice p is unknown and must be estimated from the data itself. Model selection methods offer different ways of estimating this number. Later, in this section we provide a brief overview of some of these methods.

4.3.1. Periodogram Estimators

The most popular estimator of the frequencies is the periodogram estimator. The periodogram function $I(\theta)$ of the signal $y(t)$ is defined as follows:

$$I(\theta) = \frac{1}{N}\left|\sum_{t=1}^{N} y(t)e^{i\theta t}\right|^2. \tag{4.43}$$

The main reason to use the periodogram function to estimate the frequencies and the number of components p is that $I(\theta)$ has local maxima at the true frequencies if there is no noise in the data. Therefore, if the noise variance is not high, then $I(\theta)$ can be used to detect the number of signals and also the associated frequencies. Let us look at the following example. A sample of $y(t)$ is generated using the following model equation:

$$y(t) = 3.0\cos(0.2\pi t) + 3.0\sin(0.2\pi t) + 3.0\cos(0.5\pi t) + 3.0\sin(0.5\pi t) + e(t). \tag{4.44}$$

Here $e(t)$ satisfies Assumption 2 with $\sigma^2 = 2$. The periodogram plot of the signal (4.44) is displayed in Fig. 4.7.

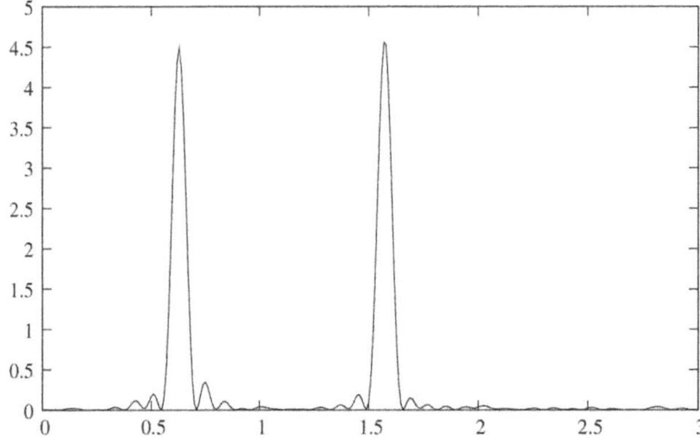

Fig. 4.7. Periodogram plot of $y(t)$ obtained from Model (4.44).

The figure clearly reveals that the number of sinusoidal components of the signal $y(t)$ is 2. Moreover, the location of the frequencies can be estimated from the periodogram function as well. However, sometimes the periodogram plot may give misleading results. To demonstrate we consider the following example. We generate a sample using the model:

$$\begin{aligned} y(t) &= 3.0\cos(0.2\pi t) + 3.0\sin(0.2\pi t) \\ &\quad + 0.25\cos(0.5\pi t) + 0.25\sin(0.5\pi t) + e(t). \end{aligned} \tag{4.45}$$

Here also $e(t)$ satisfies Assumption 2, but $\sigma^2 = 5$. In this case, the periodogram plot of $y(t)$ is shown in Fig. 4.8. The figure fails to tell us the true location of the frequencies as well as the number of components present in the signal.

Therefore, it is clear that the periodogram estimators may not always work, particularly if the sample size is not large enough or the error variance is high.

Hannan [47] and Walker [117] independently established the strong consistency and asymptotic normality results of the periodogram estimators of the frequencies based on Assumption 3. Their theoretical results justify the use of periodogram estimators at least when the number of data points N is large enough.

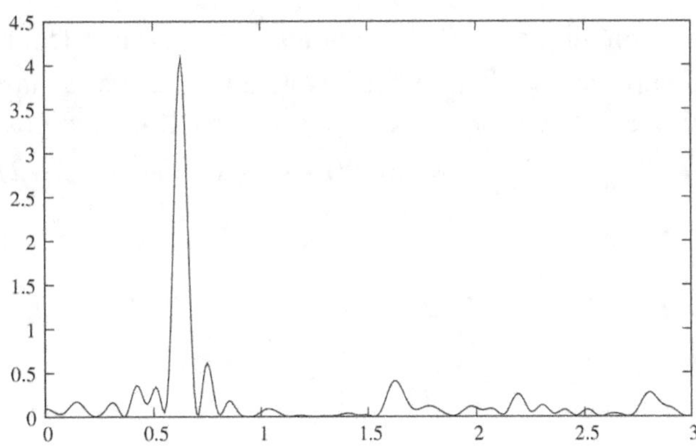

Fig. 4.8. Periodogram plot of $y(t)$ obtained from Model (4.45).

4.3.2. Least Squares Estimators

Since model (4.2) is a non-linear regression model, one natural choice of estimators is the LSEs, and they coincide with the MLEs, if the error satisfies Assumption 1. These estimators can be obtained by minimizing the error sum of squares as defined below

$$Q(\boldsymbol{\omega},\boldsymbol{\theta}) = \sum_{t=1}^{N}\left(y(t) - \sum_{k=1}^{p}\{A_k\cos(\omega_k t) + B_k\sin(\omega_k t)\}\right)^2. \tag{4.46}$$

Here $\omega = (\omega_1,\ldots,\omega_p)^{\mathrm{T}}$, $\boldsymbol{\theta} = (A_1, B_1,\ldots, A_p, B_p)^{\mathrm{T}}$. Now (4.46) can be written as

$$Q(\boldsymbol{\omega},\boldsymbol{\theta}) = (\boldsymbol{Y} - \boldsymbol{Z}(\boldsymbol{\omega})\boldsymbol{\theta})^{\mathrm{T}}(\boldsymbol{Y} - \boldsymbol{Z}(\boldsymbol{\omega})\boldsymbol{\theta}), \tag{4.47}$$

where $\boldsymbol{Y} = (y(1),\ldots,y(N))^{\mathrm{T}}$, $\boldsymbol{e} = (e(1),\ldots,e(N))^{\mathrm{T}}$ and

$$\boldsymbol{Z}(\boldsymbol{\omega}) = \begin{bmatrix} \cos(\omega_1) & \sin(\omega_1) & \ldots & \cos(\omega_p) & \sin(\omega_p) \\ \vdots & \vdots & \ddots & \vdots & \vdots \\ \cos(N\omega_1) & \sin(N\omega_1) & \ldots & \cos(N\omega_p) & \sin(N\omega_p) \end{bmatrix}.$$

For a given ω, the LSE of $\boldsymbol{\theta}$ can be obtained as

$$\widehat{\boldsymbol{\theta}}(\omega) = (\boldsymbol{Z}(\omega)^{\mathrm{T}}\boldsymbol{Z}(\omega))^{-1}\boldsymbol{Z}(\omega)^{\mathrm{T}}\boldsymbol{Y}. \tag{4.48}$$

Substituting (4.48) in (4.47) we obtain

$$R(\omega) = Q(\omega, \hat{\theta}(\omega)) = Y^{\mathrm{T}}(I - P_{Z(\omega)})Y. \tag{4.49}$$

Here $P_{Z(\omega)}$ is the projection matrix on the column space of $Z(\omega)$. Therefore, the LSE of ω can be obtained by minimizing $R(\omega)$ with respect to ω. Bresler and Macovski [12] proposed a special purpose algorithm named as iterative quadratic maximum likelihood (IQML) to minimize $R(\omega)$. We briefly describe it here. It can be observed that, using Prony's equation $I - P_{Z(\omega)} = P_{B(g)}$, where $B(g)$ is a $(N+1) \times (N+1-2p)$ matrix of rank $(N+1-2p)$ as follows:

$$B(g) = \begin{bmatrix} g_0 & 0 & \cdots & 0 \\ g_1 & g_0 & \cdots & 0 \\ \vdots & \vdots & \ddots & \vdots \\ g_{2p} & g_{2p-1} & \cdots & 0 \\ 0 & g_{2p} & \cdots & 0 \\ \vdots & \vdots & \ddots & \vdots \\ 0 & 0 & \cdots & g_0 \\ \vdots & \vdots & \ddots & \vdots \\ 0 & 0 & \cdots & g_{2p} \end{bmatrix}, \tag{4.50}$$

and $g = (g_0, g_1, \ldots, g_{2p})^{\mathrm{T}}$. Note that minimization of $Y^{\mathrm{T}}(I - P_{Z(\omega)})Y$ with respect to ω is same as the minimization of $Y^{\mathrm{T}}P_{B(g)}Y$ with respect to g. Now observe that

$$Y^{\mathrm{T}}P_{B(g)}Y = g^{\mathrm{T}}Y_D^{\mathrm{T}}(B(g)^{\mathrm{T}}B(g))^{-1}Y_D g,$$

where Y_D is an $(N-2p) \times (2p+1)$ matrix as given below:

$$Y_D = \begin{bmatrix} y(1) & \cdots & y(2p+1) \\ \vdots & \ddots & \vdots \\ y(N-2p) & \cdots & y(N) \end{bmatrix}.$$

Bresler and Macovski [12] suggested the following. If at the $k-th$ stage of the iteration the estimate of g is $g^{(k)}$, then obtain $g^{(k+1)}$ by minimizing $g^{\mathrm{T}}Y_D^{\mathrm{T}}((B(g^{(k)})^{\mathrm{T}}B(g^{(k)}))^{-1}Y_D g$ with respect to g. The iteration continues until convergence takes place. No proof of convergence has been provided by Bresler and Macovski [12]. Later Kundu [65] converted the minimization of $Y^{\mathrm{T}}P_{B(g)}Y$ as a non-linear eigenvalue problem, and provided an iterative procedure based on solving a standard non-linear eigenvalue problem. He also has provided the proof of convergence of the proposed iterative procedure.

Note that the multiple sinusoidal model (4.42) is a non-linear regression model. An extensive amount of work has been done in the statistical literature developing different properties of the LSEs for different non-linear regression model, see for example the book by Seber and Wild [111] or Bates and Watts [7] in this respect. Jennrich [58] and Wu [122] developed different sufficient conditions which ensure the consistency and asymptotic normality of the LSEs of the parameters of a nonlinear regression model. Unfortunately, the multiple sinusoidal model (4.42) does not satisfy the sufficient conditions of Jennrich [58] or Wu [122]. Hence, it is not guaranteed that the LSEs will be consistent in this case. However, in 1997, Kundu [66] provided rigorous theoretical proofs of the strong consistency and asymptotic normality of these estimators. The obtained asymptotic results were corroborated through extensive simulation studies. The following theorem reflects these results.

Theorem 1. *If the error $e(t)$ of model (4.42) satisfies Assumption 3, and for $k=1,\ldots,p$, if we denote $(\widehat{A}_k, \widehat{B}_k, \widehat{\omega}_k)$, as the LSE of $(A_k^0, B_k^0, \omega_k^0)$, then*

$$(\widehat{A}_k, \widehat{B}_k, \widehat{\omega}_k)^{\mathrm{T}} \xrightarrow{a.s.} (A_k^0, B_k^0, \omega_k^0)^{\mathrm{T}},$$

and

$$(N^{1/2}(\widehat{A}_k - A_k^0), N^{1/2}(\widehat{B}_k - B_k^0), N^{3/2}(\widehat{\omega}_k - \omega_k^0)) \xrightarrow{d} N_3(0, 2\sigma^2 c(\omega_k^0)\Sigma_k).$$

Here $c(\omega_k^0) = \left| \sum_{j=0}^{\infty} a(j)e^{-ij\omega_k^0} \right|^2$ and

$$\Sigma_k^{-1} = \frac{1}{A_k^{0^2} + B_k^{0^2}} \begin{bmatrix} A_k^{0^2} + 4B_k^{0^2} & -3A_k^0 B_k^0 & -3B_k^0 \\ -3A_k^0 B_k^0 & 4A_k^{0^2} + B_k^{0^2} & 3A_k^0 \\ -3B_k^0 & 3A_k^0 & 6 \end{bmatrix}.$$

Moreover, $(\widehat{A}_k, \widehat{B}_k, \widehat{\omega}_k)$ for $k=1,\ldots,p$ are asymptotically independent.

From the above theorem, several interesting observations can be made. It can be seen that although the multiple sinusoidal model does not satisfy some of the existing sufficient conditions for the LSEs to be consistent and asymptotically normal, still in this case the LSEs are consistent and they are asymptotically normally distributed. Also, the asymptotic variances of the linear parameters are of the order $O(N^{-1})$, whereas the asymptotic variances of the frequencies are of the order $O(N^{-3})$. This implies that the LSEs of the frequencies converge to the true frequencies much faster than the LSEs of the linear parameters. The above theorem can also be used to construct asymptotic confidence intervals of the linear parameters and frequencies. Despite these desirable statistical properties, computing the LSEs is not a trivial issue. Finding them involves solving a p dimensional optimization problem and finding the roots of a p degree polynomial.

Evidently, the problem can be numerically challenging for large values of p. An example of such a problem is that of an ECG signal. It has been observed that the number of sinusoidal components for an ECG signal can be as high as 60 and therefore finding LSEs in this case entails high computational complexity. To address this issue, Prasad, Kundu and Mitra [102] introduced a sequential estimation procedure which can be used quite effectively for large p. Moreover, the proposed sequential estimators have the same asymptotic properties as the LSEs.

4.3.3. Efficient Estimators

Nandi and Kundu [93] proposed an algorithm which is computationally very efficient. The most important feature of this algorithm is that it is computationally efficient but it produces estimators which have the same asymptotic properties as the LSEs. It is an iterative algorithm, but is known to converge in three steps when it starts with the initial estimators as the periodogram estimators. Moreover, the algorithm works even when the errors are from a stationary distribution.

The efficient algorithm uses only a fraction of the sample at the first two stages. However at the last step, the entire data set is taken into account. We describe the algorithm here for $p = 1,$, but it can be easily extended for any p using the sequential method of Prasad, Kundu and Mitra [102].

If at the j-th stage the estimator of ω is denoted by $\omega^{(j)}$, then $\omega^{(j+1)}$ is calculated as

$$\omega^{(j+1)} = \omega^{(j)} + \frac{12}{N_j} Im\left[\frac{P(j)}{Q(j)}\right], \tag{4.51}$$

where $P(j) = \sum_{t=1}^{N_j} y(t)\left(t - \frac{N_j}{2}\right)e^{-i\omega^{(j)}t}$, $Q(j) = \sum_{t=1}^{N_j} y(t)e^{-i\omega^{(j)}t}$, and N_j denotes the sample

size used at the j-th iteration. If $\omega^{(0)}$ denotes the periodogram estimator of ω over the Fourier frequencies as defined in Section 4.3.1, then the algorithm takes the following form:

Algorithm 1.

Step 1: Compute $\omega^{(1)}$ from $\omega^{(0)}$ using (4.51) with $N_1 = N^{0.8}$.

Step 2: Compute $\omega^{(2)}$ from $\omega^{(1)}$ using (4.51) with $N_2 = N^{0.9}$.

Step 3: Compute $\omega^{(3)}$ from $\omega^{(2)}$ using (4.51) with $N_3 = N$.

It has been indicated by Nandi and Kundu [93] that the choices of N_1 and N_2 are not unique and that there are several other choices for which the same asymptotic results hold. They also show that the estimators obtained by the above iterative procedure are consistent and have the same asymptotic distribution as the LSEs.

4.3.4. Super Efficient Estimators

Kundu et al. [68] proposed a modified Newton-Raphson method to obtain super efficient estimators of the frequencies of the sum of sinusoidal model. The remarkable property of this method is that it can produce estimators of the frequencies which have lower asymptotic variances than those of the LSEs. It is well known that the usual Newton-Raphson method often does not converge for this particular problem. But the authors have used a proper step factor modification which not only guarantees the convergence, but it also produces estimators which have smaller asymptotic variances than the LSEs. The method also works when the errors are from a stationary distribution.

We describe the method here for $p=1$, but one can easily use the sequential procedure to further extend the method for a general p. Let us denote

$$S(\omega) = Y^{T} Z(\omega)(Z^{T}(\omega)Z(\omega))^{-1} Z^{T}(\omega)Y, \qquad (4.52)$$

where Y and $Z(\omega)$ are same as defined before. Therefore, the LSE of ω can be obtained by maximizing $S(\omega)$ with respect to ω. The maximization of $S(\omega)$ using Newton-Raphson method can be performed as follows:

$$\omega^{(j+1)} = \omega^{(j)} - \frac{S'(\omega^{(j)})}{S''(\omega^{(j)})}, \qquad (4.53)$$

here $\omega^{(j)}$ is the estimate of ω at the j-th stage, $S'(\omega^{(j)})$ and $S''(\omega^{(j)})$ denote the first derivative and second derivative, respectively, of $S(\omega)$ evaluated at $\omega^{(j)}$. The standard Newton-Raphson algorithm is modified with a smaller correction factor as follows:

$$\omega^{(j+1)} = \omega^{(j)} - \frac{1}{4} \times \frac{S'(\omega^{(j)})}{S''(\omega^{(j)})} .$$

Suppose $\omega^{(0)}$ denotes the periodogram estimator of ω, then the algorithm can be described as follows:

Algorithm 2.

Step 1: Take $N_1 = N^{6/7}$, and calculate

$$\omega^{(1)} = \omega^{(0)} - \frac{1}{4} \times \frac{S'_{N_1}(\omega^{(0)})}{S''_{N_1}(\omega^{(0)})},$$

here $S'_{N_1}(\omega^{(0)})$ and $S''_{N_1}(\omega^{(0)})$ are same as $S'(\omega^{(0)})$ and $S''(\omega^{(0)})$, respectively, computed using a subsample of size N_1.

Step 2: With $N_{j+1} = N$, repeat

$$\omega^{(j+1)} = \omega^{(j)} - \frac{1}{4} \times \frac{S'_{N_{j+1}}(\omega^{(j)})}{S''_{N_{j+1}}(\omega^{(j)})}, \quad j = 1, 2, \ldots,$$

until a suitable stopping criterion is satisfied.

It is observed by the authors based on extensive simulations that any consecutive N_1 data points can be used at Step 1 to start the algorithm, so that the dependence structure is not destroyed, if any, and the choice of the initial estimators does not have any visible effect on the final estimators. Moreover, as it has been seen before in some of the other estimators also that the factor $6/7$ in the exponent of Step 1 is not unique and there are several other ways Step 1 can be initiated. It is observed that the iteration converges very quickly and it produces frequency estimator which has lower variance than the LSE.

4.3.5. Weighted Least Squares Estimators

The weighted least squares estimators (WLSEs) are proposed by Irizarry [55, 56] to estimate the frequencies in a musical signal. The WLSEs method proposed by the author is an extension of the standard least squares method, and the LSEs can be obtained as a special case of the WLSEs. The method has been proposed for a class of weight functions. Under the assumption that the error process is stationary, and with certain other properties, the asymptotic results of the WLSEs have been developed. The WLSEs can be obtained by minimizing the following criterion function

$$S(\boldsymbol{\omega},\boldsymbol{\theta}) = \sum_{t=1}^{N} w\left(\frac{t}{N}\right)\left(y(t) - \sum_{k=1}^{p}(A_k \cos(\omega_k t) + B_k \sin(\omega_k t))\right)^2,$$

with respect to $\boldsymbol{\omega} = (\omega_1, \ldots, \omega_p)^{\mathrm{T}}$ and $\boldsymbol{\theta} = (A_1, B_1, \ldots, A_p, B_p)^{\mathrm{T}}$. It is assumed that the weight function $w(\cdot)$ is non-negative, bounded, of bounded variation, has support on $[0,1]$, and it is such that $W_0 > 0$ and $W_1^2 - W_0 W_2 \neq 0$, where for $k = 1, 2, \ldots,$

$$W_k = \int_0^1 s^k w(s)\,ds.$$

It may be seen that $w(s) = 1$, satisfies all these conditions. Hence, the LSEs can be obtained as a special case of the WLSEs. Based on stationarity assumptions on the error components, Irizzary [55, 56] derived the consistency and asymptotic normality results of the WLSEs. The details are avoided. It may be mentioned although the theoretical properties of the WLSEs have been established, no proper efficient estimation procedure has been developed, more work is needed in that direction.

4.3.6. Estimation of the Number of Components

So far we have discussed about different methods of estimation for the frequencies and amplitudes, assuming the number of components 'p' to be known. In this section we briefly discuss some of the well known methods of estimation of p. The estimation of 'p' can be considered as a model selection problem. We consider the class of models

$$M_k = \left\{ \mu_k : \mu_k(t) = \sum_{j=1}^{k} \left[A_{j,k}^0 \cos(\omega_{j,k}^0 t) + B_{j,k}^0 \sin(\omega_{j,k}^0 t) \right] \right\}; \quad k = 1, 2, \dots \quad (4.54)$$

Based on the data $\{y(t); t = 1, \dots, N\}$ one needs to choose $\hat{p}$, an estimate of 'p', such that the model $M_{\hat{p}}$ fits the data 'best'. Therefore, any model selection procedure can be used to estimate 'p'.

One important point to observe here that the models are nested, i.e.

$$M_1 \subset M_2 \subset M_3 \subset \dots.$$

Hence, one of the natural procedures is to use a test of significance for each additional term as it is introduced in the model. Fisher [33] first considered this problem as a simple testing of hypothesis problem and it is based on the well known *likelihood ratio test*. This test uses the ratio of the maximized likelihood for the k-th term to the maximized likelihood for $(k-1)$-th term of model (4.54). If this quantity is large, it indicates that the k-th term is needed in the model, otherwise not.

The problem can be formulated as follows:

$$H_0 : p = p_0 \quad vs. \quad H_1 : p = p_1, \tag{4.55}$$

where $p_1 > p_0$. Based on the assumptions that the errors are i.i.d. normal random variables with mean zero and variance σ^2, it can be shown after some calculations, see Kundu and Nandi [73], that the likelihood ratio test takes the following form: rejects H_0, if L is large, where

$$L = \frac{\sum_{t=1}^{N}\left(y(t) - \sum_{k=1}^{P_0}\left\{ \widehat{A}_{k,P_0}\cos(\widehat{\omega}_{k,P_0}t) + \widehat{B}_{k,P_0}\sin(\widehat{\omega}_{k,P_0}t)\right\}\right)^2}{\sum_{t=1}^{N}\left(y(t) - \sum_{k=1}^{P_1}\left\{ \widehat{A}_{k,P_1}\cos(\widehat{\omega}_{k,P_1}t) + \widehat{B}_{k,P_1}\sin(\widehat{\omega}_{k,P_1}t)\right\}\right)^2}, \qquad (4.56)$$

here $\widehat{A}_{k,P_0}$, $\widehat{B}_{k,P_0}$, $\widehat{\omega}_{k,P_0}$ are MLEs of A_{k,P_0}, B_{k,P_0} and ω_{k,P_0}, respectively. Similarly, $\widehat{A}_{k,P_1}$, $\widehat{B}_{k,P_1}$, $\widehat{\omega}_{k,P_1}$ are also defined. Now to apply this likelihood ratio test in practice, one needs to obtain the exact/ asymptotic distribution of L under the null hypothesis. It seems finding the exact distribution of L is not an easy problem. Quinn and Hannan [104] provided an approximate distribution of L under certain restrictions on the form of the frequencies, which is quite complicated and may not have much practical importance.

Rao [106] proposed to use the cross validation technique to estimate the number of components in a sinusoidal model. It may be mentioned that the cross validation is a model selection technique and it can be used in a fairly general set up. The basic assumption of cross validation technique is that there exists an M, such that $1 \le k \le M$ for the models defined in (4.54). It can be described as follows. For a given k, such that $1 \le k \le M$, remove the j-th observation from $\{y(1),\ldots,y(N)\}$, and estimate '$y(j)$', say $\hat{y}_k(j)$ based on the model assumption $\boldsymbol{M}_k$ and the observation $\{y(1),\ldots,y(j-1),y(j+1),\ldots,y(N)\}$. Obtain the cross validatory error for the model $\boldsymbol{M}_k$ as

$$CV(k) = \sum_{t=1}^{N}\left(y(t) - \hat{y}_k(t)\right)^2,$$

for $k=1,\ldots,M$. Choose $\hat{p}$ as the estimate of p, if

$$\hat{p} = arg\ min\{CV(1),\ldots,CV(M)\}.$$

One major issue in implementing the cross validation method is to estimate $y(j)$ based on $\{y(1),\ldots,y(j-1),y(j+1),\ldots,y(N)\}$ and on the underlying model $\boldsymbol{M}_k$. Rao [106] did not provide any explicit method to compute $\hat{y}_k(j)$. Later Kundu and Kundu [69] provided a non-iterative method to estimate $y(j)$ based on the missing observation. The method has been further modified by Kundu and Mitra [70] and this can be used quite effectively to estimate 'p' based on cross validation method. Extensive simulations, performed by Kundu and Mitra [70], indicate that for small sample the method works quite well.

Rao [106] also proposed to use different information theoretic criteria to detect the number of components based on the assumptions that the errors are i.i.d. normal random variables with mean zero and finite variance. Based on these error assumptions, Akaike Information Criterion (AIC) takes the following form:

$$AIC(k) = N \ln R_k + 2(3k+1),$$

here R_k denotes the minimum value of

$$Q(\boldsymbol{\theta}_k) = \sum_{t=1}^{N} \left(y(t) - \sum_{j=1}^{k} \left(A_j \cos(\omega_j t) + B_j \sin(\omega_j t) \right) \right)^2, \qquad (4.57)$$

where $\boldsymbol{\theta}_k = (A_1,\ldots,A_k,B_1,\ldots,B_k,\omega_1,\ldots,\omega_k)^{\mathrm{T}}$ and the minimization of $Q(\boldsymbol{\theta}_k)$ as defined in (4.57) is performed with respect to $\boldsymbol{\theta}_k$. Here '$3k+1$' denotes the number of unknown parameters when the number of sinusoids is k. Choose $\hat{p}$ an estimate of p, where

$$\hat{p} = arg\ min\{AIC(1),\ldots,AIC(M)\}.$$

Under the same assumption, Bayesian Information Criterion (BIC) takes the following form

$$BIC(k) = N \ln R_k + \frac{1}{2}(3k+1)\ln N,$$

here R_k is same as defined above. In this case also, choose $\hat{p}$ an estimate of p, where

$$\hat{p} = arg\ min\{BIC(1),\ldots,BIC(M)\}.$$

Although, Rao [106] suggested to use different Information Theoretic Criteria to estimate p, he did not provide any practical implementation procedure particularly, for the computation of R_k. Later, Kundu [64] suggested a practically feasible method to implement for different information criterion and performed extensive simulation experiments for different models, for different error variances, and for different sample sizes. He observed that AIC does not provide consistent estimate of the model order, although for small sample sizes, performance of AIC is quite good. In general it is observed that BIC performs quite well.

In 1997, Kundu [67] suggested another estimation procedure for finding the number of components p. We explain it here in brief. Let us suppose that M denotes the maximum possible order, then for some fixed $L > 2M$, we write the data matrix A_L as follows:

$$A_L = \begin{bmatrix} y(1) & \cdots & y(L) \\ \vdots & \ddots & \vdots \\ y(N-L+1) & \cdots & y(N) \end{bmatrix}.$$

Let $\hat{\sigma}_1^2 > \ldots > \hat{\sigma}_L^2$ be the L eigenvalues of the $L \times L$ matrix $\frac{1}{N} A_L^{\mathrm{T}} A_L$. Consider,

$$KIC(k) = \hat{\sigma}^2_{2k+1} + kc(N),$$

here, $c(N) > 0,$ and satisfies the following two conditions

$$\lim_{N \to \infty} c(N) = 0 \quad and \quad \lim_{N \to \infty} \frac{c(N)\sqrt{N}}{(\ln \ln N)^{1/2}} = \infty.$$

Choose $\hat{p}$ an estimate of $p,$ where

$$\hat{p} = arg\ min\{KIC(1),\ldots,KIC(M)\}.$$

Under the assumption of i.i.d. errors, Kundu [67] proved the strong consistency of the above procedure. The probability of wrong detection has also been obtained in terms of linear combination of chi-square variables. Extensive simulations have been performed to check the effectiveness of the proposed method and to find proper $c(N)$. It is observed that $c(N) = (\ln N)^{-1/2}$ performs quite well, although no theoretical justification has been provided.

There are several other methods which are available in the literature and the interested reader is referred to the works of Sakai [110], Wang [118], Quinn [103] and the references cited therein.

4.3.7. Bayes Estimation of the Parameters and the Number of Components

Nielsen [96] proposed Bayes estimation method for the number of components and also the unknown parameters of the sinusoidal model. In this case the author considered model (4.42) where p is also unknown. It is assumed that the error random variables $e(t)$s are i.i.d. normal random variables with mean zero and variance σ^2. It is also assumed that there exists a $P,$ which is known such that the number of component $p \le P$. The following prior assumptions are made on the different parameters. The Jeffrey's prior has been assumed on $\sigma^2,$ i.e.

$$\pi(\sigma^2) \sim \left(\sigma^2\right)^{-1}.$$

For the model $M_p,$ let us use the following notations:

$$Z_p(\omega_p) = \begin{bmatrix} \cos(\omega_1) & \sin(\omega_1) & \ldots & \cos(\omega_p) & \sin(\omega_p) \\ \vdots & \vdots & \ddots & \vdots & \vdots \\ \cos(N\omega_1) & \sin(N\omega_1) & \ldots & \cos(N\omega_p) & \sin(N\omega_p) \end{bmatrix},$$

$\theta_p = (A_1, B_1, \ldots, A_p, B_p)^{\mathrm{T}},$ $\omega_p = (\omega_1, \ldots, \omega_p)^{\mathrm{T}}.$ Now the following prior assumptions are being made on $\theta_p,$ ω_p and $M_p.$

$$\pi(\boldsymbol{\theta}_p | g, \sigma^2, \omega_p, M_p) \sim N_{2p}\left(\boldsymbol{\theta}_p^0, g\sigma^2 [Z_p(\omega_p)^{\mathrm{T}} Z_p(\omega_p)]^{-1}\right).$$

Here, g and $\boldsymbol{\theta}_p^0$ are known quantities. Further, ω_p is assumed to be uniform in $[0, 2\pi]^p$, the model order p is also assumed to be uniform over $\{1, \ldots, P\}$ and they are independently distributed. The Bayes estimates of the unknown parameters cannot be obtained in closed forms. Laplace approximation may be used to obtain the approximate Bayes estimates of the unknown parameters. But in that case the credible intervals cannot be obtained. Hence, Markov Chain Monte Carlo (MCMC) method may be used to generate samples from the full conditional posterior density functions, and they can be used to compute the Bayes estimates of the unknown parameters, and also the associated credible intervals. Interested readers are referred to Nielsen [96] for more details.

So far, we have focussed on a brief review of the several sinusoidal parameter estimation methods. This was mainly a motivation for the core of this chapter which is the chirp signal model. Chirp signals are encountered in many different areas of engineering applications, particularly in radar, active sonar and also in passive sonar systems. Due to its applicability, the problem of parameter estimation of the chirp model parameters has received considerable amount of attention in the Signal Processing literature. The next two sections are devoted to one dimensional chirp model with one component and multiple components. We discuss several existing parameter estimation methods for both of them.

4.4. One Dimensional One Component Chirp Model

Recall that a one component chirp model can be written as follows:

$$y(t) = A^0 \cos(\alpha^0 t + \beta^0 t^2) + B^0 \sin(\alpha^0 t + \beta^0 t^2) + e(t); \quad t = 1, \ldots, N. \tag{4.58}$$

Here A^0, B^0 are unknown real numbers, $A^{0^2} + B^{0^2}$ is known as amplitude, α^0 is the frequency and β^0 is the frequency rate. Also, $0 < |A^0|, |B^0| < \infty$, and $0 < \alpha^0, \beta^0 < \pi$. The error $e(t)$ is an additive error with mean zero and finite variance. At this moment we assume that $e(t)$ s are i.i.d. random variables. However, a more general structure will be considered later. The problem is to estimate the unknown parameters $A^0, B^0, \alpha^0, \beta^0$ based on a sample of size N, namely $\{y(1), \ldots, y(N)\}$.

4.4.1. Least Squares Estimators

Like the sinusoidal model, a one component chirp model (4.58) is a non-linear regression model as well. Therefore, the LSEs or equivalently the MLEs when the errors are i.i.d. Gaussian random variables are the natural choice. The LSEs of the unknown parameters of model (4.58) can be obtained by minimizing the residual sums of squares, as given below:

$$Q(\boldsymbol{\theta}) = \sum_{t=1}^{N}(y(t) - A\cos(\alpha t + \beta t^2) - B\sin(\alpha t + \beta t^2))^2. \tag{4.59}$$

Here $\boldsymbol{\theta} = (A, B, \alpha, \beta)^{\mathrm{T}}$, and $\widehat{\boldsymbol{\theta}} = (\widehat{A}, \widehat{B}, \widehat{\alpha}, \widehat{\beta})^{\mathrm{T}}$ minimizes $Q(\boldsymbol{\theta})$, and it is the LSE of $\boldsymbol{\theta}^0 = (A^0, B^0, \alpha^0, \beta^0)^{\mathrm{T}}$. Now to minimize $Q(\boldsymbol{\theta})$ let us rewrite (4.59) in matrix notation.

$$Q(\boldsymbol{\theta}) = (\boldsymbol{Y} - \boldsymbol{W}(\alpha, \beta)\boldsymbol{\delta})^{\mathrm{T}}(\boldsymbol{Y} - \boldsymbol{W}(\alpha, \beta)\boldsymbol{\delta}), \tag{4.60}$$

here $\boldsymbol{Y} = (y(1), \ldots, y(N))^{\mathrm{T}}$, $\boldsymbol{\delta} = (A, B)^{\mathrm{T}}$ and the matrix $\boldsymbol{W}(\alpha, \beta)$ is

$$\boldsymbol{W}(\alpha, \beta) = \begin{bmatrix} \cos(\alpha + \beta) & \sin(\alpha + \beta) \\ \cos(2\alpha + 4\beta^2) & \sin(2\alpha + 4\beta^2) \\ \vdots & \vdots \\ \cos(N\alpha + N^2\beta) & \sin(N\alpha + N^2\beta) \end{bmatrix}. \tag{4.61}$$

Therefore, for fixed α and β,

$$\begin{bmatrix} \widehat{A}(\alpha, \beta) \\ \widehat{B}(\alpha, \beta) \end{bmatrix} = (\boldsymbol{W}^{\mathrm{T}}(\alpha, \beta)\boldsymbol{W}(\alpha, \beta))^{-1}\boldsymbol{W}^{\mathrm{T}}(\alpha, \beta)\boldsymbol{Y},$$

minimizes $Q(\boldsymbol{\theta})$. Hence, the LSEs of α and β, say $\widehat{\alpha}$ and $\widehat{\beta}$, respectively, can be obtained as

$$(\widehat{\alpha}, \widehat{\beta})^{\mathrm{T}} = arg\ min\ Q(\widehat{A}(\alpha, \beta), \widehat{B}(\alpha, \beta), \alpha, \beta). \tag{4.62}$$

Note that

$$Q(\widehat{A}(\alpha, \beta), \widehat{B}(\alpha, \beta), \alpha, \beta) = \boldsymbol{Y}^{\mathrm{T}}\left(\boldsymbol{I} - \boldsymbol{W}(\alpha, \beta)(\boldsymbol{W}^{\mathrm{T}}(\alpha, \beta)\boldsymbol{W}(\alpha, \beta))^{-1}\boldsymbol{W}^{\mathrm{T}}(\alpha, \beta)\right)\boldsymbol{Y} =$$
$$= \boldsymbol{Y}^{\mathrm{T}}(\boldsymbol{I} - \boldsymbol{P}_{\boldsymbol{W}(\alpha, \beta)})\boldsymbol{Y},$$

here $\boldsymbol{I}$ is the identity matrix of order N and

$$\boldsymbol{P}_{\boldsymbol{W}(\alpha, \beta)} = \boldsymbol{W}(\alpha, \beta)(\boldsymbol{W}^{\mathrm{T}}(\alpha, \beta)\boldsymbol{W}(\alpha, \beta))^{-1}\boldsymbol{W}^{\mathrm{T}}(\alpha, \beta)$$

is the projection matrix on the column space of $\boldsymbol{W}(\alpha, \beta)$. Therefore,

$$(\widehat{\alpha}, \widehat{\beta})^{\mathrm{T}} = arg\ max\, \boldsymbol{Y}^{\mathrm{T}}\boldsymbol{P}_{\boldsymbol{W}(\alpha, \beta)}\boldsymbol{Y}.$$

Once $\widehat{\alpha}$ and $\widehat{\beta}$ are obtained, the LSEs of A and B can be obtained as

$$\widehat{A} = \widehat{A}(\widehat{\alpha}, \widehat{\beta}) \quad and \quad \widehat{B} = \widehat{B}(\widehat{\alpha}, \widehat{\beta}),$$

respectively. It can be seen that for one component chirp model the LSEs of the unknown parameters can be obtained by solving a two dimensional optimization problem. Some of the standard numerical techniques like Newton-Raphson method or Gauss-Newton method may be used to compute them. Unfortunately, the least squares surface of α and β, namely $Y^{\mathrm{T}}(I - P_{W(\alpha,\beta)})Y$ is a highly non-linear surface. Fig. 4.9 illustrates this issue and evidently it has several local minima present on it.

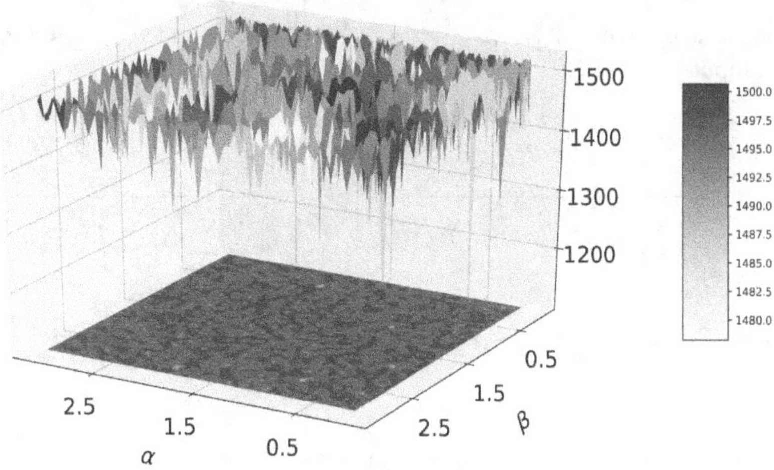

Fig. 4.9. Least Squares Surface.

Therefore, a very good set of initial values of α and β is needed for any iterative process to converge. Otherwise, the iterative process may not converge, or some times even if it converges it might converge to a local minimum rather than the global minimum.

Saha and Kay [109] proposed a Monte Carlo importance sampling procedure which does not need any initial guesses. It is a purely Monte Carlo based method, and uses the basic result of Pincus [100]. The implementation of the method is very simple and it can be described as follows. Based on the result of Pincus [100] it follows that

$$\widehat{\alpha} = \lim_{c \to \infty} \frac{\int_0^\pi \int_0^\pi \alpha \exp(cR(\alpha, \beta)) d\beta d\alpha}{\int_0^\pi \int_0^\pi \exp(cR(\alpha, \beta)) d\beta d\alpha} \quad and$$

$$\widehat{\beta} = \lim_{c \to \infty} \frac{\int_0^\pi \int_0^\pi \beta \exp(cR(\alpha, \beta)) d\beta d\alpha}{\int_0^\pi \int_0^\pi \exp(cR(\alpha, \beta)) d\beta d\alpha}, \tag{4.63}$$

where

$$R(\alpha,\beta) = Y^{\mathrm{T}} P_{W_{(\alpha,\beta)}} Y.$$

Now to compute $\hat{\alpha}$ and $\hat{\beta}$ based on (4.63) Monte Carlo importance procedure can be used. We describe the algorithm below.

Algorithm 3.

Step 1: Generate a random sample of size M, say $\{\alpha_1,\ldots,\alpha_M\}$ from uniform $(0,\pi)$. Similarly, generate another random sample of size M, say $\{\beta_1,\ldots,\beta_M\}$ also from uniform $(0,\pi)$.

Step 2: Choose a sequence $\{c_1,c_2,\ldots,\}$, such that $c_n \to \infty$, and $c_1 < c_2 < c_3 < \ldots$. For a fixed c_m, compute

$$\hat{\alpha}_{c_m} = \frac{\dfrac{1}{M}\sum_{k=1}^{M}\alpha_k \exp(c_m R(\alpha_k,\beta_k))}{\dfrac{1}{M}\sum_{k=1}^{M}\exp(c_m R(\alpha_k,\beta_k))} \quad and \quad \hat{\beta}_{c_m} = \frac{\dfrac{1}{M}\sum_{k=1}^{M}\beta_k \exp(c_m R(\alpha_k,\beta_k))}{\dfrac{1}{M}\sum_{k=1}^{M}\exp(c_m R(\alpha_k,\beta_k))}.$$

Step 3: Stop the iteration if $\left|\hat{\alpha}_{c_{m-1}} - \hat{\alpha}_{c_m}\right| < \varepsilon$ and $\left|\hat{\beta}_{c_{m-1}} - \hat{\beta}_{c_m}\right| < \varepsilon$. Otherwise change c_m to c_{m+1} and go back to Step 2.

It is important to note that the same generated sample namely $\{\alpha_1,\ldots,\alpha_M\}$ and $\{\beta_1,\ldots,\beta_M\}$ can be used in each iteration. Therefore, the proposed method is computationally very efficient. We will see later that the method can be easily generalized for multiple chirp model also.

Although, Saha and Kay [109] proposed a very efficient estimation procedure of the unknown parameters, they did not discuss any theoretical properties of the LSEs or the MLEs based on the assumption that the error are i.i.d. Gaussian random variables. Like the sinusoidal model, the one component chirp model as described in (4.58) is a non-linear regression model that does not satisfy the sufficient conditions of Jennrich [58] or Wu [122]. Therefore, it is not immediate that the LSEs will be consistent. Nandi and Kundu [91] established the consistency and asymptotic normality properties of the LSEs of the unknown parameters of the complex counterpart of this model. The results for the underlying real valued model are presented in the following theorems.

Theorem 2. *If there exists a* K, *such that* $0 < |A^0| + |B^0| < K$, $0 < \alpha^0, \beta^0 < \pi$, *and* $\sigma^2 > 0$, *then* $\hat{\theta} = (\hat{A},\hat{B},\hat{\alpha},\hat{\beta})^{\mathrm{T}}$ *is a strongly consistent estimate of* $\theta^0 = (A^0,B^0,\alpha^0,\beta^0)^{\mathrm{T}}$.

Theorem 3. *Under the same assumptions as in Theorem 2,*

$$\left(N^{1/2}(\widehat{A}-A^0), N^{1/2}(\widehat{B}-B^0), N^{3/2}(\widehat{\alpha}-\alpha^0), N^{5/2}(\widehat{\beta}-\beta^0)\right)^{\mathrm{T}} \xrightarrow{d} N_4(0, 2\sigma^2 \Sigma). \quad (4.64)$$

Here

$$\Sigma = \frac{2}{A^{0^2}+B^{0^2}} \begin{bmatrix} \frac{1}{2}(A^{0^2}+9B^{0^2}) & -4A^0B^0 & -18B^0 & 15B^0 \\ -4A^0B^0 & \frac{1}{2}(9A^{0^2}+B^{0^2}) & 18A^0 & -15A^0 \\ -18B^0 & 18A^0 & 96 & -90 \\ 15B^0 & -15A^0 & -90 & 90 \end{bmatrix}. \quad (4.65)$$

Theorem 2 establishes the consistency of the LSEs, whereas Theorem 3 states the asymptotic distribution along with the rates of convergence of the LSEs. It is important to note that when the errors are i.i.d. Gaussian random variables then the LSEs are same as the MLEs and the asymptotic variance-covariance matrix coincide with the corresponding Cramer-Rao lower bounds.

It is interesting to observe that the rates of the convergence of the LSEs of the linear parameters are significantly different than the corresponding non-linear parameters as we have observed in case of sinusoidal model parameters also in Section 4.3. The non-linear parameters are estimated more efficiently than the linear parameters for a given sample size. The variance of the frequency estimator goes to zero with the rate $O(N^{-3})$, whereas the variance of the frequency rate estimator goes to zero with the rate $O(N^{-5})$. Moreover, as $A^{0^2}+B^{0^2}$ decreases, the asymptotic variances of the LSEs of the frequency and frequency rate increase. It may be noted that the asymptotic distribution of the LSEs can be easily used to construct asymptotic confidence intervals of the unknown parameters as well as to develop testing of hypotheses.

Nandi and Kundu [91] were the first to formally establish the asymptotic properties of the LSEs of the parameters of the one component complex chirp model based on the assumptions that the additive errors are i.i.d. with mean zero and finite variance. Later, the same authors (see Kundu and Nandi [72]) obtained the theoretical results for model (4.58) for a more general class of error distributions in which the errors are assumed to be stationary. The explicit error assumption is stated below.

Assumption 4.

$$e(t) = \sum_{j=-\infty}^{\infty} a(j)X(t-j), \quad (4.66)$$

where $\{X(t)\}$ is a sequence of i.i.d. random variables with mean zero, variance σ^2 and finite fourth moment, and

$$\sum_{j=-\infty}^{\infty} |a(j)| < \infty. \tag{4.67}$$

This assumption is a standard assumption of a linear process. The stationary autoregressive process or the moving average process is a special case of the above linear process. Kundu and Nandi [72] established the following two results under this assumption.

Theorem 4. *If there exists a K, such that $0 < |A^0| + |B^0| < K$, $0 < \alpha^0, \beta^0 < \pi$, $\sigma^2 > 0$ and $e(t)$ satisfies Assumption 4, then $\widehat{\boldsymbol{\theta}} = (\widehat{A}, \widehat{B}, \widehat{\alpha}, \widehat{\beta})^{\mathrm{T}}$ is a strongly consistent estimate of $\boldsymbol{\theta}^0 = (A^0, B^0, \alpha^0, \beta^0)^{\mathrm{T}}$.*

Theorem 5. *Under the same assumptions as in Theorem 4,*

$$\left(N^{1/2}(\widehat{A} - A^0), N^{1/2}(\widehat{B} - B^0), N^{3/2}(\widehat{\alpha} - \alpha^0), N^{5/2}(\widehat{\beta} - \beta^0)\right)^{\mathrm{T}} \overset{d}{\to} N_4(0, 2c\sigma^2 \boldsymbol{\Sigma}), \tag{4.68}$$

here $\boldsymbol{\Sigma}$ is same as defined in Theorem 3, and $c = \sum_{j=-\infty}^{\infty} a^2(j)$.

Note that the original asymptotic variance-covariance matrix given by Kundu and Nandi [72] was much complicated, however the simplified form provided here was derived later by Lahiri, Kundu and Mitra [79]. Lahiri, Kundu and Mitra [79] performed extensive simulations to study the behavior of the LSEs under different error assumptions, and it is observed that the performance of the LSEs are quite satisfactory. The sample variances of the different estimators match quite well with the asymptotic variances even for moderate sample sizes.

4.4.2. Periodogram Estimator

Periodogram estimator as described in Section 4.3.1 is one of the most popular estimators for the sinusoidal model. Along the same line the periodogram estimators of the frequency and frequency rate of the one component chirp model (4.58) can be obtained. First let us define the periodogram function associated with the chirp model as follows:

$$I(\alpha, \beta) = \frac{1}{N}\left|\sum_{t=1}^{N} y(t)e^{i(\alpha t + \beta t^2)}\right|^2 =$$

$$= \frac{1}{N}\left\{\left(\sum_{t=1}^{N} y(t)\cos(\alpha t + \beta t^2)\right)^2 + \left(\sum_{t=1}^{N} y(t)\sin(\alpha t + \beta t^2)\right)^2\right\}. \tag{4.69}$$

Grover, Kundu and Mitra [43] proposed periodogram estimators (PEs) or approximate least squares estimators (ALSEs) of α and β as

$$(\widehat{\alpha}, \widehat{\beta}) = arg \ max \ I(\alpha, \beta). \tag{4.70}$$

The basic idea is same as that for the periodogram estimator of the frequency of a sinusoidal model. That is, if the error variance is small then the maximum of the periodogram function $I(\alpha, \beta)$ occurs near the true values of the frequency and frequency rate, i.e. at (α^0, β^0).

Clearly, the PEs of α and β are obtained by solving a two dimensional optimization problem. The Newton-Raphson or the Gauss-Newton method may be used to compute them. Like the least squares surface, the periodogram surface is also highly non-linear in nature as it has several local maxima. Hence, precise initial estimators are needed for any algorithm to converge to the global maximum rather than a local maximum. In concurrence with the Fourier frequencies for the sinusoidal model, to find the initial values of (α^0, β^0) one can search among the points

$$\left\{ \left(\frac{\pi k}{N}, \frac{\pi j}{N^2} \right); k = 1, \ldots, N-1, j = 1, \ldots, (N-1)^2 \right\} \text{ and select } (k_0, j_0), \text{ such that}$$

$$I\left(\frac{\pi k_0}{N}, \frac{\pi j_0}{N^2} \right) \geq I\left(\frac{\pi k}{N}, \frac{\pi j}{N^2} \right),$$

for all k, j. Alternatively, Algorithm 3 also can be used to maximize $I(\alpha, \beta)$ simply by replacing $R(\alpha, \beta)$ with $I(\alpha, \beta)$ at each step.

Once $\widehat{\alpha}$ and $\widehat{\beta}$ are obtained the estimators of A and B can be obtained as

$$\widehat{A} = \frac{1}{N} \sum_{t=1}^{N} y(t) \cos(\widehat{\alpha}t + \widehat{\beta}t^2) \quad and \quad \widehat{B} = \frac{1}{N} \sum_{t=1}^{N} y(t) \sin(\widehat{\alpha}t + \widehat{\beta}t^2). \tag{4.71}$$

Grover, Kundu and Mitra [43] established the asymptotic properties of the PEs of the unknown parameters of the one component chirp model. It has been shown that under the assumption of stationary errors as in Assumption 4, the PEs are strongly consistent and they have the same asymptotic distribution as the LSEs. Grover, Kundu and Mitra [43] performed extensive simulations to compare the performances of the ALSEs and LSEs. Based on simulation experiments it is observed that although LSEs and ALSEs have the same asymptotic variances, for small and moderate sample sizes the sample variances of the ALSEs are slightly higher than the corresponding variances of the LSEs.

4.4.3. Finite Step Algorithm

So far we have discussed about the LSEs and PEs and although both of them have some computational issues, they are the most efficient estimators in the sense that their asymptotic variances achieve the Cramer-Rao lower bound under the assumption of i.i.d. Gaussian noise. Therefore a trade-off has to be made between statistical properties and computational complexity. A numerically faster algorithm was proposed by Lahiri, Kundu and Mitra [76] that offers computational advantage and has the same convergence properties as the LSEs. They observed that if one starts with the initial guesses of α^0 and β^0 with convergence rates $O_p(N^{-1})$ and $O_p(N^{-2})$, respectively, then after four iterations, the algorithm produces an estimate of α^0 with convergence rate $O_p(N^{-3/2})$, and an estimate of β^0 with convergence rate $O_p(N^{-5/2})$. Before providing details of their algorithm, we first show how to improve the estimators of α^0 and β^0. If $\tilde{\alpha}$ is an estimator of α^0, such that for $\delta_1 > 0$, $\tilde{\alpha} - \alpha^0 = O_p(N^{-1-\delta_1})$, and $\tilde{\beta}$ is an estimator of β^0, such that for $\delta_2 > 0$, $\tilde{\beta} - \beta^0 = O_p(N^{-2-\delta_2})$, then the improved estimators of α^0 and β^0, can be obtained as

$$\tilde{\tilde{\alpha}} = \tilde{\alpha} + \frac{48}{N^2} Im\left(\frac{P_N^\alpha}{Q_N}\right), \tag{4.72}$$

$$\tilde{\tilde{\beta}} = \tilde{\beta} + \frac{45}{N^4} Im\left(\frac{P_N^\beta}{Q_N}\right), \tag{4.73}$$

respectively, where

$$P_N^\alpha = \sum_{t=1}^{N} y(t)\left(t - \frac{N}{2}\right)e^{-i(\tilde{\alpha}t + \tilde{\beta}t^2)}, \quad P_N^\beta = \sum_{t=1}^{N} y(t)\left(t^2 - \frac{N^2}{3}\right)e^{-i(\tilde{\alpha}t + \tilde{\beta}t^2)},$$

$$Q_N = \sum_{t=1}^{N} y(t)e^{-i(\tilde{\alpha}t + \tilde{\beta}t^2)}.$$

The theoretical justification of the algorithm comes from the following two theorems. One may refer to Lahiri, Kundu and Mitra [76] for the detailed proof.

Theorem 6. *If* $\tilde{\alpha} - \alpha^0 = O_p(N^{-1-\delta_1})$ *for* $\delta_1 > 0$, *then*

$$(a)\,(\tilde{\tilde{\alpha}} - \alpha^0) = O_p(N^{-1-2\delta_1}) \quad if \quad \delta_1 \le 1/4,$$
$$(b)\, N^{3/2}(\tilde{\tilde{\alpha}} - \alpha^0) \xrightarrow{d} \mathcal{N}(0, \sigma_1^2) \quad if \quad \delta_1 > 1/4,$$

where $\sigma_1^2 = \dfrac{384\sigma^2}{A^{0^2} + B^{0^2}}$.

Theorem 7. *If* $\tilde{\beta} - \beta^0 = O_p(N^{-2-\delta_2})$ *for* $\delta_2 > 0$, *then*

$$(a)\, (\tilde{\tilde{\beta}} - \beta^0) = O_p(N^{-2-2\delta_2}) \quad \text{if} \quad \delta_2 \leq 1/4,$$

$$(b)\, N^{5/2}(\tilde{\tilde{\beta}} - \beta^0) \xrightarrow{d} \mathcal{N}(0,\sigma_2^2) \quad \text{if} \quad \delta_2 > 1/4,$$

where $\sigma_2^2 = \dfrac{360\sigma^2}{A^{0^2} + B^{0^2}}.$

Now we show that starting from initial guesses $\tilde{\alpha}$, $\tilde{\beta}$ with convergence rates $\tilde{\alpha} - \alpha^0 = O_p(N^{-1})$ and $\tilde{\beta} - \beta^0 = O_p(N^{-2})$, respectively, how the above results/theorems can be used to obtain efficient estimators. It may be noted that finding initial guesses with the above convergence rates is not difficult. We can use the same initial guesses as it has been used in case of PEs in Section 4.4.2. The main idea is not to use the whole sample at the beginning, as it was originally suggested by Bai et al. [3]. A fraction of the sample is used at the beginning, and we gradually proceed towards the complete sample. With varying sample size, more and more data points are used with the increasing number of iterations. The algorithm is described below. Note that we denote the estimates of α^0 and β^0 obtained at the j-th iteration as $\tilde{\alpha}^{(j)}$ and $\tilde{\beta}^{(j)}$, respectively.

Algorithm 4.

Step 1: Choose $N_1 = N^{8/9}$. Therefore, $\tilde{\alpha}^{(0)} - \alpha^0 = O_p(N^{-1}) = O_p(N_1^{-1-1/8})$ and $\tilde{\beta}^{(0)} - \beta^0 = O_p(N^{-2}) = O_p(N_1^{-2-1/4})$. Perform steps (4.72) and (4.73). Therefore, after 1-st iteration, we have

$$\tilde{\alpha}^{(1)} - \alpha^0 = O_p(N_1^{-1-1/4}) = O_p(N^{-10/9}) \quad and \quad \tilde{\beta}^{(1)} - \beta^0 = O_p(N_1^{-2-1/2}) = O_p(N^{-20/9}).$$

Step 2: Choose $N_2 = N^{80/81}$. Therefore, $\tilde{\alpha}^{(1)} - \alpha^0 = O_p(N_2^{-1-1/8})$ and $\tilde{\beta}^{(1)} - \beta^0 = O_p(N_2^{-2-1/4})$. Perform steps (4.72) and (4.73). Therefore, after 2-nd iteration, we have

$$\tilde{\alpha}^{(2)} - \alpha^0 = O_p(N_2^{-1-1/4}) = O_p(N^{-100/81}) \quad and \quad \tilde{\beta}^{(2)} - \beta^0 = O_p(N_2^{-2-1/2}) = O_p(N^{-200/81}).$$

Step 3: Choose $N_3 = N$. Therefore, $\tilde{\alpha}^{(2)} - \alpha^0 = O_p(N_3^{-1-19/81})$ and $\tilde{\beta}^{(2)} - \beta^0 = O_p(N_3^{-2-38/81})$. Perform steps (4.72) and (4.73). Therefore, after 3-rd iteration, we have

$$\tilde{\alpha}^{(3)} - \alpha^0 = O_p(N^{-1-38/81}) \quad and \quad \tilde{\beta}^{(3)} - \beta^0 = O_p(N^{-2-76/81}).$$

Step 4: Choose $N_4 = N$ and perform steps (4.72) and (4.73). Now we obtain the required convergence rates, i.e.

$$\tilde{\alpha}^{(4)} - \alpha^0 = O_p(N^{-3/2}) \quad and \quad \tilde{\beta}^{(4)} - \beta^0 = O_p(N^{-5/2}).$$

The above algorithm can be used quite efficiently to compute the estimators in four steps which are equivalent to the LSEs. Lahiri, Kundu and Mitra [76] performed extensive simulations for different sample sizes and for different error structures, mainly to see the behavior of the proposed estimators and to compare their performances with the LSEs. It is observed that the performances of the proposed estimators are quite satisfactory even for moderate sample sizes, although the sample variances of the LSEs are slightly smaller than the corresponding variances of these estimators.

It may be mentioned that the fraction of the sample sizes which have been used in each step is not unique. It is possible to obtain equivalent estimators with different choices of N_i. Moreover, if observations are independent then the algorithm can be started with any sub-sample of size N_1. But if the data are stationary, then the sub-sample of size N_1 has to be chosen with consecutive data points. Although they are asymptotically equivalent, the finite sample performances might be different. It might be interesting to compare the finite sample performances of the different estimators by the use of Monte Carlo simulations.

4.4.4. Least Absolute Deviation Estimators

As discussed above, LSEs and PEs are the most natural choice of estimators in case of chirp model. However, despite their theoretical appeal, a widely-recognized practical difficulty with these estimators is that they are not robust. For a model to describe majority of the data and avert the influence of outliers on the results, we need robust methods of estimation. Among different robust estimators in the literature, the L_1 norm estimators or least absolute deviation (LAD) estimators are quite popular for general linear and non-linear regression problems. For reference, one may look into Dielman [23] or Dodge [28]. The two major issues associated with the LAD estimators are: (i) Computational complexity, i.e. how to compute the LAD estimators in an efficient manner, and (ii) Theoretical intractability, i.e. how to develop properties of the LAD estimators.

In the renowned paper by Charnes, Cooper and Ferguson [17], it has been shown that for a linear regression model the LAD estimators can be obtained very efficiently by solving a linear programming problem. Similar method can be extended for the non-linear regression model as well although this has not been explored yet in the literature. The LAD estimators of the unknown parameters of the one component chirp model (4.58) can be obtained by minimizing

$$R(\boldsymbol{\theta}) = \sum_{n=1}^{N} \left| y(n) - A\cos(\alpha n + \beta n^2) - B\sin(\alpha n + \beta n^2) \right|. \tag{4.74}$$

Here $\theta = (A, B, \alpha, \beta)^{\mathrm{T}}$ is same as defined before. Let us denote the LAD estimators of A^0, B^0, α^0 and β^0, by $\widehat{A}, \widehat{B}, \widehat{\alpha}$ and $\widehat{\beta}$, respectively. They can be obtained as

$$(\widehat{A}, \widehat{B}, \widehat{\alpha}, \widehat{\beta}) = \arg \min R(\theta) \, . \tag{4.75}$$

Clearly, the computation of the LAD estimators is a challenging issue. Any optimization algorithm which does not require derivative information may be used for this purpose. Alternatively, one can use the importance sampling technique as suggested by Saha and Kay [109] in Section 4.4.1 with the obvious modifications. Developing efficient LAD estimators is a challenging and still an open problem.

Another main difficulty is the derivation of the theoretical properties of the LAD estimators. Even for a linear regression model, this method requires stronger assumptions on the error distribution than those needed in case of the LSEs. Recently, Lahiri, Kundu and Mitra [78] developed the properties of LAD estimators under certain assumptions on the probability density function (PDF) of the error distribution (see Assumption 5). We present their results in the following theorems, however for the proofs of these theorems one may refer to their work.

Assumption 5. *The error random variable $\{e(t)\}$ is a sequence of i.i.d. random variables with mean zero, variance σ^2, and it has a PDF $f(x)$. The PDF $f(x)$ is symmetric and differentiable in $(0, \varepsilon)$ and $(-\varepsilon, 0)$, for some $\varepsilon > 0$, and $f(0) > 0$.*

Theorem 8. *If there exists a K, such that $0 < |A^0| + |B^0| < K$, $0 < \alpha^0, \beta^0 < \pi$, $\sigma^2 > 0$, and $\{e(t)\}$ satisfies Assumption 5, then $\widehat{\theta} = (\widehat{A}, \widehat{B}, \widehat{\alpha}, \widehat{\beta})^{\mathrm{T}}$ is a strongly consistent estimate of θ^0.*

Theorem 9. *Under the same assumptions as in Theorem 8,*

$$\left(N^{1/2}(\widehat{A} - A^0), N^{1/2}(\widehat{B} - B^0), N^{3/2}(\widehat{\alpha} - \alpha^0), N^{5/2}(\widehat{\beta} - \beta^0) \right)^{\mathrm{T}} \xrightarrow{d} N_4\left(0, \frac{1}{f^2(0)}\Sigma \right), \tag{4.76}$$

here Σ is same as defined in (4.65).

It is evident that the rates of convergence of the LAD estimators are same as that of the LSEs, however for the heavy tailed distributions, the LAD estimators are asymptotically more efficient that the LSEs. Lahiri, Kundu and Mitra [78] performed some simulations to observe the behavior of the LAD estimators mainly for Gaussian errors. They have used the Downhill Simplex Algorithm to minimize (4.74) and the asymptotic results of Theorem 9 have been used to construct confidence intervals of the unknown parameters assuming $f(0)$ is known. It is observed that the sample variances are quite close to the asymptotic variances even for moderate sample sizes. The performances of the confidence intervals based on the asymptotic distribution, are not very good for moderate sample sizes in the sense that the coverage percentages are significantly smaller than the nominal level. Although, for large sample sizes, they are quite close. More work is needed in developing efficient estimation procedure to compute LAD estimators, and also a proper methodology to construct confidence intervals mainly for small and moderate sample sizes.

4.4.5. Iterative Approach

So far all the methods we have discussed are computationally quite demanding, but all of them are shown to achieve the best convergence rates possible. In all these cases the main computational time involves finding the initial guess which requires a $O(N^3)$ search. Therefore, if N is large this can take a significant amount of time. To avoid that several heuristic methods are available in the literature. Out of many of these methods, the iterative method suggested by Ikarm, Abed-Meraim and Hua [53] (see also Ikarm, Abed-Meraim and Hua [52]) is an interesting one and we will briefly discuss this method here.

To implement the iterative method, Ikarm, Abed-Meraim and Hua [53] used the complex one parameter chirp model defined as follows:

$$y(t) = Ae^{i(\alpha t + \beta t^2)} + e(t); \quad t = 1, \ldots, N. \tag{4.77}$$

Here A is a complex number and $e(t)$ is a complex valued random variable with mean zero and finite variance. The frequency α and the frequency rate β are same as defined before. The basic idea of their proposed algorithm is the following. Suppose

$$\mu(t) = Ae^{i(\alpha t + \beta t^2)}; \quad t = 1, \ldots, N.$$

Then for any fixed integer τ, consider

$$m(t) = \mu^H(t)\mu(t + \tau) = Ce^{i(2\beta \tau t)}; \quad t = 1, \ldots, N - \tau. \tag{4.78}$$

Here $\mu^H(t)$ is the conjugate of $\mu(t)$ and $C = |A|^2 e^{i(\alpha \tau + \beta \tau^2)}$. The equation (4.78) indicates that the sequence $m(t)$ represents a sinusoidal signal with frequency $2\beta\tau$. Therefore, the frequency β can be estimated from $m(t)$ using one of the frequency estimation technique. Once β is estimated say by $\hat{\beta}$ then the signal $\mu(t)$ can be demodulated as

$$z(t) = \mu(t)e^{-i(\hat{\beta}t^2)} \approx Ae^{i(\alpha t)}.$$

Hence, α can be estimated from the demodulated signal $z(t)$ by using again the same frequency estimation technique. Note that in estimating α, it is assumed that the effect of β has been removed in the demodulated signal $z(t)$. Hence, Ikarm, Abed-Meraim and Hua [53] proposed the following iterative technique to estimate β. Once an accurate estimate of β is obtained in the first phase then in the second phase α is estimated by using the demodulation technique as described above. Let us use the subscript k as the iteration number and τ_k as the lag parameter at the k-th iteration.

The first step of the iterative algorithm depends on the following transformation

$$m_k(t) = z_{k-1}(t + \tau_{k-1})z_{k-1}^H(t), \ t = 1,\ldots,N - \tau_{k-1}, \tag{4.79}$$

where $z_{k-1}(t) = y(t)e^{-i\hat{\beta}_{k-1}t^2}$, $t = 1,\ldots,N-1$.

Here, $\hat{\beta}_{k-1}$ is the estimate of β at the $(k-1)$-th iterate. Let us start the iteration as follows. For the first iteration ($k=1$) the initial values of $\hat{\beta}_0 = 0$ and $\tau_0 = 1$ are chosen. Now from the transformed data $m_1(t) = y(t+1)y^H(t);\quad t = 1,\ldots,N-1$, obtain $\hat{\beta}_1$. At the kth ($k = 1,2,\ldots$) iteration, the demodulated signal $z_{k-1}(t)$ is considered to be a chirp signal with the same frequency as the original signal $y(t)$ but with different frequency rate $\Delta\beta_{k-1} = \beta - \hat{\beta}_{k-1}$. The parameter $\Delta\beta_{k-1}$ is estimated using the transform $m_k(t)$, say as $\hat{\Delta\beta}_{k-1}$. The estimate of β is then updated using

$$\hat{\beta}_k = \hat{\beta}_{k-1} + \hat{\Delta\beta}_{k-1}.$$

It is very clear that the selection of the lag parameter τ_k affects the accuracy in estimating β. Different τ_k s can be chosen at different iteration steps. Ikarm, Abed-Meraim and Hua [53] conducted extensive simulation experiments and they have taken fixed lag size $\tau_k = 9$. It is observed that the above method provides very good results in practice, although theoretically the properties of the estimators could not be established.

4.4.6. Bayes Estimates

Lin and Djurić [81] and recently, Mazumder [87] considered the Bayes estimators of the unknown parameters of model (4.58) when $e(t)$ s are i.i.d. normal random variables. Here we elaborate on the methodology proposed by Mazumder [87], as it is more general in nature. The following transformations and assumptions have been made for mathematical convenience.

$$A^0 = r^0 \cos(\theta^0), \ B^0 = r^0 \sin(\theta^0), \ r^0 \in (0,K], \ \theta^0 \in [0,2\pi], \ \alpha,\beta \in (0,\pi).$$

The following assumptions on prior distributions have been made on the above unknown parameters.

$$r \sim uniform(0,K), \ \theta \sim uniform(0,2\pi), \ \alpha \sim vonMises(a_0,a_1),$$

$$\beta \sim vonMises(b_0,b_1), \ \sigma^2 \sim inverse\ gamma(c_0,c_1)$$

It may be mentioned that r and θ have non-informative priors and σ^2 has a conjugate prior. In this case, 2α and 2β are circular random variables, that is why, von Misses distribution, the natural analog of the normal distribution in circular data has been considered. We denote the prior densities of r, θ, α, β, σ^2 as $[r]$, $[\theta]$, $[\alpha]$, $[\beta]$, $[\sigma^2]$, respectively, and Y as the data vector as defined in Section 4.4.1. Under the assumption that the priors are independently distributed, the joint posterior density function of r, θ, α, β, σ^2 can be obtained as follows:

$$[r,\theta,\alpha,\beta,\sigma^2 \,|\, Y] \propto [r][\theta][\alpha][\beta][\sigma^2][Y\,|\,r,\theta,\alpha,\beta,\sigma^2].$$

Now, computation of the Bayes estimates of the unknown parameters is based on the Gibbs sampling technique. For that one needs to compute conditional distribution of each parameter given all the parameters, known as the full conditional distribution and denoted by $[\cdot\,|\,...]$. These are given by

$$[r\,|\,...] \propto [r][Y\,|\,r,\theta,\alpha,\beta,\sigma^2], \quad [\theta\,|\,...] \propto [\theta][Y\,|\,r,\theta,\alpha,\beta,\sigma^2],$$

$$[\alpha\,|\,...] \propto [\alpha][Y\,|\,r,\theta,\alpha,\beta,\sigma^2], \quad [\beta\,|\,...] \propto [\beta][Y\,|\,r,\theta,\alpha,\beta,\sigma^2],$$

$$[\sigma^2\,|\,...] \propto [\sigma^2][Y\,|\,r,\theta,\alpha,\beta,\sigma^2].$$

The closed form expression of the full conditionals cannot be obtained. Mazumder [87] proposed to use the random walk MCMC technique to update these parameters, which can be easily implemented in practice. The author has used this method to predict future observation also.

4.4.7. Testing of Hypothesis

So far we have mainly talked about different estimation procedures of the unknown parameters for one component chirp model. But along with the estimation problem the associated testing of hypothesis problem is of significant practical importance. The testing of hypothesis can be very useful in identifying a source, mainly to detect whether a particular signal is coming from a known or an unknown object. Recently, Dhar, Kundu and Das [22] considered the following testing of hypothesis problem. The authors considered model (4.58) and assumed the additive errors $e(t)$ s to be i.i.d. random variables with mean zero and finite variance σ^2. In order to develop the testing procedure some more technical assumptions on the density functions of the error random variables will be specified later.

We denote the vector $\boldsymbol{\theta}^0 = (A^0, B^0, \alpha^0, \beta^0)^{\mathrm{T}}$. The testing of hypothesis can be formulated as follows. We want to test the following null hypothesis

$$H_0 : \boldsymbol{\theta} = \boldsymbol{\theta}^0 \quad vs. \quad H_1 : \boldsymbol{\theta} \neq \boldsymbol{\theta}^0. \tag{4.80}$$

We may interpret the problem as whether the observed signal is coming from a known source which emits chirp model with parameter $\boldsymbol{\theta}^0$.

Dhar, Kundu and Das [22] proposed four different tests to test the hypothesis (4.80) based on the following test statistics:

$$T_{N,1} = \| \boldsymbol{D}^{-1}(\widehat{\boldsymbol{\theta}}_{N,LSE} - \boldsymbol{\theta}^0) \|_2^2, \tag{4.81}$$

$$T_{N,2} = \| \boldsymbol{D}^{-1}(\widehat{\boldsymbol{\theta}}_{N,LAD} - \boldsymbol{\theta}^0) \|_2^2, \tag{4.82}$$

$$T_{N,3} = \| \boldsymbol{D}^{-1}(\widehat{\boldsymbol{\theta}}_{N,LSE} - \boldsymbol{\theta}^0) \|_1^2, \tag{4.83}$$

$$T_{N,4} = \| \boldsymbol{D}^{-1}(\widehat{\boldsymbol{\theta}}_{N,LAD} - \boldsymbol{\theta}^0) \|_1^2. \tag{4.84}$$

Here, $\widehat{\boldsymbol{\theta}}_{N,LSE}$ and $\widehat{\boldsymbol{\theta}}_{N,LAD}$ denote the LSE of $\boldsymbol{\theta}^0$ and LAD estimate of $\boldsymbol{\theta}^0$ as discussed in Sections 4.4.1 and 4.4.4, respectively. Further, $\|\cdot\|_2$ and $\|\cdot\|_1$ denote the Euclidean and L_1 distances, respectively, and the 4×4 diagonal matrix $\boldsymbol{D}$ is as follows:

$$\boldsymbol{D} = diag\{N^{-1/2}, N^{-1/2}, N^{-3/2}, N^{-5/2}\}.$$

All these test statistics are based on some normalized values of the distances between the estimates and the true parameter value under the null hypothesis. The authors have chosen two specific distances, but in principle, any other distance can be considered. In all these cases, it is expected that the null hypothesis should be rejected if the values of the test statistics are large.

At this point, a natural question to ask is how large is large? For this purpose we need to compute the critical value of a test for a given significance level. The hypothesis will be rejected for the given significance level, if the value of the test statistic exceeds the critical value. In practice, for a given error distribution and significance level, it is possible to compute the critical value of a test based on extensive simulations. To calculate it theoretically, the following error assumptions are made by the authors.

Assumption 6. *Let F_n be the distribution function of $y(n)$ with the probability density function $f_n(y,\boldsymbol{\theta})$, which is twice continuously differentiable with respect to $\boldsymbol{\theta}$. It is*

assumed that $\quad E\left[\dfrac{\partial}{\partial \theta_i} f_n(y,\boldsymbol{\theta})\right]_{\boldsymbol{\theta}=\boldsymbol{\theta}^0}^{2+\delta} < \infty, \quad$ *for* *some* $\delta > 0$ *and*

$E\left[\dfrac{\partial^2}{\partial \theta_i \partial \theta_j} f_n(y,\boldsymbol{\theta})\right]_{\boldsymbol{\theta}=\boldsymbol{\theta}^0}^{2} < \infty,$ *for all $n=1,2,\dots,N$. Here θ_i and θ_j, for $1 \le i,j \le 4$, are*

the i-th and j-th component of $\boldsymbol{\theta}$.

Note that the above assumption is not very unnatural. It holds for most of the well known probability density functions, e.g. normal, Laplace and Cauchy probability density functions. Based on the above assumption the critical values for all the four tests can be calculated. Since they are quite involved, they are not presented here. Interested readers are refereed to the original article of Dhar, Kundu and Das [22] for details. It can be shown that all the tests are consistent, i.e. as the sample size tends to infinity, the power of the tests goes to one. Dhar, Kundu and Das [22] performed extensive simulation experiments to study the power of these tests under different error assumptions, namely standard normal (light tailed), standard Laplace (heavy tailed) and t-distribution with 5 degrees of freedom (heavy tailed). It is observed that the tests based on least squares ($T_{N,1}$ and $T_{N,3}$) perform well when the data are obtained from a light tailed distribution like Gaussian distribution. Similarly, the tests based on least absolute deviations ($T_{N,2}$ and $T_{N,4}$) perform well when the data are obtained from standard Laplace or t distribution with 5 degrees of freedom. It is recommended that the least absolute based methodologies should be preferred when the data are likely to have influential observations/ outliers.

In this section we have mainly addressed different inferential issues associated with the one component chirp model. Now in the next section we will address the multicomponent chirp model.

4.5. One Dimensional Multicomponent Chirp Model

For real life applications, we need a more general model than a one component chirp model. Here in this section, we work with its natural generalization to a model with multiple components, defined as follows:

$$y(t) = \sum_{k=1}^{p} \{A_k^0 \cos(\alpha_k^0 t + \beta_k^0 t^2) + B_k^0 \sin(\alpha_k^0 t + \beta_k^0 t^2)\} + e(t); \quad t = 1, \ldots, N. \tag{4.85}$$

Similar to the one component model, A_k^0, B_k^0 are real numbers, $|A_k^0|^2 + |B_k^0|^2$ is the amplitude associated with the k-th component, α_k^0 and β_k^0 are the frequency and frequency rate, respectively. The additive error $e(t)$ has mean zero and the other conditions will be explicitly stated whenever required. At this moment, A_k^0, B_k^0, α_k^0 and β_k^0 are assumed to be unknown, and the number of components p is assumed to be known. Based on the sample $\{y(t); t = 1, \ldots, N\}$, the problem is to estimate the unknown parameters.

Even though all natural signals are real-valued, sometimes it might be more beneficial to use their complex counterpart. With the abuse of notations, the associated complex model can be written as follows:

$$y(t) = \sum_{k=1}^{p} A_k^0 e^{i(\alpha_k^0 t + \beta_k^0 t^2)} + e(t); \quad t = 1, \ldots, N. \tag{4.86}$$

Here, A_k^0 is a complex number, $e(t)$ is a complex valued error with mean zero. The problem remains the same, i.e. to estimate the unknown parameters based on the complex valued signal $\{y(t); t = 1, \ldots, N\}$. Most of the estimation methods which have been developed for the one component model, can be extended for the multicomponent model also. We mainly discuss the estimation procedures for the real multicomponent model (4.85) only, but all the methods can be used with the obvious modification for model (4.86) also.

Before proceeding further, let us define the following notations: $\boldsymbol{a} = (\alpha_1, \ldots, \alpha_p)^{\mathrm{T}}$, $\boldsymbol{\beta} = (\beta_1, \ldots, \beta_p)^{\mathrm{T}}$, $\boldsymbol{A} = (A_1, \ldots, A_p)^{\mathrm{T}}$, $\boldsymbol{B} = (B_1, \ldots, B_p)^{\mathrm{T}}$, and $\boldsymbol{\Gamma}_j = (A_j, B_j, \alpha_j, \beta_j)^{\mathrm{T}}$, $j = 1, \ldots, p$. The vectors $\boldsymbol{a}^0$, $\boldsymbol{\beta}^0$, $\boldsymbol{A}^0$, $\boldsymbol{B}^0$ and $\boldsymbol{\Gamma}_j^0$ are defined in the similar manner. Based on the assumption that additive error $e(t)$s are i.i.d. random variables with mean 0 and variance σ^2, the most natural estimator of the unknown parameters will be the LSEs. In the subsequent subsection, we provide the least squares estimation procedure in detail and derive the theoretical properties of the LSEs.

4.5.1. Least Squares Estimators

The LSEs of the unknown parameters of model (4.85) can be obtained by minimizing

$$Q(\boldsymbol{\Gamma}_1, \ldots, \boldsymbol{\Gamma}_p) = \sum_{t=1}^{N}\left(y(t) - \sum_{k=1}^{p}\{A_k \cos(\alpha_k t + \beta_k t^2) + B_k \sin(\alpha_k t + \beta_k t^2)\} \right)^2. \quad (4.87)$$

Note that if $e(t)$s are i.i.d. Gaussian random variables, then the LSEs become MLEs also. The minimization of (4.87) can be performed along the same line as the one component model. We can write $Q(\boldsymbol{\Gamma}_1, \ldots, \boldsymbol{\Gamma}_p)$ as follows:

$$Q(\boldsymbol{\Gamma}_1, \ldots, \boldsymbol{\Gamma}_p) = \left[\boldsymbol{Y} - \sum_{j=1}^{p} W(\alpha_j, \beta_j)\boldsymbol{\phi}_j \right]^{\mathrm{T}} \left[\boldsymbol{Y} - \sum_{j=1}^{p} W(\alpha_j, \beta_j)\boldsymbol{\phi}_j \right], \quad (4.88)$$

where the $N \times 2$ matrix $W(\alpha, \beta)$ is same as defined in (4.61) and $\boldsymbol{\phi}_j = (A_j, B_j)^{\mathrm{T}}$ is a 2×1 vector for $j = 1, \ldots, p$. The data vector $\boldsymbol{Y} = (y(1), \ldots, y(N))^{\mathrm{T}}$ is same as defined before. The LSEs of the unknown parameters can be obtained by minimizing (4.88) with respect to the unknown parameters. Now we define the $N \times 2p$ matrix $\widetilde{W}(\boldsymbol{a}, \boldsymbol{\beta})$ as

$$\widetilde{W}(\boldsymbol{a}, \boldsymbol{\beta}) = [W(\alpha_1, \beta_1) : \cdots : W(\alpha_p, \beta_p)].$$

Then, for fixed $\boldsymbol{a}$ and $\boldsymbol{\beta}$, the LSEs of $\boldsymbol{\phi}_1, \ldots, \boldsymbol{\phi}_p$, the linear parameter vectors, can be obtained as follows:

$$[\hat{\boldsymbol{\phi}}_1^{\mathrm{T}}(\alpha_1, \beta_1) : \cdots : \hat{\boldsymbol{\phi}}_p^{\mathrm{T}}(\alpha_p, \beta_p)]^{\mathrm{T}} = \left[\widetilde{W}^{\mathrm{T}}(\boldsymbol{a}, \boldsymbol{\beta})\widetilde{W}(\boldsymbol{a}, \boldsymbol{\beta}) \right]^{-1} \widetilde{W}^{\mathrm{T}}(\boldsymbol{a}, \boldsymbol{\beta})\boldsymbol{Y}.$$

It is important to note that $\widetilde{W}^{\mathrm{T}}(\alpha,\beta)\widetilde{W}(\alpha,\beta)$ is a diagonal matrix for large N by the use of Lemma 1. Therefore, $\widehat{\boldsymbol{\phi}}_j = \widehat{\boldsymbol{\phi}}_j(\alpha_j,\beta_j)$, the LSEs of $\boldsymbol{\phi}_j^0$, $j=1,\ldots,p$ can also be expressed as follows:

$$\widehat{\boldsymbol{\phi}}_j(\alpha_j,\beta_j) = \left[W^{\mathrm{T}}(\alpha_j,\beta_j)W(\alpha_j,\beta_j)\right]^{-1}W^{\mathrm{T}}(\alpha_j,\beta_j)Y .$$

Using similar techniques as in the one component case, the LSEs of α and β can be obtained as the argument maximum of

$$Y^{\mathrm{T}}\widetilde{W}(\alpha,\beta)[\widetilde{W}^{\mathrm{T}}(\alpha,\beta)\widetilde{W}(\alpha,\beta)]^{-1}\widetilde{W}^{\mathrm{T}}(\alpha,\beta)Y . \qquad (4.89)$$

The criterion function, given in (4.89), is a highly non-linear function of α and β, therefore, the LSEs of α and β, cannot be obtained in a closed form. One needs to solve a $2p$ dimensional optimization problem to compute these LSEs. Also, because of the non-linear nature of the function defined in (4.89), very good initial estimates are needed to find the global maximum. The problem becomes all the more complicated for large values of p. An alternative is to use the method of Pincus [100] to maximize (4.89), as suggested by Saha and Kay [109]. But before describing different efficient methods, we provide the properties of the LSEs.

Kundu and Nandi [72] obtained the following consistency result for the LSEs of the multicomponent model.

Theorem 10. *Suppose there exists a K, such that for $j=1,\ldots,p$, $0<|A_j^0|+|B_j^0|<K$, $0<\alpha_j^0,\beta_j^0<\pi$, α_j^0 are distinct, similarly β_j^0 are also distinct. If $e(t)$s are i.i.d. random variables with mean zero and finite variance σ^2, then $\widehat{\boldsymbol{\Gamma}}_j = (\widehat{A}_j,\widehat{B}_j,\widehat{\alpha}_j,\widehat{\beta}_j)^{\mathrm{T}}$, the LSE of $\boldsymbol{\Gamma}_j^0 = (A_j^0,B_j^0,\alpha_j^0,\beta_j^0)^{\mathrm{T}}$ is strongly consistent, for $j=1,\ldots,p$.*

They also obtained the asymptotic normality properties of $\widehat{\boldsymbol{\Gamma}}_j$ along with the consistency results, but the elements of the asymptotic variance covariance matrix were quite complex. Later Lahiri, Kundu and Mitra [79] simplified the entries of the asymptotic variance covariance matrix using Lemma 1 and the Conjecture 1. They obtained the following results.

Theorem 11. *Under the same assumptions as in Theorem 10, for $j=1,\ldots,p$,*

$$\left(N^{1/2}(\widehat{A}_j - A_j^0), N^{1/2}(\widehat{B}_j - B_j^0), N^{3/2}(\widehat{\alpha}_j - \alpha_j^0), N^{5/2}(\widehat{\beta}_j - \beta_j^0)\right)^{\mathrm{T}} \xrightarrow{d} N_4(0, 2\sigma^2\Sigma_j), \qquad (4.90)$$

where Σ_j can be obtained from the matrix Σ defined in (4.65), by replacing A^0 and B^0 with A_j^0 and B_j^0, respectively. Moreover, $\widehat{\boldsymbol{\Gamma}}_j$ and $\widehat{\boldsymbol{\Gamma}}_k$, for $j \neq k$ are asymptotically independently distributed.

Note that the consistency and asymptotic normality of the LSEs hold even when the errors are from a stationary linear process. More precisely, under Assumption 4, the asymptotic variance covariance matrix is $2\sigma^2 c\Sigma_j$, where c is same as defined in Theorem 5.

4.5.2. Sequential Least Squares Estimators

Even though, the LSEs are the most natural estimators for the multicomponent chirp model, finding the LSEs is a numerically challenging problem, particularly if p is large. To address this issue, Lahiri, Kundu and Mitra [79] proposed a sequential procedure for the multiple chirp model similar to the sequential procedure suggested by Prasad, Kundu and Mitra [102] for the multiple sinusoidal model. Based on the number theoretic Lemma 1 and the Conjecture 1, it has been shown that the LSEs and the sequential estimators are asymptotically equivalent, i.e. the sequential estimators are also strongly consistent and they have the same asymptotic distributions as the LSEs.

The sequential procedure proposed by Lahiri, Kundu and Mitra [79] can be described as follows. Let us assume the following, without loss of generality:

$$A_1^{0^2} + B_1^{0^2} > ... > A_p^{0^2} + B_p^{0^2}\ .$$

We first obtain the estimates of A_1^0, B_1^0, α_1^0 and β_1^0 by minimizing

$$Q(\mathbf{\Gamma}_1) = \left[\mathbf{Y} - \mathbf{W}(\alpha_1,\beta_1)\boldsymbol{\phi}_1\right]^{\mathrm{T}}\left[\mathbf{Y} - \mathbf{W}(\alpha_1,\beta_1)\boldsymbol{\phi}_1\right]. \tag{4.91}$$

Note that the minimization of (4.91) can be performed by solving a two dimensional optimization problem. Suppose $\widehat{\alpha}_1$, $\widehat{\beta}_1$, $\widehat{\boldsymbol{\phi}}_1 = (\widehat{A}_1,\widehat{B}_1)^{\mathrm{T}}$, are the estimates of α_1^0, β_1^0 and $(A_1^0,B_1^0)^{\mathrm{T}}$, respectively. We then obtain the new data set $\mathbf{Y}^{(1)}$ from the original data set $\mathbf{Y}$, by removing the effect of the first component as follows:

$$\mathbf{Y}^{(1)} = \mathbf{Y} - \mathbf{W}(\widehat{\alpha}_1,\widehat{\beta}_1)\widehat{\boldsymbol{\phi}}_1. \tag{4.92}$$

Then at the second step, we obtain the estimates of A_2^0, B_2^0, α_2^0 and β_2^0 by minimizing

$$Q(\mathbf{\Gamma}_2) = \left[\mathbf{Y}^{(1)} - \mathbf{W}(\alpha_2,\beta_2)\boldsymbol{\phi}_2\right]^{\mathrm{T}}\left[\mathbf{Y}^{(1)} - \mathbf{W}(\alpha_2,\beta_2)\boldsymbol{\phi}_2\right]. \tag{4.93}$$

Minimization of (4.93) can again be performed by solving a two dimensional optimization problem. Repeating this process we can obtain the estimates of $(A_3^0,B_3^0,\alpha_3^0,\beta_3^0)$, ..., $(A_p^0,B_p^0,\alpha_p^0,\beta_p^0)$, sequentially. We call these estimators as the sequential LSEs. It has been shown by Lahiri, Kundu and Mitra [79] that under the assumption of stationary errors, the sequential estimators are strongly consistent and they have the same asymptotic distribution as the LSEs. They also performed extensive simulation experiments and observed that the sequential estimators perform as good as the LSEs. Another important result is that if the process is repeated beyond p times, then the corresponding linear parameter estimates converge to zero almost surely. Hence, in practice the sequential procedure can be used to estimate the number of components in a chirp signal also.

Note that the sequential procedure is a very powerful tool. It mainly works due to the fact that as the sample size increases the two chirp components become orthogonal because of the number theoretic results described in Section 4.2. Thus, this sequential procedure can be extended for other estimators as well. For example, Grover, Kundu and Mitra [44] developed the sequential PEs. They showed that these estimators are strongly consistent and have the same asymptotic distribution as the LSEs. Through extensive simulations, it is observed that the sequential PEs perform at par with the LSEs and are computationally faster in comparison. It can be easily shown that the sequential method works for the finite step algorithm (Section 4.4.3) as well. Also, it will be interesting to develop the theoretical properties of the sequential LAD estimators and sequential testing of hypothesis for the multicomponent chirp model.

So far, most of the discussion has been on the chirp model, its applications, motivation and various methods of estimation of its parameters are briefly reviewed. The rest of the chapter, however, will deal with the more general variations of this model. For instance, in the next section, we consider a polynomial phase model which is more flexible as it allows the phase to be a time-varying polynomial of any degree.

4.6. Polynomial Phase Model

4.6.1. One Component Polynomial Phase Model

The one component polynomial phase signal was first introduced by Djurić and Kay [24] and it can be described as follows:

$$y(t) = A^0 e^{i(\alpha_1^0 t + \alpha_2^0 t^2 + \ldots + \alpha_p^0 t^p)} + e(t); \quad t = 1, \ldots, N. \tag{4.94}$$

Here also A^0 is a complex number, and $0 < \alpha_1^0, \ldots, \alpha_p^0 < \pi$ are the coefficients of the polynomial and $e(t)$s are noise random variables with mean zero and finite variance. The explicit assumptions on $e(t)$s will be mentioned later. The problem remains the same, that is, based on the observations $\{y(1), \ldots, y(N)\}$, we need to estimate the unknown parameters A^0, $\alpha_1^0, \ldots, \alpha_p^0$. It is assumed that the degree of the polynomial p is known.

The associated real valued model can be written as follows:

$$\begin{aligned}
y(t) &= A^0 \cos(\alpha_1^0 t + \alpha_2^0 t^2 + \ldots + \alpha_p^0 t^p) \\
&+ B^0 \cos(\alpha_1^0 t + \alpha_2^0 t^2 + \ldots + \alpha_p^0 t^p) + e(t); t = 1, \ldots, N.
\end{aligned} \tag{4.95}$$

In case of model (4.95), A^0, B^0 are real valued and $e(t)$s are real valued random variables with mean zero and finite variance.

One of the primary motivations for studying polynomial phase signals comes from Doppler radar applications. Although the continuous time transmitted radar signal does not have polynomial phase, the samples taken at the matched filter output of a pulsed radar system give rise to a discrete time polynomial phase signal, when the target is moving, see for example Rihaczek [108]. The polynomial coefficients are then related to the kinetic parameters of the target. The polynomial phase models are often used in analyzing synthetic aperture radar (SAR) images, see for example Porchia et al. [101].

Djurić and Kay [24] considered the MLEs of the unknown parameters based on the assumptions that the error random variables are complex valued Gaussian random variables. Hence, for model (4.94) the MLEs of the unknown parameters can be obtained by minimizing

$$Q(\boldsymbol{\theta}_p) = \sum_{t=1}^{N} \left| y(t) - A e^{i(\alpha_1 t + \ldots, \alpha_p t^p)} \right|^2, \tag{4.96}$$

with respect to $\boldsymbol{\theta}_p$, where $\boldsymbol{\theta}_p = (A_R, A_I, \alpha_1, \ldots, \alpha_p)^{\mathrm{T}}$ and $A = A_R + iA_I$. The minimization of (4.96) can be performed by solving a p dimensional optimization problem. Any standard non-linear optimization method may be used, but one needs a very good set of initial guesses to attain the global optimum solution. Alternatively, the importance sampling as described in Section 4.4.1 can be used for this purpose. Although, Djurić and Kay [24] justified the use of MLEs through Monte Carlo simulations, they did not provide any theoretical properties of the MLEs. Clearly, many desirable properties of the MLEs are not assured as the non-linear model (4.94) does not satisfy the sufficient conditions of Jennrich [58], Wu [122] or Kundu [63]. Later, Nandi and Kundu [91] established the consistency and asymptotic normality properties of the LSEs under the assumptions that the errors are i.i.d. complex valued random variables with mean zero and finite variance. To state their results, we need the following assumption.

Assumption 7. *$A^0 = A_R^0 + iA_I^0$ is an arbitrary complex number, $0 < \alpha_1^0, \ldots, \alpha_p^0 < \pi$ and $e(t)$s are i.i.d. complex valued random variables. Let us write $e(t) = e_R(t) + ie_I(t)$, where $e_R(t)$ and $e_I(t)$ are the real and imaginary parts of $e(t)$. It is assumed that*

$$E(e_R(t)) = E(e_I(t)) = 0 \quad \text{and} \quad V(e_R(t)) = V(e_I(t)) = \frac{\sigma^2}{2}, \quad \text{and} \quad e_R(t) \text{ and } e_I(t) \text{ are}$$

independently distributed.

The following results were obtained.

Theorem 12. If $e(t)$s are i.i.d. complex valued random variables satisfying Assumption 7, then the LSEs of $(A_R^0, A_I^0, \alpha_1^0, \ldots, \alpha_p^0)^{\mathrm{T}}$, say $(\hat{A}_R, \hat{A}_I, \hat{\alpha}_1, \ldots, \hat{\alpha}_p)^{\mathrm{T}}$, are strongly consistent and

$$\left(\sqrt{N}(\hat{A}_R - A_R^0), \sqrt{N}(\hat{A}_I - A_I^0), N^{\frac{3}{2}}(\hat{\alpha}_1 - \alpha_1^0), \ldots, N^{\frac{2p+1}{2}}(\hat{\alpha}_p - \alpha_p^0) \right)^{\mathrm{T}} \xrightarrow{d} N_{p+2}(\mathbf{0}, \sigma^2 \Sigma_{p+2}),$$

where Σ_{p+2} is a $(p+2) \times (p+2)$ positive definite matrix, and it is defined through its inverse as follows:

$$\Sigma_{p+2}^{-1} = \begin{bmatrix} 1 & 0 & -\dfrac{1}{2}A_I^0 & -\dfrac{1}{3}A_I^0 & \cdots & -\dfrac{1}{p+1}A_I^0 \\[2ex] 0 & 1 & \dfrac{1}{2}A_R^0 & \dfrac{1}{3}A_R^0 & \cdots & \dfrac{1}{p+1}A_R^0 \\[2ex] -\dfrac{1}{2}A_I^0 & \dfrac{1}{2}A_R^0 & \dfrac{1}{3}|A^0|^2 & \dfrac{1}{4}|A^0|^2 & \cdots & \dfrac{1}{p+2}|A^0|^2 \\[2ex] -\dfrac{1}{3}A_I^0 & \dfrac{1}{3}A_R^0 & \dfrac{1}{4}|A^0|^2 & \dfrac{1}{5}|A^0|^2 & \cdots & \dfrac{1}{p+3}|A^0|^2 \\[2ex] \vdots & \vdots & \vdots & \vdots & \ddots & \vdots \\[2ex] -\dfrac{1}{p+1}A_I^0 & \dfrac{1}{p+1}A_R^0 & \dfrac{1}{p+2}|A^0|^2 & \dfrac{1}{p+3}|A^0|^2 & \cdots & \dfrac{1}{2p+1}|A^0|^2 \end{bmatrix}.$$

Similar to periodogram estimators for the chirp model (see Section 4.4.2), one can compute them for the polynomial phase model (4.94) as well. The periodogram function for the latter can be defined as follows:

$$I(\alpha_1, \ldots, \alpha_p) = \frac{1}{N}\left| \sum_{t=1}^{N} y(t)e^{-i(\alpha_1 t + \ldots + \alpha_p t^P)} \right|^2. \tag{4.97}$$

Thus, the estimators of $\alpha_1, \ldots, \alpha_p$ can be obtained by maximizing $I(\alpha_1, \ldots, \alpha_p)$ with respect to the unknown parameters. If

$$(\hat{\alpha}_1, \ldots, \hat{\alpha}_p) = arg\ max\ I(\alpha_1, \ldots, \alpha_p),$$

then $\hat{\alpha}_1, \ldots, \hat{\alpha}_p$ are called the periodogram estimators of $\alpha_1^0, \ldots, \alpha_p^0$, respectively. Once we obtain the periodogram estimators of the phase parameters, the periodogram estimator of A^0 can be obtained as

$$\hat{A} = \frac{1}{N}\sum_{t=1}^{N} y(t)e^{-i(\hat{\alpha}_1 t + \ldots + \hat{\alpha}_p t^P)}.$$

Evidently, the periodogram estimators of the polynomial chirp model can be obtained by solving a p dimensional optimization problem, similar to the LSEs. Moreover, if the error random variables $e(t)$s satisfy Assumption 7, then it can be shown that the periodogram estimators are strongly consistent and they have the same asymptotic distribution as the LSEs. This can be proved theoretically following similar approach to that used for a chirp model by Grover, Kundu and Mitra [44]. It will be interesting to develop efficient estimation procedures to compute LSEs and ALSEs for a general polynomial of order p and compare their performances.

4.6.2. Multicomponent Polynomial Phase Model

Although one component polynomial phase model has been used quite extensively in modeling SAR data and in many other signal processing applications, in a number of practical situations such as analysis of non-stationary signal in the presence of another non-stationary jamming signal, the multicomponent model is more suitable. A multicomponent polynomial phase model in presence of additive noise can be described as follows:

$$y(t) = \sum_{k=1}^{m} A_k^0 e^{i(\alpha_{1k}^0 t + \alpha_{2k}^0 t^2 + \ldots + \alpha_{pk}^0 t^P)} + e(t); \quad t = 1, \ldots, N. \tag{4.98}$$

Here $A_1^0, \ldots, A_m^0$ are complex numbers, $e(t)$ s are complex valued random variables as before, and $0 < \alpha_{1k}^0, \ldots, \alpha_{pk}^0 < \pi$, for $k = 1, \ldots, m$. The problem remains the same, i.e. to estimate the unknown parameters namely $A_1^0, \ldots, A_m^0$ and $\alpha_{1k}^0, \ldots, \alpha_{pk}^0$, for $k = 1, \ldots, m$.

Note that a more general model than model (4.98) can be more flexible in the sense that the former can have different components with varying degrees of the polynomials. However, analytically it has the same challenges as the underlying model. Hence, we restrict to model (4.98) mainly for notational simplicity.

This problem has been considered by several authors, see for example Friedlander and Francos [37], Barbarossa, Scaglione and Giannakis [5], Ticahvsky and Handel [113], Pham and Zoubir [99] and the references cited therein. There are mainly two issues related to the multicomponent polynomial phase model. First, the interaction between the two components, often called cross-terms, give rise to undesired sinusoids in the higher order instantaneous moment. This strongly affects algorithms based on frequency estimation. Second, the principle of demodulation of mono component polynomial phase signals no longer works with multicomponent polynomial phase signals.

Pham and Zoubir [99] proposed to use the maximum likelihood method to estimate the unknown parameters when the errors are i.i.d. complex Gaussian random variables with mean zero and finite variance. The MLEs of the unknown parameters can be obtained by solving a non-linear optimization problem. If the errors are not complex Gaussian random variables but i.i.d. with mean zero and finite variance, then the least squares method can be applied. The LSEs can be obtained by minimizing

$$Q(A,\alpha) = \sum_{t=1}^{N}\left| y(t) - \sum_{k=1}^{m} A_k e^{i(\alpha_{1k}t + \alpha_{2k}t^2 + \ldots + \alpha_{pk}t^P)} \right|^2 \qquad (4.99)$$

with respect to $A = (A_1,\ldots,A_m)^{\mathrm{T}}$ and $\alpha = (\alpha_{11},\ldots,\alpha_{p1},\ldots,\alpha_{1m},\ldots,\alpha_{pm})^{\mathrm{T}}$. The theoretical properties of the MLEs or the LSEs have not yet been established and is an open problem. Therefore, it will be interesting to develop them under different error assumptions.

4.7. Random Amplitude Chirp Model

4.7.1. One Component Random Amplitude Chirp Model

So far we have discussed one component and multicomponent chirp models when the amplitude is constant. However in many applications of signal processing, particularly in different radar problems, it is more apt to use a model with time-varying amplitude. Let us consider an example of a radar illuminating a target. The transmitted signal can then be affected by two different phenomena. First, it will usually undergo a phase shift induced by the distance and relative motion between the target and the receiver. Based on some smoothness assumption on the motion, the phase shift can be modelled as quadratic function of time. The coefficients of the quadratic functions are usually dependent on the speed and acceleration of the radar, and kind of waveforms transmitted by the target. The second phenomenon is due to amplitude distortion caused either by target fluctuation or scattering of the medium. Due to these reasons, the random amplitude chirp signal (4.100) can be used quite effectively in dealing with different radar problems. Mathematically, the random amplitude chirp model can be described as follows:

$$y(t) = \alpha(t)e^{i(\theta_0^0 + \theta_1^0 t + \theta_2^0 t^2)} + e(t); \quad t = 1,\ldots,N. \qquad (4.100)$$

Here, $e(t)$ s are the additive noise random variables with mean zero and finite variance, and $\alpha(t)$ is the random time varying amplitude. The explicit structure of $\alpha(t)$ will be defined later.

Besson, Ghogho and Swami [10] made the following assumptions on the random amplitude $\alpha(t)$ and the error component $e(t)$. It is assumed that $\alpha(t)$ is a real valued stationary mixing process with non-zero mean. The error component $e(t)$ is a white complex Gaussian process with mean zero and finite variance. Moreover, $e(t)$ and $\alpha(t)$ are assumed to be independent. Recently, Nandi and Kundu [94] considered the random amplitude chirp model (4.100), and provided a theoretical justification that if the mean of the stationary mixing process $\alpha(t)$ is unknown and θ_0^0 is also unknown, then both are not identifiable. Hence, if both are present then one has to be known. Due to this, without loss of generality, we consider the model

$$y(t) = \alpha(t)e^{i(\theta_1^0 t + \theta_2^0 t^2)} + e(t); \quad t = 1,\ldots,N. \qquad (4.101)$$

The problem remains the same, i.e., estimate the frequency θ_1^0 and the frequency rate θ_2^0 based on a random sample $\{y(1),\ldots,y(N)\}$ from model (4.101). Besson, Ghogho and Swami [10] proposed the following estimators of θ_1^0 and θ_2^0. Consider the following function

$$Q(\boldsymbol{\theta}) = \frac{1}{N}\left|\sum_{t=1}^{N} y^2(t)e^{-i2(\theta_1 t+\theta_2 t^2)}\right|^2 . \tag{4.102}$$

Note that the function $Q(\boldsymbol{\theta})$ as defined in (4.102) is the periodogram function of the transformed signal $y^2(t)$ with exponent replaced by twice the usual periodogram component. The unknown parameters θ_1 and θ_2 are estimated by maximizing $Q(\boldsymbol{\theta})$. Let $\widehat{\boldsymbol{\theta}} = (\widehat{\theta}_1,\widehat{\theta}_2)^{\mathrm{T}}$ be the maximizer of $Q(\boldsymbol{\theta})$, then

$$\widehat{\boldsymbol{\theta}} = (\widehat{\theta}_1,\widehat{\theta}_2)^{\mathrm{T}} = arg\ max_{\boldsymbol{\theta}}\, Q(\boldsymbol{\theta}) . \tag{4.103}$$

In this case also, the maximization of $Q(\boldsymbol{\theta})$ can be performed by solving a two dimensional optimization problem. Once $\widehat{\theta}_1$ and $\widehat{\theta}_2$ are obtained, then the random amplitude $\alpha(t)$ at the time point t may be estimated by

$$\widehat{\alpha}(t) = Re\left\{ y(t)\times e^{-(\widehat{\theta}_1 t+\widehat{\theta}_2 t^2)} \right\} .$$

Besson, Ghogho and Swami [10] provided the Cramer-Rao lower bound, and performed extensive simulations to show that the above estimators perform very well in practice. However, they did not provide any theoretical properties of the estimators. Very recently, Nandi and Kundu [94] provided the theoretical properties of $\widehat{\theta}_1$ and $\widehat{\theta}_2$ based on the following assumptions on $\alpha(t)$ and $e(t)$.

Assumption 8. *The random amplitude $\{\alpha(t)\}$ is a sequence of i.i.d. real-valued random variables with mean μ_α, variance σ_α^2, $\mu_\alpha \neq 0$ and $\sigma_\alpha^2 > 0$. The fourth moment of $\{\alpha(t)\}$ exists.*

Assumption 9. *The additive error $\{e(t)\}$ is a sequence of complex-valued i.i.d. random variables with mean zero and variance σ^2. Write $e(t) = e_R(t) + ie_I(t)$, then $\{e_R(t)\}$ and $\{e_I(t)\}$ are i.i.d. random variables with mean 0 and variance $\dfrac{\sigma^2}{2}$, have fourth moment γ, and are independently distributed.*

Assumption 10. *$\{e(t)\}$ is assumed to be independent of $\{\alpha(t)\}$ and $0 < \theta_1^0,\theta_2^0 < \pi$.*

It may be noted that the assumptions used by Nandi and Kundu [94] in case of the random amplitude are stronger than it has been mentioned by Besson, Ghogho and Swami [10], whereas the assumptions on the error components are weaker in comparison. Based on the above assumptions, the following consistency and asymptotic normality results have been obtained. For detailed proofs, interested readers are referred to Nandi and Kundu [94].

Theorem 13. *Under Assumptions 8-10, $\widehat{\theta}_1$ and $\widehat{\theta}_2$ as defined in (4.103) are strongly consistent estimators of θ_1^0 and θ_2^0, respectively.*

Theorem 14. *Under Assumptions 8-10, as $N \to \infty$,*

$$D^{-1}(\widehat{\boldsymbol{\theta}} - \boldsymbol{\theta}^0) \xrightarrow{d} N_2(0, 4(\sigma_\alpha^2 + \mu_\alpha^2)^2 \boldsymbol{\Sigma}^{-1} \boldsymbol{\Gamma} \boldsymbol{\Sigma}^{-1}),$$

where with $C_\alpha = 8(\sigma_\alpha^2 + \mu_\alpha^2)\sigma^2 + \dfrac{1}{2}\gamma + \dfrac{1}{8}\sigma^4$,

$$D = \begin{pmatrix} N^{-\frac{3}{2}} & 0 \\ 0 & N^{-\frac{5}{2}} \end{pmatrix}, \quad \boldsymbol{\Sigma} = \frac{2(\sigma_\alpha^2 + \mu_\alpha^2)^2}{3}\begin{pmatrix} 1 & 1 \\ 1 & \dfrac{16}{15} \end{pmatrix}, \quad \boldsymbol{\Gamma} = C_\alpha \begin{bmatrix} \dfrac{1}{3} & \dfrac{1}{4} \\ \dfrac{1}{4} & \dfrac{1}{5} \end{bmatrix}.$$

From Theorem 14 it follows that the asymptotic variances of $\hat{\theta}_1$ and $\hat{\theta}_2$ are

$$Var(\widehat{\theta}_1) = \frac{93}{N^3(\sigma_\alpha^2 + \mu_\alpha^2)^2}\left[8(\sigma_\alpha^2 + \mu_\alpha^2)\sigma^2 + \frac{1}{2}\gamma + \frac{1}{8}\sigma^4 \right],$$

$$Var(\widehat{\theta}_2) = \frac{135}{N^5(\sigma_\alpha^2 + \mu_\alpha^2)^2}\left[8(\sigma_\alpha^2 + \mu_\alpha^2)\sigma^2 + \frac{1}{2}\gamma + \frac{1}{8}\sigma^4 \right],$$

respectively. In case the additive error is Gaussian with mean zero and variance σ^2, then the fourth moment is $3\sigma^4$ and the asymptotic variances reduce to

$$Var(\widehat{\theta}_1) = \frac{93}{N^3(\sigma_\alpha^2 + \mu_\alpha^2)^2}\left[8(\sigma_\alpha^2 + \mu_\alpha^2)\sigma^2 + \frac{13}{8}\sigma^4 \right],$$

$$Var(\widehat{\theta}_2) = \frac{135}{N^5(\sigma_\alpha^2 + \mu_\alpha^2)^2}\left[8(\sigma_\alpha^2 + \mu_\alpha^2)\sigma^2 + \frac{13}{8}\sigma^4 \right].$$

It should be mentioned that the asymptotic variances of $\hat{\theta}_1$ and $\hat{\theta}_2$ in case of Gaussian errors, obtained heuristically by Besson, Ghogho and Swami [10] are slightly different than the above and they are

$$Var(\widehat{\theta}_1) = \frac{96}{N^3(\sigma_\alpha^2 + \mu_\alpha^2)^2}\left[(\sigma_\alpha^2 + \mu_\alpha^2)\sigma^2 + \frac{1}{2}\sigma^4\right],$$

$$Var(\widehat{\theta}_2) = \frac{90}{N^5(\sigma_\alpha^2 + \mu_\alpha^2)^2}\left[(\sigma_\alpha^2 + \mu_\alpha^2)\sigma^2 + \frac{1}{2}\sigma^4\right].$$

Comment 1. It may be mentioned that the random amplitude sinusoidal model $y(t) = \alpha(t)e^{i\theta^0 t} + e(t)$ is a special case of model (4.101). In this case, the effective frequency does not change over time and it is constant since the frequency rate is zero. The unknown frequency can be estimated by maximizing $Q(\theta)$, where

$$Q(\theta) = \frac{1}{N}\left|\sum_{t=1}^{N} y^2(t)e^{-i2\theta t}\right|^2.$$

The consistency and asymptotic normality of the estimator of θ^0 follow along the same way.

Recall that in Section 4.6.1 we examined the polynomial phase signal model. It can be used to generalize the underlying random amplitude chirp model to define a more flexible random amplitude polynomial phase signal. Mathematically, the latter can be described as follows:

$$y(t) = \alpha(t)e^{i(\theta_1^0 t + \dots + \theta_p^0 t^p)} + e(t); \quad t = 1,\dots,N. \tag{4.104}$$

Under proper conditions on the random amplitude $\alpha(t)$ and the error random variables $e(t)$, it will be interesting to develop consistency and asymptotic normality properties of the estimators of the unknown parameters. More work is needed along that direction.

4.7.2. Multi Component Random Amplitude Chirp Model

Nandi and Kundu [94] proposed the multi component random amplitude chirp model, where instead of a single frequency and chirp rate pair, p such pairs are present. The model can be written as follows:

$$y(t) = \sum_{k=1}^{p} \alpha_k(t)e^{i(\theta_{1k}^0 t + \theta_{2k}^0 t^2)} + e(t); \quad t = 1,\dots,N. \tag{4.105}$$

Here, the number of components p is assumed to be known. The additive error $e(t)$ satisfies Assumption 9 and the random amplitudes $\alpha_1(t),\dots,\alpha_p(t)$ and the parameters satisfy the following assumptions.

Assumption 11. *The random amplitude corresponding to the k-th component $\{\alpha_k(t)\}$ is a sequence of i.i.d. real-valued random variables with mean $\mu_{k\alpha} \neq 0$, variance $\sigma_{k\alpha}^2 > 0$ and finite fourth moment, for $k = 1, \ldots, p$. Moreover, $\{\alpha_j(t)\}$ and $\{\alpha_k(t)\}$ for $j \neq k$, are independent.*

Assumption 12. *The additive error component $\{e(t)\}$ is independent of the random amplitudes $\{\alpha_1(t)\}, \ldots, \{\alpha_p(t)\}$.*

Assumption 13. *The parameters satisfy $0 < \theta_{11}^0, \theta_{21}^0, \ldots, \theta_{1p}^0, \theta_{2p}^0 < \pi$, and $(\theta_{1j}^0, \theta_{2j}^0) \neq (\theta_{1k}^0, \theta_{2k}^0)$, for $j \neq k$, $j, k = 1, \ldots, p$.*

The unknown parameters are estimated by maximizing $Q(\boldsymbol{\theta})$, for $\boldsymbol{\theta} = (\theta_1, \theta_2)^{\mathrm{T}}$, as defined in (4.102), locally. Let us write $\boldsymbol{\theta}_k = (\theta_{1k}, \theta_{2k})^{\mathrm{T}}$ and let $\boldsymbol{\theta}_k^0$ be the true value of $\boldsymbol{\theta}_k$. Let us further use the notation N_k, as a neighborhood of $\boldsymbol{\theta}_k^0$, such that for $j \neq k$, $\boldsymbol{\theta}_j^0 \notin N_k$. We estimate $\boldsymbol{\theta}_k^0$ as follows:

$$\hat{\boldsymbol{\theta}}_k = (\hat{\theta}_{1k}, \hat{\theta}_{2k})^{\mathrm{T}} = \arg\max_{(\theta_1, \theta_2) \in N_k} \frac{1}{N} \left| \sum_{t=1}^{N} y^2(t) e^{-i2(\theta_1 t + \theta_2 t^2)} \right|^2 . \tag{4.106}$$

Nandi and Kundu [94] observed that the proposed estimators have the following theoretical properties.

Theorem 15. *Under Assumptions 9 and 11 to 13, $\hat{\boldsymbol{\theta}}_k$ is a strongly consistent estimator of $\boldsymbol{\theta}_k^0$, for $k = 1, \ldots, p$.*

Theorem 16. *Under Assumptions 9 and 11 to 13, as $N \to \infty$,*

$$\boldsymbol{D}^{-1}(\hat{\boldsymbol{\theta}}_k - \boldsymbol{\theta}_k^0) \xrightarrow{d} N_2(0, 4(\sigma_{k\alpha}^2 + \mu_{k\alpha}^2)^2 \boldsymbol{\Sigma}_k^{-1} \boldsymbol{\Gamma}_k \boldsymbol{\Sigma}_k^{-1}),$$

where with $C_{k\alpha} = 8(\sigma_{k\alpha}^2 + \mu_{k\alpha}^2)\sigma^2 + \frac{1}{2}\gamma + \frac{1}{8}\sigma^4$,

$$\boldsymbol{D} = \begin{pmatrix} N^{-\frac{3}{2}} & 0 \\ 0 & N^{-\frac{5}{2}} \end{pmatrix}, \quad \boldsymbol{\Sigma}_k = \frac{2(\sigma_{k\alpha}^2 + \mu_{k\alpha}^2)}{3} \begin{pmatrix} 1 & 1 \\ 1 & \frac{16}{15} \end{pmatrix}, \quad \boldsymbol{\Gamma}_k = C_{k\alpha} \begin{bmatrix} \frac{1}{3} & \frac{1}{4} \\ \frac{1}{4} & \frac{1}{5} \end{bmatrix}.$$

Theorem 17. *Under Assumptions 9 and 11 to 13, $\boldsymbol{D}^{-1}(\hat{\boldsymbol{\theta}}_k - \boldsymbol{\theta}_k^0)$ and $\boldsymbol{D}^{-1}(\hat{\boldsymbol{\theta}}_j - \boldsymbol{\theta}_j^0)$ for $k \neq j$ are asymptotically independently distributed.*

At this point we would like to bring the reader's attention to some interesting observations: a) The rates of convergence of the LSEs of the frequencies and frequency rates are same as that for the multicomponent chirp model, b) The asymptotic distribution of the estimators depends on the parameters of $\alpha(t)$ and $e(t)$, but not the true values of the frequencies or frequency rates, c) and lastly the assumption that $\{\alpha_j(t)\}$ and $\{\alpha_k(t)\}$ are independent for $j \neq k$ is crucial in obtaining the independence of $\boldsymbol{D}^{-1}(\widehat{\boldsymbol{\theta}}_j - \boldsymbol{\theta}_j^0)$ and $\boldsymbol{D}^{-1}(\widehat{\boldsymbol{\theta}}_k - \boldsymbol{\theta}_k^0)$.

Nandi and Kundu [94] performed extensive simulations both for one component and multicomponent random amplitude chirp models. It is observed that the performances of the proposed estimators are quite good even for moderate sample sizes if $E(\alpha_k^2(t))$ is not very close to zero, for $k = 1, \ldots, p$. But as sample size increases the sample biases of the proposed estimators become negligible and sample variances become quite close to the theoretical asymptotic variances.

4.8. Sum of Sinusoidal and Chirplet Model

In this section, we will have a detailed discussion on a new model which has been recently introduced by Grover [42] in her doctoral thesis. This novel model is closely associated with the multicomponent chirp model and it has been named as 'chirp like model'. The chirp like model can be formally described as follows:

$$
\begin{aligned}
y(t) = \sum_{j=1}^{p} &\left\{ A_j^0 \cos(\alpha_j^0 t) + B_j^0 \sin(\alpha_j^0 t) \right\} \\
+ \sum_{k=1}^{q} &\left\{ C_k^0 \cos(\beta_k^0 t^2) + D_k^0 \sin(\beta_k^0 t^2) \right\} + e(t); t = 1, \ldots, N.
\end{aligned}
\tag{4.107}
$$

Here $p \geq 1$, $q \geq 1$, A_j^0s, B_j^0s, C_k^0s, D_k^0s are real numbers, α_j^0s and β_j^0s are the frequencies, and frequency rates, respectively. It has been observed through extensive simulation studies that model (4.107) behaves very much like a multicomponent chirp model (4.85), in the sense that if a signal has been generated from model (4.85) then model (4.107) fits the generated signal very well and vice versa. She also observed that if the multicomponent chirp model (4.85) fits a data set well, then the 'chirp-like-model' also fits the data set reasonably well and the two model fittings are almost indistinguishable. To illustrate the validity of her analysis, in Fig. 4.10, we show four examples of speech signal data sets. We plot the original speech signal along with the estimated signals in each of the sub-plots, where the estimated signals are obtained by fitting a chirp model and a chirp like model to the original data set.

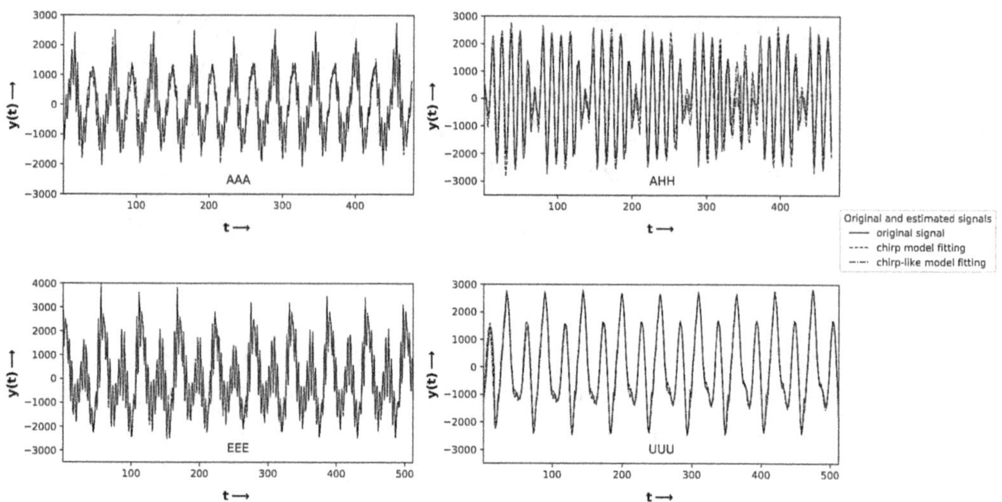

Fig. 4.10. Observed speech signals along with estimated signals using chirp and chirp-like models.

It is clear from the above figure that the two fittings for each data set are extremely similar. The major advantage of switching to the proposed model (4.107) is that it is easier to handle both analytically as well as computationally. Although the theoretical properties of different estimators are developed assuming that the number of sinusoids and chirp components, p and q are known, but in her thesis, Grover [42] has discussed their estimation when analyzing real data sets. In the following table, we report the number of components that were needed to model the speech signals using the chirp model as well as using the chirp like model. These models were fitted to the observed signals using the sequential estimation procedure that we describe in Section 4.5 for the multicomponent chirp model and in this section for the chirp like model. It is clear that in most of the cases the chirp like model needs less number of parameters than a chirp model to fit a particular data set. Moreover, we will see later that fitting a multicomponent chirp-like model is much easier compared to a multicomponent chirp model. The results, undeniably, demonstrate how using a chirp like model as an alternative to the chirp model lowers the computational cost significantly.

Table 4.2. Number of components used to fit chirp and chirp like model to the speech data sets.

Data Set	Chirp Model		Chirp-like model	
	p	Number of parameters	(p, q)	Number of parameters
AAA	9	36	(10, 1)	33
AHH	8	32	(7, 1)	24
UUU	9	36	(8, 1)	27
EEE	11	44	(14, 1)	45

It is interesting to note that the sinusoidal frequency model can be obtained as a special case of model (4.107) when $C_k^0 = D_k^0 = 0,$ for $k = 1,\ldots,q$. Similarly, a simple elementary chirp model namely

$$y(t) = C^0 \cos(\beta^0 t^2) + D^0 \sin(\beta^0 t^2) + e(t);\, t = 1,\ldots,N, \qquad (4.108)$$

can be obtained as a special case of model (4.107). Compared with the sinusoidal model, the literature on a simple elementary chirp model is rather limited. However, the interested reader is referred to articles by Casazza and Fickus [15] and Mboup and Adali [88] for references.

In the next two subsections, we briefly discuss the different estimation methods of a one component and a multicomponent chirp like model.

4.8.1. One Component Chirp Like Model

The one component chirp like model can be described as follows:

$$\begin{aligned}
y(t) = {}& A^0 \cos(\alpha^0 t) + B^0 \sin(\alpha^0 t) + \\
& + C^0 \cos(\beta^0 t^2) + D^0 \sin(\beta^0 t^2) + e(t);\, t = 1,\ldots,N.
\end{aligned} \qquad (4.109)$$

Here A^0, B^0, C^0 and D^0 are real numbers and $0 < \alpha^0, \beta^0 < \pi$, $e(t)$ s are i.i.d. random variables with mean zero and finite variance. Note that the one component chirp like model has six parameters, whereas one component chirp model has four parameters. Due to presence of additional parameters, the one component chirp like model (4.109) is expected to be more flexible than the one component chirp model (4.58).

Now the natural question is how to estimate the unknown parameters given the observed signal data $\{y(1),\ldots,y(N)\}$. The most reasonable estimators of the unknown parameters will be the traditional LSEs. The LSEs of $\boldsymbol{\theta}^0 = (A^0, B^0, \alpha^0, C^0, D^0, \beta^0)^{\mathrm{T}}$ can be obtained by minimizing

$$Q(\boldsymbol{\theta}) = \sum_{t=1}^{N} \left(y(t) - A^0 \cos(\alpha^0 t) - B^0 \sin(\alpha^0 t) - C^0 \cos(\beta^0 t^2) - D^0 \sin(\beta^0 t^2) \right)^2, \qquad (4.110)$$

with respect to the unknown parameters $\boldsymbol{\theta} = (A, B, \alpha, C, D, \beta)^{\mathrm{T}}$. For notational convenience, we write (4.110) in matrix notation as follows:

$$Q(\boldsymbol{\theta}) = (\boldsymbol{Y} - \boldsymbol{Z}(\alpha,\beta)\boldsymbol{\mu})^{\mathrm{T}} (\boldsymbol{Y} - \boldsymbol{Z}(\alpha,\beta)\boldsymbol{\mu}), \qquad (4.111)$$

here $\boldsymbol{Y} = (y(1),\ldots,y(N))^{\mathrm{T}}$, $\boldsymbol{\mu} = (A, B, C, D)^{\mathrm{T}}$ and

$$\boldsymbol{Z}(\alpha,\beta) = \begin{bmatrix} \cos(\alpha) & \sin(\alpha) & \cos(\beta) & \sin(\beta) \\ \vdots & \vdots & \vdots & \vdots \\ \cos(N\alpha) & \sin(N\alpha) & \cos(N^2\beta) & \sin(N^2\beta) \end{bmatrix}.$$

Since $\boldsymbol{\mu}$ is a linear parameter vector, we use separable regression technique of Richards [107] and obtain the LSE of $\boldsymbol{\mu}$ for a given α and β as follows:

$$\widehat{\boldsymbol{\mu}}(\alpha,\beta) = [\boldsymbol{Z}^{\mathrm{T}}(\alpha,\beta)\boldsymbol{Z}(\alpha,\beta)]^{-1}\boldsymbol{Z}(\alpha,\beta)^{\mathrm{T}}\boldsymbol{Y} . \tag{4.112}$$

Now, we denote the LSEs of A, B, C and D as $\widehat{A}(\alpha,\beta)$, $\widehat{B}(\alpha,\beta)$, $\widehat{C}(\alpha,\beta)$ and $\widehat{D}(\alpha,\beta)$, respectively, so that

$$R(\alpha,\beta) = Q(\widehat{A}(\alpha,\beta), \widehat{B}(\alpha,\beta), \alpha, \widehat{C}(\alpha,\beta), \widehat{D}(\alpha,\beta), \beta) =$$
$$= \boldsymbol{Y}^{\mathrm{T}}(\boldsymbol{I} - \boldsymbol{Z}(\alpha,\beta)[\boldsymbol{Z}^{\mathrm{T}}(\alpha,\beta)\boldsymbol{Z}(\alpha,\beta)]^{-1}\boldsymbol{Z}^{\mathrm{T}}(\alpha,\beta))\boldsymbol{Y}.$$

Hence, the LSEs of α^0 and β^0 can be obtained by minimizing $R(\alpha,\beta)$ with respect to α and β. Clearly, it is a two dimensional optimization process. As we have shown before, once the LSEs of α^0 and β^0 are obtained, say $\widehat{\alpha}$ and $\widehat{\beta}$, then the LSEs of A^0, B^0, C^0 and D^0 can be easily obtained as

$$\widehat{A} = \widehat{A}(\widehat{\alpha},\widehat{\beta}), \quad \widehat{B} = \widehat{B}(\widehat{\alpha},\widehat{\beta}), \quad \widehat{C} = \widehat{C}(\widehat{\alpha},\widehat{\beta}), \quad \widehat{D} = \widehat{D}(\widehat{\alpha},\widehat{\beta}),$$

respectively. Therefore, it is observed that although the one component chirp like model has six parameters, the LSEs of its unknown parameters can be obtained by solving a two dimensional optimization problem, similar to the one component chirp model.

The theoretical properties of $\widehat{\boldsymbol{\theta}} = (\widehat{A},\widehat{B},\widehat{\alpha},\widehat{C},\widehat{D},\widehat{\beta})^{\mathrm{T}}$, the LSE of $\boldsymbol{\theta}^0$ are stated in the following theorems. However for proofs, one may refer to Grover [42].

Theorem 18. *Under the assumptions that there exists a K such that $0 < |A^0|, |B^0|, |C^0|, |D^0| < K$ and $e(t)$s are i.i.d. random variables with mean zero and finite variance $\sigma^2 > 0$,*

$$\widehat{\boldsymbol{\theta}} \xrightarrow{a.s.} \boldsymbol{\theta}^0 .$$

Theorem 19. *Under the same set of assumptions as of Theorem 18*

$$\boldsymbol{D}(\widehat{\boldsymbol{\theta}} - \boldsymbol{\theta}^0) \xrightarrow{d} N_6(0,\sigma^2\boldsymbol{\Sigma}),$$

where $\boldsymbol{D} = diag\{\sqrt{N},\sqrt{N},N\sqrt{N},\sqrt{N},\sqrt{N},N^2\sqrt{N}\}$, $\boldsymbol{\Sigma} = \begin{bmatrix} \boldsymbol{\Sigma}^{(1)} & \boldsymbol{0} \\ \boldsymbol{0} & \boldsymbol{\Sigma}^{(2)} \end{bmatrix}$,

and

$$\Sigma^{(1)} = \frac{1}{(A^{0^2} + B^{0^2})} \begin{bmatrix} 2(A^{0^2} + 4B^{0^2}) & -6A^0 B^0 & -12B^0 \\ -6A^0 B^0 & 2(4A^0 + B^0) & 12A^0 \\ -12B^0 & 12A^0 & 24 \end{bmatrix},$$

$$\Sigma^{(2)} = \frac{1}{2(C^{0^2} + D^{0^2})} \begin{bmatrix} 4C^{0^2} + 9D^{0^2} & -5C^0 D^0 & -15D^0 \\ -5C^0 D^0 & 9C^0 + 4D^0 & 15C^0 \\ -15D^0 & 15C^0 & 45 \end{bmatrix}.$$

The theoretical analysis suggests that the behavior of the LSEs of the linear, frequency and chirp parameters are very similar with that of the one component chirp model. Finding the LSEs, however, involves solving a two dimensional optimization problem and as we have seen it in the case of chirp model, to compute accurate initial guesses we need to perform a grid search of the order $O(N^3)$. When N is large, this problem becomes computationally quite challenging. Due to this reason, Grover [42] proposed a sequential method which involves a search of the order $O(N^2 + N)$ to find the initial estimates, and which produces estimators having the same asymptotic properties as the LSEs. We call these estimators as the sequential LSEs.

The sequential procedure is described below. We first provide the following notations for this description. Let us partition the matrix $Z(\alpha, \beta)$ as follows:

$$Z(\alpha, \beta) = \begin{bmatrix} Z^{(1)}(\alpha) & Z^{(2)}(\beta) \end{bmatrix},$$

where

$$Z^{(1)}(\alpha) = \begin{bmatrix} \cos(\alpha) & \dots & \cos(N\alpha) \\ \sin(\alpha) & \dots & \sin(N\alpha) \end{bmatrix}^{\mathrm{T}} \quad and \quad Z^{(2)}(\beta) = \begin{bmatrix} \cos(\beta) & \dots & \cos(N^2\beta) \\ \sin(\beta) & \dots & \sin(N^2\beta) \end{bmatrix}^{\mathrm{T}}.$$

Similarly, the vectors

$$\mu = \begin{pmatrix} \mu^{(1)^{\mathrm{T}}} & \mu^{(2)^{\mathrm{T}}} \end{pmatrix}^{\mathrm{T}}, \quad \theta = \begin{pmatrix} \theta^{(1)^{\mathrm{T}}} & \theta^{(2)^{\mathrm{T}}} \end{pmatrix}^{\mathrm{T}}$$

where $\mu^{(1)} = (A, B)^{\mathrm{T}}$, $\mu^{(2)} = (C, D)^{\mathrm{T}}$, $\theta^{(1)} = (A, B, \alpha)^{\mathrm{T}}$ and $\theta^{(2)} = (C, D, \beta)^{\mathrm{T}}$. Similarly, μ^0 and θ^0 are also partitioned. First obtain estimators of $\mu^{(1)^0}$ and α^0 by minimizing

$$Q_1(\theta^{(1)}) = (Y - Z^{(1)}(\alpha)\mu^{(1)})^{\mathrm{T}}(Y - Z^{(1)}(\alpha)\mu^{(1)}). \tag{4.113}$$

Note that the minimization of $Q_1(\theta^{(1)})$ can be obtained by solving a one dimensional optimization problem. Let the sequential LSEs be denoted by $\widehat{\mu}^{(1)}$ and $\widehat{\alpha}$, i.e.

$$\widehat{\boldsymbol{\theta}}^{(1)} = (\widehat{\boldsymbol{\mu}}^{(1)^{\mathrm{T}}}, \widehat{\alpha})^{\mathrm{T}} = (\widehat{A}, \widehat{B}, \widehat{\alpha})^{\mathrm{T}} = arg\ min\ Q_1(\boldsymbol{\theta}^{(1)})\,.$$

Then at the second stage obtain the new data vector:

$$\boldsymbol{Y}_1 = \boldsymbol{Y} - \boldsymbol{Z}^{(1)}(\widehat{\alpha})\widehat{\boldsymbol{\mu}}^{(1)}\,.$$

Compute the estimators of $\boldsymbol{\mu}^{(2)^0}$ and β^0, by minimizing

$$Q_2(\boldsymbol{\theta}^{(2)}) = (\boldsymbol{Y}_1 - \boldsymbol{Z}^{(2)}(\beta)\boldsymbol{\mu}^{(2)})^{\mathrm{T}}(\boldsymbol{Y}_1 - \boldsymbol{Z}^{(2)}(\beta)\boldsymbol{\mu}^{(2)})\,. \tag{4.114}$$

In this case also the sequential LSEs of $\boldsymbol{\mu}^{(2)^0}$ and β^0 can be obtained by solving a one-dimensional optimization problem. An interesting problem will be to look at the performance of ALSEs of α^0 and β^0 in comparison. These can be obtained by maximizing $I_1(\alpha)$ and $I_2(\beta)$, respectively, where

$$I_1(\alpha) = \frac{1}{N}\left|\sum_{t=1}^{N} y(t)e^{-i\alpha t}\right|^2 \quad \text{and} \quad I_2(\beta) = \frac{1}{N}\left|\sum_{t=1}^{N} y(t)e^{-i\beta t^2}\right|^2\,.$$

Once the ALSEs of α^0 and β^0 are obtained, the corresponding linear parameter estimates can be obtained in explicit form. Therefore, the ALSEs also can be obtained by solving two one dimensional optimization problems, and in this case also finding the initial guesses involves search of the order $O(N^2 + N)$. It is expected that the asymptotic properties of the ALSEs should be same as the LSEs, although it has not been established yet. More work is needed along that direction.

4.8.2. Multicomponent Chirp Like Model

To model real life phenomena efficiently, we need a more general and flexible model. We consider in this section the chirp like model with multiple components (4.107) and discuss some methods for its parameter estimation. The most intuitive estimators will be the LSEs and they can be obtained by minimizing

$$Q(\boldsymbol{\theta}) = \sum_{t=1}^{N}\left(y(t) - \sum_{j=1}^{p}\{A_j\cos(\alpha_j t) + B_j\sin(\alpha_j t)\} - \sum_{k=1}^{q}\{C_k\cos(\beta_k t^2) + D_k\sin(\beta_k t^2)\}\right)^2, \tag{4.115}$$

with respect to the unknown parameters

$$\boldsymbol{\theta} = (A_1, B_1, \alpha_1, \ldots, A_p, B_p, \alpha_p, C_1, D_1, \beta_1, \ldots, C_q, D_q, \beta_q)^{\mathrm{T}}\,.$$

It can be easily seen as before, that the LSEs of $\boldsymbol{\theta}$ can be obtained by solving a $p+q$ dimensional optimization problem. Grover [42] established the asymptotic properties of

$\widehat{\boldsymbol{\theta}}$, the LSE of $\boldsymbol{\theta}^0$, the vector of true parameter values of the model. These are indicated in the following theorems.

Theorem 20. *Under the assumptions that there exists a K such that*

$$0 <\mid A_1^0 \mid,\mid B_1^0 \mid,\ldots,\mid A_p^0 \mid,\mid B_p^0 \mid,\mid C_1^0 \mid,\mid D_1^0 \mid,\ldots,\mid C_q^0 \mid,\mid D_q^0 \mid< K,$$

and $e(t)$s are i.i.d. random variables with mean zero and finite variance $\sigma^2 > 0$, then

$$\widehat{\boldsymbol{\theta}} \xrightarrow{a.s.} \boldsymbol{\theta}^0 .$$

Theorem 21. *Under the same set of assumptions as in Theorem 20,*

$$\boldsymbol{D}\left(\widehat{\boldsymbol{\theta}} -\boldsymbol{\theta}^0\right) \xrightarrow{d} N_{3(p+q)}\left(\boldsymbol{0},\sigma^2\boldsymbol{\Sigma}^{-1}\right) .$$

Here $\boldsymbol{D}=diag(\underbrace{\boldsymbol{D}_1,\ldots,\boldsymbol{D}_1}_{ptimes},\underbrace{\boldsymbol{D}_2,\ldots,\boldsymbol{D}_2}_{qtimes})$, with

$$\boldsymbol{D}_1=diag\left(\sqrt{N},\sqrt{N},N\sqrt{N}\right), \boldsymbol{D}_2 =diag\left(\sqrt{N},\sqrt{N},N^2\sqrt{N}\right),$$

$$\boldsymbol{\Sigma}=\begin{bmatrix} \boldsymbol{\Sigma}_1^{(1)} & 0 & \cdots & \cdots & \cdots & 0 \\ 0 & \ddots & 0 & \cdots & \cdots & 0 \\ \vdots & \vdots & \boldsymbol{\Sigma}_p^{(1)} & 0 & \cdots & 0 \\ 0 & \cdots & 0 & \boldsymbol{\Sigma}_1^{(2)} & \cdots & 0 \\ 0 & \cdots & \cdots & 0 & \ddots & 0 \\ 0 & \cdots & \cdots & \cdots & 0 & \boldsymbol{\Sigma}_q^{(2)} \end{bmatrix},$$

and for $j=1,\ldots,p,\ k=1,\ldots,q,$

$$\boldsymbol{\Sigma}_j^{(1)}=\begin{bmatrix} \dfrac{1}{2} & 0 & \dfrac{B_j^0}{4} \\ 0 & \dfrac{1}{2} & -\dfrac{A_j^0}{4} \\ \dfrac{B_j^0}{4} & -\dfrac{A_j^0}{4} & \dfrac{A_j^{0^2}+B_j^{0^2}}{6} \end{bmatrix}, \ \text{and} \ \boldsymbol{\Sigma}_k^{(2)}=\begin{bmatrix} \dfrac{1}{2} & 0 & \dfrac{D_k^0}{6} \\ 0 & \dfrac{1}{2} & -\dfrac{C_k^0}{6} \\ \dfrac{D_k^0}{6} & -\dfrac{C_k^0}{6} & \dfrac{C_k^{0^2}+D_k^{0^2}}{10} \end{bmatrix} .$$

Again to avoid solving $(p+q)$ dimensional optimization problem, Grover [42] proposed sequential LSEs for this model as well. The proposed sequential procedure works on the same principle as the one described in Section 4.8.1 with some obvious modifications.

Based on the number theoretic Conjecture 1, it has been shown that the LSEs and sequential LSEs are asymptotically equivalent. Note that the sequential LSEs can be obtained by solving $(p+q)$ one dimensional optimization problems, thus lowering the computational complexity involved in computing the LSEs significantly. Grover [42] performed extensive simulation experiments to observe the performances of the sequential estimators for multicomponent chirp-like model for known p and q. She has estimated first p sinusoidal components, then q chirp components. The performances of the sequential estimators are quite satisfactory in the sense that the sample variances are quite close to the asymptotic variances even for moderate sample sizes. Grover [42] fitted multicomponent chirp-like model to several data sets, where p and q are not known. She has used Bayesian Information Criterion (BIC) to estimate p and q effectively.

In case of the multicomponent chirp like model also, the ALSEs can be examined. Following a sequential procedure as described above, the ALSEs can be obtained by solving $(p+q)$ one dimensional optimization problems. Again, it is expected that the ALSEs will be equivalent asymptotically with the LSEs, although it has not been established and is an open problem. It will be interesting to see the performance of the ALSEs and to explore how they compare with the existing least squares method for this model.

4.9. Harmonic Chirp Model

The harmonic chirp model is a special case of the chirp model, where the frequencies and frequency rates instead of being arbitrary, are harmonics of a common constant fundamental frequency and common constant fundamental frequency rate, respectively. The model can be expressed mathematically as follows:

$$y(t) = \sum_{k=1}^{p} \left\{ A_k^0 \cos(k\alpha^0 t + k\beta^0 t^2) + B_k^0 \sin(k\alpha^0 t + k\beta^0 t^2) \right\} + e(t); \quad t = 1, \ldots, N. \quad (4.116)$$

Here A_k^0 s and B_k^0 s are arbitrary real numbers, $0 < \alpha^0 < \pi$ is the fundamental frequency and $0 < \beta^0 < \pi$ is the fundamental frequency rate of the observed signal $y(t)$, and $e(t)$s are real valued additive error with mean zero and finite variance.

It may be mentioned that the problem of estimating the fundamental frequency of harmonic time stationary sinusoids has several applications in speech processing, communications, radar and sonar, biomedical systems, electric power, and semiconductor devices, see for example Christensen et al. [18, 19], Eddy [31] and Li, Stoica and Li [80]. An extensive amount of work has been done in developing efficient estimators and developing their properties both in the Statistical and Signal Processing literature, see for example Quinn and Thomson [105], Irizarry [55], Nandi and Kundu [92, 95], Chang and Chen [16], Jain and Singh [57] and the references cited therein.

But in many other applications it is observed that the signal is more appropriately modeled as a sum of harmonic components of a non-stationary signal, i.e. a signal that has its frequency content varying with time. In active transmission used in tissue harmonic imaging in ultrasound or by mammals, namely bats, dolphins, whales, the signal is deliberately transmitted as a sum of harmonic linear frequency modulated chirps to increase the detectability of the source of interest, i.e. an organ in ultrasound or a prey in case of mammals, see for example Vespe, Jones and Baker [114], Kopsinis et al. [62].

Doweck, Amar and Cohen [29] considered model (4.116) when the errors are i.i.d. Gaussian random variables with mean zero and finite variance. They have discussed about the MLEs of the unknown parameters assuming the number of components p to be known, and then using BIC criterion they have estimated p as well. They have obtained the Cramer-Rao lower bound and discussed several computational issues associated with this problem. They have also used product high order ambiguity function method to estimate the unknown parameters, but did not provide any theoretical justifications. Thus, developing theoretical properties of these estimators remains an open problem.

Grover [42] considered the LSEs and ALSEs of the unknown parameters under a fairly general set of assumptions on the additive errors $e(t)$s and provided the theoretical properties of these estimators. However, before providing the exact results let us define the following notations: $\boldsymbol{\theta} = (A_1, B_1, \ldots, A_p, B_p, \alpha, \beta)^{\mathrm{T}}$. The true parameter value will be denoted by $\boldsymbol{\theta}^0$. The LSEs of the unknown parameters can be obtained by minimizing

$$Q(\boldsymbol{\theta}) = \sum_{t=1}^{N}\left(y(t) - \sum_{k=1}^{p}\left\{ A_k \cos(k\alpha t + k\beta t^2) + B_k \sin(k\alpha t + k\beta t^2)\right\}\right)^2 \quad (4.117)$$

with respect to $\boldsymbol{\theta}$. Using matrix notation (4.117) can be written as

$$Q(\boldsymbol{\theta}) = \left(\boldsymbol{Y} - \boldsymbol{Z}(\alpha, \beta)\boldsymbol{\phi}\right)^{\mathrm{T}}\left(\boldsymbol{Y} - \boldsymbol{Z}(\alpha, \beta)\boldsymbol{\phi}\right). \quad (4.118)$$

Here $\boldsymbol{Y} = (y(1), \ldots, y(N))^{\mathrm{T}}$ is the observed data vector, $\boldsymbol{\phi} = (A_1, B_1, \ldots, A_p, B_p)^{\mathrm{T}}$ is the vector of linear parameters and the matrix $\boldsymbol{Z}(\alpha, \beta)$ is defined as

$$Z(\alpha, \beta) = \begin{bmatrix} \cos(\alpha + \beta) & \sin(\alpha + \beta) & \ldots & \cos(p\alpha + p\beta) & \sin(p\alpha + p\beta) \\ \vdots & \vdots & \ldots & \vdots & \vdots \\ \cos(N\alpha + N^2\beta) & \sin(N\alpha + N^2\beta) & \ldots & \cos(pN\alpha + pN^2\beta) & \sin(pN\alpha + pN^2\beta) \end{bmatrix}. \quad (4.119)$$

Since $\boldsymbol{\phi}$ is a vector of linear parameters, for given α and β, they can be estimated using simple linear regression technique as discussed before:

$$\hat{\boldsymbol{\phi}}(\alpha, \beta) = \left[\boldsymbol{Z}^{\mathrm{T}}(\alpha, \beta)\boldsymbol{Z}(\alpha, \beta)\right]^{-1}\boldsymbol{Z}^{\mathrm{T}}(\alpha, \beta)\boldsymbol{Y}. \quad (4.120)$$

Now the LSEs of α^0 and β^0 can be obtained by minimizing

$$R(\alpha,\beta) = \left(\boldsymbol{Y} - \boldsymbol{Z}(\alpha,\beta)\widehat{\phi}(\alpha,\beta)\right)^{\mathrm{T}} \left(\boldsymbol{Y} - \boldsymbol{Z}(\alpha,\beta)\widehat{\phi}(\alpha,\beta)\right)$$
$$= \boldsymbol{Y}^{\mathrm{T}}(\boldsymbol{I} - \boldsymbol{Pz}(\alpha,\beta))\boldsymbol{Y}, \tag{4.121}$$

where $\boldsymbol{Pz}(\alpha,\beta) = \boldsymbol{Z}(\alpha,\beta)[\boldsymbol{Z}^{\mathrm{T}}(\alpha,\beta)\boldsymbol{Z}(\alpha,\beta)]^{-1}\boldsymbol{Z}^{\mathrm{T}}(\alpha,\beta)$ is the projection matrix on the column space of the matrix $\boldsymbol{Z}(\alpha,\beta)$. Let us denote the obtained estimators as $\hat{\alpha}$ and $\hat{\beta}$, respectively, i.e.

$$(\hat{\alpha},\hat{\beta})^{\mathrm{T}} = arg\ min\, R(\alpha,\beta), \tag{4.122}$$

so that the LSE of $\boldsymbol{\phi}$ becomes

$$\widehat{\phi} = \widehat{\phi}(\hat{\alpha},\hat{\beta}) = \left[\boldsymbol{Z}^{\mathrm{T}}(\hat{\alpha},\hat{\beta})\boldsymbol{Z}(\hat{\alpha},\hat{\beta})\right]^{-1}\boldsymbol{Z}^{\mathrm{T}}(\hat{\alpha},\hat{\beta})\boldsymbol{Y}. \tag{4.123}$$

Grover [42] obtained the following result on the consistency of the LSEs.

Theorem 22. *It is assumed that there exits a K, such that* $0 < |A_1^0|^2 + |B_1^0|^2 < \ldots$ $< |A_p^0|^2 + |B_p^0|^2 < K,\quad 0 < \alpha^0, \beta^0 < \dfrac{\pi}{p}$ *and the error random variables $e(t)$s satisfy Assumption 4. Then, $\widehat{\boldsymbol{\theta}}$, the LSE of $\boldsymbol{\theta}^0$, is strongly consistent i.e.*

$$\widehat{\boldsymbol{\theta}} \xrightarrow{a.s.} \boldsymbol{\theta}^0.$$

The asymptotic distribution of the LSEs also has been derived as well, and it is stated in the following theorem.

Theorem 23. *Under the same set of assumptions as in Theorem 22*

$$\boldsymbol{D}(\widehat{\boldsymbol{\theta}} - \boldsymbol{\theta}^0) \xrightarrow{d} N_{2p+2}(\boldsymbol{0}, 2c\sigma^2\boldsymbol{\Sigma}).$$

Here $\boldsymbol{D} = diag\{\underbrace{\sqrt{N},\ldots,\sqrt{N}}_{2p\,times}, N\sqrt{N}, N^2\sqrt{N}\},\ c = \sum_{j=-\infty}^{\infty} a^2(j),$ *the matrix*

$$\boldsymbol{\Sigma}^{-1} = \begin{bmatrix} \boldsymbol{I} & \boldsymbol{\Sigma}_{12} \\ \boldsymbol{\Sigma}_{21} & \boldsymbol{\Sigma}_{22} \end{bmatrix},$$

where $\boldsymbol{I}$ is a $2p \times 2p$ identity matrix, $\boldsymbol{\Sigma}_{12}$ is a $2p \times 2$ matrix, $\boldsymbol{\Sigma}_{22}$ is a $2p \times 2p$ matrix as given below:

$$\boldsymbol{\Sigma}_{21} = \begin{bmatrix} \dfrac{B_1^0}{2} & \cdots & \dfrac{pB_p^0}{2} & -\dfrac{A_1^0}{2} & \cdots & -\dfrac{pA_p^0}{2} \\[2ex] \dfrac{B_1^0}{3} & \cdots & \dfrac{pB_p^0}{3} & -\dfrac{A_1^0}{3} & \cdots & -\dfrac{pA_p^0}{3} \end{bmatrix},$$

$$\boldsymbol{\Sigma}_{12} = \boldsymbol{\Sigma}_{21}^{\mathrm{T}} \; and \; \boldsymbol{\Sigma}_{22} = \begin{bmatrix} \dfrac{\sum\limits_{k=1}^{p} k^2 (A_k^{0^2} + B_k^{0^2})}{3} & \dfrac{\sum\limits_{k=1}^{p} k^2 (A_k^{0^2} + B_k^{0^2})}{4} \\[4ex] \dfrac{\sum\limits_{k=1}^{p} k^2 (A_k^{0^2} + B_k^{0^2})}{4} & \dfrac{\sum\limits_{k=1}^{p} k^2 (A_k^{0^2} + B_k^{0^2})}{5} \end{bmatrix}.$$

Grover [42] also proposed ALSEs of the unknown parameters of the underlying model. These can be obtained by maximizing a periodogram type function $I(\alpha, \beta)$, where

$$I(\alpha, \beta) = \frac{1}{N} \left| \sum_{t=1}^{N} y(t) e^{-i(\alpha t + \beta t^2)} \right|^2 = \frac{1}{N} \boldsymbol{Y}^{\mathrm{T}} \boldsymbol{Z}(\alpha, \beta) \boldsymbol{Z}^{\mathrm{T}}(\alpha, \beta) \boldsymbol{Y}. \qquad (4.124)$$

Evidently, the maximization of $I(\alpha, \beta)$ can be obtained by solving a two dimensional optimization problem. If $\hat{\alpha}$ and $\hat{\beta}$ maximize $I(\alpha, \beta)$, i.e.

$$(\hat{\alpha}, \hat{\beta})^{\mathrm{T}} = arg\ max\ I(\alpha, \beta),$$

then the corresponding ALSE of $\boldsymbol{\phi}$ can be obtained as

$$\hat{\phi} = \frac{2}{N} \boldsymbol{Z}^{\mathrm{T}}(\hat{\alpha}, \hat{\beta}) \boldsymbol{Y}.$$

Under the same set of assumptions as in Theorem 22, Grover [42] showed that the LSEs and ALSEs are asymptotically equivalent, i.e. ALSEs are also consistent and they have the same asymptotic distribution as the LSEs. Based on extensive simulations by Grover [42] it has been observed that although LSEs and ALSEs are asymptotically equivalent, the performances of the LSEs are slightly better than the ALSEs mainly for small and moderate sample sizes in terms of lower mean squared errors.

4.10. Two Dimensional Chirp Model

So far we have mainly talked about one dimensional chirp model and some of its variants. We will now proceed to discuss about the two dimensional (2-D) chirp model in presence of additive errors and some of its variants. Mathematically a 2-D chirp model can be expressed as follows:

$$y(m,n) = \sum_{k=1}^{p} \left\{ A_k^0 \cos(\alpha_k^0 m + \beta_k^0 m^2 + \gamma_k^0 n + \delta_k^0 n^2) + B_k^0 \sin(\alpha_k^0 m + \beta_k^0 m^2 + \gamma_k^0 n + \delta_k^0 n^2) \right\}$$

$$+ e(m,n); \; m = 1,\ldots,M; \; n = 1,\ldots,N.$$

$$(4.125)$$

Here, $y(m,n)$ is the observed signal in two dimension, A_k^0 s, B_k^0 s are real numbers, α_k^0 s, γ_k^0 s are frequencies, and β_k^0 s, δ_k^0 s are frequency rates. The random component $e(m,n)$ s are real valued random variables with mean zero and finite variance. The more explicit assumptions on $e(m,n)$ s will be stated later.

The 2-D chirp model is a natural generalization of the 2-D sinusoidal model which can be described as follows:

$$y(m,n) = \sum_{k=1}^{p} \left\{ A_k^0 \cos(\alpha_k^0 m + \gamma_k^0 n) + B_k^0 \sin(\alpha_k^0 m + \gamma_k^0 n) \right\} + e(m,n);$$

$$m = 1,\ldots,M; \; n = 1,\ldots,N.$$

$$(4.126)$$

The above model has received considerable amount of attention in the signal processing literature because of its many applications in 'Multidimensional Signal Processing'. This is one of the basic models in many fields such as antenna array processing, geophysical perception, biomedical spectral analysis etc. See for example the work by Barbieri and Barone [6], Cabrera and Bose [13], Chun and Bose [20], Hua [51] and see the references cited therein. This problem also has a special scope in spectrography, and it has been studied quite rigorously by Malliavan [84, 85].

Apart from being a generalization of the 2-D sinusoidal model, naturally the above 2-D chirp model (4.125) is also an extension of the 1-D chirp model. Many 2-D signal processing applications require modeling and analysis of non-homogeneous signals. In fact, almost any application where image interpretation is required, has to face challenge of analyzing heterogeneity in the observed image. Model (4.125) and some of its variants have been used in modeling and analyzing magnetic resonance imaging (MRI), optical imaging and different texture imaging. It has been also used in modeling black and white 'gray' images, and to analyze finger print images data. This model has a wide applications in modeling SAR data and in particular Interferometric SAR data. See for example, Pelag and Porat [98], Hedley and Rosenfeld [49], Friedlander and Francos [38], Francos and Friedlander [35, 36], Cao, Wang and Wang [14], Zhang and Liu [129], Zhang, Wang and Cao [130], and see the references cited therein.

In the subsequent subsections we mainly discuss several estimation procedures and their properties for a single component 2-D chirp model. However, all the results can be easily generalized for multiple 2-D chirp model using sequential procedures. These sequential methods can be developed following the same pattern as described before for 1-D chirp model.

4.10.1. Least Squares Estimators

The one component 2-D chirp model can be written as follows:

$$y(m,n) = A^0 \cos(\alpha^0 m + \beta^0 m^2 + \gamma^0 n + \delta^0 n^2) + B^0 \sin(\alpha^0 m + \beta^0 m^2 + \gamma^0 n + \delta^0 n^2)$$
$$+ e(m,n); m = 1,\ldots,M; n = 1,\ldots,N. \tag{4.127}$$

This model can be obtained as a particular case of model (4.125) when $p = 1$. The problem is to find the estimators of the unknown parameters based on the observed data $y(m,n)$, under a suitable error assumption. At first we assume that the error components $e(m,n)$s are i.i.d. random variables with mean zero and finite variance σ^2, for $m = 1,\ldots,M$ and $n = 1,\ldots,N$. More general error assumptions will be considered in the subsequent sections.

Based on the assumptions that the errors are i.i.d. random variables the most reasonable estimators will be the conventional LSEs, and they can be obtained by minimizing the following error sums of squares, i.e.

$$Q(\boldsymbol{\theta}) = \sum_{m=1}^{M}\sum_{n=1}^{N}\left(y(m,n) - A\cos(\alpha m + \beta m^2 + \gamma n + \delta n^2) - B\sin(\alpha m + \beta m^2 + \gamma n + \delta n^2)\right)^2,$$

$$\tag{4.128}$$

where $\boldsymbol{\theta} = (A,B,\alpha,\beta,\gamma,\delta)^{\mathrm{T}}$. The LSEs of the unknown parameters $\boldsymbol{\theta}$, say $\hat{\boldsymbol{\theta}}$, can be obtained as the argument minimum of $Q(\boldsymbol{\theta})$, i.e.

$$\hat{\boldsymbol{\theta}} = (\hat{A},\hat{B},\hat{\alpha},\hat{\beta},\hat{\gamma},\hat{\delta})^{\mathrm{T}} = arg\ min\ Q(\boldsymbol{\theta}).$$

As expected, the LSEs cannot be obtained in explicit forms and one needs to use some numerical techniques to compute them. Newton-Raphson, Gauss-Newton, Genetic algorithm or simulated annealing method may be used for this purpose. Alternatively, the method suggested by Saha and Kay [109] as described in Section 4.4.1 may be used to find the LSEs of the unknown parameters. Lahiri [74], see also Lahiri and Kundu [75] in this respect, established the asymptotic properties of the LSEs under a fairly general set of conditions. It has been shown that the LSEs are strongly consistent and they are asymptotically normally distributed. The results are provided in the theorems below.

Theorem 24. *If there exists a K, such that $0 < |A^0| + |B^0| < K$, $0 < \alpha^0, \beta^0, \gamma^0, \delta^0 < \pi$, and $\sigma^2 > 0$, then $\hat{\boldsymbol{\theta}} = (\hat{A},\hat{B},\hat{\alpha},\hat{\beta},\hat{\gamma},\hat{\delta})^{\mathrm{T}}$ is a strongly consistent estimate of $\boldsymbol{\theta}^0 = (A^0, B^0, \alpha^0, \beta^0, \gamma^0, \delta^0)^{\mathrm{T}}$.*

Theorem 25. *Under the same assumptions as in Theorem 24, if we denote $\boldsymbol{D}$ as a 6×6 diagonal matrix as*

$$\boldsymbol{D} = diag\left\{M^{1/2}N^{1/2}, M^{1/2}N^{1/2}, M^{3/2}N^{1/2}, M^{5/2}N^{1/2}, M^{1/2}N^{3/2}, M^{1/2}N^{5/2}\right\},$$

then $\mathbf{D}(\widehat{A}-A^0,\widehat{B}-B^0,\widehat{\alpha}-\alpha^0,\widehat{\beta}-\beta^0,\widehat{\gamma}-\gamma^0,\widehat{\delta}-\delta^0)^{\mathrm{T}}\overset{d}{\to}N_6(0,2\sigma^2\mathbf{\Sigma}),$

where

$$\mathbf{\Sigma}^{-1}=\begin{bmatrix} 1 & 0 & \dfrac{B^0}{2} & \dfrac{B^0}{3} & \dfrac{B^0}{2} & \dfrac{B^0}{3} \\[2ex] 0 & 1 & -\dfrac{A^0}{2} & -\dfrac{A^0}{3} & -\dfrac{A^0}{2} & -\dfrac{A^0}{3} \\[2ex] \dfrac{B^0}{2} & -\dfrac{A^0}{2} & \dfrac{A^{0^2}+B^{0^2}}{3} & \dfrac{A^{0^2}+B^{0^2}}{4} & \dfrac{A^{0^2}+B^{0^2}}{4} & \dfrac{A^{0^2}+B^{0^2}}{4.5} \\[2ex] \dfrac{B^0}{3} & -\dfrac{A^0}{3} & \dfrac{A^{0^2}+B^{0^2}}{4} & \dfrac{A^{0^2}+B^{0^2}}{5} & \dfrac{A^{0^2}+B^{0^2}}{4.5} & \dfrac{A^{0^2}+B^{0^2}}{5} \\[2ex] \dfrac{B^0}{2} & -\dfrac{A^0}{2} & \dfrac{A^{0^2}+B^{0^2}}{4} & \dfrac{A^{0^2}+B^{0^2}}{4.5} & \dfrac{A^{0^2}+B^{0^2}}{3} & \dfrac{A^{0^2}+B^{0^2}}{4} \\[2ex] \dfrac{B^0}{3} & -\dfrac{A^0}{3} & \dfrac{A^{0^2}+B^{0^2}}{4.5} & \dfrac{A^{0^2}+B^{0^2}}{5} & \dfrac{A^{0^2}+B^{0^2}}{4} & \dfrac{A^{0^2}+B^{0^2}}{5} \end{bmatrix}.$$

Note that the above asymptotic result (Theorem 25) can be used to construct confidence intervals of the unknown parameters. They can be used for testing of hypothesis problem also. When the errors are assumed to be i.i.d. Gaussian errors, the above LSEs are the MLEs. In fact, the LSEs are consistent and asymptotically normally distributed even for a more general set of error assumptions. These are stated below.

Assumption 14. *The error component* $e(m,n)$ *has the following form:*

$$e(m,n)=\sum_{j=-\infty}^{\infty}\sum_{k=-\infty}^{\infty}a(j,k)X(m-j,n-k),$$

with $\displaystyle\sum_{j=-\infty}^{\infty}\sum_{k=-\infty}^{\infty}|a(j,k)|<\infty$, *where* $\{X(m,n)\}$ *is a double array sequence of i.i.d. random variables with mean zero, variance* σ^2, *and with finite fourth moment.*

Under Assumption 14, Lahiri [74] proved that the LSE of θ^0 is strongly consistent. Furthermore, the LSEs are shown to be asymptotically normally distributed and the associated result is stated in the theorem below.

Theorem 26. *Under the same assumptions as in Theorem 25 and Assumption 14,*

$$\mathbf{D}(\widehat{\boldsymbol{\theta}}-\boldsymbol{\theta}^0)\overset{d}{\to}N_6(0,2c\sigma^2\mathbf{\Sigma}),$$

where matrices $\mathbf{D}$ *and* $\mathbf{\Sigma}$ *are same as defined in Theorem 25 and* $\displaystyle c=\sum_{j=-\infty}^{\infty}\sum_{k=-\infty}^{\infty}a^2(j,k).$

4.10.2. Approximate Least Squares Estimators

As an alternative to the LSEs, Grover, Kundu and Mitra [44] considered ALSEs which can be obtained by maximizing a 2-D periodogram type function defined as follows:

$$I(\alpha,\beta,\gamma,\delta) = \frac{2}{MN}\left\{\left(\sum_{m=1}^{M}\sum_{n=1}^{N}y(m,n)\cos(\alpha m + \beta m^2 + \gamma n + \delta n^2)\right)^2 \right.$$
$$\left. + \left(\sum_{m=1}^{M}\sum_{n=1}^{N}y(m,n)\sin(\alpha m + \beta m^2 + \gamma n + \delta n^2)\right)^2\right\}. \tag{4.129}$$

The main idea about the above 2-D periodogram type function has been obtained from the periodogram estimator in case the 2-D sinusoidal model and the former is a natural extension of the latter.

The 2-D periodogram type estimators or the ALSEs of the 2-D chirp model can be obtained by the argument maximum of $I(\alpha,\beta,\gamma,\delta)$ given in (4.129) over the range $(0,\pi) \times (0,\pi) \times (0,\pi) \times (0,\pi)$. Since, the closed-form solution of this problem cannot be obtained analytically, efficient numerical methods are required to compute them. Extensive simulation experiments have been performed by Grover, Kundu and Mitra [44], and it has been observed that the Downhill-Simplex method performs quite well to compute the ALSEs, provided the initial guesses are close to the true values. It has been observed in simulation studies that although both LSEs and ALSEs involve solving four dimensional optimization problem, computational time of the ALSEs is significantly lower than the LSEs. They also show that the LSEs and ALSEs are asymptotically equivalent. Based on extensive simulation results, it is observed by Grover, Kundu and Mitra [44] that the performances of the LSEs are slightly better than the ALSEs for small and moderate sample sizes in terms of biases and lower mean squared errors.

Since the computation of the LSEs or the ALSEs is a challenging problem, several other computationally efficient methods are devised in the literature. But unfortunately in most the cases, either the asymptotic properties are unknown or they do not have the same efficiency as the LSEs or ALSEs. In the next two subsections, we summarize two estimators which have the same asymptotic efficiency as the LSEs or ALSEs and at the same time both of them can be computed more efficiently than the LSEs or the ALSEs.

4.10.3. 2-D Finite Step Efficient Algorithm

Lahiri, Kundu and Mitra [77] proposed a finite step efficient algorithm for 2-D one component chirp signal model. It is observed that if we start with the initial guesses of α^0 and γ^0 having convergence rates $O_p(M^{-1}N^{-1/2})$ and $O_p(N^{-1}M^{-1/2})$, respectively, and β^0 and δ^0 having convergence rates $O_p(M^{-2}N^{-1/2})$ and $O_p(N^{-2}M^{-1/2})$, respectively, then after four iterations the algorithm produces estimates of α^0 and γ^0 having convergence rates $O_p(M^{-3/2}N^{-1/2})$ and $O_p(N^{-3/2}M^{-1/2})$, respectively, and β^0 and δ^0 having convergence rates $O_p(M^{-5/2}N^{-1/2})$ and $O_p(N^{-5/2}M^{-1/2})$, respectively. Therefore, the efficient algorithm produces estimates with the same rates of convergence as the LSEs or the ALSEs. Moreover, it is also guaranteed that the algorithm stops after four iterations.

Before providing the algorithm in detail, we introduce some notations and preliminary results required for further development. If $\tilde{\alpha}$ is an estimator of α^0 such that $\tilde{\alpha} - \alpha^0 = O_p(M^{(-1-\lambda_{11})}N^{-\lambda_{12}})$, for some $0 < \lambda_{11}, \lambda_{12} \leq 1/2$, and $\tilde{\beta}$ is an estimator of β^0 such that $\tilde{\beta} - \beta^0 = O_p(M^{(-2-\lambda_{21})}N^{-\lambda_{22}})$, for some $0 < \lambda_{21}, \lambda_{22} \leq 1/2$, then an improved estimator of α^0 can be obtained as

$$\tilde{\tilde{\alpha}} = \tilde{\alpha} + \frac{48}{M^2} Im\left(\frac{P_{MN}^{\alpha}}{Q_{MN}^{\alpha,\beta}}\right), \tag{4.130}$$

with

$$P_{MN}^{\alpha} = \sum_{n=1}^{N}\sum_{m=1}^{M} y(m,n)\left(m - \frac{M}{2}\right)e^{-i(\tilde{\alpha}m+\tilde{\beta}m^2)}, \tag{4.131}$$

$$Q_{MN}^{\alpha,\beta} = \sum_{n=1}^{N}\sum_{m=1}^{M} y(m,n)e^{-i(\tilde{\alpha}m+\tilde{\beta}m^2)}. \tag{4.132}$$

Similarly, an improved estimator of β^0 can be obtained as

$$\tilde{\tilde{\beta}} = \tilde{\beta} + \frac{45}{M^4} Im\left(\frac{P_{MN}^{\beta}}{Q_{MN}^{\alpha,\beta}}\right), \tag{4.133}$$

with

$$P_{MN}^{\beta} = \sum_{n=1}^{N}\sum_{m=1}^{M} y(m,n)\left(m^2 - \frac{M^2}{3}\right)e^{-i(\tilde{\alpha}m+\tilde{\beta}m^2)}, \tag{4.134}$$

and $Q_{MN}^{\alpha,\beta}$ is same as defined above in (4.132).

The following two theorems provide the theoretical justification for the improved estimators. For the proofs, however, the reader is referred to the article by Lahiri, Kundu and Mitra [77].

Theorem 27. *If $\tilde{\alpha} - \alpha^0 = O_p(M^{-1-\lambda_{11}}N^{-\lambda_{12}})$ for $\lambda_{11}, \lambda_{12} > 0$, then*

$$(a)\ (\widetilde{\widetilde{\alpha}} - \alpha^0) = O_p(M^{-1-2\lambda_{11}} N^{-\lambda_{12}})\quad if\quad \lambda_{11} \leq 1/4,$$

$$(b)\ M^{3/2} N^{1/2} (\widetilde{\widetilde{\alpha}} - \alpha^0) \xrightarrow{d} N(0, \sigma_1^2)\quad if\quad \lambda_{11} > 1/4, \lambda_{12} = 1/2,$$

where $\sigma_1^2 = \dfrac{384 c \sigma^2}{A^{0^2} + B^{0^2}}$, *the asymptotic variance of the LSE of* α^0, *and* c *is same as defined in Theorem 26.*

Theorem 28. *If* $\widetilde{\beta} - \beta^0 = O_p(M^{-2-\lambda_{21}} N^{-\lambda_{22}})$ *for* $\lambda_{21}, \lambda_{22} > 0$, *then*

$$(a)\ (\widetilde{\widetilde{\beta}} - \beta^0) = O_p(M^{-2-2\lambda_{21}} N^{-\lambda_{22}})\quad if\quad \lambda_{21} \leq 1/4,$$

$$(b)\ M^{5/2} N^{1/2} (\widetilde{\widetilde{\beta}} - \beta^0) \xrightarrow{d} N(0, \sigma_2^2)\quad if\quad \lambda_{21} > 1/4, \lambda_{22} = 1/2,$$

where $\sigma_2^2 = \dfrac{360 c \sigma^2}{A^{0^2} + B^{0^2}}$, *the asymptotic variance of the LSE of* β^0, *and* c *is same as defined in the previous theorem.*

In order to find estimators of γ^0 and δ^0, one can interchange the roles of M and N. If $\widetilde{\gamma}$ is an estimator of γ^0 such that $\widetilde{\gamma} - \gamma^0 = O_p(N^{(-1-\kappa_{11})} M^{-\kappa_{12}})$, for some $0 < \kappa_{11}, \kappa_{12} \leq 1/2$, and $\widetilde{\delta}$ is an estimator of δ^0 such that $\widetilde{\delta} - \delta^0 = O_p(N^{(-2-\kappa_{21})} M^{-\kappa_{22}})$, for some $0 < \kappa_{21}, \kappa_{22} \leq 1/2$, then an improved estimator of γ^0 can be obtained as

$$\widetilde{\widetilde{\gamma}} = \widetilde{\gamma} + \frac{48}{N^2} Im\left(\frac{P_{MN}^{\gamma}}{Q_{MN}^{\gamma,\delta}}\right), \tag{4.135}$$

with

$$P_{MN}^{\gamma} = \sum_{n=1}^{N} \sum_{m=1}^{M} y(m,n)\left(n - \frac{N}{2}\right) e^{-i(\widetilde{\gamma} n + \widetilde{\delta} n^2)}, \tag{4.136}$$

$$Q_{MN}^{\gamma,\delta} = \sum_{n=1}^{N} \sum_{m=1}^{M} y(m,n) e^{-i(\widetilde{\gamma} n + \widetilde{\delta} n^2)}, \tag{4.137}$$

and an improved estimator of δ^0 can be obtained as

$$\widetilde{\widetilde{\delta}} = \widetilde{\delta} + \frac{45}{N^4} Im\left(\frac{P_{MN}^{\delta}}{Q_{MN}^{\gamma,\delta}}\right), \tag{4.138}$$

with

$$P_{MN}^{\delta} = \sum_{n=1}^{N}\sum_{m=1}^{M} y(m,n)\left(n^2 - \frac{N^2}{3}\right) e^{-i(\tilde{\gamma}n + \tilde{\delta}n^2)}, \tag{4.139}$$

and $Q_{MN}^{\gamma,\delta}$ is same as defined in (4.137).

Again, the justification for the improved estimators is provided in the following theorems but the details of the proofs can be obtained from Lahiri, Kundu and Mitra [77].

Theorem 29. *If* $\tilde{\gamma} - \gamma^0 = O_p(N^{-1-\kappa_{11}} M^{-\kappa_{12}})$ *for* $\kappa_{11}, \kappa_{12} > 0,$ *then*

$$(a)(\tilde{\tilde{\gamma}} - \gamma^0) = O_p(N^{-1-2\kappa_{11}} M^{-\kappa_{12}}) \quad if \quad \kappa_{11} \leq 1/4,$$

$$(b) N^{3/2} M^{1/2} (\tilde{\tilde{\gamma}} - \gamma^0) \xrightarrow{d} N(0,\sigma_1^2) \quad if \quad \kappa_{11} > 1/4, \kappa_{12} = 1/2.$$

Here σ_1^2 *and* c *are same as defined in Theorem 27.*

Theorem 30. *If* $\tilde{\delta} - \delta^0 = O_p(N^{-2-\kappa_{21}} M^{-\kappa_{22}})$ *for* $\kappa_{21}, \kappa_{22} > 0,$ *then*

$$(a)(\tilde{\tilde{\delta}} - \delta^0) = O_p(N^{-2-2\kappa_{21}} M^{-\kappa_{22}}) \quad if \quad \kappa_{21} \leq 1/4,$$

$$(b) N^{5/2} M^{1/2} (\tilde{\tilde{\delta}} - \delta^0) \xrightarrow{d} N(0,\sigma_2^2) \quad if \quad \kappa_{21} > 1/4, \kappa_{22} = 1/2.$$

Here σ_2^2 *and* c *are same as defined in Theorem 28.*

Now we show that starting from the initial guesses $\tilde{\alpha}$, $\tilde{\beta}$, with convergence rates $\tilde{\alpha} - \alpha^0 = O_p(M^{-1}N^{-1/2})$ and $\tilde{\beta} - \beta^0 = O_p(M^{-2}N^{-1/2})$, respectively, how the above results can be used to obtain efficient estimators, which have the same rate of convergence as the LSEs. It may be noted that finding initial guesses with the above convergence rates are not difficult. It can be obtained by finding the minimum of $Q_1(\alpha,\beta)$, where

$$Q_1(\alpha,\beta) = \sum_{m=1}^{M}\left(\sum_{n=1}^{N} y(m,n) - A\cos(\alpha m + \beta m^2) - B\sin(\alpha m + \beta m^2)\right)^2.$$

Here also the main idea is not to use the whole, but a fraction of sample at the beginning and gradually proceed towards the complete sample. The algorithm is described below. We denote the estimates of α^0 and β^0 obtained at the j-th iteration as $\tilde{\alpha}^{(j)}$ and $\tilde{\beta}^{(j)}$, respectively.

Algorithm 4.

Step 1: Choose $M_1 = M^{8/9}$, $N_1 = N$. Therefore,

$$\widetilde{\alpha}^{(0)} - \alpha^0 = O_p(M^{-1}N^{-1/2}) = O_p(M_1^{-1-1/8}N_1^{-1/2}) \ \ and$$

$$\widetilde{\beta}^{(0)} - \beta^0 = O_p(M^{-2}N^{-1/2}) = O_p(M_1^{-2-1/4}N_1^{-1/2}) \, .$$

Perform steps (4.130) and (4.133). Therefore, after 1-st iteration, we have

$$\widetilde{\alpha}^{(1)} - \alpha^0 = O_p(M_1^{-1-1/4}N_1^{-1/2}) = O_p(M^{-10/9}N^{-1/2}) \ \ and$$

$$\widetilde{\beta}^{(1)} - \beta^0 = O_p(M_1^{-2-1/2}N_1^{-1/2}) = O_p(M^{-20/9}N^{-1/2}) \, .$$

Step 2: Choose $M_2 = M^{80/81}$, $N_1 = N$. Therefore,

$$\widetilde{\alpha}^{(1)} - \alpha^0 = O_p(M_2^{-1-1/8}N_2^{-1/2}) \ \ and \ \ \widetilde{\beta}^{(1)} - \beta^0 = O_p(M_2^{-2-1/4}N_2^{-1/2}).$$

Perform steps (4.130) and (4.130). Therefore, after 2-nd iteration, we have

$$\widetilde{\alpha}^{(2)} - \alpha^0 = O_p(M_2^{-1-1/4}N_2^{-1/2}) = O_p(M^{-100/81}N^{-1/2}) \ \ and$$

$$\widetilde{\beta}^{(2)} - \beta^0 = O_p(M_2^{-2-1/2}N_2^{-1/2}) = O_p(M^{-200/81}N^{-1/2}) \, .$$

Step 3: Choose $M_3 = M$, $N_3 = N$. Therefore,

$$\widetilde{\alpha}^{(2)} - \alpha^0 = O_p(M_3^{-1-19/81}N_3^{-1/2}) \ \ and \ \ \widetilde{\beta}^{(2)} - \beta^0 = O_p(M_2^{-2-38/81}N_3^{-1/2}).$$

Again, performing steps (4.130) and (4.133), after 3-rd iteration, we have

$$\widetilde{\alpha}^{(3)} - \alpha^0 = O_p(M^{-1-38/81}N^{-1/2}) \ \ and \ \ \widetilde{\beta}^{(3)} - \beta^0 = O_p(M^{-2-76/81}N^{-1/2}) \, .$$

Step 4: Choose $M_4 = M$, $N_4 = N$, and after performing steps (4.130) and (4.133) we obtain the required convergence rates, i.e.

$$\widetilde{\alpha}^{(4)} - \alpha^0 = O_p(M^{-3/2}N^{-1/2}) \ \ and \ \ \widetilde{\beta}^{(4)} - \beta^0 = O_p(M^{-5/2}N^{-1/2}) \, .$$

Along the similar lines, interchanging the roles of M and N, we can get the algorithm corresponding to γ^0 and δ^0. Extensive simulation experiments have been carried out by Lahiri [74], and it is observed that the performance of the finite step algorithm is quite good in terms of mean squared errors (MSEs) and biases. In the initial steps, the part of the sample is selected in such a way that the dependence structure is maintained in the subsample. The MSEs and biases of the finite step algorithm are quite comparable with the performance of the corresponding LSEs. Therefore, the finite step algorithm can be used quite effectively in practice.

4.10.4. Efficient Algorithm Based on Dimension Reduction

Recently, Grover, Kundu and Mitra [45] proposed an efficient estimation method for the estimation of the unknown parameters of a 2-D chirp model (4.127). They assumed that the errors are i.i.d. random variables with mean zero and finite variance σ^2. It is a numerically efficient method and it uses the reduction of 2-D chirp model to 1-D chirp models. The main advantage of the proposed estimators is that although they can be obtained in a more computationally efficient manner, they have the same convergence rates as the LSEs.

Suppose we fix $n = n_0$, then (4.127) can be rewritten for $m = 1, \ldots, M$ as follows:

$$
\begin{aligned}
y(m, n_0) &= A^0 \cos(\alpha^0 m + \beta^0 m^2 + \gamma^0 n_0 + \delta^0 n_0^2) \\
&\quad + B^0 \sin(\alpha^0 m + \beta^0 m^2 + \gamma^0 n_0 + \delta^0 n_0^2) + e(m, n_0) \\
&= A^0(n_0) \cos(\alpha^0 m + \beta^0 m^2) + B^0(n_0) \sin(\alpha^0 m + \beta^0 m^2) + e(m, n_0).
\end{aligned}
\tag{4.140}
$$

Clearly, (4.140) represents a 1-D chirp model with $A^{0^2}(n_0) + B^{0^2}(n_0)$ as amplitudes, α^0 as the frequency parameter and β^0 as the frequency rate parameter. Here,

$$
\begin{aligned}
A^0(n_0) &= A^0 \cos(\gamma^0 n_0 + \delta^0 n_0^2) + B^0 \sin(\gamma^0 n_0 + \delta^0 n_0^2), \\
B^0(n_0) &= -A^0 \sin(\gamma^0 n_0 + \delta^0 n_0^2) + B^0 \cos(\gamma^0 n_0 + \delta^0 n_0^2).
\end{aligned}
$$

It is important to note that each column of the 2-D data matrix can be thought of as a data vector coming from 1-D chirp model with the same frequency and frequency rate. Therefore, each column of the data matrix can be used to estimate α^0 and β^0. For this purpose, least squares method may be used. Hence, the estimators of α^0 and β^0 can be obtained by minimizing the following function:

$$
R_M(\alpha, \beta, n_0) = Y_{n_0}^{\mathrm{T}} (I - P_{Z_M}(\alpha, \beta)) Y_{n_0},
$$

for each n_0. Here, $Y_{n_0} = \left[y[1, n_0], \ldots, y[M, n_0] \right]^{\mathrm{T}}$ is the n_0th column of the original data matrix, $P_{Z_M}(\alpha, \beta) = Z_M(\alpha, \beta)(Z_M(\alpha, \beta)^{\mathrm{T}} Z_M(\alpha, \beta))^{-1} Z_M(\alpha, \beta)^{\mathrm{T}}$ is the projection matrix on the column space of the matrix $Z_M(\alpha, \beta)$ and

$$
Z_M(\alpha, \beta) = \begin{bmatrix} \cos(\alpha + \beta) & \sin(\alpha + \beta) \\ \vdots & \vdots \\ \cos(M\alpha + M^2\beta) & \sin(M\alpha + M^2\beta) \end{bmatrix}.
\tag{4.141}
$$

The estimates of α^0 and β^0 can be obtained from each column of the data matrix and they can be combined thereafter. This process involves minimizing of N 2-D functions corresponding to the N columns of the matrix. To avoid solving N such optimization problems, Grover, Kundu and Mitra [45] propose to minimize the following function instead:

$$R_{MN}^{(1)}(\alpha,\beta) \;=\; \sum_{n_0=1}^{N} R_M(\alpha,\beta,n_0) = \sum_{n_0=1}^{N} \mathbf{Y}_{n_0}^{\mathrm{T}}(\mathbf{I}-\mathbf{P}_{\mathbf{Z}_M}(\alpha,\beta))\mathbf{Y}_{n_0}, \qquad (4.142)$$

with respect to α and β simultaneously and obtain $\hat{\alpha}$ and $\hat{\beta}$. This reduces the computational burden significantly. It may be mentioned that since the errors are assumed to be i.i.d., replacing these N functions by their sum is justifiable.

Similarly, we can obtain the estimates, $\hat{\gamma}$ and $\hat{\delta}$, of γ^0 and δ^0, by minimizing the following criterion function:

$$R_{MN}^{(2)}(\gamma,\delta) \;=\; \sum_{m_0=1}^{M} R_N(\gamma,\delta,m_0) = \sum_{m_0=1}^{M} \mathbf{Y}_{m_0}^{\mathrm{T}}(\mathbf{I}-\mathbf{P}_{\mathbf{Z}_N}(\gamma,\delta))\mathbf{Y}_{m_0}, \qquad (4.143)$$

with respect to γ and δ simultaneously. The data vector $\mathbf{Y}_{m_0}=\left[y[m_0,1],\ldots,y[m_0,N]\right]^{\mathrm{T}}$, is the $m_0 th$ row of the data matrix, $m_0=1,\ldots,M$, $\mathbf{P}_{\mathbf{Z}_N}(\gamma,\delta)$ is the projection matrix on the column space of the matrix $\mathbf{Z}_N(\gamma,\delta)$ and the matrix $\mathbf{Z}_N(\gamma,\delta)$ can be obtained by replacing α, β, N by γ, δ and M, respectively in the matrix $\mathbf{Z}_M(\alpha,\beta)$, defined in (4.141). Once the non-linear parameters have been estimated, the estimates of the linear parameters can be obtained as given below:

$$\begin{bmatrix} \hat{A} \\ \hat{B} \end{bmatrix} = [\mathbf{W}(\hat{\alpha},\hat{\beta},\hat{\gamma},\hat{\delta})^{\mathrm{T}}\mathbf{W}(\hat{\alpha},\hat{\beta},\hat{\gamma},\hat{\delta})]^{-1}\mathbf{W}(\hat{\alpha},\hat{\beta},\hat{\gamma},\hat{\delta})^{\mathrm{T}}\mathbf{Y}_{MN\times1}.$$

Here, $\mathbf{Y}_{MN\times1}=\left[y(1,1),\ldots,y(M,1),\ldots,y(1,N),\ldots,y(M,N)\right]^{\mathrm{T}}$ is the data vector obtained by stacking the columns of the observed data matrix on top of one another, and

$$\mathbf{W}(\alpha,\beta,\gamma,\delta)_{MN\times2} = \begin{bmatrix} \cos(\alpha+\beta+\gamma+\delta) & \sin(\alpha+\beta+\gamma+\delta) \\ \cos(2\alpha+4\beta+\gamma+\delta) & \sin(2\alpha+4\beta+\gamma+\delta) \\ \vdots & \vdots \\ \cos(M\alpha+M^2\beta+\gamma+\delta) & \sin(M\alpha+M^2\beta+\gamma+\delta) \\ \vdots & \vdots \\ \cos(\alpha+\beta+N\gamma+N^2\delta) & \sin(\alpha+\beta+N\gamma+N^2\delta) \\ \cos(2\alpha+4\beta+N\gamma+N^2\delta) & \sin(2\alpha+4\beta+N\gamma+N^2\delta) \\ \vdots & \vdots \\ \cos(M\alpha+M^2\beta+N\gamma+N^2\delta) & \sin(M\alpha+M^2\beta+N\gamma+N^2\delta) \end{bmatrix}. \qquad (4.144)$$

The following assumptions are needed to establish the asymptotic properties of the estimators.

Assumption 15. *The error random variable* $\{e(m,n)\}$ *is a double array sequence of i.i.d. random variables with mean zero, variance* σ^2 *and finite fourth order moment.*

Assumption 16. *The non-linear parameters satisfy* $0 < \alpha^0, \beta^0, \gamma^0, \delta^0 < \pi$ *and there exists a* K, *such that* $0 < A^{02} + B^{02} < K$.

The results obtained on the consistency of the proposed estimators are presented in the following theorems:

Theorem 31. *Under Assumptions 15 and 16,* $\hat{\alpha}$ *and* $\hat{\beta}$ *are strongly consistent estimators of* α^0 *and* β^0 *respectively, that is,*

$$\hat{\alpha} \xrightarrow{a.s.} \alpha^0 \text{ as } M \to \infty,$$
$$\hat{\beta} \xrightarrow{a.s.} \beta^0 \text{ as } M \to \infty.$$

Theorem 32. *Under Assumptions 15 and 16,* $\hat{\gamma}$ *and* $\hat{\delta}$ *are strongly consistent estimators of* γ^0 *and* δ^0 *respectively, that is,*

$$\hat{\gamma} \xrightarrow{a.s.} \gamma^0 \text{ as } N \to \infty,$$
$$\hat{\delta} \xrightarrow{a.s.} \delta^0 \text{ as } N \to \infty.$$

The following theorems provide the asymptotic distributions of the proposed estimators:

Theorem 33. *If Assumptions 15 and 16 are satisfied, then*

$$\mathbf{D}_1\left[(\hat{\alpha} - \alpha^0), (\hat{\beta} - \beta^0)\right]^{\mathrm{T}} \xrightarrow{d} N_2(\mathbf{0}, 2\sigma^2\mathbf{\Sigma}) \text{ as } \min\{M, N\} \to \infty.$$

Here, $\mathbf{D}_1 = diag(M^{\frac{3}{2}} N^{\frac{1}{2}}, M^{\frac{5}{2}} N^{\frac{1}{2}})$ *and*

$$\mathbf{\Sigma} = \frac{2}{A^{02} + B^{02}} \begin{bmatrix} 96 & -90 \\ -90 & 90 \end{bmatrix}. \tag{4.145}$$

Theorem 34. *If Assumptions 15 and 16 are satisfied, then*

$$\mathbf{D}_2\left[(\hat{\gamma} - \gamma^0), (\hat{\delta} - \delta^0)\right]^{\mathrm{T}} \xrightarrow{d} \mathcal{N}_2(\mathbf{0}, 2\sigma^2\mathbf{\Sigma}) \text{ as } \min\{M, N\} \to \infty.$$

Here, $\mathbf{D}_2 = diag(M^{\frac{1}{2}} N^{\frac{3}{2}}, M^{\frac{1}{2}} N^{\frac{5}{2}})$ *and* $\mathbf{\Sigma}$ *is as defined in (4.145).*

The asymptotic distributions of $(\hat{\alpha}, \hat{\beta})$ and $(\hat{\gamma}, \hat{\delta})$ are observed to be the same as those of the corresponding LSEs. Thus, we get the same efficiency as that of the LSEs without going through the process of actually computing the LSEs. Grover, Kundu and Mitra [45] performed extensive simulations when the errors are i.i.d. Gaussian random variables to compare the performances of the LSEs and the estimators obtained through dimension reduction technique. It is observed that their performances are quite comparable. Even for moderate sample sizes, the variances of the estimators obtained through dimension reduction technique reach the Cramer-Rao lower bound. Grover, Kundu and Mitra [45] also compared the computational times of the two estimators, and it is observed that even for one component model the computation time of the LSEs is more than 200 times compared to the other.

4.10.5. Two-dimensional Polynomial Phase Signal Model

A related but more general model that has received significant amount of attention in the signal processing literature is a 2-D polynomial phase signal model. This model was first introduced by Francos and Friedlander [35] and in presence of additive errors, it can be described as follows:

$$y(m,n) = A^0 \cos\left(\sum_{p=1}^{r}\sum_{j=0}^{p}\alpha^0(j,p-j)m^j n^{p-j}\right) + B^0 \sin\left(\sum_{p=1}^{r}\sum_{j=0}^{p}\alpha^0(j,p-j)m^j n^{p-j}\right) \tag{4.146}$$

$$+ e(m,n); \, m = 1,\ldots,M; \; n = 1,\ldots,N,$$

here $e(m,n)$ is the additive error with mean 0, A^0 and B^0 are non zero real numbers, and for $j = 0,\ldots,p, \; = 1,\ldots,r, \; \alpha^0(j,p-j)$ s are distinct frequency rates of order $(j,p-j)$, respectively, and lie strictly between 0 and π. Here $\alpha^0(0,1)$ and $\alpha^0(1,0)$ are called frequencies. The explicit assumptions on the error $e(m,n)$ will be provided later.

Different specific forms of model (4.146) have been used quite extensively in the literature. For example Friedlander and Francos [38] used the following 2-D polynomial phase signal model

$$y(m,n) = A^0 \cos(\alpha^0 m + \beta^0 m^0 + \gamma^0 n + \delta^0 n^2 + v^0 mn)$$

$$+ B^0 \sin(\alpha^0 m + \beta^0 m^0 + \gamma^0 n + \delta^0 n^2 + v^0 mn)$$

$$+ e(m,n); \; m = 1,\ldots,M; \; n = 1,\ldots,N,$$

to analyze finger print type data and Djurović et al. [27] used a specific 2-D cubic phase signal model due to its applications in modelling SAR data and in particular Interferometric SAR data. Further, 2-D polynomial phase signal model also has been used in modeling and analyzing MRI, optical imaging and different texture imaging. For some of the other specific applications of this model, one may refer to Djukanović and Djurović

[21], Ikram and Zhou [54], Wang and Zhou [120], Tichavsky and Handel [113], Amar, Leshem and van der Veen [2] and see the references cited therein.

Due to the wide-ranging applications of the model, several estimation procedures have been suggested in the literature. These are mostly based on the assumption that the error components are i.i.d. random variables with mean zero and finite variance. For example, Farquharson et al. [32] provided a computationally efficient estimation procedures of the unknown parameters of a polynomial phase signals. Djurović and Stanković [26] proposed the quasi maximum likelihood estimators based on the normality assumption of the error random variables. Recently, Djurović et al. [25] considered an efficient estimation method of the polynomial phase signal parameters by using a cubic phase function. Some of the refinement of the parameter estimation of the polynomial phase signals can be obtained in O'Shea [97].

Interestingly, a significant amount of work has been done in developing different estimation procedures to compute parameter estimates of the polynomial phase signal, but not much work has been done in developing the properties of the estimators. In most of the cases, the MSEs or the variances of the estimates are compared with the corresponding Cramer-Rao lower bound, without establishing formally that the asymptotic variances of the MLEs attain the lower bound in case the errors are i.i.d. normal random variables. Recently, Lahiri and Kundu [75] established formally the asymptotic properties of the LSEs of parameters in model (4.146) under certain general assumptions on the error random variables. From the results of Lahiri and Kundu [75] it can be easily seen that when the errors are i.i.d. normally distributed, the asymptotic variance of the MLEs attain the Cramer-Rao lower bound. Although, the LSEs are the most efficient estimators, finding the LSEs is a computationally challenging problem. Moreover, computationally efficient algorithm which provide estimators with the same rate of convergence as the LSEs, do not exist and developing such algorithms is an area to explore.

4.11. Conclusions

In this chapter we have provided an overview of different 1-D and 2-D chirp and related models which have received a considerable amount of attention in recent years in the Statistical Signal Processing literature. Our main aim is to introduce different models which are being extensively used in practice. These models have been used quite effectively in analyzing 1-D and 2-D non-stationary signals. We have highlighted several numerical algorithms and different sophisticated statistical tools which have been used for the efficient estimation of the unknown parameters of the underlying models. Their theoretical properties are provided and the associated problems have been discussed as well. Since the models are non-linear in nature, most of the properties are asymptotic in nature. We have discussed simulation results which are available in the literature regarding the behavior of these estimators mainly for small and moderate sample sizes. This chapter has surveyed many important and effective methods of parameter estimation in the Classical as well as Bayesian framework suggested by different researchers over the years and advocated the use of such coherent methods of estimation. Several open problems for future research are proffered in the hope that this will generate enough interests among young researchers to come forward and solve some of these challenging problems.

Acknowledgements

Part of the work of the first author has been supported by a grant from the Science and Engineering Research Board, Department of Science and Technology, Government of India.

References

[1]. T. Abatzoglou, Fast maximum likelihood joint estimation of frequency and frequency rate, *IEEE Transactions on Aerospace and Electronic Systems*, Vol. 22, 1986, pp. 708-715.

[2]. A. Amar, A. Leshem, A. J. van der Veen, A low complexity blind estimator of narrow band polynomial phase signals, *IEEE Transactions on Signal Processing*, Vol. 58, 2010, pp. 4674-4683.

[3]. Z. D. Bai, C. R. Rao, M. Chow, D. Kundu, An efficient algorithm for estimating the parameters of superimposed exponential signals, *Journal of Statistical Planning and Inference*, Vol. 110, 2003, pp. 23-34.

[4]. S. Barbarossa, V. Petrone, Analysis of polynomial-phase signals by the integrated generalized ambiguity function, *IEEE Transactions on Signal Processing*, Vol. 45, 1997, pp. 316-327.

[5]. S. Barbarossa, A. Scaglione, G. Giannakis, Product high-order ambiguity function for multicomponent polynomial-phase signal modeling, *IEEE Transactions on Signal Processing*, Vol. 46, 1998, pp. 691-708.

[6]. M. M. Barbieri, P. Barone, A two dimensional Prony's method for Spectral estimation, *IEEE Transactions on Signal Processing*, Vol. 40, 1992, pp. 2747-2756.

[7]. D. M. Bates, D. G. Watts, Nonlinear Regression Analysis and Its Applications, *John Wiley & Sons*, New York, 1988.

[8]. O. Besson, F. Castanié, On estimating the frequency of a sinusoid in autoregressive multiplicative noise, *Signal Processing*, Vol. 30, 1993, pp. 65-83.

[9]. O. Besson, P. Stoica, Sinusoidal signals with random amplitude: Least- squares estimators and their statistical analysis, *IEEE Transactions on Signal Processing*, Vol. 43, 1995, pp. 2733-2744.

[10]. O. Besson, M. Ghogho, A. Swami, Parameter estimation for random amplitude chirp signals, *IEEE Transactions on Signal Processing*, Vol. 47, 1999, pp. 3208-3219.

[11]. O. Besson, G. B. Giannakis, F. Gini, Improved estimation of hyperbolic frequency modulated chirp signals, *IEEE Transactions on Signal Processing*, 1999, Vol. 47, pp. 1384-1388.

[12]. Y. Bresler, A. Macovski, Exact maximum likelihood parameter estimation of superimposed exponential signals in noise, *IEEE Transactions on Acoustics, Speech and Signal Processing*, Vol. 34, 1986, pp. 1081-1089.

[13]. S. D. Cabrera, N. K. Bose, Prony's method for two-dimensional complex exponential modeling, Chapter 15, in Applied Control Theory (S. G. Tzafestas, Ed.), *Marcel and Dekker*, New York, NY, 1993, pp. 401-411.

[14]. F. Cao, S. Wang, F. Wang, Cross-spectral method based on 2-D cross polynomial transform for 2-D chirp signal parameter estimation, in *Proceedings of the 8th International Conference on Signal Processing (ICSP'06)*, 2006.

[15]. P. G. Casazza, M. Fickus, Fourier transforms of finite chirps, *EURASIP Journal on Applied Signal Processing*, Vol. 2006, 2006, 70204.

[16]. G. W. Chang, C. Chen, An accurate time domain procedure for harmonics and inter-harmonics detection, *IEEE Transactions of Power Del.*, Vol. 25, 2010, pp. 1787-1795.

[17]. A. Charnes, W. W. Cooper, R. Ferguson, Optimal estimation of executive compensation by linear programing, *Management Science*, Vol. 1, 1955, pp. 138-151.

[18]. M. G. Christensen, P. Stoica, A. Jakobsson, H. S. Jensen, Multi-pitch estimation, *Signal Processing*, Vol. 88, 2008, pp. 972-983.

[19]. M. G. Christensen, S. H. Jensen, A. Jakobsson, S. H. Jensen, Optimal filter designs for fundamental frequency estimation, *IEEE Signal Processing Letters*, Vol. 15, 2008, pp. 745-748.

[20]. J. Chun, N. K. Bose, Parameter estimation via signal selectivity of signal subspaces (PESS) and its applications, *Digital Signal Processing*, Vol. 5, 1995, pp. 58-76.

[21]. S. Djukanović, I. Djurović, Aliasing detection and resolving in the estimation of polynomial-phase signal parameters, *Signal Processing*, Vol. 92, 2012, pp. 235-239.

[22]. S. S. Dhar, D. Kundu, U. Das, On testing parameters of chirp signal model, *IEEE Transactions on Signal Processing*, Vol. 67, 2019, pp. 4291-4301.

[23]. T. E. Dielman, Least absolute value estimation in regression models: An annotated bibliography, *Communication in Statistics – Theory and Methods*, Vol. 4, 1984, pp. 513-541.

[24]. P. M. Djurić, S. M. Kay, Parameter estimation of chirp signals, *IEEE Transactions on Acoustics, Speech and Signal Processing*, Vol. 38, 1990, pp. 2118-2126.

[25]. I. Djurović, M. Simeunović, P. Wang, Cubic phase function: A simple solution for polynomial phase signal analysis, *Signal Processing*, Vol. 135, 2017, pp. 48-66.

[26]. I. Djurović, L. J. Stanković, Quasi maximum likelihood estimators of polynomial phase signals, *IET Signal Processing*, Vol. 13, 2014, pp. 347-359.

[27]. I. Djurović, P. Wang, C. Ioana, Parameter estimation of 2-D cubic phase signal function using genetic algorithm, *Signal Processing*, Vol. 90, 2010, pp. 2698-2707.

[28]. Y. Dodge, An introduction to L_1-norm bases statistical data analysis, *Computational Statistics and Data Analysis*, Vol. 5, 1987, pp. 239-253.

[29]. Y. Doweck, A. Amar, I. Cohen, Joint model order selection and parameter estimation of chirps with harmonic components, *IEEE Transactions on Signal Processing*, Vol. 63, 2015, pp. 1765-1778.

[30]. R. F. Dwyer, Fourth-order spectra of Gaussian amplitude modulated sinusoids, *Journal of the Acoustics Society of America*, Vol. 90, 1991, pp. 918-926.

[31]. T. W. Eddy, Maximum likelihood detection and estimation for harmonic sets, *Journal of Acoustic Society of America*, Vol. 68, 1980, pp. 149-155.

[32]. M. Farquharson, P. O'Shea, G. Ledwich, A computationally efficient technique for estimating the parameters phase signals from noisy observations, *IEEE Transactions on Signal Processing*, Vol. 53, 2005, pp. 3337-3342.

[33]. R. A. Fisher, Tests of signifcance in harmonic analysis, *Proceedings of the Royal Society of London, Series A*, Vol. 125, 1929, pp. 54-59.

[34]. D. Fourier, F. Auger, K. Czarnecki, S. Meignen, Chirp rate and instantaneous frequency estimation: Application to recursive vertical synchrosqueezing, *IEEE Signal Processing Letters*, Vol. 24, Issue 11, 2017, pp. 1724-1728.

[35]. J. M. Francos, B. Friedlander, Two-dimensional polynomial phase signals: parameter estimation and bounds, *Multidimensional Systems and Signal Processing*, Vol. 9, 1998, pp. 173-205.

[36]. J. M. Francos, B. Friedlander, Parameter estimation of 2-D random amplitude polynomial phase signals, *IEEE Transactions on Signal Processing*, Vol. 47, 1999, pp. 1795-1810.

[37]. B. Friedlander, J. M. Francos, Estimation of amplitudes and phase parameters of multicomponent signals, *IEEE Transactions on Signal Processing*, Vol. 43, 1995, pp. 917-926.

[38]. B. Friedlander, J. M. Francos, An estimation algorithm for 2-D polynomial phase signals, *IEEE Transactions on Image Processing*, Vol. 5, 1996, pp. 1084-1087.

[39]. C. E. Froberg, Introduction to Numerical Analysis, 2nd Edition, *Addison-Wesley Pub. Co.*, 1969.

[40]. D. Gabor, Theory of communication. Part 1: The analysis of information *Journal of the Institution of Electrical Engineers – Part III: Radio and Communication Engineering*, Vol. 93, 1946, pp. 429-441.

[41]. F. Gini, M. Montanari, L. Verrazzani, Estimation of chirp signals in compound Gaussian clutter: a cyclostationary approach, *IEEE Transactons on Acoustics, Speech and Signal Processing*, Vol. 48, 2000, pp. 1029-1039.

[42]. R. Grover, Frequency and frequency rate estimation of some non-stationary signal processing models, PhD Thesis, *Indian Institute of Technology*, Kanpur, India, 2019.

[43]. R. Grover, D. Kundu, A. Mitra, On approximate least squares estimators of parameters of one-dimensional chirp signal, *Statistics*, Vol. 52, 2018, pp. 1060-1085.

[44]. R. Grover, D. Kundu, A. Mitra, Asymtotic of approximate least squares estimators of parameters of two-dimensional chirp signal, *Journal of Multivariate Analysis*, Vol. 168, 2018, pp. 211-220.

[45]. R. Grover, D. Kundu, A. Mitra, An efficient methodology to estimate the parameters of a two-dimensional chirp signal model, *Multidimensional Systems and Signal Processing*, 2020.

[46]. J. Guo, H. Zou, X. Yang, G. Liu, Parameter estimation of multicomponent chirp signals via sparse representation, *IEEE Transactions on Aerospace and Electronic Systems*, Vol. 47, 2011, pp. 2261-2268.

[47]. E. J. Hannan, Non-linear time series regression, *Journal of Applied Probability*, Vol. 8, 1971, pp. 767-780.

[48]. E. J. Hannan, Time series analysis, *IEEE Transactions on Automatic Control*, Vol. 19, 1974, pp. 706-715.

[49]. M. Hedley, D. Rosenfeld, A new two-dimensional phase unwrapping algorithm for MRI images. *Magnetic Resonance in Medicine*, Vol. 24, 1992, pp. 177-181.

[50]. F. B. Hilderband, An Introduction to Numerical Analysis, *McGraw Hill*, New York, 1956.

[51]. Y. Hua, Estimating two-dimensional frequencies by matrix enhancement and matrix Pencil, *IEEE Transactions on Signal Processing*, Vol. 40, 1992, pp. 2267-2280.

[52]. M. Z. Ikram, K. Abed-Meraim, Y. Hua, Iterative parameter estimation of multiple chirp signals, *Electronics Letters*, Vol. 33, 1997, pp. 657-659.

[53]. M. Z. Ikram, K. Abed-Meraim, Y. Hua, Estimating the parameters of chirp signals: An iterative approach, *IEEE Transactions on Signal Processing*, Vol. 46, 1998, pp. 3436-3441.

[54]. M. Z. Ikram, G. T. Zhou, Estimation of multicomponent phase signals of mixed orders, *Signal Processing*, Vol. 81, 2001, pp. 2293-2308.

[55]. R. A. Irizarry, Asymptotic distribution of estimates for a time-varying parameter in a harmonic model with multiple fundamentals, *Statistica Sinica*, Vol. 10, 2000, pp. 1041-1067.

[56]. R. A. Irizarry, Weighted estimation of harmonic components in a musical sound signal, *Journal of Time Series Analysis*, Vol. 23, 2002, pp. 29-48.

[57]. S. K. Jain, S. N. Singh, Exact model order ESPRIT techniquefor harmonics and inter-harmonics estimation, *IEEE Transactions on Instrumental Measurement*, Vol. 61, 2012, pp. 1915-1923.

[58]. R. I. Jennrich, Asymptotic properties of the nonlinear least squares estimators, *Annals of Mathematical Statistics*, Vol. 40, 1969, pp. 633-643.

[59]. T. L. Jensen, J. K. Nielsen, J. R. Jensen, M. G. Christensen, S. H. Jensen, A fast algorithm for maximum likelihood estimation of harmonic chirp parameters, *IEEE Transactions on Signal Processing*, Vol. 65, 2017, pp. 5137-5152.

[60]. S. M. Kay, Modern Spectral Estimation, Theory and Application, *Prentice Hall*, Englewood Cliffs, NJ, 1988.

[61]. W. J. Kennedy Jr., J. E. Gentle, Statistical Computing, *Marcel Dekker Inc.*, New York, 1980.

[62]. Y. Kopsinis, E. Aboutanios, D. A. Waters, S. McLaughlin, Time-frequency and advanced frequency estimation techniques for the investigation of bat echolocation calls, *Journal of Acoustics Society of America*, Vol. 127, 2010, pp. 1124-1134.

[63]. D. Kundu, Asymptotic properties of the complex valued non-linear regression model, *Communications in Statistics – Theory and Methods*, Vol. 20, 1991, pp. 3793-3803.

[64]. D. Kundu, Estimating the number of signals using information theoretic criterion, *Journal of Statistical Computation and Simulation*, Vol. 44, 1992, pp. 117-131.

[65]. D. Kundu, Estimating the parameters of complex valued exponential signals, *Technometrics*, Vol. 35, 1993, pp. 215-218.

[66]. D. Kundu, Asymptotic properties of the least squares estimators of sinusoidal signals, *Statistics*, Vol. 30, 1997, pp. 221-238.

[67]. D. Kundu, Estimating the number of sinusoids in additive white noise, *Signal Processing*, Vol. 56, 1997, pp. 103-110.

[68]. D. Kundu, Z. D. Bai, S. Nandi, L. Bai, Super efficient frequency estimation, *Journal of Statistical Planning and Inference*, Vol. 141, 2011, pp. 2576-2588.

[69]. D. Kundu, R. Kundu, Consistent estimates of superimposed exponential signals when some observations are missing, *Journal of Statistical Planning and Inference*, Vol. 44, 1995, pp. 205-218.

[70]. D. Kundu, A. Mitra, Consistent method of estimating the superimposed exponential signals, *Scandinavian Journal of Statistics*, Vol. 22, 1995, pp. 73-82.

[71]. D. Kundu, S. Nandi, Determination of discrete spectrum in a random field, *Statistica Neerlandica*, Vol. 57, 2003, pp. 258-283.

[72]. D. Kundu, S. Nandi, Parameter estimation of chirp signals in presence of stationary Christensen, noise, *Statistica Sinica*, Vol. 18, 2008, pp. 187-201.

[73]. D. Kundu, S. Nandi, Statistical Signal Processing: Frequency Estimation, *Springer*, New Delhi, 2012.

[74]. A. Lahiri, Estimators of parameters of chirp signals and their properties, PhD Thesis, Department of Mathematics and Statistics, *Indian Institute of Technology*, Kanpur, India, 2011.

[75]. A. Lahiri, D. Kundu, On parameter estimation of two-dimensional polynomial phase signal model, *Statistica Sinica*, Vol. 27, 2017, pp. 1779-1792.

[76]. A. Lahiri, D. Kundu, A. Mitra, Efficient algorithm for estimating the parameters of chirp signal, *Journal of Multivariate Analysis*, Vol. 108, 2012, pp. 15-27.

[77]. A. Lahiri, D. Kundu, A. Mitra, Efficient algorithm for estimating the parameters of two dimensional chirp signal, *Sankhya, Ser. B*, Vol. 75, 2013, pp. 65-89.

[78]. A. Lahiri, D. Kundu, A. Mitra, On least absolute deviation estimator of one dimensional chirp model, *Statistics*, Vol. 48, 2014, pp. 405-420.

[79]. A. Lahiri, D. Kundu, A. Mitra, Estimating the parameters of multiple chirp signals, *Journal of Multivariate Analysis*, Vol. 139, 2015, pp. 189-205.

[80]. H. Li, P. Stoica, J. Li, Computationally efficient parameter estimation for harmonic sinusoidal signals, *Signal Processing*, Vol. 80, 2000, pp. 1937-1944.

[81]. C.-C. Lin, P. M. Djurić, Estimation of chirp signals by MCMC, in *Proceedings of the IEEE International Conference on Acoustics, Speech and Signal Processing (ICASSP'98)*, Vol. 1, 2000, pp. 265-268.

[82]. X. Liu, H. Yu, Time-domain joint parameter estimation of chirp signal based on SVR, *Mathematical Problems in Engineering*, Vol. 2013, 2013, 952743.

[83]. Y. Lu, R. Demirli, G. Cardoso, J. Saniie, A successive parameter estimation algorithm for chirplet signal decomposition, *IEEE Transactions on Ultrasonic, Ferroelectrics and Frequency Control*, Vol. 53, 2006, pp. 2121-2131.

[84]. P. Malliavan, Sur la norme d'une matrice circulate Gaussienne Serie I, *C. R. Acad. Sc. Paris t*, Vol. 319, 1994, pp. 45-49.

[85]. P. Malliavan, Estimation d'un signal Lorentzien Serie I, *C. R. Acad. Sc. Paris t*, Vol. 319, 1994, pp. 991-997.

[86]. V. Mangulis, Handbook of Series for Scientists and Engineers, *Academic Press*, New York, 1965.

[87]. S. Mazumder, Single-step and multiple-step forecasting in one-dimensional chirp signal using MCMC-based Bayesian analysis, *Communications in Statistics – Simulation and Computation*, Vol. 46, 2017, pp. 2529-2547.

[88]. M. Mboup, T. Adali, A generalization of the Fourier transform and its application to spectral analysis of chirp-like signals, *Applied and Computational Harmonic Analysis*, Vol. 32, Issue 2, 2012, pp. 305-312.

[89]. H. L. Montgomery, Ten Lectures on the Interface Between Analytic Number Theory and Harmonic Analysis, *American Mathematical Society*, 1990.

[90]. S. Nandi, D. Kundu, Estimating the fundamental frequency of a periodic function, *Statistical Methods and Applications*, Vol. 12, 2003, pp. 341-360.

[91]. S. Nandi, D. Kundu, Asymptotic properties of the least squares estimators of the parameters of the chirp signals, *Annals of the Institute of Statistical Mathematics*, Vol. 56, 2004, pp. 529-544.

[92]. S. Nandi, D. Kundu, Analyzing non-stationary signals using a cluster type model, *Journal of Statistical Planning and Inference*, Vol. 136, 2006, pp. 3871-3903.

[93]. S. Nandi, D. Kundu, A fast and efficient algorithm for estimating the parameters of sinusoidal signals, *Sankhya*, Vol. 68, 2006, pp. 283-306.

[94]. S. Nandi, D. Kundu, Random amplitude chirp model, *Signal Processing*, Vol. 168, 2020, 107328.

[95]. S. Nandi, D. Kundu, Estimating the fundamental frequency using modified Newton-Raphson algorithm, *Statistics*, Vol. 53, 2019, pp. 440-458.

[96]. J. K. Nielsen, Some new results on the estimation of sinusoids in noise, PhD Thesis, *Aalborg University*, Denmark, 2012.

[97]. P. O'Shea, On refining polynomial phase signal parameter estimates, *IEEE Transactions on Aerospace, Electronic Systems*, Vol. 4, 2010, pp. 978-987.

[98]. S. Pelag, B. Porat, Estimation and classification of polynomial phase signals, *IEEE Transactions on Information Theory*, Vol. 37, 1991, pp. 422-430.

[99]. D. S. Pham, A. M. Zoubir, Analysis of multicomponent polynomial phase signals, *IEEE Transations on Signal Processing*, Vol. 55, 2007, pp. 56-65.

[100]. M. Pincus, A closed form solution of certain programming problems, *Operation Research*, Vol. 16, 1968, pp. 690-694.

[101]. A. Porchia, S. Barbarossa, A. Scaglione, G. B. Giannakis, Auto focusing techniques for SAR Imaging using the multilag high-order ambiguity function, in *Proceedings of the IEEE International Conference on Acoustics, Speech and Signal Processing (ICASSP'96)*, Atlanta, GA, Vol. IV, 1996, pp. 2086-2089.

[102]. A. Prasad, D. Kundu, A. Mitra, Sequential estimation of the sum of sinusoidal model parameters, *Journal of Statistical Planning and Inference*, Vol. 138, 2008, pp. 1297-1313.

[103]. B. G. Quinn, Estimating the number of terms in a sinusoidal regression, *Journal of Time Series Analysis*, Vol. 10, 1989, pp. 71-75.

[104]. B. G. Quinn, E. J. Hannan, The Estimation and Tracking of Frequency, *Cambridge Universty Press*, New York, 2001.

[105]. B. G. Quinn, P. J. Thomson, Estimating the frequency of a periodic function, *Biometrika*, Vol. 78, 1991, pp. 65-74.

[106]. C. R. Rao, Some results in signal detection, in Decision Theory and Related Topics (S. S. Gupta, J. O. Berger, Eds.), *Springer*, New York, 1988, pp. 319-332.

[107]. F. S. G. Richards, A method of maximum likelihood estimation, *Journal of the Royal Statistical Society, Ser. B*, Vol. 23, 1961, pp. 469-475.

[108]. A. W. Rihaczek, Principles of High Resolution Radar, *McGraw-Hill*, New York, 1969.

[109]. S. Saha, S. M. Kay, Maximum likelihood parameter estimation of superimposed chirps using Monte Carlo importance sampling, *IEEE Transactions on Signal Processing*, Vol. 50, 2002, pp. 224-230.

[110]. H. Sakai, An application of a BIC-type method to harmonic analysis and a new criterion for order determination of an error process, *IEEE Transactions of Acoustics Speech and Signal Processing*, Vol. 38, 1990, pp. 999-1004.

[111]. G. A. F. Seber, C. J. Wild, Nonlinear Regression, *John Wiley and Sons*, New York, 1989.

[112]. P. Stoica, R. Moses, Spectral Analysis of Signals, *Prentice Hall*, Upper Saddle River, New Jersey, 2005.

[113]. P. Ticahvsky, P. Handel, Multicomponent polynomial phase signal analysis using a tracking algorithm, *IEEE Transactions on Signal Processing*, Vol. 47, 1999, pp. 1390-1395.

[114]. M. Vespe, G. Jones, C. Baker, Lessons for radar: Waveform diversity in echolocating mammals, *IEEE Transactions on Signal Processing Magazine*, Vol. 26, 2009, pp. 65-75.

[115]. I. M. Vinogradov, The Method of Trigonometrical Sums in the Theory of Numbers, (Translated from Russian, Reprint of the 1954 Translation), *Dover Publications Inc.*, Mineola, New York, USA, 2004.

[116]. B. Volcker, B. Ottersten, Chirp parameter estimation from a sample covariance matrix, *IEEE Transactions on Signal Processing*, Vol. 49, 2001, pp. 603-612.

[117]. A. M. Walker, On the estimation of a harmonic component in a time series with stationary independent residuals, *Biometrika*, Vol. 58, 1971, pp. 21-36.

[118]. X. Wang, An AIC type estimator for the number of cosinusoids, *Journal of Time Series Analysis*, Vol. 14, 1993, pp. 433-440.

[119]. P. Wang, J. Yang, Multicomponent chirp signals analysis using product cubic phase function, *Digital Signal Processing*, Vol. 16, 2006, pp. 654-669.

[120]. Y. Wang, Y. G. T. Zhou, On the use of high-order ambiguity function for multi-component polynomial phase signals, *Signal Processing*, Vol. 5, 1998, pp. 283-296.

[121]. D. R. Wehner, High-Resolution Radar, 2nd Edition, *Artech House*, Norwell, MA, USA, 1995.

[122]. C. F. J. Wu, Asymptotic theory of the nonlinear least squares estimation, *Annals of Statistics*, Vol. 9, 1981, pp. 501-513.

[123]. J. Z. Wang, S. Y. Su, Z. Chen, Parameter estimation of chirp signal under low SNR, *Science China: Information Sciences*, Vol. 58, 2015, 020307.

[124]. Y. Wu, H. C. So, H. Liu, Subspace-based algorithm for parameter estimation of polynomial phase signals, *IEEE Transactions on Signal Processing*, Vol. 56, 2008, pp. 4977-4983.

[125]. Z. Xinghao, T. Ran, Z. Siyong, A novel sequential estimation algorithm for chirp signal parameters, in *Proceedings of the IEEE Conference in Neural Networks and Signal Processing (NNSP'03)*, Nanjing, China, December 14-17, 2003, pp. 628-631.

[126]. P. Yang, Z. Liu, W.-L. Jiang, Parameter estimation of multicomponent chirp signals based on discrete chirp Fourier transform and population Monte Carlo, *Signal, Image and Video Processing*, Vol. 9, 2015, pp. 1137-1149.

[127]. D. Yaron, A. Alon, C. Israel, Joint model order selection and parameter estimation of chirps with harmonic components, *IEEE Transactions on Signal Processing*, Vol. 63, 2015, pp. 1765-1778.

[128]. H. Zhang, H. Liu, S. Shen, Y. Zhang, X. Wang, Parameter estimation of chirp signals based on fractional Fourier transform, *The Journal of China Universities of Posts and Telecommunications*, Vol. 20, Suppl. 2, 2013, pp. 95-100.

[129]. H. Zhang, Q. Liu, Estimation of instantaneous frequency rate for multicomponent polynomial phase signals, in *Proceedings of the 8th International Conference on Signal Processing (ICSP'06)*, 2006, pp. 498-502.

[130]. K. Zhang, S. Wang, F. Cao, Product cubic phase function algorithm for estimating the instantaneous frequency rate of multicomponent two-dimensional chirp signals, in *Proceedings of the Congress on Image and Signal Processing (CISP'08)*, 2008, pp. 498-502.

[131]. Y. M. Zhu, F. Peyrin, R. Goutte, The use of a two-dimensional Hilbert transform for Wigner analysis of 2-dimensional real signals, *Signal Processing*, Vol. 19, 1990, pp. 205-220.

Chapter 5
New Statistical Algorithms for 3D Signal Processing: The Detection of Floating Objects on an Agitated Sea Surface

Victor Golikov, Oleg Samovarov, Evgeniy Zhilyakov and Rafael Sanchez Lara

5.1. Introduction

Early work in search of video/infrared systems utilized algorithms that first attempted to detect the object spatially in each image, and then applied a temporal association algorithm [1, 2]. Although these algorithms were adequate for early applications in which the targets were bright compared to the background, they performed poorly with dim targets in severe real-world clutter [3, 4]. Approaches that are more recent used multiple frames to incorporate temporal as well as spatial information. The initial work performed by Reed [2] derived the filter that maximized the output SNR in the general case of noise with known auto-covariance function. A number of approaches have been developed for the detection of targets in the presence of the dominant background clutter and noise. The common detection methods include the orthogonal subspace projection and constrained energy minimization [5, 6]. Recently, Kim improved the mean subtraction filter by inserting a target enhancement and noise reduction filter called modified mean subtraction filter (M-MSF) [5, 7]. In [8], the authors develop the FC method. This method is based on the observation that the two images of the background (for example, sea surface) do not correlate, whereas the two images of the same boat do correlate.

The existing multipixel and subpixel target detection methods mainly depend on the spectral characteristics of the targets to be detected [1, 9, 10]. One kind of subpixel detection method is the linear unmixing (LU) analysis methods. These methods are based on the linear mixture model (LMM), in which a pixel's spectrum is assumed to be composed of several pure materials' spectra, called endmembers. In this model, pixels in the imagery are called mixed pixels, and the signal of a mixed pixel is the sum of the

Victor Golikov
UNACAR, Ciudad del Carmen, Mexico

different endmembers' spectral signals. Different endmembers' weighted values are assigned according to their fractions in the pixel. This model is based on the assumption that the signals acquired by the sensor are reflected directly from the ground objects' surfaces, not reflected twice or more times by surfaces on the ground. If multiple reflections are taken into consideration, the LMM has to be extended into a nonlinear model. However, the LMM is suitable in most cases, and it has clear physical meaning. Based on LMM, several spectral LU methods were developed [11-13]. Among them, the fully constrained least squares (FCLS) method is one that has been successfully used in subpixel target detection. In this method, two constrained conditions are used: each endmember's abundance is nonnegative and all the endmembers' abundances in each pixel sum to one. With these two constraining conditions, the abundance results have physical meanings and can be weighted as the percentages of different materials in each pixel. In this method, the subpixel target is assumed as an endmember in the imagery and its percentage in each pixel is figured out by the least squares estimation (LSE) method [14]. In this way, a quantitative detection result is determined.

Another kind of multipixel and subpixel target detection method is based on a statistical hypothesis test. Adaptive matched subspace detector (AMSD) is such an algorithm that formulates the target and background subspaces and uses the LMM and the generalized likelihood ratio (LR) approach to separate a probable subpixel target [15, 16]. The key factor in AMSD is the target abundance and the noise variance, which can be estimated by the maximum likelihood (ML) method, and with which a Neyman-Pearson detector can be designed to maximize the probability of detection for a certain given false alarm probability. Due to the complexity of the changes in atmospheric conditions, sensor geometry, surface defects, and films, target variability is not negligible and can have an impact on the detection result. However, by the use of the subspace model, AMSD can resolve the problem and provides a reliable rule to separate the target pixels from nontarget ones.

Adaptive cosine/coherent estimate (ACE) is another statistical hypothesis test-based method [1, 6, 10, 17]. Unlike AMSD, ACE assumes no structured background. Instead, ACE models the background as a multivariate normal distribution. ACE discards both the sum to one constraint and the nonnegative constraint. This characteristic allows it to provide a better separability between targets and background as the target abundance can be fully expressed by a large enough value.

Recently, hybrid subpixel target detection methods have been proposed, which combine the advantages of LU-based methods and statistical hypothesis test-based methods [18-20].

Hybrid detectors model the background with both physically meaningful abundance and the statistical hypothesis test. In this way, the background is better characterized and the separability of targets and background is improved. In [19], Broadwater and Chellappa used FCLS, AMSD, and ACE to develop two hybrid detectors. Compared to the aforementioned two kinds of method, the two hybrid methods proved to be less sensitive to the number of endmembers used and provided a better separability and improved receiver operating characteristic (ROC) curves.

All the aforementioned methods use the first-order statistics in the formulation of the problem of binary statistical hypotheses testing. But in the problem of detecting an object against a dynamic surface, a situation is possible when the object obscures the background and the task is to distinguish between the background and the object, which can differ in both first order and second order statistics. The most interesting case is when the spectral composition of the object is much narrower than the spectral composition of the background. In this case, it is possible to estimate the background parameters from the primary data. An example is the task of detecting a floating object on the surface of an agitated sea using a video camera. Light signals reflected from a floating object contain only low-frequency harmonics, and a signal reflected from an excited sea surface contains higher frequencies.

To remedy this situation, the new detection algorithms for subpixel objects have been recently developed [21-23]. In these papers, the authors took the modified GLRT-based approach which takes consists of taking into account that the statistical characteristics of the background under H_0 hypothesis (object is not present) are not similar to that under H_1 (object is present). This model is physically motivated. Indeed, in video systems, the target may completely cover the pixel on the fluctuating surface and, in this case, the received signal contains only target signal plus channel noise. Hence, the presence of the target removes the background clutter from the received signal. In this case, it is more appropriate to use the general likelihood ratio (GLR) approach with different background plus noise power under the two hypotheses. Specifically, each pixel contains the background-plus-noise power under the null hypothesis and the signal-plus-noise power under the alternative hypothesis only in the case of the presence of the target in this pixel. We have recently proposed a modified GLR approach associated with the hypothesis-dependent background clutter power for subpixel optical objects in the case of unstructured background clutter in [22-24] and Bo Du proposed it in [25] for the case of structured background clutter.

5.2. 3D Subpixel Target Detection

5.2.1. 3D Subpixel Target Detection Algorithms for Structured Signal Models

In the literature the detection problem with the hypothesis dependent noise power had been considered in [21] for the case of the white normal noise. In [21], Vincent, *et al.* examined specific aspects the situation when the presence of the signal of interest triggers a change in the noise power. This problem has been expanded to the case of the subpixel targets detection where the structured background power is only known under the null hypothesis and the presence of the target triggers a change in the noise power. It was presented an algorithm for subpixel detection in sequence of the images. The approach is based on the use of linear subspaces to model the target and background spectra. The idea of this approach is derived from the phenomena nature. When the target is present in a pixel, the background power will change in this pixel. Accordingly, a noncomplete knowledge of the target's size leads to the background power variation under hypotheses H_0 and H_1.

It is considered the modified adequate to the problem modified statistical test that detects the target presence and the background power change simultaneously. Both phenomena give their contributions into the detection quality. The subject of this paragraph is the design and analysis of the modified algorithm for the detection of subpixel targets with partially known spectral signature in the presence of structured background with unknown power under alternative statistical hypothesis H_1 and channel noise. It is interesting to perform comparisons of the MSD, ACE and the modified MSD (MMSD) detectors. The analytical results explain when the MMSD outperforms the MSD and ACE. Numerical simulations attest to the validity of the analysis.

The linear mixing modeling is widely used when one has to detect a partly known deterministic signal belonging to a known subspace in the presence of an additive background belonging to a known or unknown subspace and white Gaussian noise [1, 6, 26-29]. Because the background power is changed if a target is present, it is necessary to modify the MSD. It is formulated the subpixel detection problem in the case of known and unknown structured background subspace. Then we derive the GLRT for the problems at hand and the distributions under both the hypotheses. The modified MSD employs an additional term in test for the known background subspace. The additional terms should estimate the difference between background variances under the H_1 and H_0. The performance analysis is presented by developing the approximated analytical expressions. It was investigated the detection performance gains in the case of the background power variations between two hypotheses in a Gaussian environment for the proposed detectors with respect to the MSD. Here, the numerical simulations are included to verify the validity of the theoretical analysis and show that the proposed detectors could outperform the classical one. When the closed-form expressions for the false alarm probability and the probability of detection are not available we evaluate them resorting to Monte Carlo techniques. Brief conclusion ends the work.

A pixel that contains more than one material is called a mixed pixel. A mixed pixel is often represented as the linear combination of component spectra (endmembers) [1, 20]. Let L denote the number of spectral bands and let S and at denote an $L \times P$ matrix with columns that contain the orthogonal basis vectors that span the target subspace and the abundance vector of the target, respectively. A mixed pixel, denoted by an L-dimensional vector x, can be described

$$x = \boldsymbol{B}\boldsymbol{a}_{b0} + \sigma_0 \boldsymbol{n}, \text{ under } H_0,$$

$$x = \boldsymbol{S}\boldsymbol{a}_t + \boldsymbol{B}\boldsymbol{a}_{b1} + \sigma_1 \boldsymbol{n}, \text{ under } H_1, \tag{5.1}$$

where $\boldsymbol{B}$ is the $L \times Q$ matrix representing the background endmembers (Discreet Fourier components), $\boldsymbol{a}_b$ are the abundance vector of the background endmembers, $\boldsymbol{n}$ is the L-dimensional random background error vector (or error noise) accounting for lack-of-fit and noise effects [1, 30], σ^2 is the variance of the error noise, and $\sigma\boldsymbol{n} \sim N[\boldsymbol{0}, \sigma^2 \boldsymbol{I}]$. Note that the lack-of-fit effect causes a spectral tail and we assume that its variance predominates over the thermal noise [1, 30]. Subpixel material detection requires determining if a pixel spectrum x contains some amount of the target material. Our approach is based on the use of linear subspaces and LMM with a structured background (5.1). We assume that the background spectra span some subspace of the real vector space $\mathcal{R}^L$ and the target material spectrum lies in a different subspace within $\mathcal{R}^L$. We assume that

$$\boldsymbol{a}_{b1} = c\boldsymbol{a}_{b0}, \text{ and } \sigma_1 = c\sigma_0, \qquad (5.2)$$

where c is the pixel fill factor. The scale total interference under H_1 is used in this model because in the case of subpixel targets, the amount of background covered area may be different from that of a pure background pixel. The target's size determines the pixel fill factor c and practically always is unknown (see Fig. 5.1(b)). In this work, we focus on the detection of a small target in the case of unknown total interference scale factor c under hypothesis H_1.

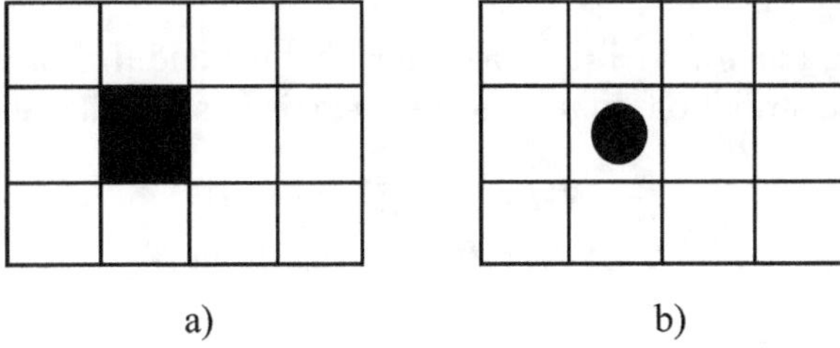

a)   b)

Fig. 5.1. Typical target detection scenario: a) Resolved target occupying a whole pixel (pixel fill factor $c = 0$); b) Subpixel target in pure background with the pixel fill factor of $c \approx 0.3$.

Healy and Slater [29] used a physical model to show that the set of sensor spectral radiance vectors for a material can be approximated accurately by a small number of orthogonal basis vectors. We will use this approach to define a target subspace. Also, we describe shortly a method for finding the subspace that is spanned by the background component spectra. It was considered a rectangular image region of size $\alpha \times \beta$ pixels with L spectral bands. It was constructed a matrix of size $L \times \alpha \beta$, where each column contains a hyperspectral pixel vector in the region. Using the singular value decomposition, it was obtained the diagonal matrix that contains the singular values λ_1, λ_2, ..., λ_L. Suppose that the eigenvectors columns (background basis vectors) are arranged so that $\lambda_1 \geq \lambda_2 \geq ... \geq \lambda_L$.

The first M background basis vectors will always be included in the background subspace to ensure that the squared error is less than a determined fraction of the total variance. It is not advantageous to include a number of basis vectors in the background subspace that reduces the squared error below the noise or error levels. However, there are situations where either we do not have sufficient information, or we are unwilling to estimate the background subspace $\boldsymbol{B}$. In these situations, we may be only interested in a certain object present in an unknown image scene and only its spectral subspace $\boldsymbol{S}$ is available. We assume that the background matrix $\boldsymbol{B}$ is unavailable, and it cannot be estimated from the data. Then, we deal with the total interference

$$\sigma_{0v}\boldsymbol{v} = \boldsymbol{B}\boldsymbol{a}_{b0} + \sigma_0\boldsymbol{n} \text{ under } H_0$$

$$\text{and } \sigma_{1v}\boldsymbol{v} = \boldsymbol{B}\boldsymbol{a}_{b1} + \sigma_1\boldsymbol{n} \text{ under } H_1 \qquad (5.3)$$

Under these assumptions, the received signal is

$$x = \begin{cases} \sigma_{0v}v \text{ under } H_0 \\ Sa_t + \sigma_{1v}v \text{ under } H_1 \end{cases}, \tag{5.4}$$

where $\sigma_{1v} = c\sigma_{0v}$, $a_{b1} = ca_{b0}$.

5.2.2. Modified Matched Subspace Detectors

The purpose of this section is to derive the modified matched subspace detector (MMSD) for known background structure to differentiate between the pixels that contain a target and the pixels that contain only background.

The structured background model (5.1) assumes that the additive noise is modeled by a multivariate normal distribution. In this case the received signal has distributions

$$x \sim \begin{cases} N[Ba_{b0}, \sigma_0 I] \text{ under } H_0 \\ N[Sa_t + Ba_{b1}, \sigma_1 I] \text{ under } H_1 \end{cases}, \tag{5.5}$$

where I is L by L identity matrix, abundance vectors a_t, a_{b0} and a_{b1} are unknown, and $\sigma_1 = c\sigma_0$ is unknown (or c is unknown). Under these assumptions, we can calculate the remaining unknown parameters using the Maximum likelihood Estimation (MLE) technique. The MMSD algorithm provides a statistical test using a Generalized Likelihood Ratio Test (GLRT). To do this, we calculate the likelihood equations for the null and alternative hypotheses as

$$L(x|H_0) = (2\pi\sigma_0^2)^{-L/2}exp\left\{-\frac{1}{2\sigma_0^2}(x - Ba_{b0})^T(x - Ba_{b0})\right\},$$

$$L(x|H_1) = (2\pi\sigma_1^2)^{-L/2}exp\left\{-\frac{1}{2\sigma_1^2}(x - Sa_t - Ba_{b1})^T(x - Sa_t - Ba_{b1})\right\} \tag{5.6}$$

Taking the derivative of the logarithm of (5.6) with respect to each of the unknown parameters and setting them equal to zero allows us to arrive at the (MLE) abundance estimates

$$\hat{a}_{b0} = (B^T B)^{-1}B^T x, \; \hat{a}_{b1} = (A^T A)^{-1}A^T x, \text{ and } \hat{a}_t = (C^T C)^{-1}C^T x, \tag{5.7}$$

and the MLE noise variance estimate under alternative hypothesis

$$\hat{\sigma}_1^2 = \frac{1}{(L-J)}\left\|P_E^\perp x\right\|^2 = \frac{1}{(L-J)}\left\|x_E^\perp\right\|^2, \tag{5.8}$$

where $x_E^\perp = P_E^\perp x$, $P_E^\perp = I - P_E$, $P_E = E(E^T E)^{-1}E^T$, $E = [S, P_S^\perp B]$ is a $L \times J$ matrix, $A = P_S^\perp B$, $C = P_B^\perp S$, $P_S^\perp = I - P_S = I - S(S^T S)^{-1}S^T$, $\hat{a}_{b0}$ is the estimate of a_{b0}, $\hat{a}_{b1}$ is the estimate of the concatenation of a_t and a_{b1}, P_S and P_B are the orthogonal projection matrices onto the target and background subspaces, and P_E is the orthogonal projection matrix onto the concatenation of the target and background (without its fraction occupied by a target) subspaces. We assume here that L >> P and substituting (5.8) and (5.7) into (5.6) provides the generalized likelihood equation under alternative hypothesis. Then, the GLR is given by

$$\frac{p(x;\hat{a},\hat{\sigma}_1^2|H_1)}{p(x|H_0)} =$$

$$= \frac{\left\{2e^{(L-J)/L}\pi(L-J)^{-1}\|x_E^\perp\|^2\right\}^{-\frac{L}{2}}}{(2\pi)^{-\frac{L}{2}}\sigma_0^{-L}exp\left\{-\frac{\|x_B^\perp\|^2}{2\sigma_0^2}\right\}} = \left[e^{(L-J)/L}\frac{\|x_E^\perp\|^2}{(L-J)\sigma_0^2}exp\left\{-\frac{\|x_B^\perp\|^2}{L\sigma_0^2}\right\}\right]^{-L/2} \qquad (5.9)$$

Taking the logarithm of the Lth root of the (5.9) and using some algebra, the GLRT is given by the following MMSD statistical test

$$T_{MMSD}(x) = \frac{\|x_t\|^2}{L\sigma_0^2} + \frac{\|x_E^\perp\|^2}{L\sigma_0^2} - ln\frac{\|x_E^\perp\|^2}{(L-J)\sigma_0^2} - \frac{L-J}{L}, \qquad (5.10)$$

where $x_t = P_C x$. The first term in (5.10) is well known MSD statistical test [1, 10, 31]

$$T_{MSD}(x) = \frac{\|x_t\|^2}{L\sigma_0^2} \qquad (5.11)$$

The test (5.11) provides such a statistical test using a GLRT in the case of the known background variance under the H_1. We can see (5.10) that the detection quality depends on relation between the target recognition contribution (the first term) and the background power change contribution (the second and third terms).

We can regulate the target presence sensitivity of detector with respect to the background power change sensitivity using the factor m. Varying the factor m we can change the relation between the first and other terms in the statistical test (5.12), i.e. between the target recognition contribution (depends on the target signature) and the background power change contribution (in the fraction of the total space outside of the target subspace). We can introduce the factor of signal detection sensitivity m and rewrite the proposed detector in the following form:

$$T_{MMSD}(x) = \frac{m\|x_t\|^2}{L\sigma_0^2} + \frac{\|x_E^\perp\|^2}{L\sigma_0^2} - ln\frac{\|x_E^\perp\|^2}{(L-J)\sigma_0^2} - \frac{L-J}{L} \overset{>H_1}{\underset{<H_0}{}} \eta, \qquad (5.12)$$

where η is the threshold. Fig. 5.2 shows the computational structure of the MMSD. We observe that the MMSD, MSD and ACE differ in the way they remove the background power from the total power.

The MMSD uses the background power estimate as the additional term in the statistics but the ACE [6, 23, 24] applies this estimate as the MSD normalizing factor. Unfortunately, such unmixing methods require the complete knowledge of the image endmembers and background subspace. On some practical occasions, obtaining this prior knowledge may not be realistic. In this problem statement, we addressed the problem of detecting targets in the error noise-dominant scenario, that is, we limited the influence of the thermal noise.

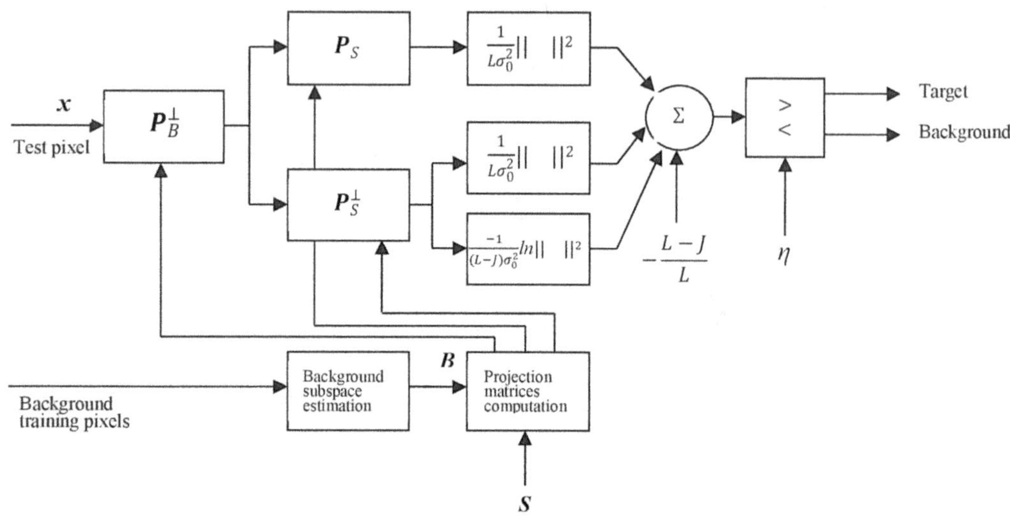

Fig. 5.2. Computational structure for the implementation of the MMSD for unknown structured background.

5.2.3. Detection Power Analysis

In the case of structured background is known to make a decision we need to compare T_{MMSD} (5.12) to a given threshold η and decide H_1 when $T_{MMSD}(x) \geq \eta$ and H_0 otherwise. Since the threshold determines both probability of detection P_D and false alarm probability P_{FA}, we need to determine the probability distribution of $T_{MMSD}(x)$. The first term of the $T_{MMSD}(x)$ (5.12) is the MSD statistical test $T_{MSD}(x)$ in the subspace orthogonal to background subspace. Since x is drawn from a real multivariate Gaussian distribution, with mean and covariance matrix $\sigma_0^2 I$, it follows immediately that

$$T_{MSD}(x) = \frac{m\|x_t\|^2}{L\sigma_0^2} \sim \begin{cases} \dfrac{m}{L}\chi^2_{J-Q}(0) & \text{under } H_0 \\[2mm] \dfrac{mc^2}{L}\chi^2_{J-Q}(\lambda_S^2) & \text{under } H_1 \end{cases}, \tag{5.13}$$

where $\lambda_S^2 = \dfrac{(Sa_t)^T P_B^\perp (Sa_t)}{\sigma_1^2}$ is a noncentrality parameter. When the pixel fill factor decreases the parameter λ_S^2 increases but threshold increases too. It is a reason of a detection performance diminution. By analogy, we observe that the second term in (5.12) $\dfrac{\|x_E^\perp\|^2}{L\sigma_0^2}$ is $\dfrac{1}{L}\chi^2_{L-J}(0)$ a central distribution under H_0 and $\dfrac{c^2}{L}\chi^2_{L-J}(0)$ under H_1. Since the third term $ln\dfrac{\|x_E^\perp\|^2}{(L-J)\sigma_0^2}$ is nonlinear function, the deriving the exact distribution it is very difficult. In order to come up with manageable expressions, we thus investigate an asymptotic approach, assuming that the number of samples L is large. The approximation is based on well-known convergence the central chi-square distribution to a Gaussian distribution. It is possible to obtain the following approximation for the chi-square distribution:

$$\frac{\|x_E^\perp\|^2}{(L-J)\sigma_0^2} \sim \begin{cases} N\left(1, \frac{2}{L-J}\right) \text{ under } H_0 \\ N\left(c^2, \frac{2c^4}{N-J}\right) \text{ under } H_1 \end{cases} \tag{5.14}$$

For high value of L it is easy to obtain the approximation

$$\frac{\|x_E^\perp\|^2}{L\sigma_0^2} - \ln\frac{\|x_E^\perp\|^2}{(L-J)\sigma_0^2} - 1 \approx \frac{1}{2}\left[\frac{\|x_E^\perp\|^2}{(L-J)\sigma_0^2} - 1\right]^2 \tag{5.15}$$

We used the latter approximation along with (5.15) and obtained that

$$\frac{\|x_E^\perp\|^2}{L\sigma_0^2} - \ln\frac{\|x_E^\perp\|^2}{(L-J)\sigma_0^2} - 1 \sim \begin{cases} \frac{1}{L-J}\chi_1^2(0) \text{ under } H_0 \\ \frac{c^4}{L-J}\chi_1^2\left(\frac{(L-J)(c^2-1)^2)}{c^4}\right) \text{ under } H_1 \end{cases} \tag{5.16}$$

Since the first and second terms are independent, the asymptotic distribution of $T_{MMSD}(x)$ is given by as follows:

$$T_{MMSD}(x) \sim \begin{cases} \frac{1}{L-J}\chi_1^2(0) + \frac{m}{L}\chi_{J-Q}^2(0) \; H_0 \\ \frac{c^4}{L-J}\chi_1^2\left(\frac{(L-J)(c^2-1)^2}{c^4}\right) + \frac{mc^2}{L}\chi_{J-Q}^2(\lambda_S^2) \; H_1 \end{cases} \tag{5.17}$$

The distributions derived above enable one to obtain the received operating characteristics (ROC). In order to come up with exploitable expressions, we examine a further approximation ($L \gg J$, $L \gg P$, $L \gg Q$) to (5.17):

$$T_{MMSD}(x) \sim \begin{cases} \frac{m}{L}\chi_{J-Q+1}^2(0) \text{ under } H_0 \\ \frac{mc^2}{L}\chi_{J-Q+1}^2(\lambda_S^2 + \lambda_1^2) \text{ under } H_1 \end{cases}, \tag{5.18}$$

where $\lambda_1^2 = \frac{(L-J)(c^2-1)^2}{c^4}$. This expression holds for large L and $c \approx 1$. We verified that the pdf of (5.18) is close to those of (5.12) for parameter c between 0.5 and 1. Furthermore, through extensive Monte Carlo simulations, we checked that the pdf of (5.18) matches the exact pdf of (5.12) for $L \geq 200$ and $P, Q < 20$, $J < 40$. We provided a qualitative analysis of the differences between classical test T_{MSD} and the proposed test T_{MMSD} in presence of the pixel fill factor c variation (the noise and the background power variation). The T_{MSD} depends only on the projection on the subspace orthogonal to the background subspace; the T_{MMSD} depends on the projections on the two subspaces: on the subspace orthogonal to the background subspace (first term in (5.12)) and on the white noise subspace (second and third terms in (5.12)). Noticing (15, 11 and 8) that T_{MMSD} can be rewritten as

$$T_{MMSD}(x) = T_{MSD}(x) + \frac{1}{2}\left[\frac{\hat{\sigma}_1^2 - \sigma_0^2}{\sigma_0^2}\right]^2, \tag{5.19}$$

we observe that T_{MMSD} modifies T_{MSD} by adding a corrective factor proportional to the square of the noise power variation between H_1 and H_0. This correction increases the probability of detection. One can notice that the only information used by T_{MMSD} to modify T_{MSD} is the power in the noise subspace. One can then expect the different behavior of the proposed detector, each time the estimated noise power is different from the expected one. When the target is present, the background and error noise powers will decrease. The projection onto the subspace orthogonal to the background subspace will decrease and the projection onto the noise subspace will increase (5.19). The proposed detector could recover a part of the energy having moved from one subspace to the other and try to maintain the test performance. The distribution derived above enable one to obtain the ROC, using the complementary cumulative distribution of a noncentral $\chi_n^2(\lambda)$ random variable with n degrees of freedom and noncentral parameter λ:

$$C(\eta,\, n,\, \lambda) = \int_{\eta}^{\infty} T_{\chi_n^2(\lambda)}(x)\, dx \tag{5.20}$$

We can obtain the threshold η, false alarm probability P_F and probability of detection P_D in such form:

$$\eta = C^{-1}(P_F, n, \lambda),\ P_F = C(L\,\eta, L\text{-}Q\text{+}1,0),\ P_D = C(\eta c^{-2}, L - Q + 1, \lambda_S^2 + \lambda_1^2)\ (5.21)$$

Based on these expressions, the detection probability depends on three factors: the pixel fill factor, the degree of freedom and noncentrality parameter. When the target is present, the pixel fill factor decreases and, hence, the bottom limit of the integral (5.20) increases. It leads to reduction of the detection probability. In contrast, at the pixel fill factor reduction, the noncentrality parameter increases, and that leads to improvement of detectability.

In general, as a result of the action of two opposite factors (the first and the third parameters in (5.21)), the detection probability can decrease or increase in the case of the reduction of the fill factor. These changes of detection probability depend on the target and background dimensions. Note that the noncentral parameter of the T_{MMSD} more than the noncentral parameter of T_{MSD} on the value λ_1^2. This value depends on the pixel fill factor c and the subspaces (S and B) degrees of freedom. When the target appears in the pixel the fill factor c decreases, hence, the chi-square noncentral parameter (5.18) increases on the value λ_1^2 that causes the augment of the detection probability.

The asymptotic approach, assuming that the number of samples L is large, was based (5.10) on convergence the central chi-square distribution to a Gaussian distribution. In our case the second term of the $T_{MMSD}(x)$ (5.12) has the noncentral chi-square distribution:

$$\frac{\|x_\perp\|^2}{(L-P)\sigma_0^2} \sim \begin{cases} \frac{1}{L-P}\chi_{L-P}^2(\lambda_B^2) \text{ under } H_0 \\ \frac{c^2}{L-P}\chi_{L-P}^2(\lambda_B^2) \text{ under } H_1 \end{cases}, \tag{5.22}$$

where $\lambda_B^2 = \dfrac{(Ba_{b0})^T P_S^{\perp}(Ba_{b0})}{\sigma_1^2}$. Unfortunately, the third term (5.12) $ln\,\dfrac{\|x_{\perp}\|^2}{(L-P)\sigma_0^2}$ is nonlinear function and the asymptotic approach is not applicable because the noncentral chi-square distributions (5.22) do not converge to a Gaussian distribution. In this case we will apply Monte Carlo simulations to obtain the ROC.

5.2.4. Numerical Illustrations

The aim of this section is threefold. First, we assess the validity of the theoretical formulas ((5.18), (5.20), (5.21) for T_{MMSD}) by comparing it with the actual receiver operating characteristics obtained through Monte Carlo simulations. We analyze the MMSD detection performance with respect to the MSD and ACE in the case of different pixel fill factor. Second, since closed-form expressions for the P_F and the probability of detection P_D are not available for the T_{MMSD} we evaluated them resorting to standard Monte Carlo techniques based on $100/P_F$ and $100/P_D$ independent trials, respectively. In order to limit the computational burden, we set $P_F = 10^{-3}$. We assess the performances of the MMSD and the MSD (5.13) in the case of the structured background power variation under hypothesis H_1. Third, we compare the performance gain of the proposed detectors (5.11) and (5.12) with respect to the MSD under the different pixel fill factor.

In this section, we consider the target and background abundance vectors $a_t = [1,1,\ldots,1]^T$ and $a_{b0} = [1,1,\ldots,1]^T$, number of measurements $L = 200$, the target and the background subspace dimensions $P = 11$, $Q = 10$, respectively. The target subspace matrix S is not orthogonal to the background subspace matrix B. We have used the orthogonal $L \times P$ and $L \times Q$ Fourier matrices to represent the target S and background B subspaces in the simulations. The some fraction of the target and background spectral components has been the common.

Figs. 5.3-5.5 illustrate the relation between the detection probability P_D and the input signal-to-noise ratio (SNR$_{in}$) under the chosen system constraint resulting from the 10^6 Monte Carlo trials. In Fig. 5.3 we have plotted the approximated analytical performance given by (5.18), (5.21) for MMSD (dashed line for simulation results and continuous for analytical results). One can see that the closed form expressions (5.18), (5.21) give a relatively precise approximation for $0.8 < c < 1$ and $0.6 < P_D < 1$. Notice that in the case of $c = 1$ the mean of the second term in the detector statistic (5.19) is equal to zero and, hence, the performances of the T_{MMSD} and T_{MSD} are approximately equal. Using Fig. 5.3, we can compare the detection probability behavior for weak targets and varying c. In this case the first term in (5.19) is very small and the detection probability is almost completely determined by the second term (see curves for SNR$_{in} < 22$ dB and c = 0.8). As we can see in Fig. 5.4, the MMSD has the better performance than the MSD and ACE for $0.5 < c < 0.7$.

Fig. 5.5 shows that the MSD considerably outperforms the MSD and ACE in the case of the background power deviation under the alternative hypothesis. In Fig. 5.6, we display the theoretical (5.18), (5.21) detection probability of the MMSD as a function of the false alarm probability (ROC). We can see (Fig. 5.6) that the quality of the MMSD outperforms the MSD considerably for different values of the false alarm probability. Changing the factor of signal detection sensitivity m, it is possible to increase or reduce the detection performance and the detection gain. The choice of the m leads to the trade-off between the detection sensitivity to the target signature and to the background power deviation under the alternative hypothesis. For high m the first term (5.19) determines the major contribution to the detection statistics and, hence, the detection sensitivity to the target signature is high. In the case of low m the detection sensitivity to the background power deviation is high. The MMSD or MMSRD detection gain with respect to MSD or ACE can be defined as the incremental of SNR_{in} required that to obtain under the same instance about the background and the target models, the same detection probability $P_D = 0.9$. Fig. 5.7 illustrates the MMSD detection gain factor as a function of the pixel fill factor for different values of the signal detection sensitivity factor m. One can see that the detection gain factor depends on detection sensitivity factor m.

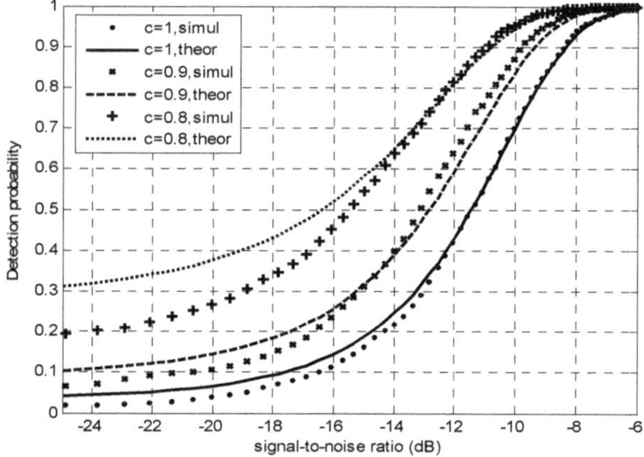

Fig. 5.3. Detection probability of the MMSD, versus SNR_{in} for varying pixel fill factor c, $m = 1.5$, $L = 200$, $P = 11$, $Q = 10$, $P_F = 10^{-3}$. Theoretical and simulation results.

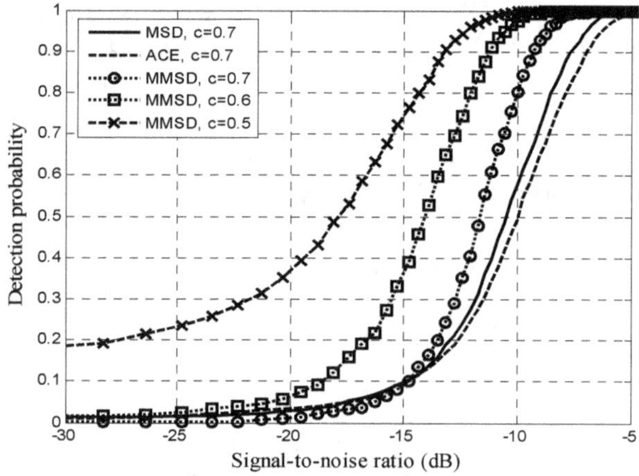

Fig. 5.4. Detection probability of the MMSD, MSD, and ACE versus SNR_{in} for varying pixel fill factor c, $m = 7$, $L = 200$, $P = 11$, $Q = 10$, $P_F = 10^{-3}$. The background matrix $\boldsymbol{B}$ is known. Simulation results.

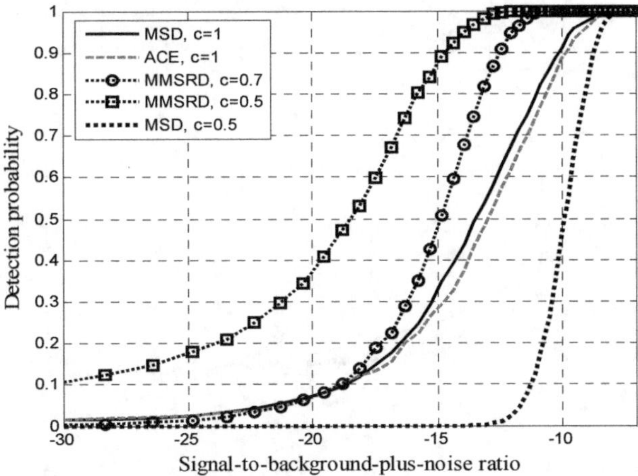

Fig. 5.5. Detection probability of the MSD, ACE and MMSD versus signal-to-background-plus-noise ratio for varying pixel fill factor c, $m = 4.4$, $L = 200$, $P = 11$, $Q = 10$, BNR $= 4$, $P_F = 10^{-3}$. The background matrix $\boldsymbol{B}$ is unknown. Simulation results.

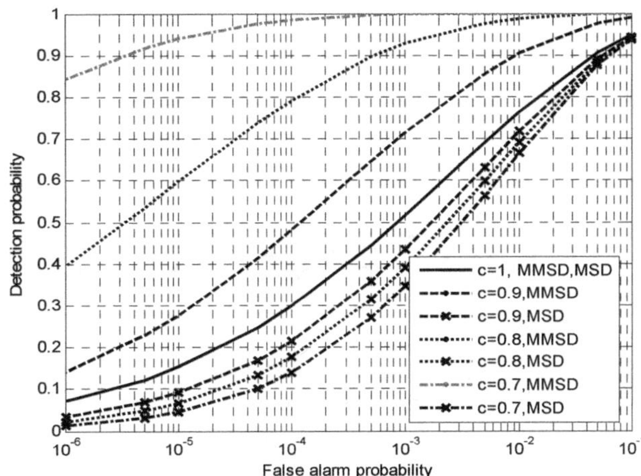

Fig. 5.6. Detectors comparison for $m = 1.25$, $L = 200$, $P = 11$, $Q = 10$, SNR$_{in}$ = -9 dB for varying pixel fill factor c. Theoretical results.

When m decreases the detection gain factor increases and can achieve 29 dB. To illustrate the behavior of the detection gain factor of T_{MMSD}, we display the Fig. 5.8. Since the MMSD does not eliminate the energy within the background subspace we introduce the background to noise ratio factor (BNR). The BNR is the relation between background and noise powers. It is clearly that the gain factor of the T_{MMSD} depends on BNR and achieves the large value (24 dB) in the case of the small c. At the decreasing of the m the sensitivity to the background power change of both proposed detectors will be increased. In this case, these detectors can have the unnecessary high sensitivity with respect to the background power changing.

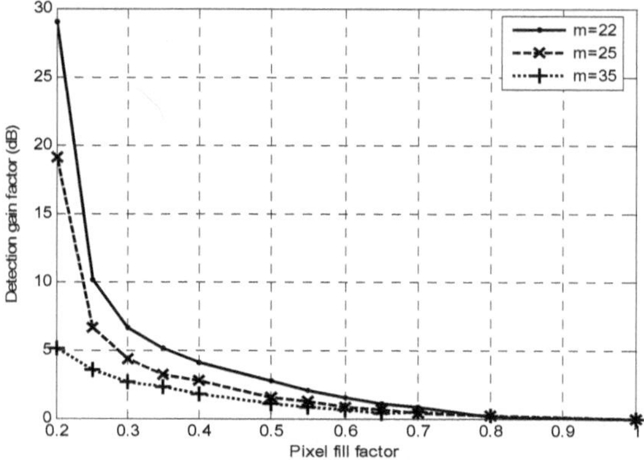

Fig. 5.7. Detection gain factor of T_{MMSD} vs. pixel fill factor c, $L = 200$, $P = 11$, $Q = 10$, $\text{SNR}_{in} = 0.024$, $P_F = 10^{-3}$. The background matrix $\boldsymbol{B}$ is known. Simulation results.

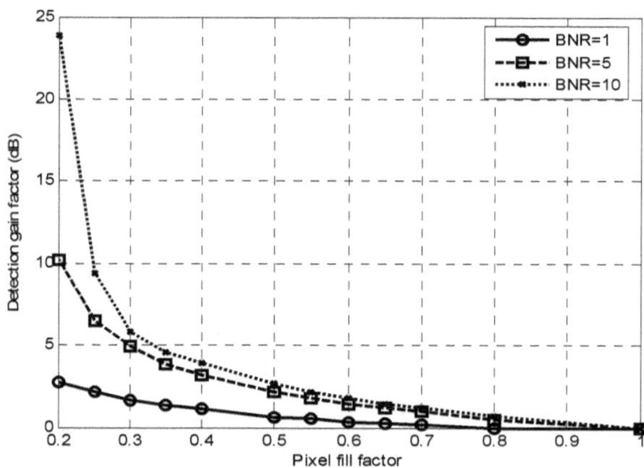

Fig. 5.8. Detection gain factor of T_{MMSRD} vs. pixel fill factor c, $L = 200$, $P = 11$, $Q = 10$, $P_F = 10^{-3}$. The background matrix $\boldsymbol{B}$ is unknown. Simulation results.

The natural variation of the background power can be equal to the background power change associated with the target presence and, hence, the false alarm probability can increase. The natural reason of this phenomenon is a possible instability of the background power.

5.2.5. Conclusion

In this section, we have proposed the asymptotic optimum target detection algorithms (with background subspace known or not) in the presence of the structured background and noise. We derived the GLRT for the case where the structured background plus noise power is only known under the null hypothesis. The proposed detectors modify the MSD by adding the corrective term proportional to the square of the background power variation. Numerical simulations attest to the validity of the theoretical analysis and show that this new detectors could considerably outperform the classical one.

248

Our approach implicitly assumes that HSI data of the targets and background can be approximated by a small number of orthogonal basis vectors [2]. Comparing the MMSD and MSD, we note that the MMSD outperforms the MSD and the MMSD is robust with respect to the background subspace B and also is less dependent on the relation between background error vector and thermal noise.

5.3. 3D Multipixel Target Detection

5.3.1. 3D Multipixel Target Detection Algorithms for Unstructured Signal Models

The goal is to design and assess the adaptive scheme to detect unknown, relatively small low contrast targets in a given sequence of images with a random homogeneous background. We use video spatial-temporal patches, called bricks, to determine whether the observed brick contains a target in the presence of the random background. We assume that some or all pixel areas in the analyzed brick can be covered by the target under hypothesis H_1. These pixels contain the target signal plus channel noise, and the remaining pixels contain the background clutter plus channel noise. The variance of each pixel of the brick under H_1 is equal to the background variance plus channel noise variance or only channel noise variance. We focus attention on a situation when the presence of the target in the pixel changes the pixel noise covariance matrix. In this case, the pixel noise is equal to the channel Gaussian white noise. It is assumed that the optical background clutter is the predominant noise factor and is modeled as a correlated Gaussian noise. For simplicity, all the image data under H_0 is assumed to be a zero-mean Gaussian random vector. Specifically, the local mean of each pixel has been subtracted from the original pixel value. The whole data set is divided into primary and secondary data sets. The primary data set (all pixels in the observed brick) consists of a real pixel-vectors, x_j, $N \times 1$, $j = 1, \ldots, L$. The secondary data set consists of G bricks (GL pixel-vectors or GLN pixels), does not contains any targets and exhibits the same structure of the background covariance structure. Under H_0, i.e. the background-plus-channel-noise hypothesis, we have $x_j = \sigma_{0,j} n_j$, where the noise pixel-vector $n_j \sim N[0, R]$, the R is normalized covariance matrix of the background clutter plus channel noise for the homogeneous environment, and $\sigma_{0,j}$ is the variance of the background plus channel noise. Under H_1, i.e. the signal-plus-background-plus-channel-noise hypothesis, we have $x_j = s_j + \sigma_{1,j} n_j$, where $s_j = H\theta_j$ is the deterministic signal of interest that belongs to a known subspace $\langle H \rangle$ of size p [32], H is an orthogonal $N \times p$ target mode matrix, p is the rank of a projection matrix [32] $P_s = H(H^H H)^{-1} H^H$, θ_j is the unknown amplitude vector, and $j = 1, \ldots, U$. The noise pixel-vector of the primary data under H_1, $\sigma_{1,j} n_j$, is scaled by $\sigma_{1,j}$, where we have $\sigma_{1,j} = \sigma_{0,j}$ for the pixel-vectors without target ($n_j \sim N[0, R]$); and $\sigma_{1,j} = \sigma_{ch,j}$ for the pixel-vectors with target ($n_j \sim N[0, I]$), where $\sigma_{ch,j}$ is the standard deviation of the channel noise and I is identity matrix.

We consider the problem of detecting a multi-pixel optical target in the sequence of N digital images with a random homogeneous Gaussian background and channel noise. We assume that the multi-pixel target may be present completely or partially anywhere in the brick of size L of the pixel-vectors. We assume that the area of the observed brick is partially covered by the target, and the presence of the target changes the background-plus-noise power in the pixels covered by the target. Let $j = 1, \ldots, U$ be the subset of integers indexing the pixel-vectors (U is unknown), which may contain a partially unknown object under the H_1 hypothesis and $j = U+1, \ldots, L$ be the subset of the pixel-vectors, which do not contain the object under the H_1. The secondary data set consists of GL pixel-vectors x_j, $N{\times}1$, $j = L+1,2, \ldots,(G+1)L$, which are assumed to have the background clutter and white channel noise components only. In this case, we assume that $x_j{\sim}N[0, \sigma^2 R]$ are independent and identically distributed Gaussian vectors and that x_j are independent of the primary data x_j. Note that we assume the homogeneous Gaussian environment and assume that the secondary data set is available to estimate both the normalized covariance matrix R and $\sigma^2 = \sigma_{0,j}^2$.

We develop a hypothesis test that distinguishes the signal-plus-noise hypothesis (H_1) from the background-plus-noise hypothesis (H_0). Then, the hypotheses are expressed as

$$\begin{cases} H_0: x_j = c_j + n_j, j = 1,2,\ldots,(G+1)L, \\ H_1:\begin{cases} x_j = s_j + n_j, j = 1,\ldots,U, \\ x_j = c_j + n_j, j = U+1,\ldots,(G+1)L \end{cases} \end{cases} \tag{5.23}$$

Our approach may be evaluated against: (i) the well-known statistic GLRT of Kelly [33] that assumes the same noise scaling in both primary and secondary data ($\sigma = \sigma_0 = \sigma_1 = 1$), (ii) the ASD ($\sigma = \sigma_0 = \sigma_1 \neq 1$) [34], and (iii) the constant false alarm ASD [35] that assumes $\sigma \neq \sigma_0 = \sigma_1$. The statistic test ACE assumes that the variance σ_0^2 is estimated using the primary data. In contrast to these papers, we assume that $\sigma_{0,j}$ may be estimated using the secondary data set, but $\sigma_{1,j}$ may be estimated by primary data and may be different with respect to $\sigma_{0,j}$. It is well known that GLRT is obtained by inserting maximum-likelihood (ML) estimates for unknown parameters into the likelihood ratio. Following the generally accepted approach, we derive the GLRT by considering the joint probability density function (*pdf*) of the measurement and the secondary data set. This paragraph is devoted to the derivation of the two-step detector. In particular, we first derive the GLRT based on primary data, assuming that the variance ($\sigma_{0,j}^2 = \sigma^2$) and normalized covariance matrix (R) of the background plus channel noise are known. Fully adaptive detector is obtained by substituting the unknown matrix by the sample covariance matrix based on secondary data only.

Step 1. Under the hypotheses H_1 and H_0, *pdf* of the $(G+1)L$ vectors may be written as

$$p_{1,x_1,\cdots,x_{(G+1)L}}\left(x_1,\cdots,x_{(G+1)L}\big|R,\sigma_{1,j}^2,\theta_j,U,H_1\right)=$$

$$= \frac{(2\pi)^{-N(G+1)L/2}}{\|I\|^{U/2}\|R\|^{[(G+1)L-U]/2}\left[\prod_{j=1}^{U}\sigma_{ch,j}^2\,\prod_{j=1+U}^{N(G+1)L}\sigma_{1,j}^2\right]^{N/2}}\,exp\left[-\frac{1}{2}tr(T_1 + R^{-1}T_2)\right], \tag{5.24}$$

$$p_{0,x_1,\cdots,x_{(G+1)L}}\left(x_1,\cdots,x_{(G+1)L}\big|R,\sigma_{0,j}^2,H_0\right) =$$

$$= \frac{(2\pi)^{-N(G+1)L/2}}{\|R\|^{(G+1)L/2}\left[\prod_{j=1}^{N(G+1)L}\sigma_{0,j}^2\right]^{N/2}}\,exp\left[-\frac{1}{2}tr(R^{-1}T_0)\right], \tag{5.25}$$

where $\| \ \|$ and $tr()$ denote the determinant and the trace of a square matrix, respectively,

$$T_1 \ = \ \sum_{j=1}^{U} \sigma_{ch,j}^{-2}(x_j - H\theta_j)(x_j - H\theta_j)^T, \tag{5.26}$$

$$T_2 \ = \ \sum_{j=U+1}^{L} \sigma_{1,j}^{-2} x_j \, x_j^T + \sum_{j=L+1}^{(G+1)L} \sigma^{-2} x_j \, x_j^T, \tag{5.27}$$

$$T_0 = \sum_{j=1}^{L} \sigma_{0,j}^{-2} x_j \, x_j^T + \sum_{j=L+1}^{(G+1)L} \sigma^{-2} x_j \, x_j^T \tag{5.28}$$

The GLR statistic for the problem at hand is

$$L \ = \ \frac{\displaystyle\max_{\sigma_{1,j}^2, \theta_j, U} p_{1,x_1,\cdots x_{(G+1)L}}\left(x_1, \cdots, x_{(G+1)L} \middle| R, \sigma_{1,j}^2, \theta_j, U, H_1\right)}{p_{0,x_1,\cdots x_{(G+1)L}}\left(x_1, \cdots, x_{(G+1)L} \middle| R, \sigma_{0,j}^2, H_0\right)} \tag{5.29}$$

where the numerator is maximized by independent varying $\sigma_{1,j}^2$, θ_j and U. Using the primary data set, we obtain two MLE of the $\sigma_{1,j}^2$: the first estimate for the subset $j = 1, \ldots, U$, where the possible object is present (in his case the pixel noise is equal to channel noise), and the second estimate for the subset $j = U+1, \ldots, L$, where the possible object is not present (in this case the pixel noise is equal to the background-plus-channel noise). Then, the MLE of the $\sigma_{1,j}^2$ has two solutions [24]:

$$\widehat{\sigma}_{1,j}^2 \ = \ \widehat{\sigma}_{ch,j}^2 \ = \ \frac{x_j^T P_s^{\perp} x_j}{N-p} \text{ for } j = 1, \ldots, U, \tag{5.30}$$

$$\text{and } \widehat{\sigma}_{1,j}^2 \ = \ \frac{x_j^T x_j}{N}, \text{ for } j = U+1, \ldots, L, \tag{5.31}$$

where p is the rank of the target subspace $\langle H \rangle$, $P_s^{\perp} = I - P_s$ is $N \times N$ orthogonal projection matrix onto the subspace orthogonal to the signal subspace $\langle H \rangle$, and P_s is $N \times N$ orthogonal projection matrix onto the signal subspace $\langle H \rangle$. The MLE of the vector θ_j has an explicit solution [24], when the target is present ($j = 1, \ldots, U$):

$$\widehat{\theta}_j = (H^H H)^{-1} H^H x_j \tag{5.32}$$

Substitutions of θ_j (5.32) and $\widehat{\sigma}_{1,j}^2$ (5.30) into the (5.26) yield

$$tr(T_1) \ = \ tr\left(\sum_{j=1}^{U} \frac{x_j P_s^{\perp} x_j^T}{\widehat{\sigma}_{ch,j}^2 (N-p)}\right) = NU, \tag{5.33}$$

and $\widehat{\sigma}_{1,j}^2$ (5.31) into the (5.27) yield

$$tr(R^{-1} T_2) \ = \ tr\left[R^{-1}\left(N \sum_{j=U+1}^{L} \frac{x_j x_j^T}{x_j^T x_j} + \sum_{j=L+1}^{(G+1)L} \sigma^{-2} x_j \, x_j^T\right)\right] \tag{5.34}$$

Maximizing the numerator over θ_j and $\sigma_{1,j}^2$, the GLR statistic can be recast

$$L = \max_{U} \frac{\|R\|^{U/2}\left(\prod_{j=1}^{U}\sigma_{0,j}^2\right)^{N/2} exp\left\{tr\left[\frac{R^{-1}}{2}\left(\sum_{j=1}^{L}\sigma_{0,j}^{-2}x_j x_j^T + \sum_{j=L+1}^{(G+1)L}\sigma^{-2}x_j x_j^T\right)\right]\right\}}{\left(\prod_{j=1}^{U}\frac{x_j^T P_S^\perp x_j}{N-p}\right)^{N/2} exp\left[tr\left(R^{-1}\frac{N}{2}\sum_{j=U+1}^{L}\frac{x_j x_j^T}{x_j^T x_j} + \frac{R^{-1}}{2}\sum_{j=L+1}^{(G+1)L}\sigma^{-2}x_j x_j^T\right)+\frac{NU}{2}\right]} \quad (5.35)$$

We use a straightforward maximization in (5.35). Since $L(G+1) > pN$, it follows immediately that the maximum of L with respect to U is obtained by replacing the unknown parameter U by known L. Then, the GLR statistic can be rewritten as

$$L = \frac{\|R\|^{L/2}\left(\prod_{j=1}^{L}\sigma_{0,j}^2\right)^{N/2} exp\left\{tr\left[\frac{R^{-1}}{2}\left(\sum_{j=1}^{L}\sigma_{0,j}^{-2}x_j x_j^T + \sum_{j=L+1}^{(G+1)L}\sigma_{0,j}^{-2}x_j x_j^T\right)\right]\right\}}{\left(\prod_{j=1}^{L}\frac{x_j^T P_S^\perp x_j}{N-p}\right)^{N/2} exp\left[tr\left(\frac{R^{-1}}{2}\sum_{j=L+1}^{(G+1)L}\sigma^{-2}x_j x_j^T\right)+\frac{NL}{2}\right]} \quad (5.36)$$

Step 2. In the case of the homogeneous environment, we can make the detector fully adaptive by plugging the maximum likelihood estimate of R and σ^2 based on the secondary data x_j,

$$j = L+1, \ldots, (G+1)L, \text{ i.e., } \widehat{\sigma}_{0,j}^2 = \widehat{\sigma}^2 = \frac{1}{NGL}\sum_{j=L+1}^{(G+1)L} x_j^T x_j,$$

$$\widehat{R} = \frac{\frac{1}{NGL}\sum_{j=L+1}^{(G+1)L} x_j x_j^T}{\widehat{\sigma}^2} \quad (5.37)$$

Following this guidance, we receive

$$L = \frac{\|\widehat{R}\|^{L/2} exp\left[\frac{N}{2\widehat{\sigma}^2}\left(\sum_{j=1}^{L} x_j^T \widehat{R}^{-1} x_j - L\right)\right]}{\left(\prod_{j=1}^{L}\frac{x_j^T P_S^\perp x_j}{\widehat{\sigma}^2(N-p)}\right)^{N/2}} \quad (5.38)$$

Taking the logarithm of the $N/2$-th root of L, the proposed GLRT-based detector, named multipixel adaptive subspace detector (MASD), is given by the following decision rule:

$$\sum_{j=1}^{L}\left(\frac{1}{\widehat{\sigma}^2} x_j^T \widehat{R}^{-1} x_j - ln\frac{x_j^T P_S^\perp x_j}{(N-p)\widehat{\sigma}^2}\right) \underset{<H_0}{\overset{>H_1}{}} t, \quad (5.39)$$

where t is the threshold. One can see from (5.39) that the detection quality depends on the relation between the object energy contribution (the first term) and the background power change contribution (the second term). Obviously, the first term contribution may be insufficient to detect the low-contrast object. Note that the second term does not depend on the object contrast and its return energy. The contribution of the second term depends on relation between the channel noise variance estimate $\widehat{\sigma}_{ch,j}^2 = \frac{x_j P_S^\perp x_j^T}{(N-p)}$ in the each pixel-vector and background-plus-channel-noise variance $\widehat{\sigma}^2$. The second term contribution achieves the high level when $\widehat{\sigma}_{ch,j}^2 \ll \widehat{\sigma}^2$, number of pixels occupied by object and number of frames within the brick are large. We can adjust the background power change sensitivity of detector with respect to the target presence sensitivity using a factor m. Varying the factor m, we can adapt the background power change to the detection performance. We can introduce the factor of signal detection sensitivity m and rewrite the decision rule in the following form:

$$T_{MASD} = \sum_{j=1}^{L} \left(\frac{m}{\hat{\sigma}^2} x_j^T \hat{R}^{-1} x_j - ln \frac{x_j^T P_s^{\perp} x_j}{(N-p)\hat{\sigma}^2} \right) \underset{<H_0}{\overset{>H_1}{}} t \qquad (5.40)$$

Note that the first term $T_{AED} = \sum_{j=1}^{L} x_j^T \hat{R}^{-1} x_j$ is the known classical adaptive energy detector (AED), and the second term $T_{ACHD} = \sum_{j=1}^{L} ln \frac{x_j^T P_s^{\perp} x_j}{(N-p)\hat{\sigma}^2}$ is the adaptive background power change detector (ACHD) that depends on the data projection on the orthogonal subspace. In the following section, we evaluate performance of the MASD and assess it against the two recently proposed M-MSF and FC detectors and classical ASD [34]:

$$T_{ASD} = \sum_{j=1}^{L} x_j^T \hat{R}^{-1} H \left(H^H \hat{R}^{-1} H \right)^{-1} H^H \hat{R}^{-1} x_j \underset{<H_0}{\overset{>H_1}{}} t_1 \qquad (5.41)$$

5.3.2. Detection Performance and Numerical Illustrations

In the following, we assess the performance of the proposed detector and compare it to the ASD by utilizing the standard Monte Carlo counting techniques. Specifically, in order to evaluate the threshold needed to ensure a pre-assigned value of the false alarm probability (P_F) and the probability of detection (P_D), we resort to $100/P_F$ and 10^3 independent trials, respectively. We model the background and channel noise as Gaussian process. We assume that the background normalized covariance matrix is exponentially shaped, i.e., $R = [r_{i,k}] = \left[\rho^{|i-k|} \right]$, where ρ is the one-lag correlation coefficient. We assume that the optical observed brick is partially covered by the target, and the presence of the target in any pixel reduces the total noise power to the channel noise power. We introduce the target fill factor $b = U/L$ and the mismatched factor $(MF = \rho_{ac} - \rho_d)$ of the actual ρ_{ac}; design ρ_d one-lag background correlation coefficient; and then analyze, as the detection performance depends on b and MF. The primary data is modeled as the deterministic target signals with the uniform amplitude vector $\theta = [1,1, ...,1]^T$ and with p low-frequency independent basis modes of the matrix H. This matrix is a Vandermonde matrix with discrete complex exponential elements. The secondary data model is a sequence of the G subimages (frames), where each frame is represented as the L, independent, identically distributed correlated Gaussian vectors. We investigate the detectability of the detector at different parameters: the target fill factor $b = U/L$, the background-to-noise ratio (BNR), the signal-to-background ratio (SBR), brick size NL, secondary data size GL, the mismatched factor MF, and the sensitivity factor m. The MASD performance is compared to the ASD performance. In Fig. 5.9, the P_D is shown as a function of SBR for several b values, given secondary data size $GL = 5000$. These curves show that, for a fixed SBR, increasing of the target fill factor enhances the performance of both detectors. One can see that the MASD performance is improved more than ASD. The analysis of the (5.40) shows (Fig. 5.9) that the second term in (5.40), which is based on the ratio between background powers in the null and alternative hypotheses, is the cause of improving the detection quality. An important parameter of any detector is the gain factor (GF) defined as the horizontal displacement at $P_D = 0.9$ between different curves in the Figs. 5.9-5.13. Fig. 5.9 shows that the *GF* of the MASD, with respect to ASD, depends on target fill factor *b*. One can see that $GF = 5$ dB at the $b = 0.3$, and for larger $b = 0.7$ the gain factor is increased to $GF = 13$ dB. The performance of the MASD depends on BNR too. In Fig. 5.10, one can see that the MASD performance is improved when the BNR is decreased. For example, when the *BNR* is decreased from $BNR = 100$ to $BNR = 20$, the gain factor $GF = 3.5$ dB.

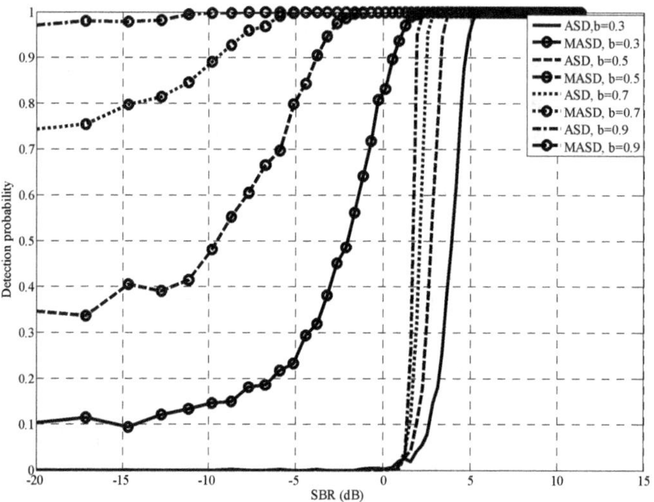

Fig. 5.9. Probability of detection versus SBR for proposed detector MASD and the ASD at different *b*, $L = 10$, $MF = 0$, $N = 10$, $BNR = 100$, $p = 3$, $GL = 5000$, $m = 1$, $\rho_{ac} = \rho_d = 0.8$, and $P_F = 10^{-3}$. Simulation results.

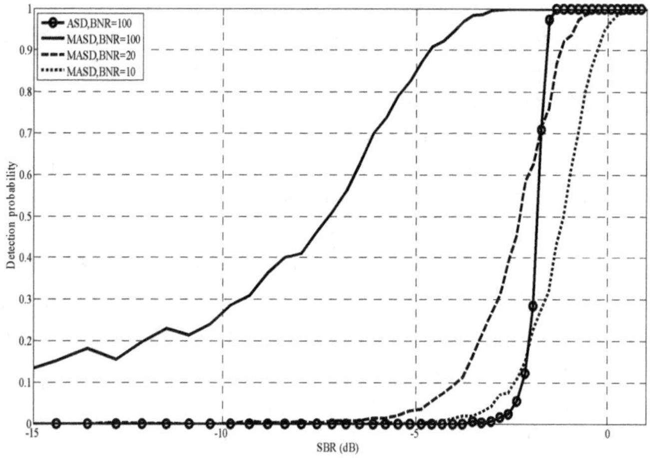

Fig. 5.10. Probability of detection versus SBR for proposed detector MASD and the ASD at different BNR, $L = 10$, $MF = 0$, $b = 0.9$, $N = 10$, $p = 3$, $GL = 1000$, $m = 1$, $\rho_{ac} = \rho_d = 0.8$, $P_F = 10^{-3}$. Simulation results.

Fig. 5.11 shows that the MASD performance depends on secondary dataset size *GL*. When the *GL* = 10000, the gain factor *GF* = 3 dB with respect to *GL* = 1000. Fig. 5.12 shows that the MASD performance depends on mismatched factor *MF*. The mismatch between actual and designed one-lag background correlation coefficients ρ_{ac} and ρ_d decreases the performance of MASD. In Fig. 5.12, the gain factor *GF* = 2.5 dB in the case of *MF* = 0.2 for different pixel fill factors. The factor of signal detection sensitivity *m* changes the detectability of the MASD. Fig. 5.13 shows that changing *m* from 1 to 0.9 improves (*GF* = 2 dB) the performance of the MASD.

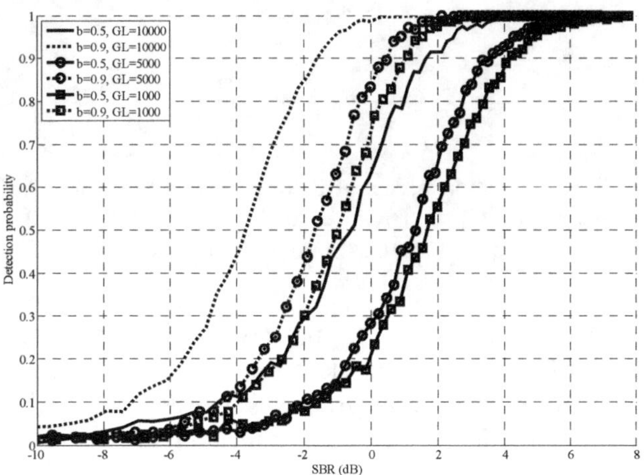

Fig. 5.11. Probability of detection versus SBR for proposed detector MASD at different sizes of the secondary dataset *GL* and target fill factor *b*, BNR = 10, *MF* = 0, *L* = 10, *N* = 10, *p* = 3, $m = 1, \rho_{ac} = \rho_d = 0.75$, $P_F = 10^{-3}$. Simulation results.

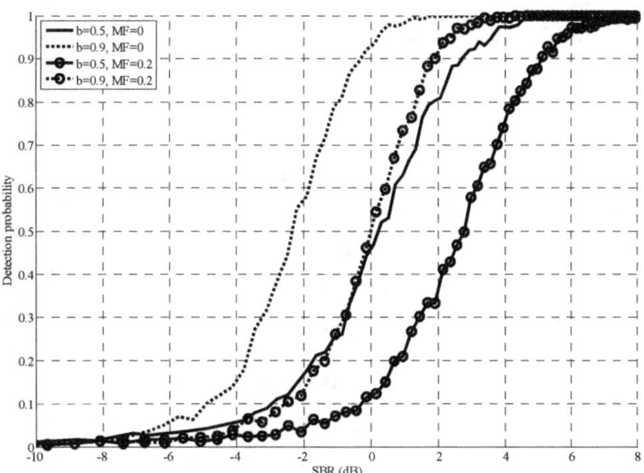

Fig. 5.12. Probability of detection versus SBR for proposed detector MASD at different mismatched factors *MF* and target fill factors *b*, BNR = 10, *N* = 10, *L* = 10, *p* = 3, *GL* = 10000, $m = 1$, $P_F = 10^{-3}$. Simulation results.

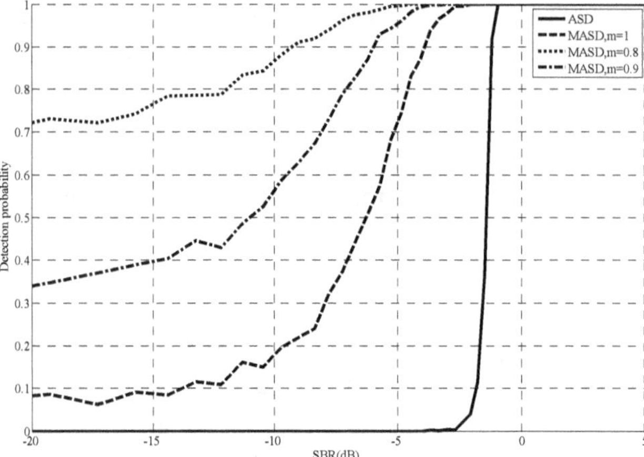

Fig. 5.13. Probability of detection versus SBR for proposed detector MASD and the ASD at different sensitive factors m, $MF = 0$, BNR $= 100$, $L = 10$, $b = 0.9$, $N = 10$, $p = 3$, $GL = 10000$, and $P_F = 10^{-3}$. Simulation results.

We compare the MASD and ASD performances for different target fill factors and successive images using the experimental video sequences of different floating objects on a sea surface. The performance evaluation of detection algorithms in practice is challenging due to the limitations imposed by the limited amount of target data. As a result, the establishment of accurate detection probability curves is quite difficult. We consider a signal at output of the video camera composed of $N = 10$ and 25 successive images (frames). Each pixel in the images is represented by 8-bits, that is, each pixel has 256 gray levels. The digital camera (30 frames/sec) is located around 8 m above the sea surface to realize the surface sea monitoring which must give a detection distance less than 600 m (to separate the sky from the sea surface). At the first stage, we have experimentally evaluated the target mode subspace size p (the number of the considerable Fourier components) for three small floating objects: the gray ball, the yellow plastic container and the swimmer in a white hat. In all experiments with different objects, the presented altitudes of the sea waves are less than 0.3 m. The maximum frequency of the Fourier spectrum is equal to 2 Hz for all types of the targets and 12 Hz for sea surface. These values are obtained by using the secondary data set $GL = 5000$. The number of the considerable target spectrum components p depends on N: $p = 3$ for $N = 10$ and $p = 7$ for $N = 25$. Then we estimate the average and the variance σ_{c+n}^2 of the background data from the observed sea surface by using 5000 successive images. Both detectors, ASD and MASD, use a sequence of optical images, which are first preprocessed by removing the average of the background. The estimates of the parameters p and σ_{c+n}^2 are used in ASD and MASD. In order to validate the MASD and ASD, three gray scale image sequences are collected as test samples: the image sequence with floating gray ball, the yellow plastic and the image sequence with the swimmer. The videos are obtained at a distance of about 200 m from the targets and contain no less than 10000 images. The set of experiments illustrates the separability between each object type and background for each detector. The figures for each target type are shown in Figs. 5.14-5.17. Each figure for gray ball (see Figs. 5.12-5.14) contains three subfigures: (a) The first image of the video, (b) The vertical bars for $N = 10$, and (c) the vertical bars for $N = 25$. The bars show the range of detection values under H_0 and H_1 for ASD and MASD. Ideally, these bars should not overlap, indicating that the target is completely separable from background. In the case where overlaps do occur, a number of targets is missed, or false alarms occur. One can evaluate as detection performance depends on target fill factors. Figs. 5.14(b), 5.15(b), and 5.16(b) illustrate that the same ball is detected at different brick sizes: $L = 16$ pixels (Fig. 5.14), $L = 36$ pixels (Fig. 5.15), and $L = 64$ pixels (Fig. 5.16). The MASD detected almost without errors at $N = 10$ and $N = 25$ and observed brick size $L = 16$. When the brick size is increased ($L = 36$ and 64), the detection quality is decreased for the same target. The maximum detection performance is achieved when the object occupies all brick pixels L. Minimum size of objects may be known. If we choose the spacial brick size L equal to minimum size of objects, the probability to miss the small object is minimized. Minimum of the number of images in analyzed sequence depends on spacial size, contrast of the object, dynamic parameters of the sea surface and determined detection quality. Figs. 5.14-5.16 illustrate that the MASD detection performance significantly surpasses the ASD performance. Figs. 5.17 and 5.18 illustrate the separability between the plastic container (Fig. 5.17(a)), swimmer (Fig. 5.18(a)) and background for each detector.

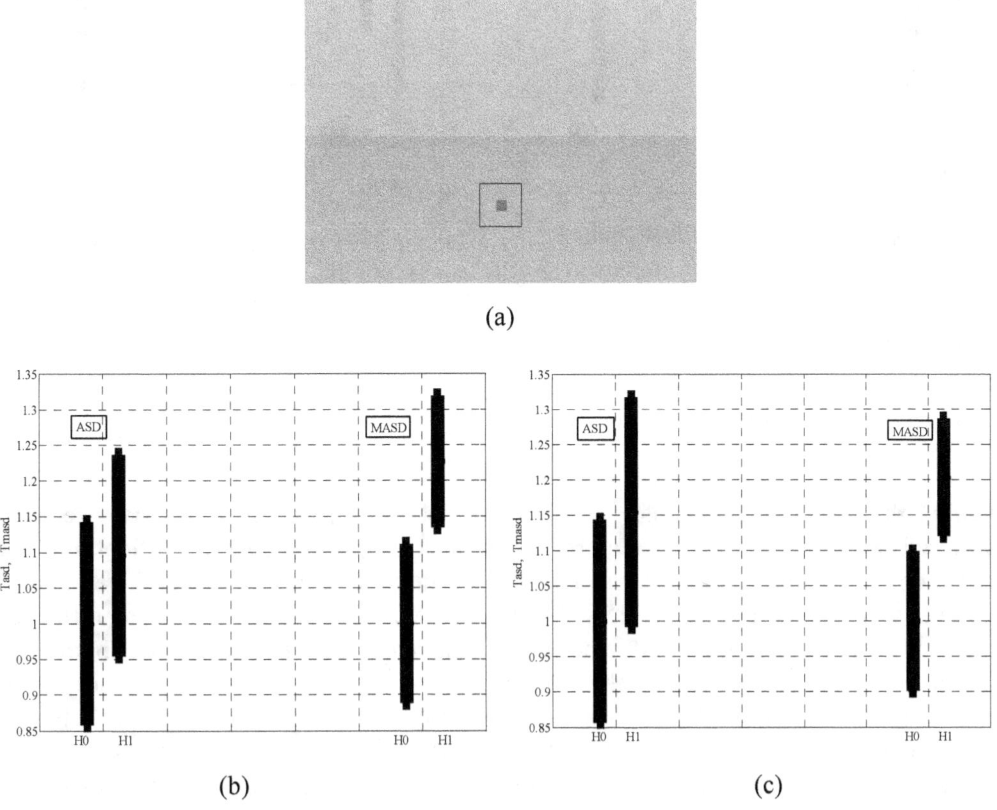

(a)

(b)   (c)

Fig. 5.14. The first image with the ball (a) into the observed bricks (the square): of size 4×4×10 (b) and 4×4×25 (for (c)). The gray ball separability analysis for MASD and ASD ($N = 10$ (b) and $N = 25$ (c)). Real experiment results.

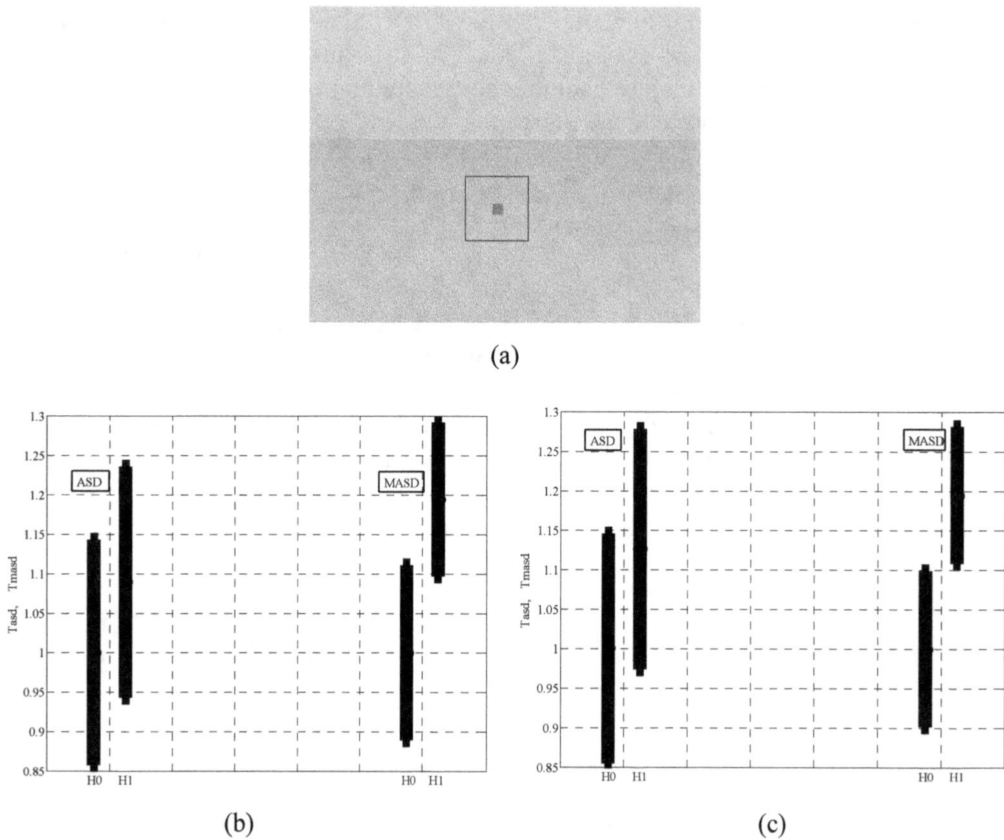

Fig. 5.15. The first image with the ball (a) into the observed bricks (the square): of size 6×6×10 (b) and 6×6×25 (for (c)). The gray ball separability analysis for MASD and ASD ($N = 10$ (b) and $N = 25$ (c)). Real experiment results.

Figs. 5.17(b) and 5.18(b) show that the MASD detectability significantly outperforms the ASD.

Finally, the detection performances of T_{MASD} and T_{ASD} are compared to two recently proposed M-MSF and FC detectors [7, 8] using real datasets (the floating blue low-contrast metallic container). In Fig. 5.19, we show the frames number 77, 80 and 83 of the observed brick of size 15×15×100. It is apparent from Fig. 5.20 that the reflection intensity from the floating metallic container practically does not change during a short time-interval between frames. On the contrary, the intensity of the reflections from the sea surface changes quickly. The proposed detector uses this distinction in the best way. For real images, the performances of the four detectors may be shown via the receive operation characteristics curves (Fig. 5.21). The results show that T_{MASD} significantly outperforms T_{ASD}, M-MSF and FC detectors. All the experiment results show that the proposed algorithm T_{MASD} can detect small targets in the presence of intensive background clutter more effectively.

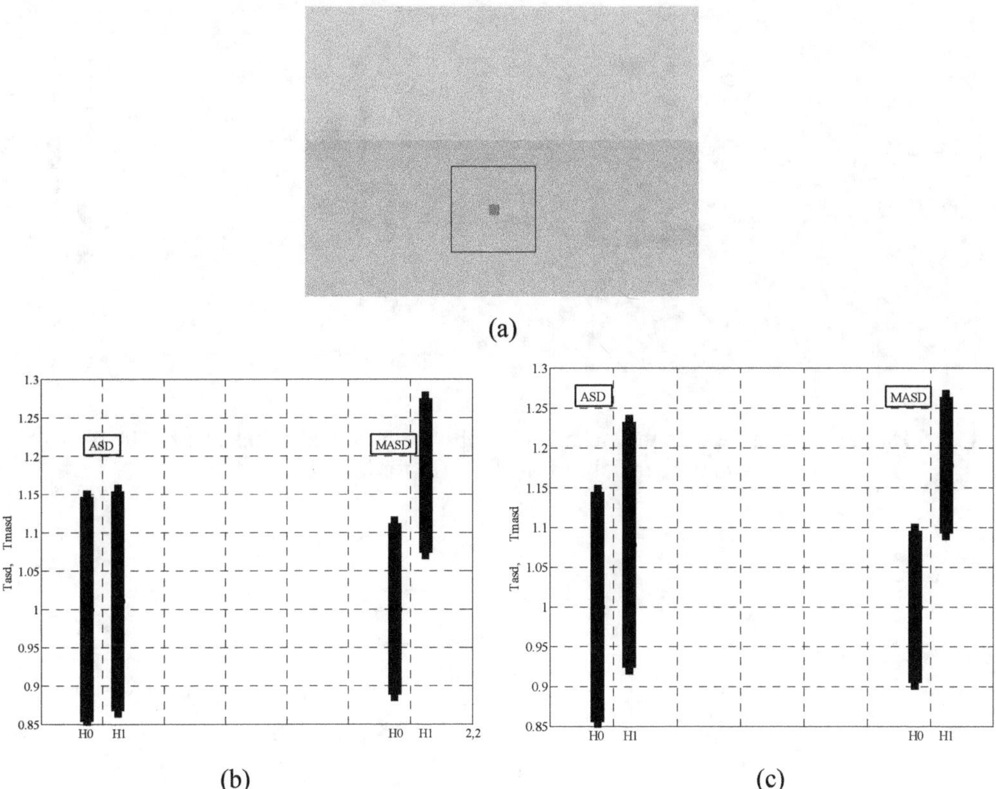

Fig. 5.16. The first image with the ball (a) into the observed bricks (the square): of size 8×8×10 (b) and 8×8×25 (for (c)). The gray ball separability analysis for MASD and ASD ($N = 10$ (b) and $N = 25$ (c)). Real experiment results.

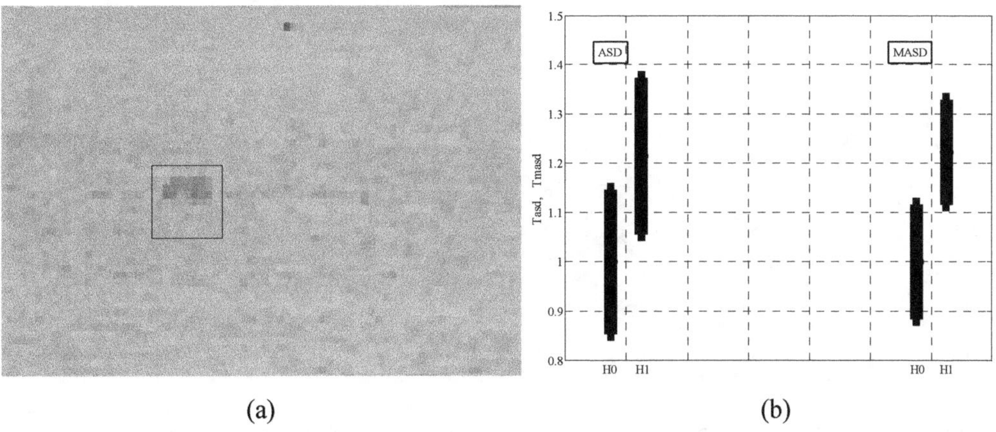

Fig. 5.17. The first image with the plastic container (a) into the observed bricks of size 10×10×100. The plastic container separability analysis (b) for MASD and ASD.

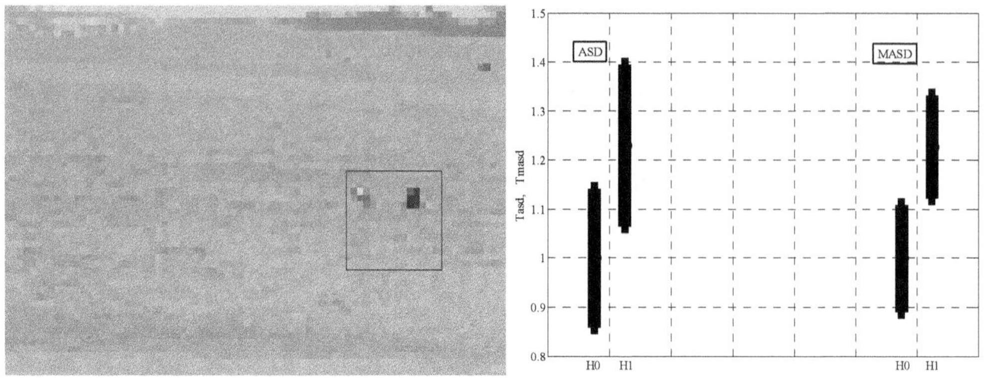

Fig. 5.18. The first image with the swimmer (a) into the observed bricks of size 10×10×100. The swimmer separability analysis (b) for MASD and ASD.

Fig. 5.19. Images number 77, 80 and 83 of the low-contrast metalic container with marked observed region.

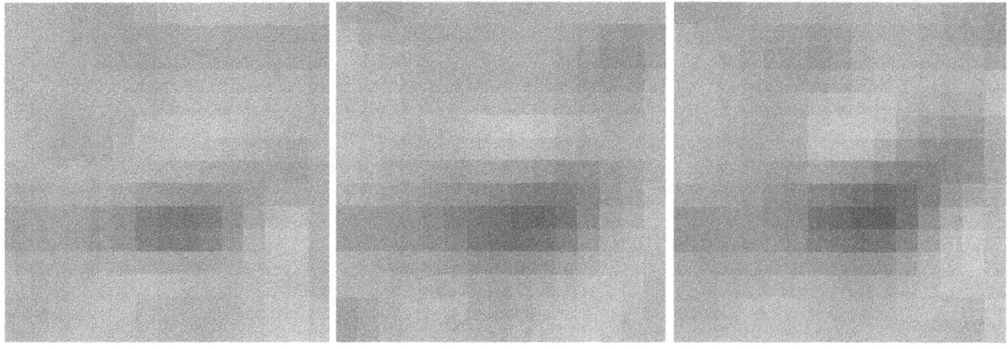

Fig. 5.20. The marked observed region of the images number 77, 80 and 83 with the floating metalic container.

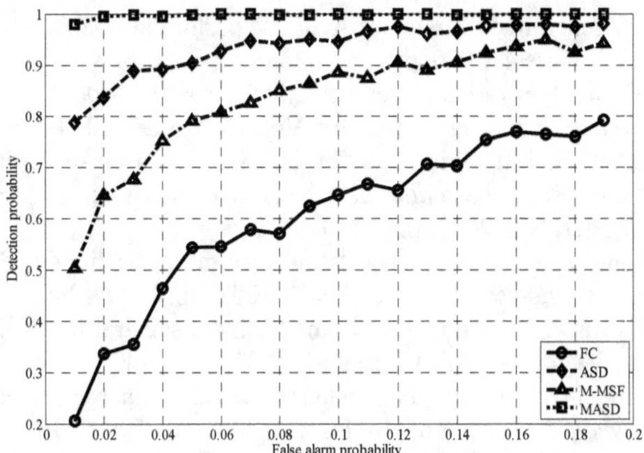

Fig. 5.21. Receive operation characteristics of the four different detectors for the floating metallic container using the observed brick of size 15×15×100. Real dataset.

5.3.3. Conclusion

We present an effective and adaptive detector of multi-pixel objects on the sea surface. The proposed detector offers two principal benefits. The first is that the proposed detector is able to detect no-contrast objects on the sea surface. The second advantage is that the proposed detector is slightly sensitive to background covariance matrix change and size of the secondary dataset. We propose the modified 3-D GLR approach based on the analysis of the certain set of the pixel-vectors in the sequence of images to detect the multi-pixel target with unknown position, size and shape. The numerical simulations and experimental results show that the proposed adaptive detector considerably outperforms the classical one and the recently proposed M-MSF and FC detectors. The MASD detection efficiency is different under different scenarios and depends on background-to-noise ratio, signal-to-background-plus-noise ratio, target fill factor, and number of images in analyzed sequence.

References

[1]. D. Manolakis, Ch. Siracusa, G. Shaw, Hyperspectral subpixel target detection using the linear mixing model, *IEEE Trans. Geosci. Remote Sensing*, Vol. 39, Issue 7, 2001, pp.1392-1408.

[2]. I. S. Reed, R. M. Gagliardi, H. M. Shao, Application of three-dimensional filtering to moving target detection, *IEEE Trans. Aerosp. Electron. Syst.*, Vol. AES-19, Issue 6, 1982, pp. 898-905.

[3]. X, Ji. L. Geng, V. Sun, Y. Zhao, CEM: More bands, better performance, *IEEE Geosci. Remote Sens. Lett.*, Vol. 11, Issue 11, 2014, pp.1876-1880.

[4]. J. Liu, Y. Feng, W. Liu, D. Orlando, H. Li, Training data assisted anomaly detection of multi-pixel targets in hyperspectral imagery, *IEEE Trans. Signal Processing*, Vol. 68, Issue 5, 2020, pp. 3022-3032.

[5]. S. Kim, High-speed incoming infrared target detection by fusion of spatial and temporal Detectors, *Sensors*, Vol. 15, Issue 4, 2015, pp. 7267-7293.

[6]. R. S. Raghavan, A generalized version of ACE and performance analysis, *IEEE Trans. Signal Processing*, Vol. 68, Issue 4, 2020, pp. 2574-2585.

[7]. S. Kim, J. Lee, Small infrared target detection by region-adaptive clutter rejection for sea-based infrared search and track, *Sensors*, Vol. 14, Issue 7, 2014, pp. 13210-13242.

[8]. A. Kadyrov, H. Yu, H. Liu, Ship detection and segmentation using image correlation, in *Proceedings of the IEEE International Conference on Systems, Man, and Cybernetics (SMC'13)*, Manchester, UK, 2013, pp. 3119-3126.

[9]. H. Ren, C.-I Chang, Automatic spectral target recognition in hyperspectral imagery, *IEEE Trans. Aerosp. Electron. Syst.*, Vol. 39, Issue 4, 2003, pp. 1232-1249.

[10]. D. Manolakis, G. Shaw, Detection algorithms for hyperspectral imaging applications, *IEEE Signal Processing Magazine*, Vol. 19, Issue 1, 2002, pp. 29-43.

[11]. B. Thai, G. Healy, Invariant subpixel material detection in hyperspectral imagery, *IEEE Trans. Geosci. Remote Sensing*, Vol. 40, Issue 3, 2002, pp. 599-608.

[12]. Q. Du, C.-I Chang, A signal-decomposed and interference annihilated approach to hyperspectral target detection, *IEEE Trans. Geosci. Remote Sens.*, Vol. 42, Issue 2, 2004, pp. 892-906.

[13]. S. Jia, Y. Qian, Constrained nonnegative matrix factorization for hyperspectral unmixing, *IEEE Trans. Geosci. Remote Sens.*, Vol. 47, Issue 1, 2009, pp. 161-173.

[14]. J. C. Harsayi, C. I. Chang, Hyperspectral image classification and dimensionality redaction: An orthogonal subspace projection approach, *IEEE Trans. Geosci. Remote Sensing*, Vol. 32, Issue 4, 1994, pp. 779-785.

[15]. N. Keshava, J. F. Mustard, Spectral unmixing, *IEEE Signal Process. Mag.*, Vol. 19, Issue 1, 2002, pp. 44-57.

[16]. J. Settle, N. A. Drake, Linear mixing and estimation of ground cover proportions, *Int. J. Remote Sens.*, Vol. 14, Issue 6, 1993, pp. 1159-1177.

[17]. E. A. Ashton, A. Schaum, Algorithms for the detection of subpixel targets in multispectral imagery, *Photogramm. Eng. Remote Sens.*, Vol. 64, Issue 7, 1998, pp. 723-731.

[18]. D. Heinz, C.-I Chang, M. L. G. Althouse, Fully constrained least squares-based linear unmixing, in *Proceedings of the Int. Geosci. Remote Sens. Symposium*, Hamburg, Germany, 1999, pp. 1401-1403.

[19]. J. Broadwater, R. Chellappa, Hybrid detector for subpixel targets, *IEEE Trans. Pattern Anal. Mach. Int.*, Vol. 29, Issue 11, 2007, pp. 1891-1903.

[20]. L. Zhang, Bo Du, Y. Zhong, Hybrid detectors based on selective endmembers, *IEEE Trans. Geosci. Remote Sensing*, Vol. 48, Issue 6, 2010, pp. 2633-2646.

[21]. F. Vincent, O. Besson, C. Richard, Matched subspace detection with hypothesis dependent noise power, *IEEE Trans. Signal Process.*, Vol. 56, Issue 11, 2008, pp. 5713-5718.

[22]. V. Golikov, O. Lebedeva, A. Castillejos-Moreno, V. Ponomaryov, Performance of the matched subspace detector in the case of subpixel targets, *IEICE Trans. Fund.*, Vol. E94-A, Issue 2, 2011, pp. 826-828.

[23]. V. Golikov, O. Lebedeva, Adaptive detection of subpixel targets with hypothesis dependent background power, *IEEE Signal Processing Letters*, Vol. 20, Issue 8, 2013, pp.751-754.

[24]. V. Golikov, M. Rodriguez-Blanco, O. Lebedeva, Robust multipixel matched subspace detection with signal-dependent background power, *Journal of Applied Remote Sensing*, Vol. 10, Issue 1, 2016, 015006.

[25]. B. Du, Y. Zhang, L. Zhang, L. Zhang, A hypothesis independent subpixel target detector for hyperspectral images, *Signal Process.*, Vol.110, Issue 5, 2015, pp. 244-249.

[26]. S. Kraut, L. Scharf, L. McWorther, Adaptive subspace detectors, *IEEE Trans. Signal Process.*, Vol. 49, Issue 1, 2001, pp. 1-16.

[27]. Z. Wang, M. Li, H. Chen, L. Zuo, P. Zhang, Y. Wu, Adaptive detection of a subspace signal in signal-dependent interference, *IEEE Trans. Signal Processing*, Vol. 65, Issue18, 2017, pp. 4812-4820.

[28]. M. Lodhi, W. U. Bajwa, Detection theory for union of subspaces, *IEEE Trans. Signal Processing*, Vol. 66, Issue 24, 2018, pp. 6347-6362.

[29]. G. Healey, D. Slater, Models and methods for automated material identification in hyperspectral imagery acquired under unknown illumination and atmospheric conditions, *IEEE Trans. Geosci. Remote Sensing*, Vol. 37, Issue 6, 1999, pp. 2706-2717.

[30]. P. Bajorski, Analytical comparison of the matched filter and orthogonal subspace projection detectors for hyperspectral images, *IEEE Trans. Geosci. Remote Sensing*, Vol. 40, Issue 7, 2007, pp. 2394-2402.

[31]. S. Kraut, L. Scharf, L. McWorther, Adaptive subspace detectors, *IEEE Trans. Signal Process.*, Vol. 49, Issue 1, 2001, pp. 1-16.

[32]. S. Kim, J. Lee, Small Infrared target Detection by region-adaptive clutter rejection for sea-based infrared search and track, *Sensors,* Vol. 14, Issue 7, 2014, pp. 13210-13242.

[33]. E. J. Kelly, An adaptive detection algorithm, *IEEE Trans. on Aerospace and Electronic Systems*, Vol. AES-22, Issue 1, 1986, pp. 115-127.

[34]. S. Kraut, L. Scharf, L. McWorther, Adaptive subspace detectors, *IEEE Trans. Signal Process.*, Vol. 49, Issue 1, 2001, pp. 1-16.

[35]. J. Settle, N. A. Drake, Linear mixing and estimation of ground cover proportions, *Int. J. Remote Sens.*, Vol. 14, Issue 6, 1993, pp. 1159-1177.

Chapter 6
Sleep Scoring Based on Biomedical Signals: A Survey and a New Algorithm

Mahtab Vaezi and Mehdi Nasri

6.1. Sleep

Sleep is essential for human survival. It is a natural state that exists in humans, animals and even invertebrates and directly affects people's health [1-3]. Sleep is a conscious act in which the brain does not respond to external stimuli and is sensitive to internal stimuli. Therefore, the brain is active in sleep mode and it affects the health of the human body. Because the brain is sensitive to the internal processes during sleep, sleep testing is one of the ways to diagnose a variety of mental and brain related diseases. Sleep is different from a coma because it is an unconscious act and one cannot get out of it [3]. Sleep is a voluntary action that increases the level of melatonin in the body and causes sleep to be altered in certain brain patterns [4]. During sleep, changes in brain, muscle, eye, heart, and respiratory activity are marked and these mood changes are necessary for mental and physical health [5].

Sleep is still a mystery at the moment, but it can be partially measured by the fact that it can be evaluated by vital signals [2]. It is estimated that about 50 million Americans suffer from sleep disorders, but no effective research has been done to study sleep [6]. Sleep and lack of sleep can have a profound effect on the learning process and memory with direct effects on synapses [7]. Enough sleep can also causes as much metabolic secretion from the brain [8]. Inadequate sleep can also have many negative effects on the cardiovascular system, the immune system with decrease in white blood cells, reduces growth hormone release, reduced ability to perform mental and mathematical calculations and one's body emotions. Protein production in body fluids also often occurs during deep sleep [3, 9, 10]. People can improve their health by changing their sleep habits. Sleep, consists of three parts, the need for sleep, the ability to sleep and the opportunity to sleep [4]. Fig. 6.1 shows these three major areas of sleep, graphically [4].

Mehdi Nasri
Department of Biomedical Engineering, Khomeinishahr branch, Islamic Azad University, Isfahn, Iran

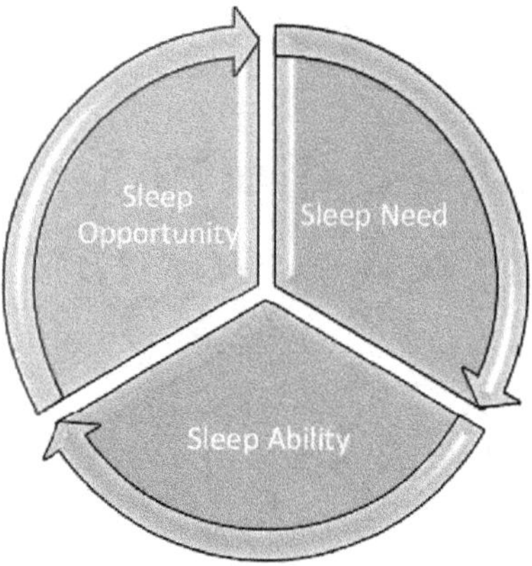

Fig. 6.1. Three major areas of sleep.

6.1.1. Need for Sleep

The amount of needed sleep varies depending on the genetic factors, and body structure of the individual. Unfortunately, there is no measurement indicator for needing sleep, but on average, every person needs 7 to 8 hours of sleep per day. This amount of need for sleep is not offset by behaviors such as staying awake for a few days and spending a lot of time for sleep on the weekend, and if the body does not pay attention to this physiological need, it will become ill [4].

6.1.2. Sleep Ability

The approximate number of hours a person can sleep during the day is call the ability to sleep. The ability to sleep can affect by physical and mental factors. For example, stress causes one to fall asleep one day, and the cold to make one sleep more than normal [4].

6.1.3. Sleep Opportunity

Sleep opportunity is called time that a person spends in bed, this time includes sleeping and doing things before fall sleep, such as reading a book before sleep. Sleep opportunity is fully conscious and the individual devotes time to sleep consciously and is strongly influenced by external factors [4].

6.2. Physiological Factors Affecting Sleep

As it was said, chemical and neurological activities in the brain cause sleep. This section examines these physiological factors in detail. The parts of the brain that make a person sleep in the body are: brainstem, hypothalamus, thalamus, cerebrum, neuro modulators and the autonomic nervous system. Fig. 6.2 show six brain parts that make a person sleep.

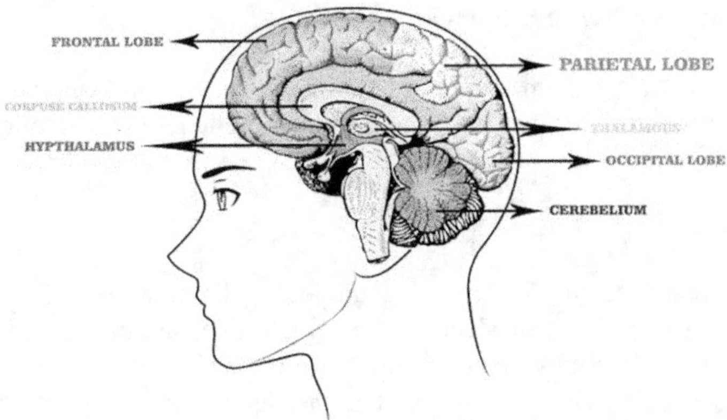

Fig. 6.2. Six brain part that make a person sleep.

6.2.1. The Brainstem

The brainstem is responsible for vital operations in the body, and in relation to sleep, controls the amount of brain activity during sleep and wake by producing substances such as serotonin, norepinephrine, and dopamine [11].

6.2.2. The Hypothalamus

Hypothalamus first by regulating the brain superchiasmatic central nucleus or nuclei (SCN), regulates the body's physiological hours of the day. Then, by adjusting the increased secretion of Histamine and Orexin, it wakes up throughout the day and reduces the secretion at night so that person can sleep [11].

6.2.3. The Thalamus

Thalamus activity helps to understand what is happening around you, including understanding events and emotions. When a person goes to sleep, the thalamus does not move these senses from the cortex to the inner part of the brain, so no processing within the brain is performed and the individual goes to sleep without knowing about the environment [11].

6.2.4. Cerebrum

The cerebrum has no specific element for sleep, but the activity of billions of neurons in this area controls sleep operations.

6.2.5. Neuromodulators

The six main and effective neural modulators of sleep and wakefulness are Dopamine, Histamine, Norepinephrine, Acetylcholine, Serotonin and Orexin.

6.2.6. The Autonomic Nervous System

The autonomic nervous system comprises sympathetic and parasympathetic nerves and helps maintain sleep during sleep by reducing sympathetic nervous activity.

6.3. Sleep Stages

During sleep, there are many physiological and behavioral activities in the body. Generally, sleep can be studied in sleep continuity and sleep architecture. Sleep continuity is defined separately for each person and includes general sleep tests including bedtime, total hours of sleep per day, and so on. Regular sleep monitoring can diagnose many diseases, including sleep apnea. Sleep architecture is a physiological study of sleep by vital signals and sleep stages is monitored and many diseases can be diagnosed by examining this area of sleep [4].

Therefore, many sleep-related illnesses such as apnea, mental illnesses such as depression, and brain diseases such as epilepsy can be diagnosed through sleep study. In general, it is necessary to diagnose different diseases through sleep by separating the different stages of sleep and examining them. A sleep period consists of several stages. These stages are from the beginning of the sleep to deep sleep.

The body's vital signals are used to monitor sleep and detect different stages of sleep. One of the sleep-related signal recorders is polysomnography (PSG) recording. PSG recording includes Electroencephalography (EEG), Electrooculography (EOG) and chin electromyography (EMG) [12]. Fig. 6.3. Shows how the PSG electrodes are used to record vital body signals.

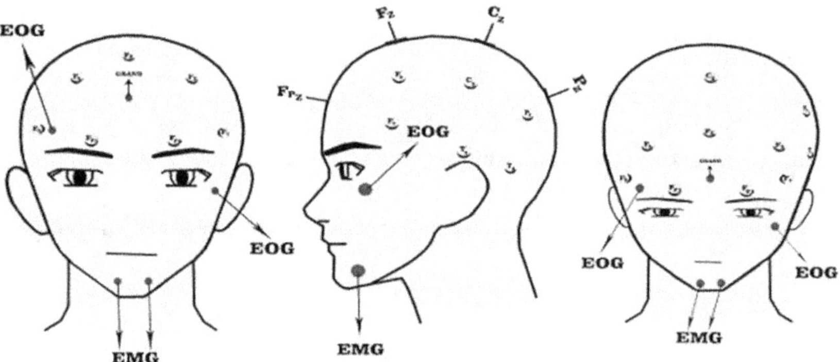

Fig. 6.3. How the PSG electrodes are used to record vital body signals.

PSG is a complete record of the body's vital signals during sleep, but due to the large number of leads that bind to the patient during sleep which causes stress, it increases the artefacts caused by motion sickness, and it is not suitable for long-term use. In recent years, much research has been done on the recording of less vital and useful sleep signals

[13-15]. The researchers concluded that by recording the ECG signal and extracting the HRV, they could perform sleep studies with high accuracy and at the lowest cost. HRV measures the activity of the autonomic nervous system and different levels of sleep can be detected by increasing and decreasing the activity of this system [16, 17]. Both the PSG record and the ECG of the individual during sleep can be used to investigate sleep characteristics and identify different stages of sleep to identify different diseases.

Researches on sleep stages began in 1968, and Rechtschaffen and kales first presented the world standard sleep stages (R&K) [18]. According to R&K standard, a sleep period consists of six stages, respectively, wake, stage 1 (S1), stage 2 (S2), stage 3 (S3), stage 4 (S4) and the rapid eye movement stage (REM). These six stages fall into two major groups REM and NREM (wake, S1, S2, S3, and S4). Fig. 6.4 shows six stages of sleep with brain signal (electroencephalography (EEG)).

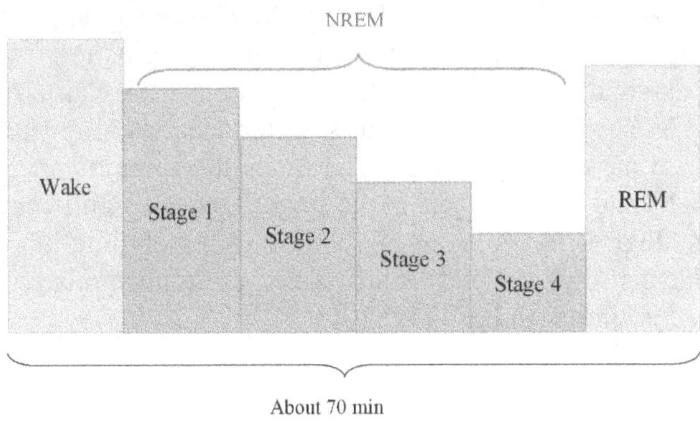

Fig. 6.4. Six stage of sleep.

In 2007, the American academy of sleep medicine (AASM) presents a new global standard for sleep stages. According to AASM manual, whole sleep period includes five stages: wake, stage 1, stage 2, stage SWS (stage 3 + stage 4) and REM [19]. In AASM standard stage 3 and stage 4 are referred to as SWS stage. The distinction between S3 and S4 is clinically meaningless [20]. To label sleep stages, every 30 seconds of the desired vital signal recording is considered as an epoch. Then, using the specific features of each stage, each epoch is identified as one stage.

6.3.1. Wake Stage

This stage usually begins at the beginning of sleep when the person is still awake. The alpha brain wave is very prominent in the wake stage, and by studying the EEG, it is possible to detect the wake stage by detecting the alpha wave. In wake stage, the EMG of the chin, due to muscle contraction has a higher amplitude than other stages. EOG at this stage is also associated with high amplitude and sustained artefacts of rapid eye movements [20].

6.3.2. Stage 1

Stage 1 of sleep is called light sleep and usually a short time of sleep, and the time it takes for stage 1 to fall sleep increases with age. In S1 in addition to the alpha wave, theta wave activity and sometimes beta appears in the EEG. The EMG of the chin also declines in frequency domain in S1. In stage 1, dreams begin, but because of light sleep, one may wake up at this stage [20].

6.3.3. Stage 2

Stage 2 is the middle stage of sleep, and the EEG of this stage will increase in amplitude compared to the previous stages. Theta wave activity is also evident in EEG. In S2 EEG, K complexes and spindle appear due to minimal response of the brain to external stimuli. K complexes may be cause by the brain's very poor response to sudden sound in the surroundings during sleep. Sometimes it has an intrinsic origin, and if we know that it is not responding to external stimuli, we can find out that it is the cause of a brain disorder. Spindles that are relate to the central nervous system reactions also usually appear at this stage, and as the age increases, the number of times the spindle appears in the EEG of sleep decreases. EOG and EMG signals in the stage 2 are not significantly different from other stages. Physiological functions of the body in S2, including blood pressure, brain metabolism, cardiac activity, and body temperature, are reduce relative to stage 1. EOG activity at this stage is similar to EEG [12, 20].

6.3.4. Stage 3

EEG of this stage have slow waves with high amplitude, and is the most relaxing and enjoyable phase of sleep. As the age decreases, the duration of this stage decrease. Sour Growth Hormone is often secreted during the SWS stage [20].

6.3.5. REM

REM stage usually occurs 90-120 minutes after S1, and despite being the deepest phase of sleep, the brain acts like an awakening state and responds to the internal stimuli. This is why REM is called the active sleep stage. In the EEG of REM stage, the alpha wave is known and brain waves are low amplitude. Unlike other stages of sleep, in the REM stage, physiological activities including blood pressure, brain metabolism, body temperature, etc. increases [12, 20].

Due to the similarity of the brain signal of some stages and the difficulty of stage recognition by EEG signal, the PSG recording uses a combination of three vital signals to improve the accuracy of the identification of sleep stages. No other signals are need if the technician or computer algorithm is capable of detecting EEG sleep stages. Table 6.1 shows brain frequencies and brain waves during sleep stages.

Table 6.1. Different brain frequencies during different sleep stages.

Feature	Frequency	Amplitude	Stages
Alpha (α)	8-13 Hz	20-60 μv	Wake, S1 and REM
Beta (β)	+13 Hz	2-20 μv	Wake
Theta (θ)	4-8 Hz	50-75 μv	S1, S2, S3, S4
Delta (δ)	0-4 Hz	+ 75 μv	S3, S4

6.4. The Importance of Sleep Study

Lack of sleep and poor sleep can affect the amplification of many diseases, and some diseases are detectable and controlled by sleep study. That is why sleep studding is so important, both to diagnose diseases, and to prevent or improve some diseases [21]. In the following, we examine the relationship between some diseases and sleep.

6.4.1. Sleep and Epilepsy

The first studies on the relationship between sleep and convulsion began in Aristotle's time, and researchers found a link between them. Internal factors affecting sleep activity and epilepsy are common, such as brainstem, hypothalamus, thalamus and cerebrum, which greatly influence the two activities. Anticonvulsants directly affect sleep, and changes in sleep rhythm stimulate seizures [21].

Lack of sleep, stimulates seizures and increases the time of the SWS stage in a person with epilepsy [22]. Sleep disorders in children, increases the likelihood of seizures and decrease the ability of the child to learn and memory, simultaneously. For this reason, the sleep of children with epilepsy should be checked [23]. If a person with epilepsy has obstructive sleep apnea, by diagnosing apnea and treating it the likelihood of a seizure in the person is significantly reduced, and this is only possible with sleep tests [24]. Highlights of the sleep stages features of people with epilepsy includes reduced REM sleep time and increased wakefulness after waking up [25].

6.4.2. Sleep and ADHD

In recent years, studies of the relationship between sleep and attention-deficit/hyperactivity disorder (ADHD) have increased dramatically. Sleep onset insomnia is one of the sleep-related diseases that children with ADHD suffer from, and it reduces learning in children. One of the signs of ADHD in children is the lack of attention to education and poor learning [26, 27].

6.5. Automatic Sleep Stages Classification

In the past, a human technician (physician and nurse) had a sleep test, with a visual inspection of vital signals recorded by a person during night-time sleep. Due to the high volume of sleep data (6 hours of sleep per night), we usually see human errors in the sleep data analysis. The process of visual inspection of sleep stages is complex, time-consuming and susceptible to human errors. It is also very costly due to the use of technician in the sleep review process. Occasionally, there are urgent cases that needed to be diagnosed quickly and manual examination, due to a time-consuming process, may not be effective in these situations [28]. In recent years, due to the importance of sleep monitoring as well as advances in technology, much researches has done to automate the classification of sleep stages using computers. Automatic identification of sleep stages can dramatically reduce sleep study time and provide reliable results.

The automatic classification of sleep stages is performed in two general ways:

- Classic methods;

- Deep learning neural network methods.

In both classical and deep learning methods, we first need a set of appropriate sleep databases to perform processing operations. In the next section, we will review some important ones.

6.5.1. Sleep Databases

Vital data in research may be available in three ways. The researcher has recorded the signals from a set of individuals, the researcher used global and available data, or the researcher has simulated the data. If the researcher has recorded the signals itself, some kind of noises may be merged with the original signal, so the received signal must be pre-processed. For example, Radha and et *al.* record sleep data from 292 patients in 584 nights and present in 2020 [16]. Tataraidze and *et al.* record sleep data from 685 healthy patients with RIP sensor and present it in 2017 [29]. Hwang and *et al.* record sleep data from 25 patients with the bed sensor and present it in 2016 [30].

Available databases will also be very useful for initial simulations of automatic classification of sleep stages algorithms in terms of time and cost. Using the public available databases, you will not have problems such as recording the signal, and it is very easy to work with them. These databases are reviewed in the following section.

6.5.1.1. Available Sleep Databases

Up to March 2020, there are seven popular sleep databases, which we will discuss in the following sections.

Sleep EDF database [expanded]

This collection contains 197 PSG sleep information. The first database of this series was published in 2002, and contains sleep information of 7 people. Complete information on this collection was presented in 2018 for Cink Challenge. The 2018 collection includes EEG (Fpz-Cz and Pz-Oz), EOG and chin EMG with 100 Hz sample rate (some records contain respiration and body temperature) for per person in day-night period. Previous researches has shown that EEG channels are better than C4-A1/C3-A2, which is why

Fpz-Cz/Pz-Oz channels are used in this database [31]. The PSG file of this database is available in EDF format. Every 30 seconds of sleep (considered an epoch) is labelled by a skilled technician according to the R&K standard. Therefore, the sleep annotation of this database is wake (0), S1 (1), S2 (2), S3 (3), S4 (4) and REM (5). Movement time in EDF database annotated by 6. Label files of this database for per person is named Hypnogram.edf [32, 33]. Table 6.2 shows total number of epochs in the sleep EDF database.

Table 6.2. Total number of epochs in sleep EDF database.

Stage	**Wake**	**REM**	**S1**	**S2**	**S3**	**S4**
Epoch	65951	21522	96132	13039	25835	222479

The PSG data in this database are in two categories: SC* (Sleep Cassette) and ST* (Sleep Telemetry). Number of 153 PSG data are available with SC* suffix. There is 20-hour PSG recording of 78 people and two nights record for per person in SC* category. From person 13, 36 and 52 there is one night of PSG recording. This data was record in 1987-1991 to examine the effect of age on sleep of healthy Caucasians aged 25-101 years [34, 35].

ST* files include 9 hours PSG record in hospital from 22 persons and two-night PSG for each person. This database was developed in 1994 to investigate the effect of temazepam on people in the Caucasus [36]. This database is available in the physionet website[1].

CAP Sleep Database

Cyclic Alternating Pattern (CAP) is an alternate and periodic pattern in EEG and occurs during NREM sleep. CAP consists of two cycles, A and B, which brain activity occurs in the A cycle and brain deactivation occurs in the B cycle. The CAP cycle may take less than one minute. Complete information on the CAP cycle is available in the references [37, 38]. Fig. 6.5 shows CAP cycle in the stage 2.

CAP indicates the brain's response to any internal or external factors during sleep, and therefore may be a sign of instability in sleep. The high number of CAP cycles can observe in the EEG of patients with epilepsy, insomnia, and sleep disturbances. For this reason, this database is suitable for the detection of such diseases through sleep [40-42].

The CAP database includes EEG, EOG (2 channels), EMG of the sub mentalis muscle, bilateral anterior tibial EMG, respiration signals (airflow, abdominal and thoracic effort and SaO2) and EKG recordings, from 16 healthy person, 40 patients diagnosed with NFLE, 22 affected by RBD, 10 with PLM, 9 insomniac, 5 narcoleptic, 4 affected by SDB and 2 by bruxism. In total, this database contains a PSG record of 108 peoples at the

---

[1] https://physionet.org/content/sleep-edfx/1.0.0/

Ospedale Maggiore Sleep Disorders Center, Parma, Italy [32, 43]. CAP database is available in the physionet website[1].

Labelling of the CAP cycle according to Terzano's reference atlas was done in REMlogic ™ software (Embla). The record scores are available in.txt file and includes sleep stage (W = wake, S1-S4 = sleep stages, R = REM, MT = body movements) and body position (Left, Right, Prone, or Supine; not recorded in some subjects).

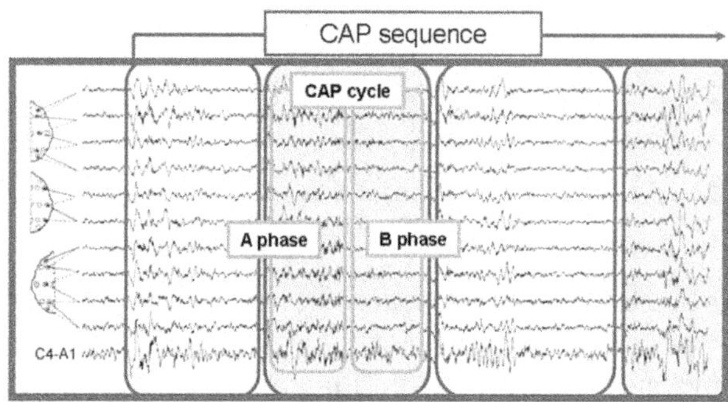

Fig. 6.5. CAP cycle in the stage 2 [39].

St. Vincent's University Hospital / University College Dublin Sleep Apnea Database

St. Vincent's University Hospital / University College Dublin Sleep Apnea Database (UCDDB) database contains PSG recordings of 21 males and 4 females (25 in total) healthy, with no history of disease and drug use and presented in 2011. The PSG includes EEG (C3-A2, C4-A1), EOG (left and right), submental EMG, ECG (modified lead V2), oxygen saturation (finger pulse oximeter), oro-nasal airflow, ribcage movements, body position, abdomen movements (uncalibrated strain gauges) and snoring (tracheal microphone) with 128 Hz sampling rate during 6 hours recordings [32]. The UCDDB database is available in the physionet website[2].

Professional experts according to R&K standard, perform sleep scoring, so that for every 30 second of the signal the stages may include wake (0), REM (1), stage 1 (2), stage 2 (3), stage 4 (5) and artefact (6). There are also four types of sleep apnea annotated in this database that include Hypopnea (HYP), Central (C), Obstructive (O) and Mixed (M).

MIT-BIH Polysomnographic Database

MIT-BIH Polysomnographic database (SLPDB) includes sleep PSG recordings from 18 persons in Boston's Beth Israel Hospital Sleep Laboratory. PSG recordings of this

---

[1] https://physionet.org/content/capslpdb/1.0.0/

[2] https://physionet.org/content/ucddb/1.0.0/

database contain ECG, EEG, EOG, EMG, invasive blood pressure, respiration and cardiac stroke volume signals [44].

The annotation of this database done by a skilled technician and includes sleep stages labels according to R&K standard and sleep apnea. Sleep stages annotations are contain wake (W), stage 1 (1), stage 2 (2), stage 3 (3), stage 4 (4), REM (R) and movement time (MT). Sleep apnea annotations of MIT-BIH Polysomnographic Database are contain hypopnea (H), hypopnea with arousal (HA), obstructive apnea (OA), obstructive apnea with arousal (X), central apnea (CA) and central apnea with arousal (CAA) [32, 44]. The SLPDB database is available in the physionet website[1].

Wrist worn wearable sleep database

Wrist worn wearable sleep database was recorded in 2017-2019 and includes acceleration and heart rate (bpm, measured from photo plethysmography) by Apple Watch from 31 persons during sleep. The purpose of this database is to make good use of smartwatches, and given the ease of clock signal recording, the algorithms presented on this database can be very useful. Subjects were both record for PSG in the laboratory for 8 hours and their accelerometers and heart rate recorded by Apple Watch. Annotation of this database is wake (0), S1 (1), S2 (2), S3 (3), S4 (4) and REM (5) [32, 45]. This database is also available in the physionet[2] website.

Sleep Bioradiolocation Database

Sleep Bioradiolocation Database (SLEEPBRL) by using a bio radar at the Remote Sensing Laboratory of Bauman Moscow State Technical University, examines the sleep of 31 people without taking any medicine. This bio radar is continuous wave type with stepped frequency modulation, quadrature receiver, 3.6-4.0 GHz frequency range and 50 Hz sampling rate. The recording data from this bio radar, annotation according to AASM manual by skilled technician and contain wake (w), stage 1 (1), stage 2 (2), stage 3 (3) and REM (R). Bio radar is a remote control device and has no contact with the person while sleeping [32, 46]. Bioradiolocation database is available in the physionet[3] website.

Sleep Heart Health Study PSG Database (SHHS)

The purpose of the database is to evaluate the health of sleep in the heart, to investigate the effect of sleep breathing on cardiovascular disease, stroke and hypertension. SHHS1 contain PSG from 3441 persons and SHHS2 contain PSG from 3295 persons. In both SHHS1 and SHHS2, PSG recording includes EEG (C3/A2-C4/A1) and EMG with 125 sampling rate, EOG with 50 Hz sampling rate, thoracic and abdominal excursions, airflow, finger-tip pulse oximetry, ECG with 125 Hz sampling rate for SHHS2 and 250 Hz for SHHS1, heart rate, body position and ambient light. SHHS annotation is

---

[1] https://physionet.org/content/slpdb/1.0.0/
[2] https://physionet.org/content/sleep-accel/1.0.0/
[3] https://physionet.org/content/sleepbrl/1.0.0/

available in Hypnogram format, according to R&K standard (such as sleep EDF database) [32, 47-50].

Different databases are used according to the researcher's needs. The number of people in the database is also important for research. For example, in deep learning research studies use large numbers of databases. Table 6.3. Summarizes all available and free available databases for automatic sleep stages classification. After selecting the appropriate database for automatic classification of sleep stages and sleep scoring, intelligent models should be used for automatic sleep scoring and it is done in both classic and deep learning methods.

Table 6.3. Summarizes all available and free databases for automatic sleep stages classification.

Database	Signal	Comment
EDF	EEG, EOG, chin EMG, respiration, body temperature	79 SC, 22 ST
CAP	EEG, EOG, EMG, EKG, respiration signals	16 healthy, 40 patients
UCDDB	EEG, EOG, ECG, EMG	21 male, 4 female
SLPDB	ECG, EEG, EOG, EMG, invasive blood pressure, respiration, cardiac stroke volume signal	18 persons
ACCEL	heart rate, accelerometer	31 persons
SLEEPBRL	Bio radio signal	31 persons
SHHS	EEG, EOG, EMG, ECG, thoracic, abdominal excursions, airflow, finger-tip pulse oximetry	6441 persons

6.5.2. Automatic Sleep Scoring by Classic Methods

Sleep signal processing like other signal processing algorithms, involves the basic stages of signal processing. In the classical signal processing algorithms, the signal is first selected, and the signal pre-processing operations are performed to prepare the signal ready for processing. Afterwards, appropriate features are extracted from the signal. To optimize the algorithms, it is necessary to select the optimal features, which is done by intelligent algorithms in the feature selection stage. Finally, according to the optimized signal features, the desired signal classification operation is performed. Fig. 6.6 shows a block diagram of the classical signal processing steps.

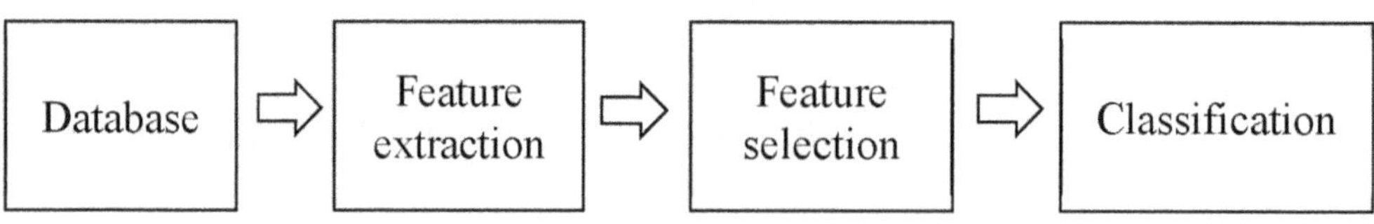

Fig. 6.6. Block diagram of the classical signal processing steps.

6.5.2.1. Feature Extraction

In relation to the automatic classification of sleep stages, five main groups of features have been used to extract the feature so far which include statistical feature, frequency feature, non-linear indicate (chaotic feature), entropy features and line-distance base feature. In addition, considering that the main EEG or ECG signals are considered for sleeping, along with the other signals, some features in each of the five groups are only used for one of the EEG or ECG signals, which will be mentioned in the description of that feature. It should be noted that some sources combine chaotic and entropy features to calculate the complexity of the signal [16, 51], but in this study, because of the many features of each group of entropy and chaos and the comprehensiveness of each group, we fall into two separate categories.

Statistical Features

The first and most common group of features that used in the classification of sleep stages and sleep scoring is statistical features [7, 52-57]. These features use time series or signal statistics. The most important statistical features of mean variance are skewness and kurtosis. In addition, one of the statistical features that is used only for EEG signal is called Hjorth parameter.

Mean and Variance

Mean and variance calculate according to Equation (6.1)-(6.2) respectively. The variance calculates the dispersion around the mean point.

$$\mu = \frac{1}{N}\sum_{i=1}^{N} x(n),$$

$$(6.1)$$

$$\sigma^2 = \frac{1}{N}\sum_{i=1}^{N} (x(n) - \mu)^2$$

$$(6.2)$$

In Equation (6.1) and (6.2), N refers to total number of time series sample and $x(n)$ is time series.

Skewness and Kurtosis

Skewness, examines symmetry or asymmetry of the data. Kurtosis is a statistical operator used to describe statistical distribution. Skewness and kurtosis calculate according to Equation (6.3)-(6.4) respectively [58, 59].

$$x_s = \frac{1}{N}\sum_{i=1}^{N} \left(\frac{x(n) - \mu}{\sigma} \right)^3,$$

$$(6.3)$$

$$x_K = \frac{1}{N}\sum_{i=1}^{N} \left(\frac{x(n) - \mu}{\sigma} \right)^4$$

$$(6.4)$$

In Equation (6.1) and (6.2), N refers to total number of time series sample and $x(n)$ is time series.

Hjorth Parameter

Hjorth parameters present for the first time by Bo Hjorth in 1970 and include tree parameters: activity, mobility and complexity [60, 61]. These parameters are statistical indicators commonly used in analyzing EEG signals for feature extraction [56]. The activity parameter indicates the power of signal. The dynamics parameter represents the mean frequency or standard deviation ratio of the power spectrum. The complexity parameter indicates the change in frequency. The complexity parameter compares the signal similarity to a sinusoidal wave. If the signal is quite similar to a sinusoidal, it reaches 1. Activity, mobility and complexity are calculated according to Equation (6.5)-(6.7).

$$Hjorth\_activity = \sigma_x^2, \tag{6.5}$$

$$Hjorth\_mobility = \frac{\sigma_{x'}}{\sigma_x}, \tag{6.6}$$

$$Hjorth\_complexity = \left(\frac{\sigma_{x''}}{\sigma_{x'}}\right) \div \left(\frac{\sigma_{x'}}{\sigma_x}\right) \tag{6.7}$$

In Equation (6.2)-(6.3), $\sigma_{x'}$ refer to standard deviation of x(n) signal derivative and $\sigma_{x''}$ refer to standard deviation of the second derivative of the x(n) signal.

Root Mean Square of the Successive Differences

Root Mean Square of the successive differences (RMSSD) is a time-domain feature that examines changes in heart rate and calculated according to Equation (6.8) [62]. RMSSD is just use for ECG analysis.

$$RMSSD = \sqrt{\frac{1}{N-2}\sum_{i=3}^{N}[RR_i - RR_{i-1}]^2}, \tag{6.8}$$

In Equation (6.8), RR_i refer to i^{th} R-R in signal and N is the total number of signal sample.

Percentage of Successive RR Differences

Percentage of successive RR differences (pNN50) Computation is the consecutive pair of R-R peaks occurring in more than 50 milliseconds. pNN50 just use for ECG analysis. Fig. 6.7 shows the R-R peak in ECG signal.

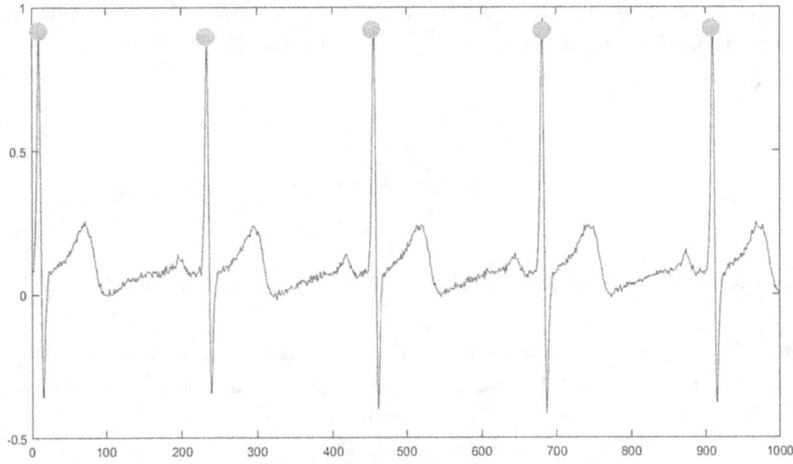

Fig. 6.7. R-R peak in ECG signal.

Frequency Analysis

The second category of features that use for sleep scoring is frequency analysis features [63, 64]. These features transmit the signal to the frequency domain and analyze their frequency dimension. The frequency dimension will display a signal to us that is not detectable in the time dimension. Frequency features used to sleep scoring are divided into two categories: Frequency domain analysis and Time-frequency analysis. Frequency analysis is often used for EEG signals. Due to the complexities of the EEG signal, time-domain analysis is not sufficient for this signal and needs to be investigated in the frequency domain as well [65-67].

Short-time Fourier Transform

Frequency domain analysis is based on the Fourier transform and usually use for EEG signal analysis. One of the most widely used features in this class is short-time Fourier transform (STFT) to determine signal complexity based on frequency function [51]. STFT is calculate according to Equation (6.9) [68].

$$STFT\left[x(n)\right](m,\omega) \equiv X(m,\omega) = \sum_{n=-\infty}^{\infty} x(n)\omega(n-m)e^{-j\omega n} \qquad (6.9)$$

In Equation (6.9), *x(n)* is a time series, ω is continuous and *m* is discrete.

Power Spectral Density

The second feature in frequency domain analysis is power spectral density (PSD). PSD represents the energy at each frequency and usually use for EEG signal analysis. This feature is obtained from the fast Fourier transform of the signal and displays the energy changes as a function of frequency. The frequency function determines at what frequencies the signal changes more and at what frequencies it changes less. Using the A method is one of the simplest ways to calculate a PSD [69, 70]. PSD is calculate according to Equation (6.10).

$$PSD = \frac{\left|FFT(x(n))\right|^2}{w_0 \cdot Q}$$

(6.10)

In Equation (6.10), $x(n)$ is a time series, Q and w_0 are constant values.

Because the EEG signal is a dynamic and time-varying signal, in EEG sleep research, each sleep phase has its own frequency component. Because of this, time-frequency analysis is used for sleep scoring. Three features in time-frequency analysis, is most common for sleep stages analysis that include Wigner-Ville distribution (WVD), wavelet transform (WT) and Hilbert-Huang transform (HHT) [51].

Wigner-ville Distribution

WVD was presents by Sen *et al.* in 2014 [71]. The PSD studies the signal energy distribution in the time-frequency domain and is very suitable for investigating dynamic and time-varying signals such as EEG [72]. Discrete WVD is calculated according to Equation (6.11).

$$WVD_x(n,\frac{m\pi}{M}) = 2 \sum_{k=L+1}^{L-1} p(k)X(n+k)X^*(n-k)e^{-jkm2\pi/M}$$

(6.11)

In Equation (6.11), mπ/M refer to circular frequency, $p(k) = \omega(k)\omega\cdot(k)$. $\omega(k)$ is the time window function at center that it can be Haining window, Gaussian window, Hamming window etc.

Wavelet Transform

The most common feature in time-frequency domain categories is wavelet transform. WT provides us with good frequency information from the sub-band of signal or time series. In order to overcome the limitations of Fourier analysis to obtain information in both time and frequency ranges, another type of conversion, called wavelet transform can be used and wavelet transform emphasizes locations and scales [73, 74]. The signals in the wavelet transform are repeatedly parsed and passed through low pass filters to obtain the system wavelet at a logarithmic frequency resolution. The decomposition process in wavelet transform is performed by the mother wavelet function which can be of various types such as Haar, Daubechies-4, Coiflet-6 and … [75]. The discrete wavelet transform passes the signal through the low pass filter and the output represents the partial coefficients, in addition to the coefficients passing through the high pass filter which also provide approximate coefficients [55, 70]. Discrete WT is calculated according to Equation (6.12) [51].

$$WT = \sum_{R} x(n) \frac{1}{\sqrt{a}} \Psi_{j,k(n)} \tag{6.12}$$

In Equation (6.12), $x(n)$ is time series and Ψ refer to mother wavelet function.

Hilbert-Huang transform

HHT present for the first time in 1998 by Tsai *et al.* and include two-step, calculate empirical mode decomposition (EMD) and Hilbert analysis [76]. HHD is calculated according to Equation (6.13)-(6.19). The EMD signal must first be calculated. Next, the Hilbert transform is performed for each of the IMFs. Then the next step is the Hilbert analysis [51].

$$X(t) = \sum_{i=1}^{n} c_i(t) + r_n, \tag{6.13}$$

$$H\left[c_i(t)\right] = \frac{1}{\pi} P \int_{-\infty}^{+\infty} \frac{c_i(\tau)}{t-\tau} d\tau, \tag{6.14}$$

$$\omega(t) = \frac{d\theta_i(t)}{dt}, \tag{6.15}$$

$$\theta_i(t) = \tan^{-1}\left(\frac{H\left[c_i(\tau)\right]}{c_i(\tau)}\right), \tag{6.16}$$

$$a_i(t) = \sqrt{c_i^2(t) + H^2\left[c_i(t)\right]}, \tag{6.17}$$

$$H(\omega,t) = \mathrm{Re}\left[\sum_{i=1}^{n} a_i(t) e^{j \int \omega_i(t) dt}\right], \tag{6.18}$$

$$HHT = \int_{0}^{T} H(\omega,t) dt \tag{6.19}$$

In Equation (6.13)-(6.19), $c_i(t)$ is the ith IMF component and r_n is the residual component. Equations (6.14)-(6.17) are for Hilbert transform and Equations (6.18)-(6.19) is for calculate HHT.

Chaotic features

The third category of features used to detect sleep stages is nonlinear indicate or chaotic features. These features on the signals describe the basic concepts of nonlinear dynamics and the measurement of signal complexity and stability. Moreover, in recent years, it has attracted the attention of researchers for sleep scoring by EEG or ECG [56, 57, 77]. The most prominent chaotic features are Poincaré plot, Lyapunov exponent, fractal dimension calculation in various ways, including Higuchi, Petrosian and Katz.

Poincaré Plot

Statistically, Poincaré plot shows the correlation between consecutive intervals graphically and present for the first time by Henri Poincare [77]. The Poincaré plot is often used to examine the ECG signal and is generally in Fig. 6.8.

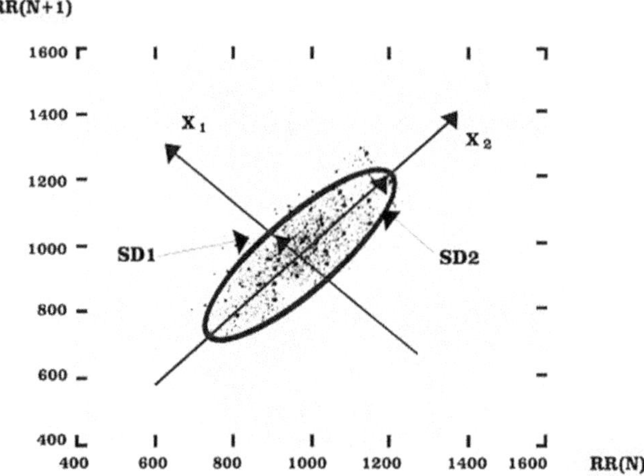

Fig. 6.8. An example of a Poincaré diagram in general.

After drawing the Poincaré plot, we can calculate the criteria for the width, length, and area of the Poincaré diagram, and obtain three features. The length of the first main diameter (SD1) indicates short-term changes in the signal and calculated according to Equation (6.20).

$$SD_1 = \frac{1}{N}\sum_{i=1}^{N}(D_i^1)^2 \qquad (6.20)$$

In Equation (6.20), D_i^1 represents i-th distance from the main signal line and N represents the number of signal samples.

The length of the second diameter of the Poincaré plot indicates long-term changes in the signal and it is calculated according to Equation (6.21).

$$SD^2 = \frac{1}{N}\sum_{i=1}^{N}(D_i^2)^2 \qquad (6.21)$$

Area of Poincare plot is calculated according to Equation (6.22).

$$area = \pi \times SD_1 \times SD_2 \qquad (6.22)$$

Petrosian Fractal Dimension

Statistically the fractal dimension of that signal can measure the complexity of a signal with details of a pattern. One of the methods of calculating the fractal dimension is the Higuchi method. Petrosian Fractal Dimension (PFD) is a combination of Higuchi method and Petrosin algorithm for fast calculation of fractal dimension of a signal [78]. PFD is calculated according to Equation (6.23) [56].

$$PFD = \frac{\log N}{\log N + \log N + 0.4M} \tag{6.23}$$

In Equation (6.23), N denotes the number of signal samples and M denotes the number of signal changes in the signal derivative.

Normalized Line Length

Another method of calculating the fractal dimension is the use of the Katz method [79]. Esteller *et al.* introduced the Normalized Line Length (NLL) feature for faster calculation of the fractal dimension by the Katz method and is calculated according to Equation (6.24) [80].

$$NLL(n) = \frac{1}{M} \sum_{m=n-N}^{n} |x(m) - x(m-1)| \tag{6.24}$$

In Equation (6.24), N denotes the length of the window, n denotes the number of samples, M denotes the number of moving and changing windows and x(n) is one epoch of signal (in sleep study x(n) is 30s of signal). In Memar et al. study about sleep scoring $M = 1$ [56].

The Largest Lyapunov Exponent

Lyapunov's view is a quantity that determines the degree of convergence or divergence of a dynamic system [81]. The largest Lyapunov exponent (LLE) is estimated as the mean rate of separation of the nearest neighbors [82]. For sleep study, a small amount of data method is used for calculate LLE. LLE is calculated according to Equation (6.25).

$$\lambda = \lim_{t \longrightarrow \alpha} \frac{1}{t} \ln \frac{|\Delta x(x_0, t)|}{|\Delta x_0|} \tag{6.25}$$

Hurst Exponent

The Hurst exponent is used to evaluate correlation and self-similarity, and usually used for EEG study [51, 83]. Hurst exponent is calculated according to Equation (6.26).

$$Hurst = \frac{\log(R / S)}{\log(T)} \tag{6.26}$$

In Equation (6.26), T refers to the length of time the sample signal lasts, and R/S is refer to scaling range of sample signal. The closer the value of H is to zero, the more unstable the signal is [51].

Entropy Features

The fourth category of features used in the classification of sleep stages are entropy-based features. These features investigate the complexity and perturbation of the EEG signal in various ways. Statistically the concept of entropy for a system is realized when the system is not separate and the states of the available states of the system are not possible. In this case, the entropy types can be used to check the complexity of the signal. For sleep scoring seven type of entropy feature are used that include normalized spectral entropy, Shannon entropy, Tsallis entropy, Kolmogorov-Sinai entropy, fuzzy entropy, sample entropy and multi-scale entropy.

Normalized Spectral Entropy

After calculating the Fourier transform of the signal, one can obtain the normalized power spectral of the signal and then enter the normalized spectral entropy (NSE) [84, 85]. NSE is calculated according to Equation (6.27).

$$NSE = \frac{1}{log\,(N_f\,)} \sum_{f=f_1}^{f_2} S\,(f\,)(log\,\frac{1}{S\,(f\,)}) \qquad (6.27)$$

In Equation (6.27), F_1 and F_2 are the high and low frequencies of the power spectrum, S refer to power density of signal and N_f is Frequency ranges between high and low frequencies.

Shannon Entropy

Reiny entropy or Shannon entropy (SE) in statistics is defined as the index of diversity and can be considered as a measure of distribution entropy [86]. SE is calculated according to Equation (6.28).

$$SE = \frac{1}{1-M} log(\sum_{i=1}^{n} P_i^m) \qquad (6.28)$$

In Memar *et al.* study about sleep scoring $m = 2$ [56]. m must be a number greater than 0 and $m \neq 1$.

Boltzmann's Entropy

Boltzmann's entropy present for the first time by Ludwig Boltzmann in 1872 and was presented to investigate the entropy of the ideal gas. Boltzmann's entropy is calculated according to Equation (6.29).

$$Boltzmann\_entropy = k \ln(W) \tag{6.29}$$

In Equation (6.29), k is Boltzmann constant and $k = 1.38065 \times 10^{-23}$ w refer to Thermodynamic probability.

Tsallis Entropy

In terms of mathematical calculations and physical states, Tsallis entropy is similar to Boltzmann's entropy. Tsallis entropy is calculated according to Equation (6.30) [87].

$$Tsallis\_entropy = \frac{k(1 - \sum_{i=1}^{m} p_i^q)}{q-1}, q \in R \tag{6.30}$$

In Equation (6.30), m is the number of sample signal and k is Boltzmann constant. Usually in sleep study $k = 1$ and $q = 2$.

Kolmogorov-Sinai Entropy

Kolmogorov-Sinai entropy (K-S) is the matrix entropy of the dynamic system and is an important measure for random testing of a time series. For a regular series, the K-S entropy is zero, positive for a series of noises, but infinite for a random signal [88, 89]. K-S entropy is calculated according to Equation (6.31).

$$K = \lim_{n \to \infty} \sum_{n}^{N-1} (S_{n+1} - S_n) = \lim_{n \to \infty} \frac{1}{N_\tau} [S_N - S_0] \tag{6.31}$$

In Equation (6.31), N refers to total number of signal sample and S is entropy of signal in limited time.

Sample Entropy

Sample entropy (SampEn) is one of the most common types of entropy for detecting diseases from biological signals. SampEn present for the first time by He *et al.* and is the improved of approximation entropy and it is used to calculate signal complexity [90]. SampEn is calculated according to Equation (6.32).

$$SampEn = -\log \frac{A}{B}, A = d[X_{m+1}(i), X_{m+1}(j) \le r, B = d[X_m(i), X_m(j) \le r \tag{6.32}$$

In Equation (6.32), A and B are number of template vector, m refers to length of template vector and d is distance function.

Fuzzy Entropy

Fuzzy entropy is also calculated to calculate the signal complexity, and Chen *et al.* for the first time, improved SampEn and present fuzzy entropy [91]. Fuzzy entropy is very suitable for use on EMG signals, but it also provides a good response for studying sleep. Fuzzy entropy is calculated according to Equation (6.33) [51].

$$FuzzyEn(m,n,r) = \ln \phi^m(n,r) - \ln \phi^{m+1}(n,r),$$

$$\phi^m(n,r) = \frac{1}{N-m}\sum_{i=1}^{N-m}\frac{1}{N-m+1}\sum_{j=1}^{N-m}D_{ij}^m, D_{ij}^m = \exp(-(d_{ij}^m)^n / r) \tag{6.33}$$

In Equation (6.33), d_{ij}^m refer to maximum value of difference in distance between x_i^m and x_j^m, N refer to total number of signal sample, m is the length of sample vector and r is tolerance.

For sample and fuzzy entropy, the selection of appropriate r and m values can influence the desired output. Sleep studies can identify sleep stages by calculating the signal complexity at each stage. For example, the complexity of the wake-up phase is greater than the other stages, and in NREM sleep, the signal complexity is greatly reduced [51].

Distance-based Features

The fifth category of features used to detect sleep stages are distance-based features. These features distinguish between different data's based on the different distances that exist between each individual data element. Among these features are the maximum and minimum data distances and *log* root sum of sequential variation.

Maximum-Minimum Distance

Maximum-Minimum Distance (MMD) calculates the distance between the highest and lowest points in each sub band based on the Pythagorean distance [92]. MMD is calculated according to Equation (6.34)-(6.35).

$$D_k = \sqrt{(\Delta x_k^2 + \Delta y_k^2)}, \tag{6.34}$$

$$MMD = \sum_{k=1}^{N} D_k \tag{6.35}$$

In Equation (6.34) and (6.35), x denotes the distance difference of two points in the time axis, y denotes the distance difference in the domain axis, and N denotes the number of windows on the signal.

Log Root Sum of Sequential Variations

Log Root Sum of Sequential Variations (LRSSV) to measure the successive variations between signal samples are presented and calculated according to Equation (6.36) [56].

$$LRSSV = log\sqrt{(\sum_{n=1}^{N-1}(x(n) - x(n-1))^2)} \tag{6.36}$$

In Equation (6.36), N denotes the number of signal samples.

Table 6.4 lists all the features used to automatic classification of sleep stages. Not all of these features are needed in sleep research, and according to the algorithm, some of these features are used to identify different stages of sleep.

Table 6.4. List of sleep features used in sleep stage classification.

Feature category	Feature	Equation
Statistical features	Mean	$\mu = \dfrac{1}{N}\sum\limits_{i=1}^{N} x(n)$
	Variance	$\sigma^2 = \dfrac{1}{N}\sum\limits_{i=1}^{N}(x(n)-\mu)^2$
	Skewness	$x_s = \dfrac{1}{N}\sum\limits_{i=1}^{N}\left(\dfrac{x(n)-\mu}{\sigma}\right)^3$
Statistical features	Kurtosis	$x_K = \dfrac{1}{N}\sum\limits_{i=1}^{N}\left(\dfrac{x(n)-\mu}{\sigma}\right)^4$
	Hjorth parameter	$Hjorth\_mobility = \dfrac{\sigma_{x'}}{\sigma_x},$ $Hjorth\_complexity = \left(\dfrac{\sigma_{x''}}{\sigma_{x'}}\right) \div \left(\dfrac{\sigma_{x'}}{\sigma_x}\right),\ Hjorth\_activity = \sigma_x^2$
	RMSSD	$RMSSD = \sqrt{\dfrac{1}{N-2}\sum\limits_{i=3}^{N}[RR_i - RR_{i-1}]^2}$
Frequency analysis	Short-time FT	$STFT\left[x(n)\right] = \sum\limits_{n=-\infty}^{\infty} x(n)\omega(n-m)e^{-j\omega n}$
	Power spectral density	$PSD = \dfrac{\left\|FFT(x(n))\right\|^2}{w_0 \cdot Q}$
	Wigner-ville distribution	$WVD_x(n,\dfrac{m\pi}{M}) = 2\sum\limits_{k=L+1}^{L-1} p(k)X(n+k)X^*(n-k)e^{-jkm2\pi/M}$
	Wavelet transform	$WT = \sum\limits_{R} x(n)\dfrac{1}{\sqrt{a}}\Psi_{j,k(n)}$
	Hilbert-Huang transform	$HHT = \int\limits_{0}^{T} H(\omega,t)dt$
Chaotic features	Poincare plot	$area = \pi \times SD_1 \times SD_2$
	Petrosian Fractal Dimension	$PFD = \dfrac{\log N}{\log N + \log N + 0.4M}$

Feature category	Feature	Equation
	Normalized Line Length	$$NLL(n) = \frac{1}{M} \sum_{m=n-N}^{n} |x(m) - x(m-1)|$$
	largest Lyapunov exponent	$$\lambda = \lim_{t \longrightarrow \alpha} \frac{1}{t} \ln \frac{|\Delta x(x_0, t)|}{|\Delta x_0|}$$
	Hurst exponent	$$Hurst = \frac{\log(R/S)}{\log(T)}$$
Entropy features	Normalized spectral entropy	$$NSE = \frac{1}{\log(N_f)} \sum_{f=f_1}^{f_2} S(f)(\log \frac{1}{S(f)})$$
	Shannon entropy	$$SE = \frac{1}{1-M} \log(\sum_{i=1}^{n} P_i^m)$$
	Boltzmann's entropy	$$Boltzmann\_entropy = k \ln(W)$$
	Tsallis entropy	$$Tsallis\_entropy = \frac{k(1 - \sum_{i=1}^{m} p_i^q)}{q-1}, q \in R$$
	Kolmogorov-Sinai entropy	$$K = \lim_{n \to \infty} \frac{1}{N_\tau}[S_N - S_0]$$
	Sample entropy	$$SampEn = -\log \frac{A}{B}$$
	Fuzzy entropy	$$FuzzyEn(m,n,r) = \ln \phi^m(n,r) - \ln \phi^{m+1}(n,r)$$
Distance-based features	MMD	$$MMD = \sum_{k=1}^{N} D_k$$
	LRSSV	$$LRSSV = \log \sqrt{(\sum_{n=1}^{N-1}(x(n) - x(n-1))^2)}$$

After reviewing the appropriate features for sleep analysis, we will examine the feature selectors. In automatic classification, it is necessary to identify the superior features after extracting the features.

6.5.2.2. Feature Selection

Human expertise, often needed to convert raw data into useful feature sets, can be complemented by automated feature extraction methods and operations on useful features. In order to automatically process signals, we first need a set of features that represent the properties of a particular signal, and can clearly distinguish the desired signal from other signals.

The number of features is a key factor that determines the size of the hypothesis space that contains all the hypotheses that can be derived from the data. Linear increase in the number of features results in an exponential increase in the hypothesis space. Therefore, the selection of superior features can effectively reduce the hypothesis space by eliminating inefficient and unnecessary features. Feature selection is essentially a task of removing inappropriate or excessive features.

Feature selectors have two general stages, named search and selection mechanisms. Each search and selection algorithm can be done by different methods. There are three categories for search algorithms that are complete search, stochastic search and heuristic search methods. Moreover, there are five categories for selection types that are filter based, wrapper based, embedded selection, hybrid selection and feature selection algorithms according to behavior of living [93]. First, we will review three categories of search algorithms, and then investigate different feature selection methods.

Complete Search

This complete search method is a comprehensive algorithm, and at first a complete search method starts with all the features. Afterwards, at each step a feature is removed to finally achieve a set of efficient features.

Random Search

This method usually provides a smaller number of subsets than the original set of 2^n by setting the maximum number of iterations (n refers to the number of features). The optimality of the selected subset depends on the available constant values. Each random production method requires some parameter values. Assigning appropriate values to these parameters is an important step in achieving the desired results.

Heuristics Search

In each iteration of this category, all remaining features that are still selected will be nominated, and the number of subsets increases with each step. This method has high computational speed and produces good results.

After reviewing the three available methods for searching feature selectors, we move on to the description of selection methods. There are five general methods for selection in feature selectors, which are: filter based, wrapper based, embedded selection, hybrid selection method and feature selection according to behavior of living. We will examine these selection methods

Filter Based Selection

Filter-based methods are used separately from the classifier to identify the pattern and work on the data without any particular parameter dependency. This method is very suitable for use on extensive data. Filter based methods uses a filter to apply to the feature matrix, which can have different types. In the filter based methods, first by using the filter, the feature vector dimming operation is performed, and a superior feature vector is created and then the this vector can be classified [94, 95]. Cho *et al.* use normalized mutual information feature selection method that is a filter based feature selection for sleep scoring [96]. Fig. 6.9 shows the flowchart of a filter-based selection method.

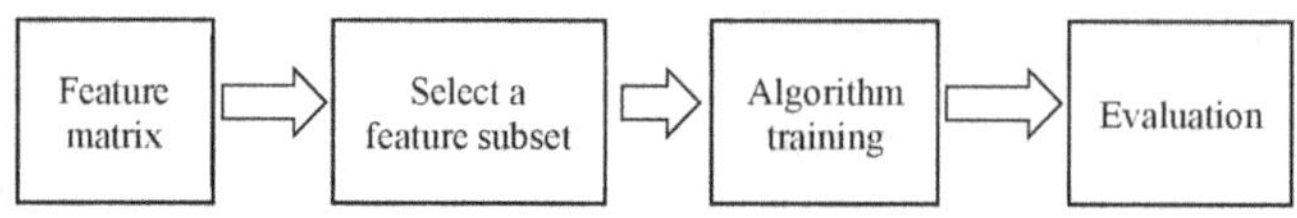

Fig. 6.9. Flowchart of filter-based selection method.

The nature of the general filter-based methods has made them universally. However, in recent studies, the types of filter-based methods have been less accurate than other methods. One of the filter-based feature selectors is the Laplacian selector, which works for labelled and unlabelled data based on the Laplacian graph [97, 98].

Wrapper-based Selection

In wrapper-based methods, in order to evaluate the features, the classification error is calculated using the selected features, and the set of features are considered as the superior features which results in the least error in the classification stage. In the Wrapper-based method, the main feature vector is used and simultaneously in a dual rotational process, the feature dimension reduction is performed and the superior data are classified [99]. Seifpour *et al.* use a wrapper-based feature selection to classify six sleep stages [100]. Fig. 6.10 show the flowchart of a wrapper-based selection method.

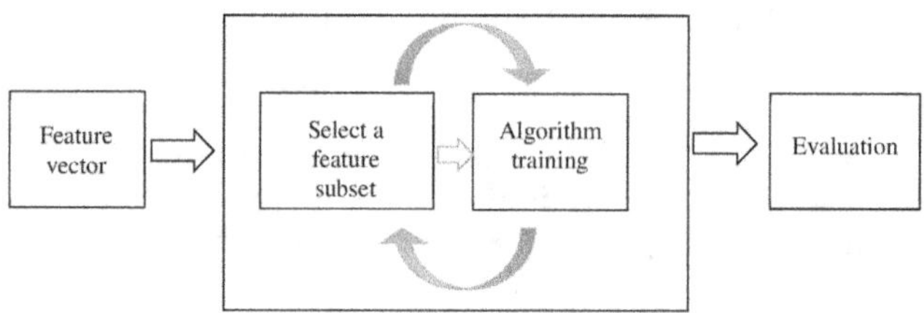

Fig. 6.10. Flowchart of wrapper-based selection method.

Wrapper-based methods create a set of custom features that are tailored to some specific task and some specific system, which limits this. In addition, this method is only applicable to small datasets and will face computational weakness in the face of large amounts of data.

Embedded Selection

Embedded selection methods are a combination of two previous methods, and is presented to solve the weaknesses of the two previous ones. Similar to wrapper-based methods, this method is also dependent on the type of classifier, and will therefore create problems in the process of selecting the superior features. Rahman *et al.* classified six sleep stages by EOG signal using neighborhood component analysis as feature selector, that is an embedded one [101]. Fig. 6.11 shows the flowchart of an embedded based feature selection method.

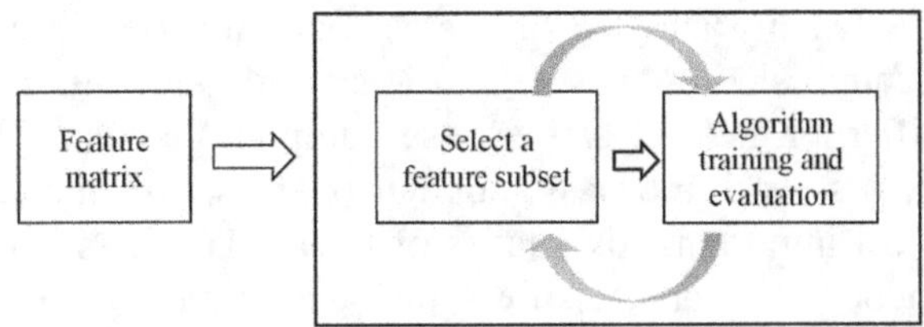

Fig. 6.11. Flowchart of an embedded based selection method.

Hybrid Selection Method

Hybrid selection methods are a combination of the two methods of filter-based and wrapper-based selection algorithms [102, 103]. Fig. 6.12 shows the flowchart of a hybrid selection method.

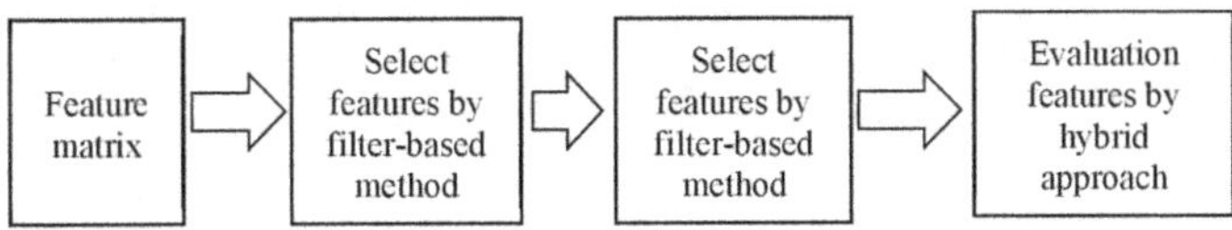

Fig. 6.12. Flowchart of hybrid selection method.

Selection Methods Inspired from Nature

This category of feature selection is based on the survival process of organisms. For example, the ant colony algorithm is one of these selectors and works on the ant process of finding food [104, 105]. Genetic algorithm optimization is another example of these selectors that works based on the evolutionary process of organisms and the Darwin principle [106]. Yundong *et al.* uses genetic algorithm feature selection for sleep scoring in 2010 [107].

Once the top features are selected, it is necessary to use the appropriate classifications to identify the different stages of sleep. Therefore, in the next section, we will examine the different types of classifiers for automatic classification of sleep stages.

6.5.2.3. Classifier

An algorithm that implements data classification is known as a classifier. The term classifier also refers to the mathematical function performed by a classification algorithm that assigns the input data to a specified class. In pattern recognition, data categorization is divided into two different types: classification of supervised data by classifiers and grouping of unsupervised data by clustering methods. Classifiers often use labelled data and assign new input data to one of the categories. Grouping of unsupervised data, known as clustering operations, use unlabelled data and divide the input data into different categories and each category considered as an output.

Because sleep data is usually labelled one, classifiers are used in connection with the automatic classification of sleep stages. Classifiers are divided into two classes of linear and nonlinear classifiers. Linear classifiers use complex mathematical relationships to classify feature vectors, and are less popular because of the complexity of the relationships. The most important advantages of linear classifiers is their simplicity and computational efficiency. Linear classifiers take less time to train, use and achieve comparable accuracy to nonlinear classifiers. Due to the simplicity of computation, linear classifiers cannot properly classify all feature vectors and their applications are limited. The types of classifiers used to classify the features of a signal, including the sleep signal are: artificial neural networks (ANN), Naïve Bayes Classifier (NBC), k-nearest neighbor (KNN), AdaBoost, Bagging, RF, denoising auto encoder (DA), support vector machine (SVM), decision tree (DA), linear discriminant analysis (LDA) and random forest (RF). Recent researches on the automatic classification of sleep stages by classic method with the type of used classifier in Table 6.5 are shown.

Table 6.5. Recent research on the automatic classification of sleep stages by classic method with the type of used classifier.

Method	Type of classifier	Signal	Number of sleep stages	Accuracy
Memar [56]	RF	EEG	5	98.44 %
Liang [108]	DT	EEG, EOG, EMG	6	86.68 %
Phan [66]	KNN	EEG	6	94.4 %
Ronzhina [109]	ANN	EEG	6	76.7 %
Hassan [54]	AdaBoost	EEG	6	84.3 %
Yu Zhang [110]	LDA	EEG	5	75.6 %
Yücelbaş [111]	RF	ECG	3	76 %
Adnane [112]	SVM	ECG	2	79.9 %
Hassan [113]	SVM	EOG	6	91.7 %
Geng [114]	SVM	HRV	3	77.9 %

Automatic classification of sleep stages is done in two ways: classic and deep learning methods. So far, all the steps of the classical method have been studied. In the next section, deep learning methods are introduced.

6.5.3. Automatic Sleep Scoring by Deep Learning Method

Deep learning neural networks are new area of machine learning, and there are a variety of applications in different engineering problems in recent years. They are an advanced type of simple elementary neural networks, and allow for signal processing in greater number of hidden layer than other neural networks including perceptron. Due to the use of more hidden layers in this type of neural networks, these networks can process large data. Due to the high volume of sleep data that results from recording the signal during several hours of sleep, in recent years the use of deep learning neural networks for automated sleep studies has received much attention from researchers [115-118].

Deep learning overcomes the deficiencies of the classical methods with automatic learning. That is, deep learning methods require only a partially processed data set, rather than manually editing data (classic method). Afterwards, attributes are extracted, and feature vectors are classified. Thus, the engineering burden of features has been transferred from humans to computers, enabling users of machine learning to use deep learning networks for their research or applications, especially in image analysis and medical signal processing applications.

In addition to automatic processing of all signal processing steps, deep training networks are not noise sensitive. In sleep data, there is a high likelihood of noise contamination due to prolonged recording time and individual movements during sleep and this has been the second reason researchers have been using deep learning networks in recent years. In addition, deep training networks need high GPUs to run because of the high volume of computing, and these computers are readily available today.

The theory of deep learning networks was introduced for the first time in year 1980, and has made many advances up to date. The types of deep learning networks, from the first type to the most advanced ones used today, are as follows: Feed-Forward Neural Networks, Deep Boltzmann machine, Stacked auto encoder and Convultional neural network (CNN), respectively [119].

CNN is an advanced type of perceptron neural network with more hidden layer. The learning stage is very strong in the convolutional networks and easily recognizes the signal pattern, which is why they are used for any kind of signal and image use. In general, it consists of four general layers: input layer, convolutional layer (middle layer), pooling layer and dropout layer, respectively. The raw signal enters the input layer, and the desired output, which is the desired class, is provided after passing the fourth layer. Fig. 6.13 shows a view of a convulousional neural network (one dimensional).

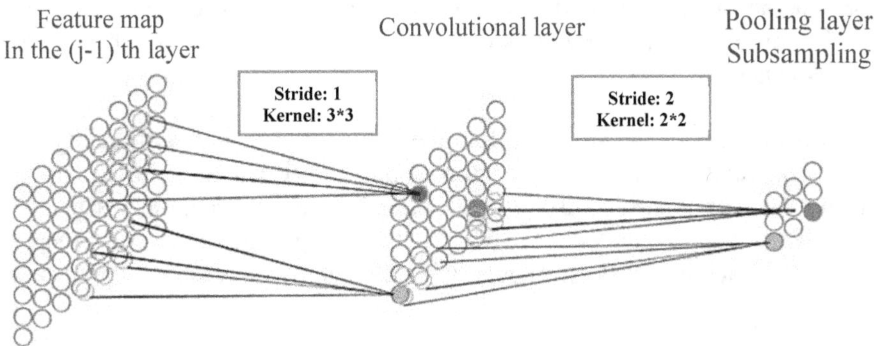

Fig. 6.13. Shows a view of a convulousional neural network (one dimensional).

Input Layer

This layer is responsible for receiving the signal. In CNN, data can be entered as one-dimensional, two-dimensional, or three-dimensional. Two-dimensional data is usually used for image processing applications.

Convolutional Layer

In the convolutional layer, the pattern learning of the signal is performed on the basis of some of the filters and the canonical calculations of the signal. This layer itself contains several substrates. Usually the number of middle layers is equal to four layers. The first layer is a filter, also known as the feature identification layer or kernel. With each filter, CNN only can learn one pattern or feature, so it is usually designed to have between 20-100 filters to train the network for more features. The second, third, and fourth layers, like the first layer, are different types of filters for pattern learning. This step is similar to the feature extraction and optimum feature selection in the classical method, except that in deep training networks this is done automatically.

Pooling Layer

This layer is used to collect the output of the previous layers and classify the data. It is very important to determine the dimension of this layer. For example, if the dimension of this layer is selected 3, it means that the output matrix size of this layer is only one third of the input matrix of this layer. In some CNN networks a middle pooling layer is placed after the second and third layers.

Dropout Layer

Dropout layer is usually used to set up convolutional networks and gives us the desired output as well as noise-related operations in this layer. This layer is used during training and not during testing. The weights given to the samples during the training make the network unable to accept new data and the connections between the layers become too much. Dropout randomly converts some of these weights to zero and by this way the network is balanced.

Convolutional networks, despite the complexity of training and adaptability to big data, have low computational speed and require sophisticated hardware to run, limiting their use. In the last few years, studies on the classification of sleep stages using cannulas networks have been conducted to investigate them. A review of recent research into the automatic classification of sleep stages using deep learning neural networks are shown in Table 6.6.

After reviewing two general classical methods and deep learning method to study the sleep stages, we present a new method for automatic classification of sleep stages.

Table 6.6. A review of recent research into the automatic classification of sleep stages using deep learning neural networks.

Method	Network	Hidden layer	Signal	Sleep stages	Accuracy
Supratak [120]	CNN	12	EEG	5	82 %
Zhang [121]	CU-CNN	4	EEG	5	87 %
Sors [122]	CNN	14	EEG	5	87 %
Qiao [123]	CNN	7	ECG	4	75.4 %
Wei [124]	CNN	4	ECG	3	77 %

6.5.4. New Method for Sleep Scoring

An EEG signal is a signal that contains useful information to check for sleep, and in spite of more complex recording than other signals, it has more useful information. We can achieve good results for classifying different stages of sleep by using EEG signal in proposed method. Deep learning methods are not sensitive to noise and perform all the steps of the signal automatically but now these networks are more complex and less accurate than the classical methods.

Section 6.5.4 presents a classic method using single-channel EEG for automatic classification of sleep stages. In the proposed algorithm, statistical, entropy and distance-based features are first extracted from EEG signal. In the last step, to improve the classification accuracy of the different stages of sleep, we combine the accuracy of three different classifiers named SVM, KNN and DA by Dempster-Shafer algorithm. The results of several different classifications can be combined with different methods, one of this method is Dempster-Shafer.

Like other signal processing methods, the proposed method includes three main parts of database, feature extraction, and classification, which are examined in these steps. Fig. 6.14 show the block diagram of proposed method.

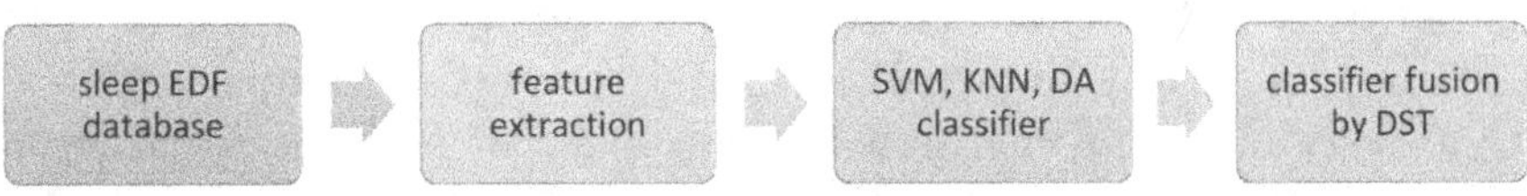

Fig. 6.14. Block diagram of proposed method.

6.5.4.1. Database

The proposed method uses sleep-EDF database that is introduced in Section 6.5.1.1. This database includes PSG signals from 197 persons. In the proposed method, we used single channel EEG (Pz-Oz channel used according to [125]) from 100 persons for automatic classification of sleep stages.

6.5.4.2. Feature Extraction

In the proposed algorithm, statistical, entropy and distance-based features are first extracted from EEG signal. Table 6.7 shows all used features in the proposed method.

Table 6.7. All features that used in proposed method.

Feature category	Type of feature
Statistical features	Mean, variance, skewness, kurtosis, Hjorth parameters
Entropy features	normalized spectral entropy, Shannon entropy, Boltzmann entropy
Distance-based features	Maximum minimum distance

6.5.4.3. Classification

Three different classifiers are used for automatic classification of sleep stages are: SVM (quadratic kernel), KNN (medium KNN) and DA (fine tree). Initially, data classification operations are performed with all three classifier and finally, the results of all three classifier are combined by Dempster-Shafer theory [126, 127].

6.5.4.4. Dempster-Shafer Classifier Fusion

Data theory is inspired by the function of the human brain. Because the brain simultaneously combines different information, including eye, ear and mouth information. Sometimes we can not get complete information for signal processing from one source, so using several information sources at the same time can increase the accuracy of signal processing.

Dempster–Shafer theory (DST) was first proposed by Dempster in 1967 and promoted by Glenn Shafer [128, 129]. DST is based on mathematical evidence and according to this theory, information from different sources can be combined. This combination can be done at the data level, at the feature level or at the classifier level [130]. There are three important functions in DST theory that are central to computation. These three functions are: basic belief assignment (m or BBL), belief function (Bel) and plausibility function (Pl). The basic belief assignment is a function of belief mapping is evidence of the occurrence of A. Bel and Pl represents the upper and lower limits of a probability. m, Bel and Pl are calculated according to Equation (6.37)-(6.39).

$$m : 2^\theta \rightarrow [0,1] \, if \sum_{A \subseteq \Theta} m(A) = 1 \_ and \_ m(\phi) = 0, \tag{6.37}$$

$$Bel(A) = \sum_{B \subseteq A} m(B), \tag{6.38}$$

$$Pl(A) = 1 - Bel(\bar{A}) = \sum_{B \cap A = \odot} m(B) \tag{6.39}$$

6.5.4.5. Results

In the proposed method, a set of statistical, entropy and distance-based features are used. After extracting the feature, data classification operations are performed by three classifiers SVM, KNN and DA. Then, to improve the results of automatic classification of sleep stages, the results of three classifications are combined using DST. The proposed method is simulated in MATLAB 2018a.

Evaluation Criteria

Accuracy, sensitivity and specificity are evaluation criteria that calculated in proposed method according to Equation (6.40)-(6.42).

$$Sensitivity = \frac{T_{pos}}{T_{pos} + F_{neg}},\tag{6.40}$$

$$Specificity = \frac{T_{neg}}{T_{neg} + F_{pos}},\tag{6.41}$$

$$Accuracy = \frac{T_{pos} + T_{neg}}{T_{neg} + F_{neg} + T_{pos} + F_{pos}}\tag{6.42}$$

In these equations, T_{pos} is correctly identified target class and F_{pos} is incorrectly identified target class. Also T_{neg} refers to correctly identified non-target class and F_{neg} refers to incorrectly identified non-target class. Another evaluation criterion is called the kappa coefficient that Introduced in 1960 by Cohen and calculate agreement between two or more cases. Kappa coefficient calculated according to Equation (6.43).

$$k = 1 - \frac{1 - p_o}{1 - p_e}\tag{6.43}$$

In Equation (6.43), p_0 relative observed agreement among rates and p_e is hypothetical probability of chance agreement. The most ideal case for the kappa coefficient is 1.

Table 6.8 show accuracy, kappa, sensitivity and specificity for sex sleep stages according to R&K standard by SVM, KNN and DA classifier. Specificity and sensitivity values are expressed as average for 6 stages of sleep. As shown in Table 6.8, accuracy, kappa, sensitivity and specificity increase after classifier fusion by DST and DA is the weakest classifier for proposed method.

Stimulation Results

The confusion matrix (based on epoch) of automatic sleep classification by SVM, KNN and DA classifier is shown by in Tables 6.9-6.11. As shown in Table 6.9, the best classifier for automatic classification of sleep stages is SVM, but despite its superiority over other classifier, it does not give us acceptable accuracy and as can be seen in all three tables, the detection accuracy of stage 2 is very low.

After automatically classifying the sleep stages with three different classifiers, in order to increase the classification accuracy, we combine the classification results with DST method. Table 6.12 show confusion matrix (based on accuracy) of sleep scoring after classifier fusion by DST. As can be seen, the classification accuracy of each stage of sleep has increased, especially the proposed method provides acceptable accuracy in the diagnosis of stage 2.

Fig. 6.15. Show the boxplot of features in proposed method. The size of the box for each feature indicates the scatter of values for that feature. So the smaller a box feature, the more suitable it is. Therefor mean, NSE, Shanonn_e, Boltzman_e and MMD are the best features for proposed method according to Fig. 6.15.

Table 6.8. Accuracy, kappa, sensitivity and specificity foe sex sleep stages according to R&K standard by SVM, KNN and DA classifier.

Classifier	Accuracy	Kappa	Sensitivity	Specificity
SVM	91.9 %	72.96 %	42.5 %	97.97 %
KNN	90.9 %	70.53 %	30.02 %	97.00 %
DA	89.6 %	67.22 %	30.2 %	97.2 %
Proposed method	**97.94 %**	**91.29 %**	**51.69 %**	**98.65 %**

Table 6.9. Confusion matrix of automatic classification of six sleep stages by SVM classifier.

	Wake	4841	8	56	0	2	26	98 %
	REM	43	17	75	0	0	51	46 %
True Class	S1	29	6	429	3	0	57	69 %
	S2	1	0	13	3	1	0	30 %
	S3	2	0	1	4	6	0	67 %
	S4	27	6	77	0	0	182	58 %
		Wake	REM	S1	S2	S3	S4	accuracy
				Prediction class				

Table 6.10. Confusion matrix of automatic classification of six sleep stages by KNN classifier.

	Wake	4843	6	60	1	0	23	97 %
	REM	55	14	69	0	0	48	27 %
True Class	S1	46	13	482	1	0	45	66 %
	S2	1	0	15	0	2	0	0 %
	S3	2	0	2	1	8	0	80 %
	S4	41	18	100	0	0	133	53 %
		Wake	REM	S1	S2	S3	S4	accuracy
				Prediction class				

Table 6.11. Confusion matrix of automatic classification of six sleep stages by DA classifier.

<table>
<tr><td rowspan="8">True Class</td><td>Wake</td><td>4788</td><td>35</td><td>80</td><td>2</td><td>0</td><td>28</td><td>97 %</td></tr>
<tr><td>REM</td><td>51</td><td>18</td><td>76</td><td>0</td><td>0</td><td>41</td><td>17 %</td></tr>
<tr><td>S1</td><td>49</td><td>23</td><td>440</td><td>3</td><td>1</td><td>71</td><td>63 %</td></tr>
<tr><td>S2</td><td>2</td><td>0</td><td>10</td><td>4</td><td>2</td><td>0</td><td>33 %</td></tr>
<tr><td>S3</td><td>1</td><td>0</td><td>1</td><td>3</td><td>8</td><td>0</td><td>73 %</td></tr>
<tr><td>S4</td><td>32</td><td>29</td><td>88</td><td>0</td><td>0</td><td>143</td><td>51 %</td></tr>
<tr><td></td><td>Wake</td><td>REM</td><td>S1</td><td>S2</td><td>S3</td><td>S4</td><td>accuracy</td></tr>
<tr><td colspan="8" align="center">Prediction class</td></tr>
</table>

Table 6.12. Confusion matrix (based on accuracy) of sleep scoring after classifier fusion by DST.

<table>
<tr><td rowspan="8">True Class</td><td>Wake</td><td>99 %</td><td>0</td><td>0</td><td>0</td><td>0</td><td>1 %</td></tr>
<tr><td>REM</td><td>19 %</td><td>42 %</td><td>16 %</td><td>0</td><td>0</td><td>21 %</td></tr>
<tr><td>S1</td><td>0.2 %</td><td>0.3 %</td><td>98 %</td><td>0</td><td>0</td><td>0.6 %</td></tr>
<tr><td>S2</td><td>0</td><td>0</td><td>34 %</td><td>37 %</td><td>28 %</td><td>0</td></tr>
<tr><td>S3</td><td>0</td><td>0</td><td>0</td><td>0.7 %</td><td>99 %</td><td>0</td></tr>
<tr><td>S4</td><td>0.5 %</td><td>3 %</td><td>6 %</td><td>0</td><td>0</td><td>90 %</td></tr>
<tr><td></td><td>Wake</td><td>REM</td><td>S1</td><td>S2</td><td>S3</td><td>S4</td></tr>
<tr><td colspan="7" align="center">Prediction class</td></tr>
</table>

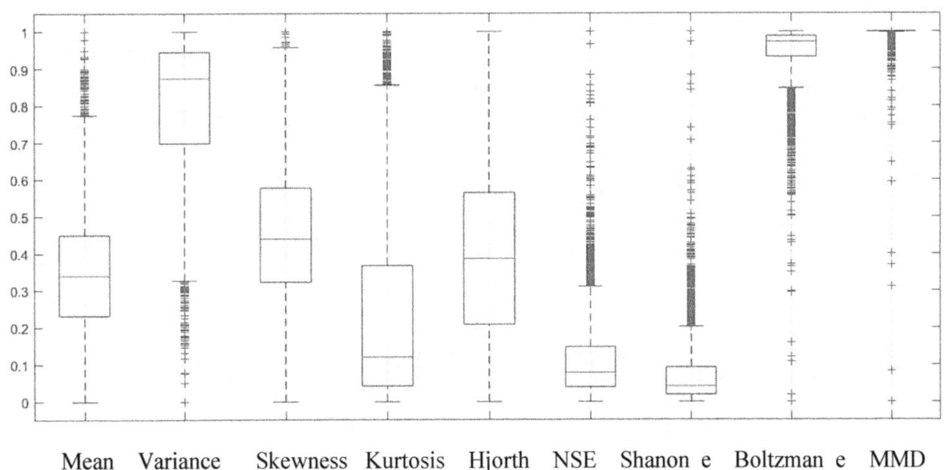

Fig. 6.15. Show the boxplot of features in proposed method.

6.5.5. Discussion New Method

In the proposed method, single channel EEG from sleep EDF database was used to identify six stages of sleep. In the next step, feature extraction operations were performed using statistical, entropy and distance-based features and the extracted features were classified into three classifier SVM, KNN and DA. The proposed method does not use the feature selector and all features are used for classification. Then, to increase the accuracy of the automatic classification of sleep stages, we combined the results of three different classifications with each other for classifier fusion by DST method. It was noted that the accuracy of classification of sleep stages using the combined DST, increased by an average of 4-5 % for each stage of sleep. Also, the overall accuracy of the classification reached 97.97 %, which is an increase of 6 % compared to the accuracy of the best type of classification (SVM) in proposed method. Therefore, using classifier fusion instead of feature selectors can increase the accuracy of classifying different stages of sleep.

6.6. Conclusion and Future Works

In this chapter, at first, the nature of sleep and its various stages were fully studied. Then, the importance of automatic review of sleep stages was discussed, and due to this importance, we reviewed various databases and methods for automatic classification of sleep stages. Regarding the databases, seven public available databases for sleep study which contains different vital biological signals were reviewed. Afterwards, two methods of automatic classification of sleep stages include the classical and the deep learning algorithms are reviewed. Automatic classification of sleep stages in the classical method includes three stages of feature extraction, feature selection and classification, all of which have been thoroughly examined. In the classical method, the feature selection step may be omitted and replaced instead classifier fusion, which has been explored in the introduction of a new method for automatic sleep monitoring. In order to automatically classify the sleep stages with deep learning networks, all types of deep learning networks and the structure of each were examined, separately. Finally, a new method was introduced for automatic classification of sleep stages, which is a classic method. In the proposed method, statistical, entropy and distance-based features are used for feature selection, and SVM, KNN and DA as classifier. DST method was used for classifier fusion to increase the classification accuracy and six sleep stages according to R&K standard were categorized with 98.96 % accuracy. Due to the importance of investigating sleep disorders and identifying these disorders through the correct identification of sleep stages, it is necessary to do more work on automatic methods of classifying sleep stages. In the future, we plan to research deep learning networks to find networks that classification sleep stages with the highest accuracy in the shortest time. Deep learning networks now have lower computational speed and accuracy compared to classical methods.

References

[1]. N. M. Gage, B. J. Baars, Fundamentals of Cognitive Neuroscience, 2nd Edition, *Elsevier*, 2018, pp. 393-435.

[2]. S. C. Chong, L. Xin, L. J. Ptáček, Y.-H. Fu, Disorders of sleep and circadian rhythms, in Handbook of Clinical Neurology, Vol. 148, *Elsevier*, 2018, pp. 531-538.

[3]. S. Sanei, J. A. Chambers, EEG Signal Processing, *John Wiley & Sons*, England, 2007.

[4]. A. S. Tubbs, H. K. Dollish, F. Fernandez, M. A. Grandner, The basics of sleep physiology and behavior, in Sleep and Health, *Elsevier*, 2019, pp. 3-10.

[5].   O. Faust, H. Razaghi, R. Barika, E. J. Ciaccio, U. R. Acharya, A review of automated sleep stage scoring based on physiological signals for the new millennia, *Computer Methods Programs in Biomedicine,* Vol. 176, July 2019, pp. 81-91.

[6].   H. R. Colten, B. M. Altevogt (Eds.), Sleep Disorders and Sleep Deprivation, *National Academies Press,* US, 2006.

[7].   T. Abel, R. Havekes, J. M. Saletin, M. P. Walker, Sleep, plasticity and memory from molecules to whole-brain networks, *Current Biology,* Vol. 23, Issue 17, 2013, pp. R774-R788.

[8].   L. Xie *et al.*, Sleep drives metabolite clearance from the adult brain, *Science,* Vol. 342, Issue 6156, 2013, pp. 373-377.

[9].   K. Eckel-Mahan, P. Sassone-Corsi, Metabolism and the circadian clock converge, *Physiological Reviews,* Vol. 93, Issue 1, 2013, pp. 107-135.

[10]. F. Portaluppi, R. Tiseo, M. H. Smolensky, R. C. Hermida, D. E. Ayala, F. Fabbian, Circadian rhythms and cardiovascular health, *Sleep Medicine Reviews,* Vol. 16, Issue 2, 2012, pp. 151-166.

[11]. M. M. Steriade, R. W. McCarley, Brainstem Control of Wakefulness and Sleep, *Springer Science & Business Media,* 2013.

[12]. R. B. Berry, R. Brooks, C. E. Gamaldo, S. M. Harding, C. Marcus, B. V. Vaughn, The AASM Manual for the Scoring of Sleep and Associated Events: Rules, Terminology and Technical Specifications, *American Academy of Sleep Medicine,* Darien, Illinois, 2012.

[13]. S. J. Redmond, C. Heneghan, Cardiorespiratory-based sleep staging in subjects with obstructive sleep apnea, *IEEE Transactions on Biomedical Engineering,* Vol. 53, Issue 3, 2006, pp. 485-496.

[14]. T. Willemen *et al.*, An evaluation of cardiorespiratory and movement features with respect to sleep-stage classification, *IEEE Journal of Biomedical Health Informatics,* Vol. 18, Issue 2, 2013, pp. 661-669.

[15]. P. Fonseca, X. Long, M. Radha, R. Haakma, R. M. Aarts, J. Rolink, Sleep stage classification with ECG and respiratory effort, *Physiological Measurement,* Vol. 36, Issue 10, 2015, pp. 2027-2040.

[16]. M. Radha *et al.*, Sleep stage classification from heart-rate variability using long short-term memory neural networks, *Scientific Reports,* Vol. 9, Issue 1, 2019, 14149.

[17]. U. R. Acharya, K. P. Joseph, N. Kannathal, L. C. Min, J. S. Suri, Heart rate variability, in Advances in Cardiac Signal Processing, *Springer,* 2007, pp. 121-165.

[18]. A. Rechtschaffen, A Manual for Standardized Terminology, Techniques and Scoring System for Sleep Stages in Human Subjects, *Brain Information Service,* 1968.

[19]. C. Iber, The AASM Manual for the Scoring of Sleep and Associated Events: Rules, Terminology and Technical Specifications, *American Academy of Sleep Medicine,* Westchester, IL, 2007.

[20]. R. K. Malhotra, A. Y. Avidan, Sleep stages and scoring technique, in Atlas of Sleep Medicine, *Elsevier,* 2014, pp. 77-99.

[21]. D. A. Nita, S. K. Weiss, Sleep and epilepsy, in Sleep in Children with Neurodevelopmental Disabilities, *Springer,* 2019, pp. 227-240.

[22]. A. J. Rowan, R. J. Veldhuisen, N. J. Nagelkerke, Comparative evaluation of sleep deprivation and sedated sleep EEGs as diagnostic aids in epilepsy, *Electroencephalography Clinical Neurophysiology,* Vol. 54, Issue 4, 1982, pp. 357-364.

[23]. S. Chan, T. Baldeweg, J. Cross, A role for sleep disruption in cognitive impairment in children with epilepsy, *Epilepsy Behavior,* Vol. 20, Issue 3, 2011, pp. 435-440.

[24]. B. V. Vaughn, F. C. O'Neill, R. Beach, J. A. Messenheimer, Improvement of epileptic seizure control with treatment of obstructive sleep apnoea, *Seizure-European Journal of Epilepsy,* Vol. 5, Issue 1, 1996, pp. 73-78.

[25]. M. N. Shouse, A. M. da Silva, M. Sammaritano, Circadian rhythm, sleep, and epilepsy, *Journal of Clinical Neurophysiology,* Vol. 13, Issue 1, 1996, pp. 32-50.

[26]. K. B. Van der Heijden, M. G. Smits, E. J. V. Someren, W. Boudewijn Gunning, Idiopathic chronic sleep onset insomnia in attention-deficit/hyperactivity disorder: A circadian rhythm sleep disorder, *Chronobiology International,* Vol. 22, Issue 3, 2005, pp. 559-570.

[27]. S. Miano *et al.*, Sleep phenotypes in attention deficit hyperactivity disorder, *Sleep Medicine,* Vol. 60, 2019, pp. 123-131.

[28]. N. A. Collop, Scoring variability between polysomnography technologists in different sleep laboratories, *Sleep Medicine,* Vol. 3, Issue 1, 2002, pp. 43-47.

[29]. A. Tataraidze, L. Anishchenko, L. Korostovtseva, M. Bochkarev, Y. Sviryaev, S. Ivashov, Estimation of a priori probabilities of sleep stages: A cycle-based approach, in *Proceedings of the 39th Annual International Conference of the IEEE Engineering in Medicine and Biology Society (EMBC'17)*, 2017, pp. 3745-3748.

[30]. S. H. Hwang, Y. J. Lee, D. U. Jeong, K. S. Park, Unconstrained sleep stage estimation based on respiratory dynamics and body movement, *Methods of Information in Medicine,* Vol. 55, Issue 6, 2016, pp. 545-555.

[31]. B. Van Sweden, B. Kemp, H. Kamphuisen, E. Van der Velde, Alternative electrode placement in (automatic) sleep scoring (f pz-cz/p z-oz versus c4-at), *Sleep,* Vol. 13, Issue 3, 1990, pp. 279-283.

[32]. A. L. Goldberger *et al.*, PhysioBank, PhysioToolkit, PhysioNet: Components of a new research resource for complex physiologic signals, *Circulation,* Vol. 101, Issue 23, 2000, pp. e215-e220.

[33]. B. Kemp, A. H. Zwinderman, B. Tuk, H. A. Kamphuisen, J. J. Oberye, Analysis of a sleep-dependent neuronal feedback loop: The slow-wave micro continuity of the EEG, *IEEE Transactions on Instrumentation Measurement,* Vol. 47, Issue 9, 2000, pp. 1185-1194.

[34]. M. Mourtazaev, B. Kemp, A. Zwinderman, H. Kamphuisen, Age and gender affect different characteristics of slow waves in the sleep EEG, *Sleep,* Vol. 18, Issue 7, 1995, pp. 557-564.

[35]. B. Kemp, A model-based monitoring of human sleep stages, *Biol. Cybern.,* Vol. 57, Issue 6, 1987, pp. 365-378.

[36]. B. Kemp, A. Janssen, M. Roessen, A digital telemetry system for ambulatory sleep recording, *Sleep-Wake Research in The Netherlands,* Vol. 4, 1993, pp. 129-132.

[37]. M. Terzano, D. Mancia, M. Salati, G. Costani, A. Decembrino, L. Parrino, The cyclic alternating pattern as a physiologic component of normal NREM sleep, *Sleep,* Vol. 8, Issue 2, 1985, pp. 137-145.

[38]. L. Parrino, R. Ferri, O. Bruni, M. G. Terzano, Cyclic alternating pattern (CAP): The marker of sleep instability, *Sleep Medicine Reviews,* Vol. 16, Issue 1, 2012, pp. 27-45.

[39]. M. G. Terzano *et al.*, Atlas, rules, and recording techniques for the scoring of cyclic alternating pattern (CAP) in human sleep, *Sleep Medicine,* Vol. 2, Issue 6, 2001, pp. 537-553.

[40]. L. Parrino, P. Halasz, C. A. Tassinari, M. G. Terzano, CAP, epilepsy and motor events during sleep: The unifying role of arousal, *Sleep Medicine Reviews,* Vol. 10, Issue 4, 2006, pp. 267-285.

[41]. M. G. Terzano, L. Parrino, M. C. Spaggiari, V. Palomba, M. Rossi, A. Smerieri, CAP variables and arousals as sleep electroencephalogram markers for primary insomnia, *Clinical Neurophysiology,* Vol. 114, Issue 9, 2003, pp. 1715-1723.

[42]. L. Parrino, M. Boselli, G. P. Buccino, M. C. Spaggiari, G. Di Giovanni, M. G. Terzano, The cyclic alternating pattern plays a gate-control on periodic limb movements during non-rapid eye movement sleep, *Journal of Clinical Neurophysiology,* Vol. 13, Issue 4, 1996, pp. 314-323.

[43]. M. G. Terzano *et al.*, Atlas, rules, and recording techniques for the scoring of cyclic alternating pattern (CAP) in human sleep, *Sleep Medicine,* Vol. 3, Issue 2, 2002, pp. 187-199.

[44]. Y. Ichimaru, G. Moody, Development of the polysomnographic database on CD-ROM, *Psychiatry Clinical Neurosciences,* Vol. 53, Issue 2, 1999, pp. 175-177.

[45]. O. Walch, Y. Huang, D. Forger, C. Goldstein, Sleep stage prediction with raw acceleration and photoplethysmography heart rate data derived from a consumer wearable device, *Sleep,* Vol. 42, Issue 12, 2019, zsz180.

[46]. A. Tataraidze, L. Korostovtseva, L. Anishchenko, M. Bochkarev, Y. Sviryaev, S. Ivashov, Bioradiolocation-based sleep stage classification, in *Proceedings of the 38th Annual International Conference of the IEEE Engineering in Medicine and Biology Society (EMBC'16),* 2016, pp. 2839-2842.

[47]. S. F. Quan *et al.*, The sleep heart health study: Design, rationale, and methods, *Sleep,* Vol. 20, Issue 12, 1997, pp. 1077-1085.

[48]. B. K. Lind, J. L. Goodwin, J. G. Hill, T. Ali, S. Redline, S. F. Quan, Recruitment of healthy adults into a study of overnight sleep monitoring in the home: Experience of the sleep heart health study, *Sleep Breathing,* Vol. 7, Issue 1, 2003, pp. 13-24.

[49]. S. Redline *et al.*, Sleep heart health research group methods for obtaining and analyzing unattended polysomnography data for a multicenter study, *Sleep,* Vol. 21, Issue 7, 1998, pp. 759-767.

[50]. C. W. Whitney *et al.*, Reliability of scoring respiratory disturbance indices and sleep staging, *Sleep,* Vol. 21, Issue 7, 1998, pp. 749-757.

[51]. D. Zhao, Comparative analysis of different characteristics of automatic sleep stages, *Computer Methods Programs in Biomedicine,* Vol. 175, July 2019, pp. 53-72.

[52]. A. R. Hassan, M. I. H. Bhuiyan, Automatic sleep scoring using statistical features in the EMD domain and ensemble methods, *Biocybernetics and Biomedical Engineering,* Vol. 36, Issue 1, 2016, pp. 248-255.

[53]. A. R. Hassan, M. I. H. Bhuiyan, Computer-aided sleep staging using complete ensemble empirical mode decomposition with adaptive noise and bootstrap aggregating, *Biomedical Signal Processing and Control,* Vol. 24, 2016, pp. 1-10.

[54]. A. R. Hassan, M. I. H. Bhuiyan, An automated method for sleep staging from EEG signals using normal inverse Gaussian parameters and adaptive boosting, *Neurocomputing,* Vol. 219, 2017, pp. 76-87.

[55]. B. K. Kanoje, A. S. Shingare, Automatic sleep stage detection of an EEG signal using an ensemble method, *International Journal of Advanced Research in Computer Engineering & Technology,* Vol. 3, Issue 8, 2014, pp. 2717-2724.

[56]. P. Memar, F. Faradji, A novel multi-class EEG-based sleep stage classification system, *IEEE Transactions on Neural Systems and Rehabilitation Engineering,* Vol. 26, Issue 1, 2018, pp. 84-95.

[57]. R. Sharma, R. B. Pachori, A. Upadhyay, Automatic sleep stages classification based on iterative filtering of electroencephalogram signals, *Neural Computing Applications,* Vol. 28, Issue 10, 2017, pp. 2959-2978.

[58]. K. V. Mardia, Measures of multivariate skewness and kurtosis with applications, *Biometrika,* Vol. 57, Issue 3, 1970, pp. 519-530.

[59]. D. Joanes, C. Gill, Comparing measures of sample skewness and kurtosis, *Journal of the Royal Statistical Society: Series D,* Vol. 47, Issue 1, 1998, pp. 183-189.

[60]. B. Hjorth, EEG analysis based on time domain properties, *Electroencephalography, Clinical Neurophysiology,* Vol. 29, Issue 3, 1970, pp. 306-310.

[61]. B. Hjorth, Time domain descriptors and their relation to a particular model for generation of EEG activity, in *Proceedings of the Symposium of Merck'sche Gesellschaft für Kunst und Wissenschaft, Kronberg/Taunus (CEAN'75),* 1975, pp. 3-8.

[62]. J. Malik, Y.-L. Lo, H.-t. Wu, Sleep-wake classification via quantifying heart rate variability by convolutional neural network, *Physiological Measurement,* Vol. 39, Issue 8, 2018, 085004.

[63]. L. Fraiwan, K. Lweesy, N. Khasawneh, H. Wenz, H. Dickhaus, Automated sleep stage identification system based on time–frequency analysis of a single EEG channel and random forest classifier, *Computer Methods Programs in Biomedicine,* Vol. 108, Issue 1, 2012, pp. 10-19.

[64]. L. Fraiwan, K. Lweesy, N. Khasawneh, M. Fraiwan, H. Wenz, H. Dickhaus, Time frequency analysis for automated sleep stage identification in full-term and preterm neonates, *Journal of Medical Systems,* Vol. 35, Issue 4, 2011, pp. 693-702.

[65]. P. Puranik, R. Kshirsagar, S. Motdhare, Elementary time frequency analysis of EEG signal processing, *EAI Endorsed Transactions on Pervasive Health Technology,* Vol. 4, Issue 14, 2018, 155081.

[66]. S. Pan, J. Liu, Minimal wave speed of traveling wave fronts in delayed Belousov-Zhabotinskii model, *Electronic Journal of Qualitative Theory of Differential Equations,* Vol. 2012, Issue 90, 2012, pp. 1-12.

[67]. B. Weiss, Z. Clemens, R. Bódizs, Z. Vágó, P. Halász, Spatio-temporal analysis of monofractal and multifractal properties of the human sleep EEG, *Journal of Neuroscience Methods,* Vol. 185, Issue 1, 2009, pp. 116-124.

[68]. L. Zhou, Z. Sun, W. Wang, Learning to short-time Fourier transform in spectrum sensing, *Physical Communication,* Vol. 25, 2017, pp. 420-425.

[69]. P. Welch, The use of fast Fourier transform for the estimation of power spectra: a method based on time averaging over short, modified periodograms, *IEEE Transactions on Audio Electroacoustics,* Vol. 15, Issue 2, 1967, pp. 70-73.

[70]. M. Prucnal, A. G. Polak, Effect of feature extraction on automatic sleep stage classification by artificial neural network, *Metrology and Measurement Systems,* Vol. 24, Issue 2, 2017, pp. 229-240.

[71]. B. Şen, M. Peker, A. Çavuşoğlu, F. V. Çelebi, A comparative study on classification of sleep stage based on EEG signals using feature selection and classification algorithms, *Journal of Medical Systems,* Vol. 38, Issue 3, 2014, 18.

[72]. Y. Chen, Y. Zhao, W. Zhao, L. Zhao, A comparative study of preconditioners for GPU-accelerated conjugate gradient solver, in *Proceedings of the IEEE 10th International Conference on High Performance Computing and Communications & IEEE International Conference on Embedded and Ubiquitous Computing,* 2013, pp. 628-635.

[73]. N. Kumar, K. Alam, A. H. Siddiqi, Wavelet transform for classification of EEG signal using SVM and ANN, *Biomedical Pharmacology Journal,* Vol. 10, Issue 4, 2017, pp. 2061-2069.

[74]. C. Yücelbaş, Ş. Yücelbaş, S. Özşen, G. Tezel, S. Küççüktürk, Ş. Yosunkaya, Automatic detection of sleep spindles with the use of STFT, EMD and DWT methods, *Neural Computing Applications,* Vol. 29, Issue 8, 2018, pp. 17-33.

[75]. A. Petrosian, New classes of hybrid Hadamard-wavelet transforms for signal-image processing, in *Proceedings of the Second Joint 24th Annual Conference and the Annual Fall Meeting of the Biomedical Engineering Society Engineering in Medicine and Biology,* Vol. 1, 2002, pp. 153-154.

[76]. F.-F. Tsai, S.-Z. Fan, Y.-S. Lin, N. E. Huang, J.-R. Yeh, Investigating power density and the degree of nonlinearity in intrinsic components of anesthesia EEG by the Hilbert-Huang transform: an example using ketamine and alfentanil, *PloS ONE,* Vol. 11, Issue 12, 2016, e0168108.

[77]. M. Brennan, M. Palaniswami, P. Kamen, Do existing measures of Poincare plot geometry reflect nonlinear features of heart rate variability?, *IEEE Transactions on Biomedical Engineering,* Vol. 48, Issue 11, 2001, pp. 1342-1347.

[78]. A. Petrosian, Kolmogorov complexity of finite sequences and recognition of different preictal EEG patterns, in *Proceedings of the Eighth IEEE Symposium on Computer-Based Medical Systems (CBMS'95)*, 1995, pp. 212-217.

[79]. M. J. Katz, Fractals and the analysis of waveforms, *Computers in Biology Medicine,* Vol. 18, Issue 3, 1988, pp. 145-156.

[80]. R. Esteller, J. Echauz, T. Tcheng, B. Litt, B. Pless, Line length: an efficient feature for seizure onset detection, in *Proceedings of the 23rd Annual International Conference of the IEEE Engineering in Medicine and Biology Society (EMBC'01)*, 2001, Vol. 2, pp. 1707-1710.

[81]. G. Boeing, Visual analysis of nonlinear dynamical systems: chaos, fractals, self-similarity and the limits of prediction, *Systems,* Vol. 4, Issue 4, 2016, 37.

[82]. A. P. Guerrero, G. E. Paredes, Linear and Non-linear Stability Analysis in Boiling Water Reactors: The Design of Real-time Stability Monitors, *Woodhead Publishing,* 2018.

[83]. S. Tzouras, C. Anagnostopoulos, E. McCoy, Financial time series modeling using the Hurst exponent, *Physica A: Statistical Mechanics Its Applications,* Vol. 425, 2015, pp. 50-68.

[84]. T. Inouye *et al.*, Quantification of EEG irregularity by use of the entropy of the power spectrum, *Electroencephalography Clinical Neurophysiology,* Vol. 79, Issue 3, 1991, pp. 204-210.

[85]. M. Sabeti, S. Katebi, R. Boostani, Entropy and complexity measures for EEG signal classification of schizophrenic and control participants, *Artificial Intelligence in Medicine,* Vol. 47, Issue 3, 2009, pp. 263-274.

[86]. A. Rényi, On measures of entropy and information, in *Proceedings of the Fourth Berkeley Symposium on Mathematical Statistics and Probability,* Vol. 1, 1961, pp. 547-561.

[87]. A. Plastino, M. Rocca, G. Ferri, Dimensional regularization of Renyi's statistical mechanics, *Physica A: Statistical Mechanics its Applications,* Vol. 505, 2018, pp. 794-804.

[88]. V. Latora, M. Baranger, Kolmogorov-Sinai entropy rate versus physical entropy, *Physical Review Letters,* Vol. 82, Issue 3, 1999, 520.

[89]. Y. Sinai, Kolmogorov-Sinai entropy, *Scholarpedia,* Vol. 4, Issue 3, 2009, 2034.

[90]. X. He, C. Shao, Y. Xiong, A non-parametric symbolic approximate representation for long time series, *Pattern Analysis Applications,* Vol. 19, Issue 1, 2016, pp. 111-127.

[91]. Z. Mu, J. Hu, J. Min, EEG-based person authentication using a fuzzy entropy-related approach with two electrodes, *Entropy,* Vol. 18, Issue 12, 2016, 432.

[92]. K. A. I. Aboalayon, M. Faezipour, W. S. Almuhammadi, S. Moslehpour, Sleep stage classification using EEG signal analysis: a comprehensive survey and new investigation, *Entropy,* Vol. 18, Issue 9, 2016, 272.

[93]. R. Dwivedi, R. Kumar, E. Jangam, V. Kumar, An ant colony optimization based feature selection for data classification, *International Journal of Recent Technology and Engineering,* Vol. 7, Issue 5S4, 2019, pp. 35-40.

[94]. H. Liu, R. Setiono, A probabilistic approach to feature selection-a filter solution, in *Proceedings of the 20th International Conference on Machine Learning (ICML'96)*, 1996, Vol. 96, pp. 319-327.

[95]. L. Yu, H. Liu, Feature selection for high-dimensional data: A fast correlation-based filter solution, in *Proceedings of the 20th International Conference on Machine Learning (ICML'03)*, 2003, pp. 856-863.

[96]. D. Cho, B. Lee, Optimized automatic sleep stage classification using the normalized mutual information feature selection (NMIFS) method, in *Proceedings of the 39th Annual International Conference of the IEEE Engineering in Medicine and Biology Society (EMBC'17)*, 2017, pp. 3094-3097.

[97]. X. He, D. Cai, P. Niyogi, Laplacian score for feature selection, in *Proceedings of the 18th International Conference on Neural Information Processing Systems (NIPS'05)*, 2005, pp. 507-514.

[98]. Z. Zhao, H. Liu, Spectral feature selection for supervised and unsupervised learning, in *Proceedings of the 24th International Conference on Machine learning (ICML'07)*, 2007, pp. 1151-1157.

[99]. R. Kohavi, G. H. John, Wrappers for feature subset selection, *Artificial Intelligence,* Vol. 97, Issues 1-2, 1997, pp. 273-324.

[100]. S. Seifpour, H. Niknazar, M. Mikaeili, A. M. Nasrabadi, A new automatic sleep staging system based on statistical behavior of local extrema using single channel EEG signal, *Expert Systems with Applications,* Vol. 104, 2018, pp. 277-293.

[101]. M. M. Rahman, M. I. H. Bhuiyan, A. R. Hassan, Sleep stage classification using single-channel EOG, *Computers in Biology Medicine,* Vol. 102, 2018, pp. 211-220.

[102]. H.-H. Hsu, C.-W. Hsieh, M.-D. Lu, Hybrid feature selection by combining filters and wrappers, *Expert Systems with Applications,* Vol. 38, Issue 7, 2011, pp. 8144-8150.

[103]. M.-C. Lee, Using support vector machine with a hybrid feature selection method to the stock trend prediction, *Expert Systems with Applications,* Vol. 36, Issue 8, 2009, pp. 10896-10904.

[104]. M. Dorigo, G. Di Caro, Ant colony optimization: A new meta-heuristic, in *Proceedings of the Congress on Evolutionary Computation (CEC'99),* 1999, Vol. 2, pp. 1470-1477.

[105]. M. Dorigo, M. Birattari, T. Stutzle, Ant colony optimization, *IEEE Computational Intelligence Magazine,* Vol. 1, Issue 4, 2006, pp. 28-39.

[106]. D. S. Weile, E. Michielssen, Genetic algorithm optimization applied to electromagnetics: A review, *IEEE Transactions on Antennas Propagation,* Vol. 45, Issue 3, 1997, pp. 343-353.

[107]. Y. Ji, X. Bu, J. Sun, Z. Liu, An improved simulated annealing genetic algorithm of EEG feature selection in sleep stage, in *Proceedings of the Asia-Pacific Signal and Information Processing Association Annual Summit and Conference (APSIPA'16),* 2016, pp. 1-4.

[108]. S.-F. Liang, C.-E. Kuo, Y.-H. Hu, Y.-S. Cheng, A rule-based automatic sleep staging method, *Journal of Neuroscience Methods,* Vol. 205, Issue 1, 2012, pp. 169-176.

[109]. M. Ronzhina, O. Janoušek, J. Kolářová, M. Nováková, P. Honzík, I. Provazník, Sleep scoring using artificial neural networks, *Sleep Medicine Reviews,* Vol. 16, Issue 3, 2012, pp. 251-263.

[110]. Y. Zhang, B. Wang, J. Jing, J. Zhang, J. Zou, M. Nakamura, A comparison study on multidomain EEG features for sleep stage classification, *Computational Intelligence Neuroscience,* Vol. 2017, 2017, 4574079.

[111]. Ş. Yücelbaş, C. Yücelbaş, G. Tezel, S. Özşen, Ş. Yosunkaya, Automatic sleep staging based on SVD, VMD, HHT and morphological features of single-lead ECG signal, *Expert Systems with Applications,* Vol. 102, 2018, pp. 193-206.

[112]. M. Adnane, Z. Jiang, Z. Yan, Sleep-wake stages classification and sleep efficiency estimation using single-lead electrocardiogram, *Expert Systems with Applications,* Vol. 39, Issue 1, 2012, pp. 1401-1413.

[113]. M. M. Rahman, M. I. H. Bhuiyan, A. R. Hassan, Sleep stage classification using single-channel EOG, *Computers in Biology Medicine,* Vol. 102, 2018, pp. 211-220.

[114]. D. Geng, J. Zhao, J. Dong, X. Jiang, Comparison of support vector machines based on particle swarm optimization and genetic algorithm in sleep staging, *Technology Health Care,* 2019, Vol. 27, Issue 4, pp. 1-9.

[115]. O. Yildirim, U. B. Baloglu, U. R. Acharya, A deep learning model for automated sleep stages classification using PSG signals, *International Journal of Environmental Research Public Health,* Vol. 16, Issue 4, 2019, 599.

[116]. M. Sokolovsky, F. Guerrero, S. Paisarnsrisomsuk, C. Ruiz, S. A. Alvarez, Deep learning for automated feature discovery, classification of sleep stages, *IEEE/ACM Transactions on Computational Biology,* 2019.

[117]. H. Phan, F. Andreotti, N. Cooray, O. Y. Chén, M. De Vos, Joint classification and prediction CNN framework for automatic sleep stage classification, *IEEE Transactions on Biomedical Engineering,* Vol. 66, Issue 5, 2018, pp. 1285-1296.

[118]. Y. Yang, X. Zheng, F. Yuan, A study on automatic sleep stage classification based on CNN-LSTM, in *Proceedings of the 3rd International Conference on Crowd Science and Engineering (ICCSE'18),* 2018, pp. 1-5.

[119]. D. Shen, G. Wu, H.-I. Suk, Deep learning in medical image analysis, *Annual Review of Biomedical,* Vol. 19, 2017, pp. 221-248.

[120]. A. Supratak, H. Dong, C. Wu, Y. Guo, DeepSleepNet: A model for automatic sleep stage scoring based on raw single-channel EEG, *IEEE Transactions on Neural Systems and Rehabilitation Engineering,* Vol. 25, Issue 11, 2017, pp. 1998-2008.

[121]. J. Zhang, Y. Wu, Complex-valued unsupervised convolutional neural networks for sleep stage classification, *Computer Methods Programs in Biomedicine,* Vol. 164, 2018, pp. 181-191.

[122]. A. Sors, S. Bonnet, S. Mirek, L. Vercueil, J.-F. Payen, A convolutional neural network for sleep stage scoring from raw single-channel EEG, *Biomedical Signal Processing and Control,* Vol. 42, 2018, pp. 107-114.

[123]. Q. Li, Q. Li, C. Liu, S. P. Shashikumar, S. Nemati, G. D. Clifford, Deep learning in the cross-time frequency domain for sleep staging from a single-lead electrocardiogram, *Physiological Measurement,* Vol. 39, Issue 12, 2018, 124005.

[124]. R. Wei, X. Zhang, J. Wang, X. Dang, The research of sleep staging based on single-lead electrocardiogram and deep neural network, *Biomedical Engineering Letters,* Vol. 8, Issue 1, 2018, pp. 87-93.

[125]. S.-F. Liang, C.-E. Kuo, Y.-H. Hu, Y.-H. Pan, Y.-H. Wang, Automatic stage scoring of single-channel sleep EEG by using multiscale entropy and autoregressive models, *IEEE Transactions on Instrumentation Measurement,* Vol. 61, Issue 6, 2012, pp. 1649-1657.

[126]. J. C. Principe, T.-G. Chang, S. Gala, A. J. Tome, Information processing models for sleep staging, *Expert Systems with Applications,* Vol. 6, Issue 4, 1993, pp. 399-409.

[127]. M. A. A. Siddiqui, Fusion of ECG/EEG for improved automatic seizure detection using Dempster Shafer theory of evidence, PhD Thesis, *King Fahd University of Petroleum and Minerals,* Saudi Arabia, 2011.

[128]. A. P. Dempster, Upper and lower probabilities generated by a random closed interval, *The Annals of Mathematical Statistics,* Vol. 39, Issue 3, 1968, pp. 957-966.

[129]. G. Shafer, A Mathematical Theory of Evidence, *Princeton University Press,* 1976.

[130]. L. Hou, N. W. Bergmann, Induction motor fault diagnosis using industrial wireless sensor networks and Dempster-Shafer classifier fusion, in *Proceedings of the 37th Annual Conference of the IEEE Industrial Electronics Society (IECON'11),* 2011, pp. 2992-2997.

Chapter 7
Statistical Learning of Phase Measuring Deflectometry for Defect Detection and Classification on Specular Cab Body Surfaces

N. Pya Arnqvist, E. Lindahl and J. Yu

7.1. Introduction

Competition in the automotive industries has reached the highest level recently. Due to the competition, automated and intelligent manufacturing process is highly desired for the increase of the production efficiency, the product quality as well as the decrease of the labor costs. Nowadays automatic inspection plays a significant role in industrial quality management and defect detection is an important factor on quality control process. The location, nature and formation of the defective regions on the painted surface can be revealed through automatic inspection. Hence automatic inspection is of crucial importance to maintain the quality of the painted surface.

Despite a highly automated computerized system among many processes, the paint quality inspection process is still mainly performed manually in most worldwide automotive manufacturers. Human vision and touch are still important for the inspection of blemishes and bumps on the final painted surface, but they are susceptible to inconsistency due to [1]:

- Unavoidable human error;

- The limited time for the inspection of each vehicle body;

- Some defects are tough to observe due to their micrometer size or less-accessible location on the product;

- Some defects are only visible in some particular viewing angles;

Natalya Pya Arnqvist, Department of Mathematics and Mathematical Statistics, Umeå University, Umeå, Sweden;

- Light conditions.

In this investigation at Volvo GTO paint shop, the issue on difficulties encountered from inspecting highly reflected painted surface is especially crucial for defects of small sizes, which are only visible when using directional light or viewed at a certain angle.

Manual quality inspections and root cause analysis are intensive, time-consuming, expensive and do not necessarily minimize the consequences of the defects. An automated paint quality control can lead to higher quality products and at the same time can reduce the costs and time waste caused by defects.

Among those solutions developed for tackling defect detection problems during the last decade, deflectometry-based detection on specular cab body surfaces has shown its reliability and accuracy [2, 3]. In automotive industry, deflectometry- and vision-based technologies combined with image fusion [4-6] have been investigated, for instance Ford [7], Mercedes-Benz [8], and Opel [9]. Statistical learning, combined with machine colour-based vision and image analysis, has shown its power and potential in image segmentation and pattern recognition [10-12], e.g. recognizing defects on painted vehicle bodies [13, 14].

In this chapter, we present a statistical learning approach for defect detection on painted cab body surfaces, covering image acquisition, feature extraction and defect detection and classification. As the inspection of specular surfaces inflicts special challenges, deflectometry technique using reflected sinusoidal fringe is applied to capture images of the considered surfaces. The main objective in developing this new approach is, in addition to providing accurate and reliable classification, to furnish uncertainty estimation in the classification results.

7.2. Data Acquisition

At the paint shop in the Volvo cab plant in Umeå, fully automated robots and a group of color technicians work together to make sure each cab exits with the best smooth, high quality finish and perfect color available on the market. Before the cabs are shipped to the main assembly lines, the paint shop in Umeå takes care of their paintings. Every cab from the body shop goes through multiple stages in the paint shop, and a computerized tracking system monitors every cab through each stage. The result is a classic finish that is consistent, durable, and tough.

The color bank of the plant offers a wide range of colors to the customers. There are more than 800 different colors to choose from. A mixed customized color can be provided as well if necessary. Hundreds of cabs are painted at the paint shop every day [15]. Paint pumps up automatically from a selected bucket and only the exact amount of paint needed is used. There is an advanced system especially created to ensure this process is flawless. The paint shop has been carefully structured and designed to take all environmental aspects into account and one of its top priorities is a reduction of the amount of solvent used. The first three layers are corrosion protectants, followed by a primer to ensure better adhesion and durability for the paint system, as well as ensuring the correct underlying color. The next layer is either a base coat followed by a clear coat, or a pigmented topcoat directly on top of the primer [15]. These later coats provide the final color. The final steps in the paint shop are human visual inspection and touch to ensure there are no missed spots, irregularities, bumps, scratches or uneven surfaces.

Acquisition of image data and annotation has been conducted through a pilot system installed at the paint shop in the Volvo cab plant, described as follows.

7.2.1. Object

Three different surfaces are targeted:

1) A part of the luggage lid for the FH cabs. The size of this object is approximately 77 cm^2 × 30 cm^2 which represents about 48 % of the whole luggage lid;

2) A part of the luggage lid together with a small part of the area above it for the FM Small cabs. The size of this object is approximately 25 cm^2 × 30 cm^2;

3) A part of the luggage lid together with a small part of the area above the luggage lid for the FM Long cabs. The size of this object is approximately 62 cm^2 × 30 cm^2.

7.2.2. Camera and Screen Calibration

Fig. 7.1 displays the setup of our pilot system used in this work. The setup is composed by a 55-inch Sony screen and two Fujinon HF16SA1 cameras. The dimensions of the images taken from each of the two cameras are 2464 × 2052 with the sizes of around 5 megabits (Mb) each. The goal is to combine the images from both cameras to form one big image with the size of 9.6 Mb and the dimensions of 4928 × 2056. The area of our object has increased considerably by using this setup with two cameras. The Sony screen has about 124 cm of width, 72 cm of height and 139 cm of diagonal. The screen's resolution is 3840 × 2160 (4K display).

Fig. 7.1. Setup of the pilot system.

The most important geometrical measurements for this setup are as follows.

- The angle of the screen is 4 degrees.

- The angle of the camera is 25 degrees.

- The distance from the center of the screen to the FH cab is 60 cm.

- The vertical distance from each of both cameras to the screen is 55 cm.

- The distance from each of both cameras to the FH cab is 65 cm.

The camera and the screen work in opposite direction. The closer the screen is positioned, the bigger the reflection becomes and the further the camera is located the bigger the image taken.

7.2.3. Sinusoidal Pattern

The idea is to project sinusoidal patterns onto a screen and observe the reflection of those patterns through the specular surface under the test. The reflection of this patterns on the surface is captured by the camera. Any variations of the surface lead to distortions of the patterns as shown in Fig. 7.2.

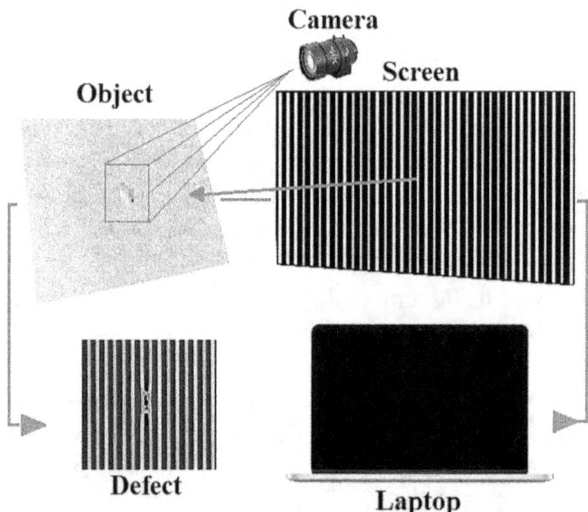

Fig. 7.2. Outline of the deflectometry-based image acquisition process.

The sinusoidal pattern is generated in a mathematical software Matlab, version R2017b. The function of this pattern is shown in the equation below.

$$I = B + A * \sin(2\pi f(x) + \theta), \qquad (7.1)$$

where A is the amptitude, B is the bias or offset, f is the frequency and θ is the phase [16]. The frequency f controls the thickness of the strips while the phase θ shifts the pattern in x or y direction, by $\pi/2$ at each step. I can be directed in the two-orthogonal directions x or y of the screen.

Phase Measuring Deflectometry (PMD) is one of the techniques utilizing a reflected sinusoidal pattern to obtain surface local slope, which can be further processed to calculate surface 3D shape. Two setups based on PMD have been evaluated previously, monocular PMD and stereo PMD [16, 17]. The monocular PMD based gradient fields and curvature maps turned out to be efficient and accurate for a simple solution.

7.2.4. Defect Type

There are many different types of defects that can occur on painted vehicle bodies. The most common defect types are dirt and crater, thereby the focus of this chapter is only on these two defects. Dirt can be described as a small bump deposited in, on, or under the painted surface, whereas crater looks like a circular low spot or bowl-shaped cavity on the painted surface. It should be noted that dirt is more common than crater on the cabs. It is almost impossible to distinguish dirt from crater just by observing the captured images, therefore, human eyes and touch are still vital for this task.

7.2.5. Image Acquisition and Annotation

An image acquisition and annotation Graphical User Interface (GUI), displayed in Fig. 7.3 below, was created and implemented in Matlab to facilitate image capturing and image annotation parts.

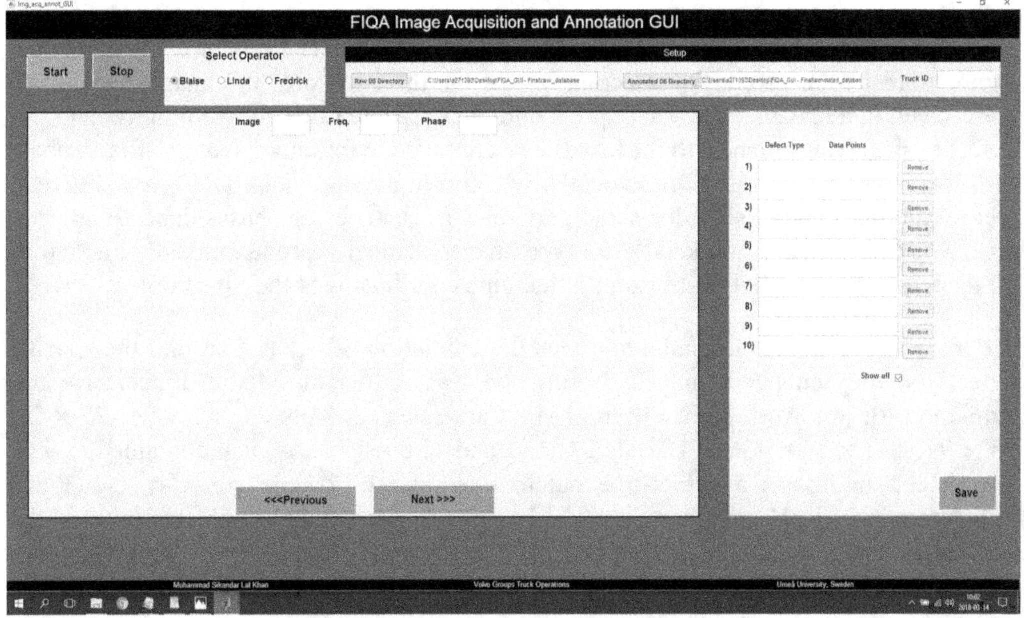

Fig. 7.3. Image acquisition and annotation GUI.

There are two sets of parameters as listed below, that need to be selected in advance.

1) Sinusoidal parameters:

   a) Pattern: horizontal and vertical patterns are available;

   b) Frequency: four different frequencies (8, 16, 32 and 64) are available;

   c) Phase: the horizontal pattern can be shifted four times by $\pi/2$ at each step $(0, \frac{\pi}{2}, \pi, \frac{3\pi}{2})$.

2) Camera parameters:

   a) Exposure: the best measure for the exposure was found to be 25000.

Only horizontal sinusoidal pattern is used here which has four different frequencies (8, 16, 32 and 64) with each frequency having four different phases $(0, \frac{\pi}{2}, \pi, \frac{3\pi}{2})$. So that in total there are 16 channels for each object.

When a new object is detected in the system, sixteen images are automatically taken by the image acquisition and annotation GUI. After the image capturing, the object's surface is inspected manually with care and under good lighting conditions. The whole process is accomplished in less than fifteen seconds per object.

7.2.6. Annotation

If there was a presence of any defect type on the surface, it was annotated by clicking on the spot on the image where the defect was situated. It was important to click exactly on the defect. The image was annotated as either dirt, crater or non-defect. Non-defect is a type of the painted surface that is defect free. For all the defect-free objects, two patches of non-defect were saved. By selecting one of the three options, the image coordinates were saved in a text-file. The same procedure was repeated if there were more than one defect on the given surface. All the text-files were saved in the annotated text-file database directory. The text-files and the channels were linked by the object ID's. Certain defects were sometimes only visible for some particular frequencies or phases depending on its size, type or location. A marked defect type on one channel was automatically marked on all the remaining channels with exactly the same coordinates of the given object.

The next step after image annotation was the extraction of patches around the specific defects on the annotated images. A function was written in Matlab to perform this extraction. In this work eight different sizes are selected: 31×31, 51×51, 71×71, 91×91, 111×111, 131×131, 151×151 and 171×171. A patch contained exactly one defect, or it was a defect-free patch. Examples of patches for dirt, crater and non-defect in 91×91 sizes are shown in Fig. 7.4. There are 11175 objects in total. For each object, $16 \times 8 = 128$ patches of different window sizes are extracted for each annotated defect. If, for example, an object has three annotated defects (two dirts and one crater), there will be stored in total $16 \times 8 \times 3 = 384$ patches.

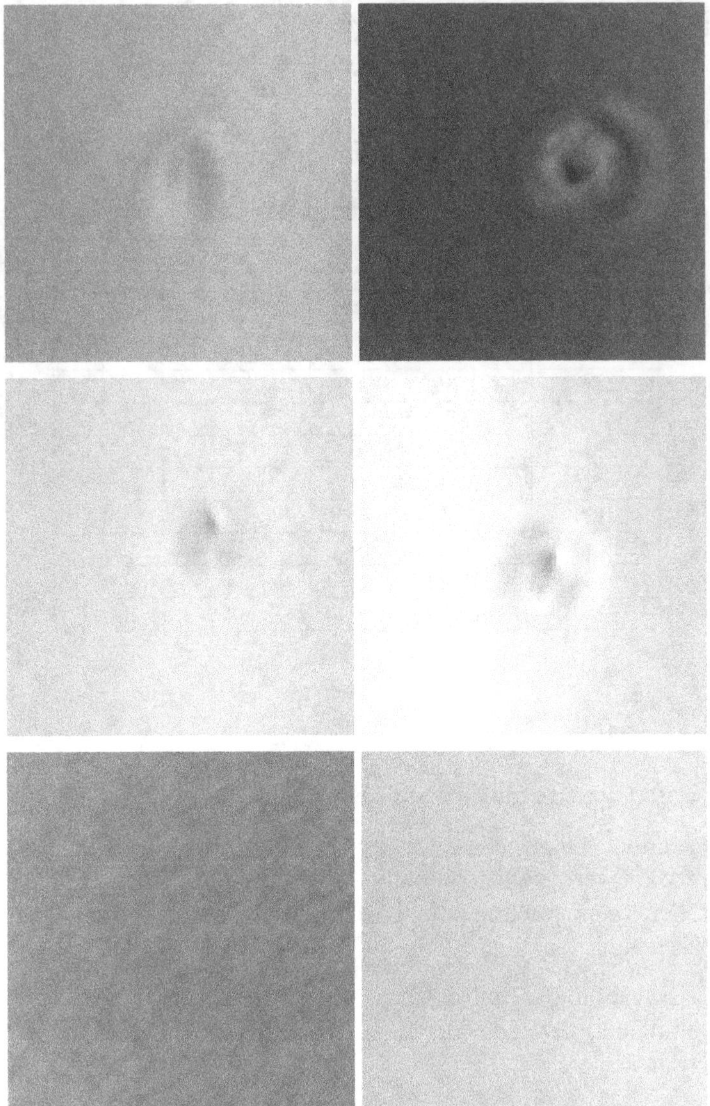

Fig. 7.4. Examples of patches for crater (top row), dirt (middle row) and non-defect (bottom row) in 91 × 91 sizes.

7.2.7. Dataset

The whole dataset consists of sixteen channels. Thus, sixteen different subsets of patches can be created from those channels. The patches close to the edges of the images were excluded from the whole dataset to withdraw the edge's effect. Each subset contains 18433 patches of eight different window sizes, which include 4234 patches of dirts, 372 patches of craters and 13827 defect-free patches. 1170 patches close to the edges were removed from each subset. All the sixteen channels are described in Table 7.1 below.

Table 7.1. Description of the channels. Each channel forms a subset of 18433 patches.

Channel	Channel's name	Frequency	Phase
1	$8 - 0$	8	0
2	$8 - \pi/2$	8	$\pi/2$
3	$8 - \pi$	8	π
4	$8 - 3\pi/2$	8	$3\pi/2$
5	$16 - 0$	16	0
6	$16 - \pi/2$	16	$\pi/2$
7	$16 - \pi$	16	π
8	$16 - 3\pi/2$	16	$3\pi/2$
9	$32 - 0$	32	0
10	$32 - \pi/2$	32	$\pi/2$
11	$32 - \pi$	32	π
12	$32 - 3\pi/2$	32	$3\pi/2$
13	$64 - 0$	64	0
14	$64 - \pi/2$	64	$\pi/2$
15	$64 - \pi$	64	π
16	$64 - 3\pi/2$	64	$3\pi/2$

7.2.8. Response Variables

Two different classification problems were considered in this work:

1) Defect detection (binary classification problem): two classes were considered. Craters and dirts were grouped into one class representing defects and labelled as 1's. The non-defects were put into the second class and labelled as 0's. Thus, the response variable takes value 0 or 1.

2) Defect classification (3-class classification problem): three different classes, crater, dirt and non-defect, were considered for this problem which led to three values of the response variable.

7.3. Feature Vectors

Six different types of features were used in this work, such as histogram of oriented gradients, local binary pattern, 2D wavelet transform, features based on variabilities and smoothing features based on P-splines and their effective degrees of freedom. The following subsections present brief descriptions of these features.

7.3.1. HOG: Histogram of Oriented Gradients

HOG is a type of feature descriptor used in computer vision and image processing for the purpose of object detection. HOG became widely used first in 2005 when [18] introduced their work. The idea behind the HOG descriptor is that local object appearance and shape within an image can be explained by the distribution (histograms) of intensity gradients [19]. Gradients of an image are useful because the magnitude of gradients is large around edges and corners. There are five steps in computing the HOG descriptors which are briefly discussed as follows:

1) Preprocessing. The sixteen channels shown in Table 7.1 were analyzed. Each patch was divided into cells of 30×30 pixels. Each patch included then $3 \times 3 = 9$ non-overlapping cells in total;

2) Computation of the image's gradients. An image's gradient is a measure of the change in pixel values along the x and y directions around each pixel. The horizontal and vertical gradients were calculated by filtering the image with the two kernels (or discrete derivative masks): horizontal $[-1\ 0\ 1]$ and vertical $\begin{bmatrix} -1 \\ 0 \\ 1 \end{bmatrix}$.

At every pixel, the gradient has a magnitude and an orientation. The gradient orientation gives you the normal to the edge (perpendicular to the edge), and the gradient magnitude gives you the strength of the edge. The magnitude and orientation of the gradient vectors ∇f were computed as

$$\text{Magnitude: } \|\nabla f\| = \sqrt{f_x'^2 + f_y'^2},$$

$$\text{Orientation: } \theta = \arctan(f_y'/f_x'),$$

where f_x' and f_y' are derivatives with respect to x and y directions correspondingly.

Histograms of the gradients for each cell. The next step after the calculation of the gradients was to create a histogram of gradients over each cell. Each histogram contained six bins with the orientations ranging between 0 and π. Hence each bin had a width of $\pi/6$ degrees. These types of histogram with the range between 0 to π are called unsigned gradients because a gradient and its negative are represented by the same numbers. Each pixel within the cell gives a weighted vote for one of the six bins based on the values found in the gradient computation. Nine histograms were created at the end. It has been shown that the nine bins histogram gives better results than the 18 bins histograms (also called signed gradients with angles between $0 - 2\pi$) for human detection.

Block normalization. Block normalization was the final step. The normalization performed at block level reduces the effect of illumination and shadowing more than performed at cell level. Each block was represented by one cell. Consequently, each patch was represented by a total of nine non-overlapping blocks with the size of 30×30 pixels each. Every block with one histogram divided into six bins, was concatenated to form a feature vector of length 54. The normalization of this feature vector was performed by dividing the feature vector with its magnitude, based on the L_2-norm:

$$f = \frac{V}{\sqrt{\|V\|_2^2 + \epsilon^2}},$$

where V is the feature vector and ϵ is some small constant.

HOG feature vector. There were:

- Three positions for each block in the y direction;

- Three positions for each block in the x direction;

- Each block contained exactly one histogram with six bins.

The length of the entire feature vector for each patch was then $3 \times 3 \times 6 = 54$. The dimensions of each subset fed to the classifiers were 1895×55 with the labels included. The HOG features were extracted with the function `HOG()` from the package `OpenImageR` in R.

7.3.2. LBP: Local Binary Pattern

The LBP operator was introduced as a gray scale invariant texture measure, derived from a general definition of texture in a local neighborhood [20]. An image texture is a set of metrics calculated in image processing designed to quantify the perceived texture of an image. Image texture gives us information about the spatial arrangement of color or intensities in an image or selected region of an image [21].

The LBP feature vector is computed as following:

a) For each pixel i in the patch, the eight neighbors (left-top, left-middle, left-bottom, right-top, etc.) are examined to see if their intensity is greater than that of i. The results from the eight neighbors are used to construct an eight-digit binary number $p_1\, p_2\, p_3\, ...\, p_8$, where $p_i = 0$ if the intensity of the ith neighbor is less than to that of i and 1 otherwise. The LBP code is obtained by converting the eight-digit binary number to decimal [22], as shown in Fig. 7.5.

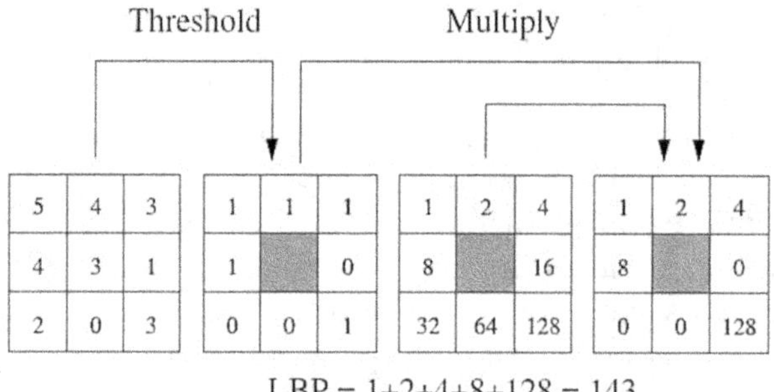

Fig. 7.5. An example of LBP code calculation.

b) Compute a histogram of the frequency of each LBP code occurring. This histogram is used to represent the texture of the patch and can be seen as a 256-dimensional feature vector.

318

c) The normalization of the histogram leads to the LBP feature vector.

d) A so-called uniform pattern [23] can be used to reduce the length of the LBP feature vector and implement a simple rotation invariant descriptor. An LBP is called uniform if the binary pattern contains at most two 0-1 or 1-0 transitions. For example, 0000010 (2 transitions) is a uniform pattern whereas 11001001 (4 transitions) is not so. This histogram will then have 58 separate bins for every uniform pattern while all non-uniform patterns will be assigned to one single bin. The length of the feature vector is then reduced from 256 to 59.

The length of the entire feature vector obtained from the LBP operator was 7921. The dimension of each subset was 1895×7922 with the labels included. The LBP feature vector was extracted with the function lbp() from the package wvtool in R. The most important properties of the LBP operator are its tolerance against illumination and image rotation changes plus its computational simplicity [22].

7.3.3. 2D DWT: 2D Discrete Wavelet Transform

Since early 90's, wavelets have become popular in many different fields such as astronomy, approximation theory, signal processing, numerical analysis, statistics, mathematics among others [24]. The wavelet transform was developed as an alternative to the short time Fourier transform to overcome problems related to its frequency and time resolution properties. The main advantages of wavelet transform are to provide the time-frequency representation and the possibility of multi-resolution analysis.

Nowadays, the usage of wavelets has increased in signal and image processing due to their multi-resolution concept [25]. They have been used for feature extraction, denoising, compression, face recognition, and image super-resolution. Discrete wavelet transform (DWT) uses filter banks to perform the wavelet analysis, i.e., it decomposes the signal into wavelet coefficients from which the original signal can be reconstructed again.

The 2D wavelet decomposition of an image is performed by applying one dimensional DWT along the rows of the image first, and then, the results are decomposed along the columns (see Fig. 7.6(a)). The wavelet coefficients are then constructed by taking the tensor product of a horizontal 1D wavelet and a vertical 1D wavelet. This leads to four types of 2D wavelets [26]:

$$\Phi(x, y) = \phi_h(x) \cdot \phi_{v(y)}, \Psi^v(x, y) = \psi_h(x) \cdot \phi_v(y),$$

$$\Psi^h(x, y) = \phi_h(x) \cdot \psi_v(y), \Psi^d(x, y) = \psi_h(x) \cdot \psi_v(y),$$

where Φ is known as the father wavelet function and Ψ as the mother wavelet function. The 2D wavelet family has then one father wavelet function and three mother wavelet functions. The father wavelet function is good at representing the smooth and the mother wavelet functions can capture the details (vertical, horizontal and diagonal details).

The decomposition of a 2D wavelet results in four sub-bands images as shown in Fig. 7.6(a) and referred to as low-low (LL), low-high (LH), high-low (HL), and high-high

(HH). LL_1 is the approximation/smooth image at the first level. LH_1, HL_1 and HH_1 capture respectively the vertical, horizontal and diagonal details at the first level. The next level of wavelet coefficients is obtained by decomposing further the sub-image LL_1 alone. Fig. 7.6 (b) shows the resulting two levels wavelet decomposition. Similarly, to obtain further decomposition, the approximation image LL_2 will be used. This process continues repeatedly until some final scale is reached.

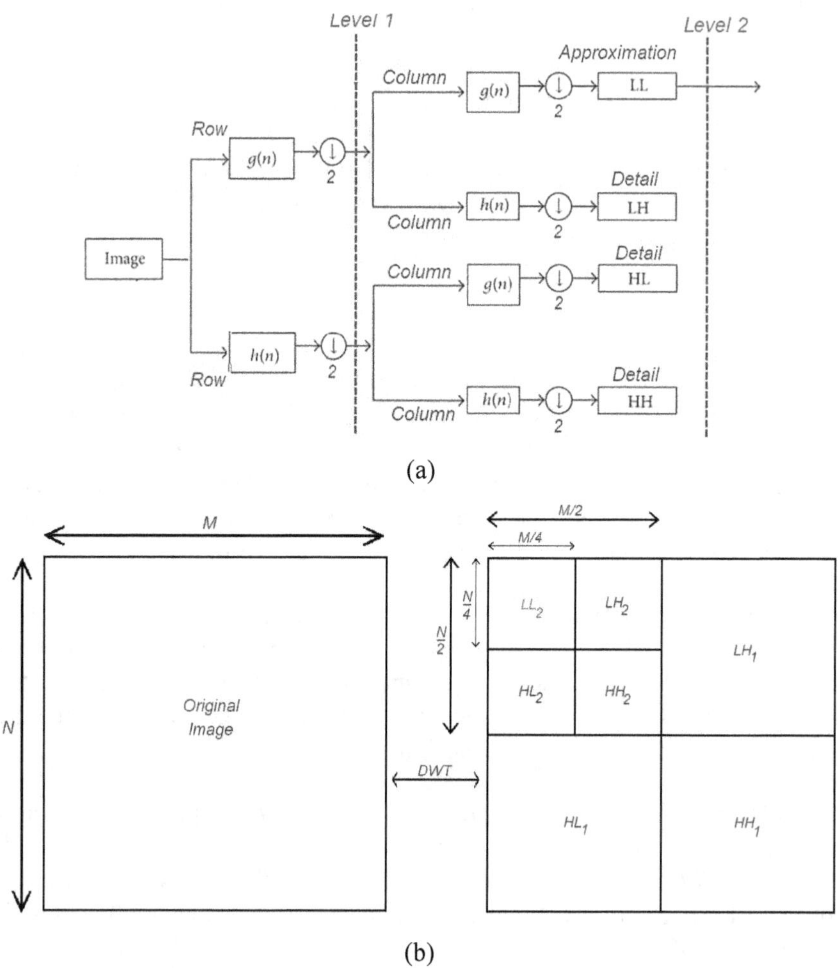

Fig. 7.6. (a) The first level in a multi-resolution pyramid decomposition of 2D DWT. $h(n)$ and $g(n)$ are respectively the high pass filter and lower pass filter; (b) The decomposition result of two multi-resolution levels.

There are two main categories in the discrete wavelet family, orthogonal and bi-orthogonal wavelets. The main difference is that the bi-orthogonal wavelets are symmetric, not orthogonal and do not introduce phase shifts in the coefficients [26]. Three types of orthogonal wavelets were used here:

- **Haar:** The Haar wavelet is a square wave. It has compact support, i.e. it is zero outside a finite interval. It is the only compact orthogonal wavelet which is symmetric and unlike other wavelets it is not continuous [27].

- **Daublets:** These are the first type of continuous orthogonal wavelet with compact support and are quite asymmetric.

- **Symmlets:** They also have compact support but are constructed to be as least asymmetric as possible. Both Daublets and symmlets were discovered by Ingrid Daubechies [28].

All the patches were resized from the size of 91×91 to the size of 88×88 to facilitate the calculations. The decomposition of each patch of size 88×88 at $k = 3$ levels of resolution results in $3k + 1 = 10$ sub-images including one approximation sub-image LL_3 and nine sub-images of details. The size of each of the four sub-images at the third level is $\frac{88}{2^3} \times \frac{88}{2^3}$, i.e. 11×11 due to the down-sampling or decimation by a factor 2 which is performed after each pass through filters.

The function `denoise.2d()` from the package `waveslim` in R was used to perform simple denoising of the patches for the 2D DWT. The chosen arguments in the denoising function were the following:

- `wf`: The name of the wavelet filter to use in the decomposition. `Haar` for Haar wavelet, `la8` for Symmlets wavelet of length L = 8 and `d4` for Daublets wavelet with L = 4;

- `j`: The depth of decomposition j = 3 was used;

- `method`: The character string describing the threshold applied. `Universal` method was applied;

- `H`: Hurst parameter to indicate spectral scaling. The white noise where H = 0.5 was used;

- `noise.dir`: The number of directions `nd` to estimate background noise standard deviation. The default nd = 3, which produces a unique estimate of the background noise for each spatial direction, was used;

- `rule`: The hard thresholding rule was applied.

7.3.4. Smoothing Features and Variabilities

7.3.4.1. EDF Smoothing Features

The smoothing features based on effective degrees of freedom (EDF) have been introduced in [14] where readers can find details. The EDF features capture the irregularities in the reflected sinusoidal patterns caused by defects. Here smoothing pixel values row-wise using penalized cubic regression splines is applied to reproduce the projected sinusoidal patterns. The irregularities of the smoothed patters are then picked up by the EDF of the fitted splines.

Given an image patch of size $m \times m$, we model each row of the intensity values by a semiparametric Gaussian model,

$$y_{ij} = f_i(t_j) + \epsilon_j, i, j = 1, \ldots, m, \tag{7.2}$$

where y_{ij} denotes the value of the pixel (i,j) in the patch, $t_j = j, f_j$ is an unknown smooth function, and ϵ_j's are white Gaussian noises with variance σ_j^2. There are several alternatives for choosing univariate penalized regression spline to estimate $f_j(t)$. Since their performance were similar in the final results, we choose cubic regression splines here due to their sufficient flexibility and efficiency. This spline can be written in the following form [29],

$$f_i(t) = \sum_{k=1}^{b} \alpha_{ik} \phi_k(t),$$

where $\phi_k(t)$ are known basis functions, α_{jk} are unknown coefficients to be estimated, and b is the number of basis functions used, which controls the degree of model smoothness. In practice, for controlling model smoothness it is commonly used by adding 'wiggliness' penalty to the least squares fitting objective., while as keeping b fixed at a sufficiently large size to avoid over smoothing. Thus, we estimate α_{jk} by minimising the following penalized regression objective

$$\|\mathbf{y}_i - \mathbf{\Phi}\boldsymbol{\alpha}_i\|^2 + \lambda_i \int_{y_*}^{y^*} f_i''(t)^2 dt, \tag{7.3}$$

where $\mathbf{y}_i = (y_{i1}, \ldots, y_{im})^T, \mathbf{\Phi} = \left(\phi_k(t_j)\right)$ is the model matrix evaluating spline basis functions at the observations, $\boldsymbol{\alpha}_i = (\alpha_{i1}, \ldots, \alpha_{ib})^T, \lambda_i$ is a smoothing parameter that controls the balance between smoothness of the fitted curve f_i and data fit, and y_* and y^* are the end knots of the cubic spline function. As shown in [38], the penalty term in (7.3) can be represented as $\int_{y_*}^{y^*} f_i''(t)^2 dt = \boldsymbol{\alpha}_i^T \mathbf{A} \boldsymbol{\alpha}_i$, where $\mathbf{A}$ is the penalty matrix of known components for the basis [29]. The value of the smoothing parameter λ_i can be determined through a generalized cross validation score

$$GCV_f = \frac{m \sum_{j=1}^{m} (y_{ij} - \hat{f}_{ij})^2}{(m - tr(\mathbf{S}_i))^2},$$

where $\mathbf{S}_i = \mathbf{\Phi}(\mathbf{\Phi}^T\mathbf{\Phi} + \lambda_i\mathbf{\Phi})^{-1}\mathbf{\Phi}^T$ is a model's influence matrix, and $\eta_i = tr(\mathbf{S}_i)$ is then its EDF.

It is reasonable to expect that the distortion of the pattern caused by defects results in some irregularities in the obtained smooth functions. Larger values of η_i indicate that the curves are more irregular, i.e. the wigglier. Hence, the EDF η can be treated as a highly plausible feature, and the combination of m scaled EDFs, obtained from smoothing the image row-wise, forms our proposed feature vector,

$$\zeta = (\zeta_1, \ldots, \zeta_m)^T, \text{where } \zeta_i = \frac{\eta_i}{\max_i \eta_i}$$

The reason for applying the feature scaling is twofold, implying that it enables better discrimination between classes in some cases and also reduces training time of classification algorithms. For a fair comparison of the means of each class, three hundred images were randomly selected within each class for the means calculations. The feature vector means can be viewed as class centroids similar to those in k-means clustering, serving as a prototype of each class. Note that a very distinct contrast between class means supports the idea of extracting EDFs [14].

7.3.4.2. Smoothing Features Based on P-splines

Since the locations of the defects and their magnitudes on those annotated patches are not known, smoothing the column means of the grey scale pixel values is suggested as an alternative way for uncovering distortions in the sinusoidal pattern. We consider the same model as in the previous subsection, model (7.2). But P-splines [30] are used here instead due to flexibility and local support of the basis functions,

$$f(t_j) = \sum_{k=1}^{b} \alpha_k B_k(t_j),$$

where $B_k(t)$ are B-spline basis functions of the third order, α_k are unknown coefficients to be estimated, and b is the number of basis functions used. P-splines use B-spline basis usually defined on evenly spaced knots, with a difference penalty applied to the coefficients α_k to control function smoothness. To estimate α_k the penalized regression fitting problem is minimized [29]

$$\|y - B\alpha\|^2 + \lambda P, \qquad (7.4)$$

where $y = (y_i, \ldots, y_m)^T$, $B = \left(B_k(t_j)\right)$ is the model matrix evaluating B-splines at the observations, $\alpha = (\alpha_1, \ldots, \alpha_b)^T$, and λ is a smoothing parameter that controls the smoothness of the fitted curve. P is the penalty applied to α

$$P = \sum_{k=1}^{b-1} (\alpha_{k+1} - \alpha_k)^2 = \alpha^T P_0^T P_0 \alpha,$$

where

$$P_0 = \begin{bmatrix} -1 & 1 & 0 & \cdot & \cdot \\ 0 & -1 & 1 & \cdot & \cdot \\ 0 & 0 & -1 & \cdot & \cdot \\ & \cdot & \cdot & \cdot & \cdot \\ & \cdot & \cdot & \cdot & \cdot \end{bmatrix}$$

Thus, the penalized least squares estimator of α is

$$\hat{\alpha} = (B^T B + \lambda P_0^T P_0)^{-1} B^T y$$

The smoothness selection, the choice of λ is made by a generalized cross validation score as above. Again, the same number of the basis dimension, b, is applied here. The

$\boldsymbol{\alpha} = (\alpha_1, \dots, \alpha_b)^T$ coefficients carry all information about the smooth function, so the estimated $\widehat{\boldsymbol{\alpha}}$ are employed as the representative features that express the changes in the sinusoidal patterns.

7.3.4.3. Variabilities

When smoothing data on the column mean gray scale values, we might expect losing some information on variability of the values by averaging them. So to keep that information, we suggest to use column standard deviations, $s_{(c)j}$, of the pixel values as features, where

$$s_{(c)j}^2 = \frac{1}{m}\sum_{i=1}^{m}(y_{ij} - \bar{y}_{(c)j})^2, j = 1, \dots, m,$$

y_{ij} is the value of pixel (i, j) and $\bar{y}_{(c)j}$ is the mean value of the jth column.

As different paint colors and various finishing paint processes on the cab surface are known to affect the gray scale pixel values, the pixel values are standardized before smoothing by subtracting the overall mean of the pixels of the patch and dividing by the overall standard deviation.

7.4. Learning Defect Classifiers

Support vector machine (SVM) and probabilistic k-NN were applied to solve the binary classification problem of this work. Probabilistic k-NN was applied for 3-class problem. These classifiers are briefly described in the following subsections.

7.4.1. Support Vector Machine

7.4.1.1. MMC: Maximal Margin Classifier

One way of solving a two-class classification problem is to find a hyperplane that separates the classes in a feature space. A hyperplane in d dimensions is a flat affine subspace of dimension $d - 1$. The equation for a hyperplane is defined as the following [31].

$$\{x: f(x) = x^T\boldsymbol{\beta} + \beta_0 = 0\}, \tag{7.5}$$

where $\boldsymbol{\beta}$ is a unit vector: $\|\boldsymbol{\beta}\| = \sum_{j=1}^{d}\beta_j = 1$. A hyperplane is a line in $d = 2$ dimensions and it is a plane in $d = 3$ dimensions. In $d > 3$, the hyperplane is a $(d - 1)$ dimensional flat subspace that can be hard to visualize.

For any point $x = (x_1, \dots, x_d)^T$ in d-dimensional space for which Equation (7.5) holds, this point lies on the hyperplane. Otherwise, i.e. when $f(x) > 0$ or $f(x) < 0$, this point lies then on one of the two sides of the hyperplane [32]. A classification rule induced by $f(x)$ is

$$G(x) = sign \left[f(x) \right]$$

In general, if the data can be perfectly separated with a hyperplane, then there will in fact exist an infinite number of such hyperplanes. Finding the optimal separating hyperplane also known as the maximal margin hyperplane is the problem to be solved. Suppose we have n training observations $x_1, x_2, \ldots, x_n \in \mathbb{R}^d$ associated with labels $l_1, l_2, \ldots, l_n \in \{-1, 1\}$. We are looking for the optimal hyperplane among all that makes the biggest gap or margin between the two classes, 1 and -1 [32]. An example of a maximal margin can be seen in Fig. 7.7. The optimal hyperplane is the solution to the following optimization problem.

$$\max_{\beta_0, \beta} M , \tag{7.6}$$

subject to $\|\beta\| = 1, l_i\left(x_i^T \beta + \beta_0\right) \geq M, \forall\, i = 1, \ldots, n.$, where M represents the width of the hyperplane margin and the optimization problem chooses β_0 and β to maximize the margin M. The constraints in the optimization problem (7.6) ensure that each observation is on the correct side of the hyperplane and at least a distance M from the hyperplane. This can be rephrased as a convex optimization problem (quadratic with linear inequality constraints) [31]. We will come back to this in the next subsection.

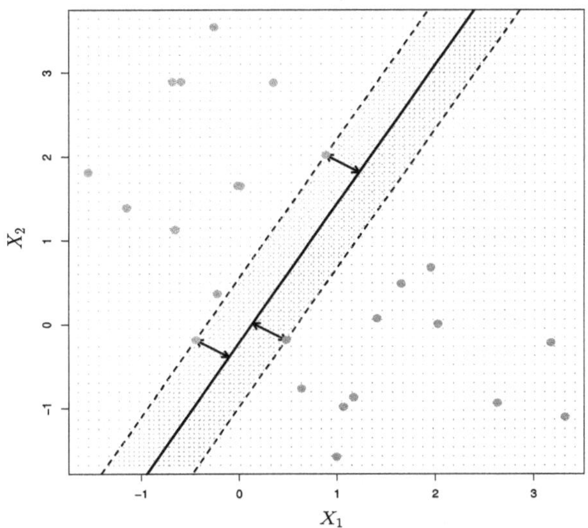

Fig. 7.7. The blue and purple dots represent the two different classes perfectly separated by the maximal margin shown as a black solid line. The distance from the solid line to either of the dashed line indicates the margin [32].

7.4.1.2. SVC: Support Vector Classifier

The SVC is an extension of the MMC. Suppose that we are dealing with two classes that overlap in feature space as seen in Fig. 7.8. In this case, the optimization problem in (7.6) has no solution with $M > 0$, i.e. no maximal margin hyperplane exists in this case. One way to deal with the overlap problem is to allow for some points to be on both, the wrong side of the margin and the wrong side of the hyperplane. Observations on the wrong side of the hyperplane are misclassified by the SVC in order to have a better overall classification. This optimization problem can be written as follows [31]

$$\max_{\beta_0, \beta, \xi} M \,, \tag{7.7}$$

subject to $\|\beta\| = 1, l_i(x_i^T \beta + \beta_0) \geq M(1 - \xi_i), \xi_i \geq 0, \sum_{i=1}^{n} \xi_i \leq C, \forall\, i = 1, \dots, n,$

where [32] $M = 1/\|\beta\|$ is the width of the margin; $\xi = \{\xi_1, \dots, \xi_n\}$ are the slack variables that allow individual observations to be on the wrong side of the margin or the hyperplane. These variables tell us where the ith observation is located relative to the hyperplane and relative to the margin. If $\xi_1 = 0$, the ith observation is on the correct side of the margin. If $\xi_i > 0$, the ith observation is on the wrong side of the margin. If $\xi_i < 0$, the ith observation is on the wrong side of the hyperplane. C is a nonnegative cost tuning parameter that bounds the sum of the slack variables. It determines the number and severity of the violations to the margin and the hyperplane that we will tolerate. If $C = 0$ then $\xi_1 = \xi_2 = \dots = \xi_n = 0$, the optimization problem becomes exactly the same as the maximal margin optimization in (7.7). If $C > 0$, no more than C observations are on the wrong side of the hyperplane. In practice, C is chosen via cross validation and controls the bias-variance trade-off. A small C leads to narrow margins that are rarely violated and this amounts to a classifier that is highly fit to the data which may have low bias but high variance and vice-versa.

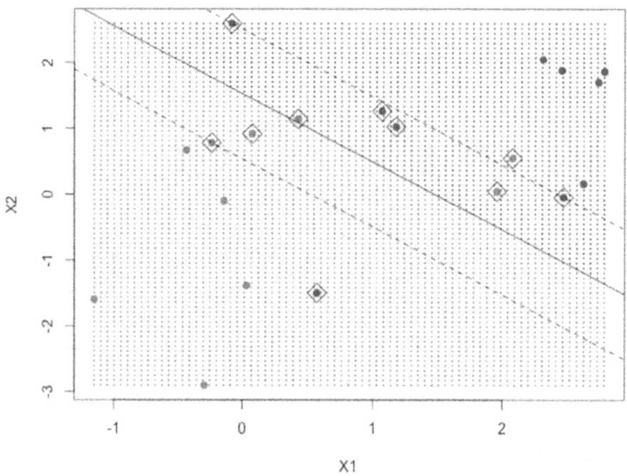

Fig. 7.8. Example with two classes (blue and red) that overlap in feature space. The hyperplane is shown as a solid line and the margins are as dashed lines. Blue observations: five observations are on the correct side of the margin, two observations are on the margin, two observations are on the wrong side of the margin and one observation is on the wrong side of the hyperplane. Red observations: five observations are on the correct side of the margin, one observation is on the margin, one observation is on the wrong side of the margin and three observations are on the wrong side of the hyperplane. The diamond-shaped blue and red points are the support vectors.

The observations that lie strictly on the correct side of the margin do not affect the SVC. This is an attractive property for the SVC. The SVC is only influenced by the support vectors which are the observations on the wrong side of their margins. The diamond-shaped blue and red observations in Fig. 7.8 are the support vectors in that case. A large value of the cost tuning parameter C will lead to many support vectors as the result of the wide margins.

The constraints of the optimization problem in (7.8) can be rephrased as a convex optimization problem (quadratic with linear inequality constraints). A quadratic programming solution using Lagrange multipliers can be used to solve this optimization problem. The Lagrange primal function is

$$L_P = \frac{1}{2}\|\boldsymbol{\beta}\|^2 +$$

$$+C\sum_{i=1}^{n}\xi_i - \sum_{i=1}^{n}\alpha_i\left[l_i\left(\boldsymbol{x}_i^T\boldsymbol{\beta} + \beta_0\right) - (1 - \xi_i)\right] - \sum_{i=1}^{n}u_i\xi_i \qquad (7.8)$$

After minimizing the Lagrange primal function in (7.8) w.r.t. the parameters $\boldsymbol{\beta}, \beta_0$ and $\boldsymbol{\xi}$ and setting the respective derivatives to zero, we arrive at a Lagrange dual problem as given below in (7.9) by substituting the parameters into the Lagrange primal function [12].

$$L_D = \sum_{i=1}^{n}\alpha_i - \frac{1}{2}\sum_{i=1}^{n}\sum_{j=1}^{n}\alpha_i\,\alpha_j l_i l_j \boldsymbol{x}_i^T \boldsymbol{x}_j \,, \qquad (7.9)$$

The solution of (7.9) is expressed in terms of fitted Lagrange multipliers $\hat{\alpha}_i$

$$\widehat{\boldsymbol{\beta}} = \sum_{i=1}^{n}\hat{\alpha}_i l_i \boldsymbol{x}_i,$$

with $\hat{\alpha}_i > 0$ only for those observations i for which the second constraints in (7.8) are fulfilled (support vectors). Suppose that S is the collection of indices of these support vectors, we have then

$$\hat{f}(\boldsymbol{x}) = \boldsymbol{x}^T\widehat{\boldsymbol{\beta}} + \hat{\beta}_0 = \sum_{i\in S}\hat{\alpha}_i l_i \boldsymbol{x}^T \boldsymbol{x}_i + \hat{\beta}_0$$

7.4.1.3. SVM: Support Vector Machine

The SVM is an extension of the SVC that results from enlarging the feature space in a specific way using basis expansions such as polynomials, splines or kernels. The idea is to enlarge the feature space in order to adapt a non-linear boundary between the classes.

The procedure that solves this problem is almost identical to the previous one, the only difference is that here we need to select the basis functions $h_j(\boldsymbol{x}), j = 1, ..., q$. Briefly, we will fit the SVC using input features $\boldsymbol{h}(\boldsymbol{x}_i) = \left(h_1(\boldsymbol{x}_i), h_2(\boldsymbol{x}_i), ..., h_q(\boldsymbol{x}_i)\right)$, $i = 1, ..., n$, and produce the nonlinear function [31]

$$\hat{f}(\boldsymbol{x}) = \boldsymbol{h}(\boldsymbol{x})^T\widehat{\boldsymbol{\beta}} + \hat{\beta}_0 \qquad (7.10)$$

The classifier is $G(\boldsymbol{x}) = sign[f(\boldsymbol{x})]$ as before. The optimization problem and its solution are presented in a special way that only involves the input features via inner products. The Lagrange dual in (7.8) has then the form [31]

$$L_D = \sum_{i=1}^{n} \alpha_i - \frac{1}{2} \sum_{i=1}^{n} \sum_{j=1}^{n} \alpha_i \, \alpha_j l_i l_j \langle h(x_i), h(x_j) \rangle \,, \qquad (7.11)$$

and the solution function can be written as

$$f(x) = h(x)^T \beta + \beta_0 = \sum_{i=1}^{n} \alpha_i l_i \langle h(x_i), h(x_j) \rangle + \beta_0 \qquad (7.12)$$

The expressions (7.11) and (7.12) involve h(x) only through inner products. This implies that the computation of the inner products in the transformed space requires only knowledge of the kernel function

$$K(x, x') = \langle h(x), h(x') \rangle \qquad (7.13)$$

Replacing the kernel function (7.13) in (7.12) gives

$$f(x) = \sum_{i=1}^{n} \alpha_i l_i K(x, x') + \beta_0$$

The three popular choices of the kernel function are:

a) p^{th} degree polynomial: $K(x, x') = (1 + \langle x, x' \rangle)^p$;

b) Radial basis: $K(x, x') = \exp(-\gamma \|x - x'\|^2)$;

c) Neural network: $K(x, x') = \tanh(\kappa_1 \langle x, x' \rangle + \kappa_2)$.

An example of the radial basis kernel is shown in Fig. 7.9 as a black solid line. The blue line is the Bayes decision boundary. The SVC will certainly perform worse than the radial basis kernel on these data because of the presence of the non-linear decision boundary.

For the SVM with more than two classes, the most popular approaches are so called "one-versus-one" and "one-versus-all".

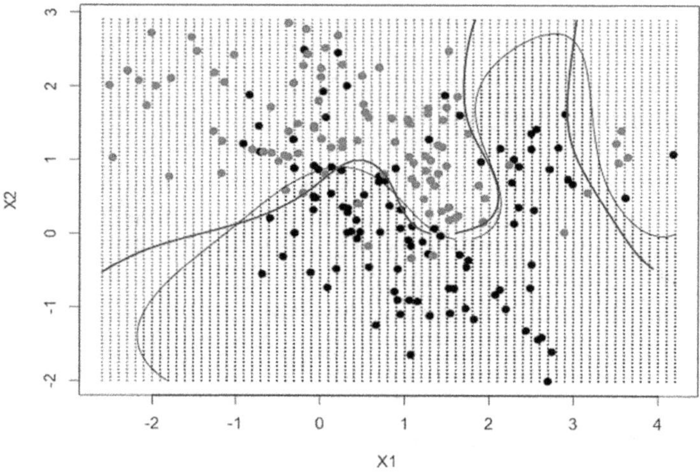

Fig. 7.9. Example of a two-class classification problem using a radial basis kernel on simulated data. The radial basis kernel with $C = 5$ and $\gamma = 0.5$ is shown as a black solid line and the Bayes decision boundary as a blue solid line. The radial basis kernel is close to the Bayes decision boundary in this case.

7.4.2. Probabilistic k-NN

The standard k-NN classifier is one of the most straightforward nonparametric methods to classifying objects on the basis of the feature vectors which are considered as points belonging to certain class, in the feature space $\mathbb{R}^d$. With the k-NN rule, class prediction is performed by finding the k nearest (in some distance metric) points and assigning the most frequent label. Despite its simplicity, the performance of k-NN shown on numerous classification tasks signifies that it continues to be a competitive classification method in machine learning and statistics [33, 34]. In order to assess the uncertainty in the assigned class labels, probabilistic alternatives of the k-NN method have been studied (see [35-37] among others). Here we use the one proposed by [38], which offers a straightforward and yet theoretically neat k-NN classifier. The probabilistic k-NN has shown good classification performance and at the same time it allows employing uncertainty measures associated with the assigned class labels, which other machine learning techniques are lacking. We describe shortly this approach below and refer the details to [38].

Given a dataset $\{(\boldsymbol{x}_1, l_1), (\boldsymbol{x}_2, l_2), \dots, (\boldsymbol{x}_n, l_n)\}$, where $l_i \in \{C_1, \dots, C_K\}$ is a class label associated with the feature vector $\boldsymbol{x}_i \in \mathbb{R}^d$. In the case of our current study, we have three classes ($K = 3$): $C_1 =$ defect-free, $C_2 =$ crater, and $C_3 =$ dirt. For a new feature vector $\boldsymbol{x} \in \mathbb{R}^d$, whose class label is unknown and to be predicted, denote the minimum Euclidean distance in the feature space from $\boldsymbol{x}$ to the data points of class C_k by

$$D(\boldsymbol{x}, C_k) = \min_{i:l_i = C_k} \|\boldsymbol{x} - \boldsymbol{x}_i\|$$

Then the conditional probability density for class C_k can be estimated by [30]

$$\hat{f}_k(\boldsymbol{x}) = \frac{1}{n_k V(D(\boldsymbol{x}, C_k))},$$

where n_k is the number of data points in class C_k, $V(r)$ is the volume of the ball in the feature space $\mathbb{R}^d$ with a centre at $\boldsymbol{x}$ and a radius of r. Following posterior class probabilities, we are able to assign proper probabilities to each class as follows [38]

$$p_k = \hat{f}(C_k|\boldsymbol{x}) = \frac{D^{-d}(\boldsymbol{x}, C_k)}{\sum_{j=1}^{K} D^{-d}(\boldsymbol{x}, C_j)}, k = 1, \dots, K$$

Therefore, the class belonging probability vector for a new point with feature vector $\boldsymbol{x}$ is obtained as

$$\boldsymbol{p}(\boldsymbol{x}) = [p_1, p_2, \dots, p_K]$$

7.5. Performance Evaluation

The performance of a classifier can be summarized in a confusion matrix as the one shown in Table 7.2 below.

The performance of a classifier can be evaluated with many different metrics. Different fields have different preferences for specific metrics due to different goals. In this work, the metrics used are the misclassification error rate (MER), false positive rate (FPR) and false negative rate (FNR). Some of the conventional evaluation metrics derived from the quantities in Table 7.2 are presented in Table 7.3 below.

Table 7.2. Confusion matrix for a binary classifier.

		Predicted class		Total
		Non-defect (0)	Defect (1)	
True class	Non-defect (0)	True Negative (TN)	False positive (FP)	n_0 = TN+FP
	Defect (1)	False Negative (FN)	True positive (TP)	n_1 = FN+TP

Table 7.3. Evaluation metrics.

Name	Acronym	Expression	Synonyms
True Negative Rate	TNR	TN/ n_0	Specificity
False Positive Rate	FPR	FP/ n_0 = 1-TNR	Type I error
False Negative Rate	FNR	FN/ n_1	Type II error
True Positive Rate	TPR	TP/ n_1 = 1-FNR	Sensitivity
Misclassification Error Rate	MER	1-(TP+TN)/(n_0+ n_1)	1-Accuracy

We also used probability-based metrics introduced in [14] which are similar to the conventional metrics such as:

Probability based misclassification error rate:

$$\text{probMER} = \frac{1}{n}\sum_{i=1}^{n} PM_i$$

Probability based false negative rate:

$$\text{probFNR} = \frac{1}{n_1}\sum_{i=1}^{n_1} PFN_i$$

Probability based false positive rate:

$$\text{probMER} = \frac{1}{n_0}\sum_{i=1}^{n_0} PFP_i,$$

where PM_i is a probability of misclassification which can be found as $PM_i = 1 - P(l_i = 0|\boldsymbol{x}_i)$ if the patch has a defect and $PM_i = P(l_i = 1|\boldsymbol{x}_i)$ otherwise. The probability of false negative is defined for the patches with defects as $PFN_i = 1 - P(l_i = 1|\boldsymbol{x}_i)$, while the probability of false positive is defined for the patches without defects as $PFP_i = P(l_i = 1|\boldsymbol{x}_i)$.

Note that it is straigforward to extend these criteria to the multi-class classification tasks by applying one-against-all principle.

Moreover, the vectors of posterior probabilities allow the classification quality to be judged at patch level, by using the uncertainty measure via Shannon entropy [30]:

$$E(\boldsymbol{x}) = -(p_1 \log p_1 + p_2 \log p_2 + \cdots + p_K \log p_K),$$

which provides us the uncertainty measure of classification for the whole dataset by its average entropy

$$E = -\frac{1}{n}\sum_{i=1}^{n} E(\boldsymbol{x}_i)$$

7.6. Results and Discussion

Here we present the results from the comparative analysis of the binary and 3-class classification problems, using cross-validation. Five different seeds were used to randomly split the collected data into five different training and test sets for SVM. 70 % of the data were included into the training sets and 30 % into the test sets for every split. Ten different seeds were used for probabilistic k-NN. All experiments were carried out in R 3.4.4 environment (R Core Team, 2018). The R library e1071 [39] was used to train a support vector machine, and knnx.dist() function of the R package FNN [40] to calculate the Euclidean distances of one-nearest neighbors.

The following ten feature vectors were constructed for classifier training.

1) EDF feature vector:

$$\text{EDF}_j = \zeta_j = (\zeta_{1,j}, \dots, \zeta_{m,j})^T, j = 1, 2, \dots, 16,$$

where $\zeta_{i,j}$ is the scaled EDF of the ith smooth of the patch pixel values for the jth channel.

2) HOG feature vector:

$$\text{HOG}_j = (\gamma_{1,j}, \dots, \gamma_{54,j})^T,$$

where $\gamma_{i,j}$ is the HOG coefficient for the corresponding j^{th} channel, $i = 1, \dots, 54$.

3) LBP-HOG feature vector:

$$\text{LBP-HOG}_j = (\varphi_{1,j}, \dots, \varphi_{54,j})^T,$$

where $\varphi_{i,j}$ is the LBP-HOG coefficient for the corresponding j^{th} channel.

4) HOG and LBP-HOG feature vector:

$$\text{HOG.LBP-HOG}_j = (\text{HOG}_j, \text{LBP-HOG}_j)^T,$$

where HOG_j and LBP-HOG_j are respectively the HOG and LBP-HOG feature vectors from above. The dimension of this vector is 108.

5) Feature vector based on the HOG applied on the denoised patches from Haar wavelet:

$$\text{HOG-Haar}_j = (\omega_{1,j}, \ldots, \omega_{54,j})^T,$$

where $\omega_{i,j}$ is the HOG coefficient for the corresponding j^{th} channel.

6) Feature vectors based on the HOG applied on the denoised patches from Daublets wavelet:

$$\text{HOG-Daublets}_j = (\vartheta_{1,j}, \ldots, \vartheta_{54,j})^T$$

The same idea as Haar wavelet in (e) was used here. 54 HOG coefficients were extracted in this case as well.

7) Feature vectors based on the HOG applied on the denoised patches from Symmlets wavelet:

$$\text{HOG-Symmlets}_j = (\delta_{1,j}, \ldots, \delta_{54,j})^T$$

Same concept as with Haar wavelets, 54 HOG coefficients extracted here.

8) Feature vectors based on column standard deviations:

$$\text{col.std}_j = (s_{(c)1,j}, \ldots, s_{(c)m,j})^T,$$

where $s_{(c)k,j}$ is the kth column standard deviation of the pixel values for the jth channel.

9) Feature vectors based on P-splines and column standard deviations:

$$\text{P-splines.col.std}_j = (\beta_j^T, s_{(c)j}^T)^T,$$

where β_j is a vector of the estimated spline coefficients, $s_{(c),j}$ is an m-dimensional vector of the column standard deviations of the pixel values for the jth channel.

10) Feature vectors based on column standard deviations and HOG feature descriptors jointly:

$$\text{col.std.HOG}_j = (s_{(c)j}^T, HOG_j)^T$$

7.6.1. Defect Detection Problem

We first present the results for defect detection, i.e. a binary task of classifying patches into defect or non-defect. Figs. 7.10 and 7.11 illustrate examples of the columns means by class of the channel "16-π-90" for the HOG and EDF feature vectors for two classes, respectively.

The performances of SVM and k-NN for binary classification problem using the probability-based metrics, with the feature vectors based on HOG, LBP, DWT, smoothing and variability are presented below in Tables 7.4 and 7.5.

Note that the results on binary classification using conventional evaluation metrics are similar and can be found in [14].

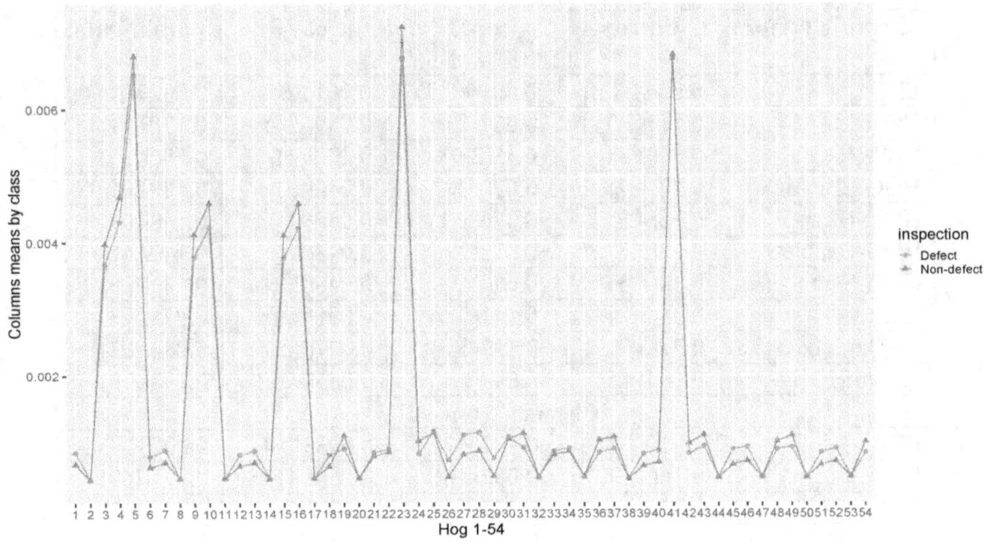

Fig. 7.10. HOG 2-class feature vector, column means by class, channel "16-π-90".

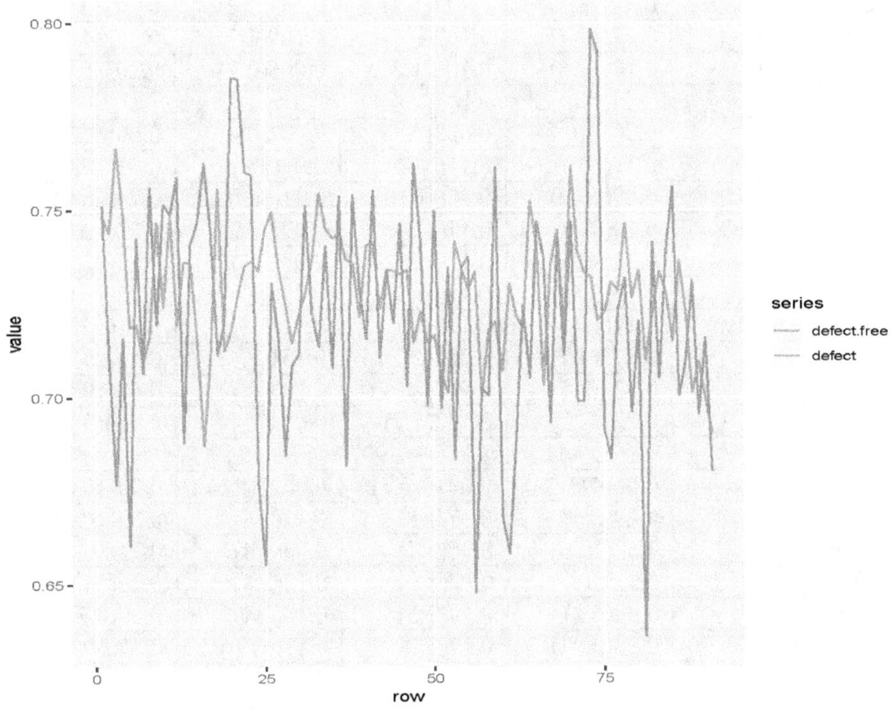

Fig. 7.11. EDF 2-class feature vector, column means by class, channel "16-π-90".

333

Table 7.4. Performance results of the SVM classifier for the binary classification problem using the probability-based metrics. The results are means of five runs (the standard errors are in brackets with a scale 10^{-3}).

Feature vector	Entropy	probMER	probFPR	probFNR
Channel "8-π-90"				
EDF	0.0044(4.6)	0.0016(0.2)	0.0013(0.3)	0.0027(1.1)
HOG	0.165(3.7)	0.087(3.0)	0.059(2.4)	0.17(9.7)
LBP-HOG	0.52(2.3)	0.35(2.0)	0.23(3.4)	0.70(15.0)
HOG:LBP-HOG	0.17(4.7)	0.09(2.8)	0.06(1.9)	0.18(9.0)
HOG-Haar	0.15(4.0)	0.076(2.1)	0.05(2.2)	0.15(1.0)
HOG-Daublets	0.13(3.2)	0.072(2.6)	0.048(1.6)	0.14(9.6)
HOG-Symmlets	0.14(5.5)	0.07(2.2)	0.048(2.8)	0.14(10.0)
col.std	0.254(3.1)	0.141(2.6)	0.095(1.9)	0.278(7.5)
P-splines:col.std	0.21(3.7)	0.117(1.6)	0.077(2.5)	0.234(9.2)
col.std:HOG	0.184(1.5)	0.099(2.4)	0.065(1.9)	0.198(8.9)
Channel "16-π-90"				
EDF	0.010(1.1)	0.0043(0.5)	0.0033(0.4)	0.0073(1.5)
HOG	0.16(3.0)	0.08(3.0)	0.06(2.4)	0.17(10.0)
LBP-HOG	0.52(10.0)	0.35(7.7)	0.23(7.0)	0.70(10.0)
HOG:LBP-HOG	0.16(3.0)	0.087(3.0)	0.058(2.3)	0.17(8.7)
HOG-Haar	0.15(3.0)	0.08(3.2)	0.06(2.9)	0.16(0.01)
HOG-Daublets	0.14(3.4)	0.08(1.6)	0.05(2.3)	0.16(0.01)
HOG-Symmlets	0.14(3.5)	0.08(2.0)	0.05(2.3)	0.15(8.9)
col.std	0.331(3.8)	0.197(1.2)	0.131(2.2)	0.391(2.9)
P-splines:col.std	0.278(2.4)	0.162(0.7)	0.109(1.3)	0.322(4.1)
col.std:HOG	0.181(2.6)	0.101(1.2)	0.066(2.2)	0.203(5.9)
Channel "32-π-90"				
EDF	0.0073(0.8)	0.003(0.7)	0.002(0.4)	0.0059(2.4)
HOG	0.19(4.0)	0.11(3.4)	0.07(2.0)	0.21(10.0)
LBP-HOG	0.49(40.0)	0.34(16.0)	0.21(36.0)	0.74(46.0)
HOG:LBP-HOG	0.20(6.4)	0.11(3.3)	0.073(3.2)	0.22(12.0)
HOG-Haar	0.21(3.0)	0.12(3.2)	0.08(3.7)	0.23(13.0)
HOG-Daublets	0.19(3.8)	0.11(3.1)	0.07(4.0)	0.22(13.0)
HOG-Symmlets	0.19(1.4)	0.11(3.2)	0.07(3.8)	0.21(16.0)
col.std	0.373(4.7)	0.227(2.1)	0.151(3.4)	0.451(9.77)
P-splines:col.std	0.333(3.5)	0.200(2.2)	0.133(2.7)	0.400(9.89)
col.std:HOG	0.221(2.2)	0.127(1.7)	0.084(2.1)	0.254(3.7)
Channel "64-π-90"				
EDF	0.0067(1.39)	0.0019(0.5)	0.0023(0.6)	0.001(0.6)
HOG	0.29(6.0)	0.17(1.8)	0.12(1.3)	0.35(10.0)
LBP-HOG	0.53(10.0)	0.35(4.7)	0.23(9.4)	0.73(10.0)
HOG:LBP-HOG	0.3(5.0)	0.18(3.9)	0.12(1.8)	0.36(14.0)
HOG-Haar	0.32(6.2)	0.19(2.5)	0.13(1.4)	0.39(8.7)
HOG-Daublets	0.28(2.03)	0.17(5.9)	0.11(17.0)	0.37(30.0)
HOG-Symmlets	0.30(5.0)	0.18(2.0)	0.12(1.8)	0.35(12.0)
col.std	0.441(2.7)	0.280(1.1)	0.187(2.2)	0.555(3.3)
P-splines:col.std	0.420(5.4)	0.266(1.5)	0.178(2.1)	0.527(2.5)
col.std:HOG	0.305(3.4)	0.185(1.3)	0.125(2.4)	0.361(6.1)

Table 7.5. Performance results of the k-NN classifier for the binary classification problem using the probability-based metrics. The results are means of ten runs (the standard errors are in brackets with a scale 10^{-3}).

Feature vector	Entropy	probMER	probFPR	probFNR
Channel "8-π-90"				
EDF	$1.4 \cdot 10^{-6}(1.7 \cdot 10^{-3})$	$1.7 \cdot 10^{-7}(2.3 \cdot 10^{-4})$	$2.2 \cdot 10^{-7}(3.2 \cdot 10^{-4})$	$3.4 \cdot 10^{-8}(8.2 \cdot 10^{-5})$
HOG	0.074(2.9)	0.22(3.5)	0.12(3.9)	0.55(8.9)
LBP-HOG	0.27(3.3)	0.37(3.9)	0.26(3.2)	0.70(9.6)
HOG:LBP-HOG	0.15(3.3)	0.36(4.6)	0.24(3.5)	0.72(10.0)
HOG-Haar	0.06(2.0)	0.21(3.7)	0.10(4.2)	0.52(8.6)
HOG-Daublets	0.07(2.0)	0.22(2.7)	0.11(4.0)	0.54(10.0)
HOG-Symmlets	0.07(1.6)	0.22(3.8)	0.11(5.0)	0.53(10.0)
col.std	0.081(1.5)	0.143(2.3)	0.116(3.7)	0.225(6.5)
P-splines:col.std	0.035(1.4)	0.144(1.9)	0.055(2.4)	0.406(8.1)
col.std:HOG	0.041(1.8)	0.183(12.0)	0.157(15.0)	0.261(7.4)
Channel "16-π-90"				
EDF	$1.9 \cdot 10^{-5}(3.4 \cdot 10^{-2})$	$6.4 \cdot 10^{-6}(1.6 \cdot 10^{-2})$	$7.1 \cdot 10^{-6}(2.1 \cdot 10^{-2})$	$4.5 \cdot 10^{-6}(7.9 \cdot 10^{-3})$
HOG	0.085(1.4)	0.24(5.1)	0.13(5.5)	0.57(14.0)
LBP-HOG	0.26(2.6)	0.35(2.4)	0.24(2.9)	0.69(7.0)
HOG:LBP-HOG	0.14(2.6)	0.34(2.8)	0.22(3.3)	0.71(7.7)
HOG-Haar	0.08(2.1)	0.23(4.0)	0.12(4.3)	0.56(7.8)
HOG-Daublets	0.09(3.6)	0.25(4.6)	0.13(4.4)	0.60(7.1)
HOG-Symmlets	0.09(3.1)	0.24(2.4)	0.13(2.9)	0.58(10.0)
col.std	0.065(1.1)	0.198(2.5)	0.128(3.3)	0.406(12.0)
P-splines:col.std	0.041(1.5)	0.194(5.2)	0.083(4.9)	0.521(13.0)
col.std:HOG	0.037(1.8)	0.190(5.0)	0.101(4.3)	0.452(11.0)
Channel "32-π-90"				
EDF	$9.3 \cdot 10^{-6}(1.4 \cdot 10^{-2})$	$2.1 \cdot 10^{-6}(4.1 \cdot 10^{-3})$	$2.3 \cdot 10^{-6}(5.5 \cdot 10^{-3})$	$1.5 \cdot 10^{-6}(2.4 \cdot 10^{-3})$
HOG	0.12(2.1)	0.28(3.2)	0.15(4.3)	0.65(12.0)
LBP-HOG	0.27(1.9)	0.37(3.0)	0.25(3.6)	0.72(8.0)
HOG:LBP-HOG	0.14(1.7)	0.36(3.3)	0.23(4.2)	0.74(8.8)
HOG-Haar	0.11(1.8)	0.26(3.4)	0.15(2.5)	0.62(10.0)
HOG-Daublets	0.12(3.0)	0.28(2.9)	0.16(2.9)	0.64(7.5)
HOG-Symmlets	0.12(3.3)	0.28(3.9)	0.16(4.4)	0.64(8.7)
col.std	0.077(1.9)	0.229(4.0)	0.135(5.4)	0.509(0.01)
P-splines:col.std	0.063(2.3)	0.239(3.5)	0.119(4.4)	0.596(11.0)
col.std:HOG	0.051(1.9)	0.232(3.3)	0.133(4.9)	0.523(0.01)
Channel "64-π-90"				
EDF	$1.0 \cdot 10^{-8}(2.2 \cdot 10^{-5})$	$8.2 \cdot 10^{-10}(1.8 \cdot 10^{-6})$	$2.7 \cdot 10^{-10}(7.9 \cdot 10^{-7})$	$2.4 \cdot 10^{-9}(5.1 \cdot 10^{-6})$
HOG	0.16(3.4)	0.32(3.7)	0.19(6.5)	0.69(7.3)
LBP-HOG	0.29(3.6)	0.37(3.0)	0.26(4.0)	0.71(12.0)
HOG:LBP-HOG	0.16(2.5)	0.36(4.0)	0.24(3.9)	0.73(0.01)
HOG-Haar	0.15(2.9)	0.31(2.3)	0.18(3.8)	0.67(8.6)
HOG-Daublets	0.15(2.5)	0.33(5.2)	0.20(7.0)	0.70(6.4)
HOG-Symmlets	0.16(3.0)	0.33(4.2)	0.21(5.6)	0.70(7.0)
col.std	0.132(2.4)	0.300(3.8)	0.193(4.2)	0.615(8.1)
P-splines:col.std	0.089(2.2)	0.301(4.4)	0.180(5.5)	0.660(11.0)
col.std:HOG	0.101(1.8)	0.310(3.5)	0.210(4.4)	0.606(13.0)

In comparison of the detection ability (binary classification problem), among those ten different types of feature vectors, EDF outperformed all the others, no matter the choice of classifier (SVM or *k*-NN) and images channels. Its probabilistic error rate and entropy are almost zero, which suggest that the approach is very certain in decision making. Between the two classifiers SVM and *k*-NN, the latter gave better performance. Therefore, we choose *k*-NN together with the feature vectors EDF and HOG (next best performance after EDF) to further investigate their abilities for not only detecting defects from non-defects but also distinguishing defects between dirt and crater, i.e. the 3-class classification problem, as follows.

7.6.2. Three-class Classification Problem

Here we present the results of our statistical learning approach for detecting the most common types of defects such as dirt and crater. Figs. 7.12 and 7.13 display the columns means by class of the channel "16-π-90" for the HOG and EDF 3-class feature vectors, respectively.

The performances of *k*-NN for 3-class classification problem, with the feature vectors based on HOG and EDF are presented in Tables 7.6 (conventional metrics) and Table 7.7 (probability-based metrics).

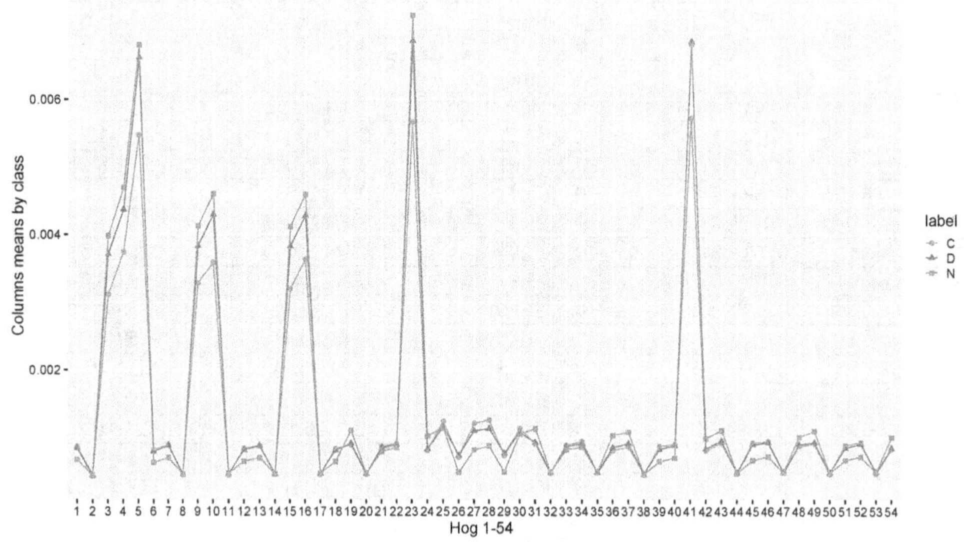

Fig. 7.12. HOG 3-class feature vector, column means by class, channel "16-π-90".

As we can see from Table 7.6, our probabilistic *k*-NN together with EDF demonstrated that almost 100 % of defect detection and 0 % of false alarms are achieved. Equivalently, nearly all patches with both crater and dirt were correctly classified, and no defect-free patch was predicted as defective. Contrary to the EDF, the feature vector based on HOG has much lower capability of defect detection, accounting for more than 10 % of false alarms and higher than 50 % of non-detections. The performance results of the HOG are worsening for the datasets of patches that correspond to channels with larger values of the frequency parameter, whereas the EDF performs equally well for all presented datasets. Consistently, Table 7.7 tells us that the probabilistic k-NN classifier using the feature vector of EDF produce very certain classified labels.

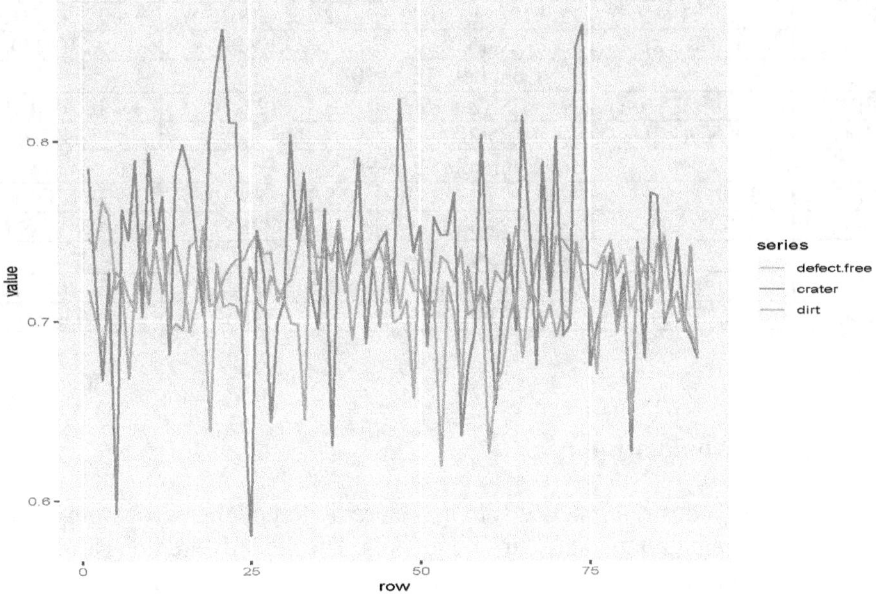

Fig. 7.13. EDF 3-class feature vector, column means by class, channel "16-π-90".

Table 7.6. Performance results of the k-NN classifier for 3-class classification problem using the conventional metrics. The results are means of ten runs (the standard errors are in brackets).

Feature vector	MER	FPR	FNR
Channel "8-π-90"			
EDF	0(0)	0(0)	0(0)
HOG	0.24(0.005)	0.11(0.005)	0.54(0.011)
Channel "16-π-90"			
EDF	0(0)	0(0)	0(0)
HOG	0.25(0.003)	0.12(0.004)	0.58(0.008)
Channel "32-π-90"			
EDF	0(0)	0(0)	0(0)
HOG	0.28(0.005)	0.15(0.005)	0.65(0.012)
Channel "64-π-90"			
EDF	0(0)	0(0)	0(0)
HOG	0.32(0.005)	0.18(0.005)	0.67(0.01)

Table 7.7. Performance results of the k-NN classifier for 3-class classification problem using the probability-based metrics. The results are means of ten runs (the standard errors are in brackets with a scale 10^{-3}).

Feature vector	Entropy	probMER	probFPR	probFNR
Channel "8-π-90"				
EDF	$1.5\cdot10^{-6}(1.6\cdot10^{-3})$	$1.9\cdot10^{-7}(2.1\cdot10^{-4})$	$2.2\cdot10^{-7}(2.8\cdot10^{-4})$	$1.1\cdot10^{-7}(2.3\cdot10^{-4})$
HOG	0.078(3.2)	0.24(5.0)	0.12(5.0)	0.61(12.0)
Channel "16-π-90"				
EDF	$2.2\cdot10^{-5}(3.7\cdot10^{-2})$	$9.5\cdot10^{-6}(2.5\cdot10^{-2})$	$1.3\cdot10^{-6}(2.6\cdot10^{-3})$	$3.4\cdot10^{-5}(9.9\cdot10^{-2})$
HOG	0.089(1.2)	0.25(2.5)	0.13(2.9)	0.64(7.9)
Channel "32-π-90"				
EDF	$1.3\cdot10^{-5}(1.6\cdot10^{-2})$	$2.7\cdot10^{-6}(4.3\cdot10^{-3})$	$2.9\cdot10^{-6}(5.9\cdot10^{-3})$	$1.9\cdot10^{-6}(3.9\cdot10^{-3})$
HOG	0.12(3.2)	0.29(3.9)	0.15(4.7)	0.69(7.6)
Channel "64-π-90"				
EDF	$2.5\,10^{-8}(3.1\cdot10^{-5})$	$2.1\cdot10^{-9}(2.6\cdot10^{-6})$	$6.2\cdot10^{-10}(1.9\cdot10^{-6})$	$6.3\cdot10^{-9}(9.1\cdot10^{-6})$
HOG	0.17(2.3)	0.33(4.2)	0.19(4.7)	0.73(7.5)

7.7. Real-time Visualization

An essential step in industrial practice is to make prediction on the new coming images of cabs, based on the trained statistical learning models. The defective regions for every cab's object are revealed through a real-time visualization GUI. The system is also able to tell the nature of the defects with properly estimated uncertainty.

Fig. 7.14 shows an example of how real-time prediction results for FH cabs are presented. What follows is a step by step description of the real-time visualization process that we built at the paint shop:

a) Sixteen images are automatically taken for each cab that passes through our system;

b) Only the image taken with the channel "16-00" is selected for further analysis;

c) A non-overlapping moving window of size 51×51 divides the selected image into 4158 patches. The edges are excluded subsequently;

d) Features are extracted from each one of those patches;

e) For every patch, the probabilistic classifier predicts the estimated probabilities of each class. The decision threshold is set to 85 %, a value above that threshold indicates defected patches while a value below indicates defect-free patches. The cross-entropy is also computed to evaluate the quality of our model;

f) The existing Volvo standard is applied for all the cabs with more than two defects.

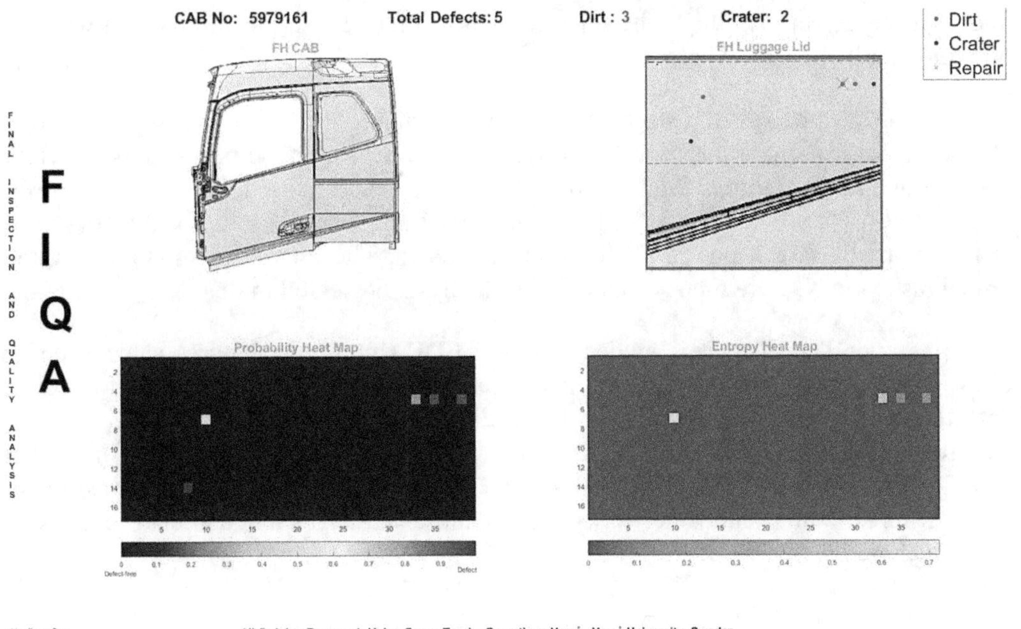

Fig. 7.14. Example of how real-time prediction results for FH cabs are presented. Top left: FH cab's image of the left side. Top right: Luggage lid for FH cab. Bottom left: Probability heat map. Bottom right: Entropy heat map.

7.8. Conclusions

Two types of problems, namely defect detection (binary problem) and defect classification (three-class problem) on the painted surface of the cab luggage lid, have been considered. The images from three different types of surfaces were collected. Rather than taking the natural images of the surface, the sinusoidal patterns were sent onto a 55-inch Sony screen and the reflections of those patterns on the surface were captured by two Fujinon cameras. Any variations of the surface due to defect, e.g., dirt or crater, resulted in the pattern distortion. In order to help with differentiating between the patches with and without defect, the gray scale pixel values of the obtained raw patches were transformed into a set of features. Together with extraction of the valuable information for the purpose of classification, the transformation reduces the dimensionality of the input data. Ten types of features were analyzed in this work, such as histogram of oriented gradients, local binary pattern, 2D wavelet transform, features based on variabilities and smoothing features based on P-splines and EDF. The derived features were converted into a vector for further use in a learning defect classifier.

Four channels were used for the data analysis. These channels have the same phase and correspond to channels 3, 7, 11, and 15 in Table 7.1. The selection of the phase π was random since it was previously known that the misclassification error rate was not affected by the phase, in contrast to the frequency which could have an impact on the performance results. The selected patch's window size was 91×91 due to its signal strength which increases the defect visibility. The selected size facilitated also the comparison with the previous results.

The preliminary analysis has been carried out to compare the performances of different classifiers such as neural networks, probabilistic k-nearest neighbors, support vector machine and random forest. Regardless of the feature vector chosen, SVM revealed itself to be the best classifier in terms of prediction accuracy, whereas the probabilistic k-NN came out as the fastest one in those previous studies. These two classifiers, SVM and probabilistic k-NN, were then selected for the analysis presented in this work.

From the overall results, it is evident that the EDF smoothing features achieve better classification performance with near zero misclassification error rate, both for defect detection and defect classification problems, in comparison with the other considered features used in k-NN and/or SVM. Besides, k-NN classifier is much faster than most commonly used classification algorithms for surface defect detection, such as SVM, random forest and neural networks, since those having computational expensive training phases.

In addition, the probability-based performance evaluation metrics have been proposed as alternatives to the conventional metrics. When an image is classified as containing a defect or non-defect, there is always a risk for erroneous classification. The usage of the probability-based metrics allows for uncertainty estimation of the predictive performance of a classifier.

It was time-consuming and expensive to collect the data and many challenges emerged during this process. Hundreds of data were thrown away due to unavailable information caused by some reported bugs from the image acquisition and annotation GUI. According to the size of the mouse click error from the operators, some defects could be hidden. Sometimes when the click error was big, some patches (often with smaller window sizes) did not even include these defects. The click error tended to be bigger when dealing with smaller defects because it was easier to miss them. The defect's hiding could as well be triggered by their micro meter size or by darker colours. Another vital and previously known issue was to separate craters from dirts during annotation. Touching the surface was still crucial to tell if it is a bump or a hole.

The developed real-time pilot system for defect detection and classification is currently running at Umeå Volvo plant and the results look promising despite the limited time for cab inspection. Developing such type of system is a complex process, where various areas of expertise have to be involved.

Acknowledgements

This work is a part of FIQA project, supported by the Strategic Vehicle Program Research and Innovation (FFI) at VINNOVA – Sweden's Innovation Agency (Reg. No. 2015-03706) and Volvo Group Trucks Operations.

References

[1]. J. Molina, J. E. Solanes, L. Arnal, J. Tornero, On the detection of defects on specular car body surfaces, *Robotics and Computer-Integrated Manufacturing*, Vol. 48, 2017, pp. 263-278.

[2]. Brilliant Insights – Deflectometry for the Inspection of Specular Surfaces, *Fraunhofer-IOSB*, 2017.

[3]. S. Kammel, F. P. León. Deflectometric measurement of specular surfaces, *IEEE Transactions on Instrumentation and Measurement*, Vol. 57, Issue 4, 2008, pp. 763-769.

[4]. E. R. Fotsing, A. Ross, E. Ruiz, Characterization of surface defects on composite sandwich materials based on deflectrometry, *NDT & E International*, Vol. 62, 2014, pp. 29-39.

[5]. F. P. León, S. Kammel, Inspection of specular and painted surfaces with centralized fusion techniques, *Measurement*, Vol. 39, Issue 6, 2006, pp. 536-546.

[6]. T. Stathaki, Image Fusion: Algorithms and Applications, *Academic Press*, 2011.

[7]. L. Armesto, J. Tornero, A. Herraez, J. Asensio, Inspection system based on artificial vision for paint defects detection on cars bodies, in *Proceedings of the IEEE International Conference on Robotics and Automation (ICRA'11)*, Shanghai, China, 9-13 May 2011, pp. 1-4.

[8]. Micro-Epsilon, Automated Surface Inspection, https://www.micro-epsilon.com/download/products/dat--reflectCONTROL-PSS-8005-D--en.pdf

[9]. J. Santolaria, J. Velázquez, D. Samper, J. Aguilar, I. Escursell, Sistema de inspección de defectos Opel España, 2016.

[10]. C. M. Bishop, Pattern Recognition and Machine Learning, *Springer*, 2007.

[11]. O. Severino Jr, A. Gonzaga, A new approach for colour image segmentation based on colour mixture, *Machine Vision and Application*, Vol. 24, Issue 3, 2013, pp. 607-618.

[12]. J. Yu, M. Ekström, Multispectral image classification using wavelets: A simulation study, *Pattern Recognition*, Vol. 36, Issue 4, 2003, pp. 889-898.

[13]. D. Maestro-Watson, J. Balzategui, L. Eciolaza, N. Arana-Arexolaleiba, Deep learning for deflectometric inspection of specular surfaces, in *Proceedings of International Joint Conference SOCO'18-CISIS'18-ICEUTE'18*, 2018, pp. 280-289.

[14]. N. Pya Arnqvist, B. Ngendangenzwa, E. Lindahl, L. Nilsson, J. Yu, Efficient surface finish defect detection using reduced rank spline smoothers and probabilistic classifiers, *Econometrics and Statistics*, 2020, in press.

[15]. N. Townsend, True colours, https://www.volvotrucks.com/en-en/news-stories/magazine-online/2017/jul/cab-paint-shop.html

[16]. Y. Mi, Specular surface inspection based on phase measuring deflectometry, MD Thesis, *Umeå University*, 2017.

[17]. M. S. L. Khan, S. ur Réhman, Computer Vision Approach Towards Final Inspection Quality Analysis, Research Report I, *FIQA*, 2018.

[18]. N. Dalal, B. Triggs, Histograms of oriented gradients for human detection, in *Proceedings of the IEEE Computer Society Conference on Computer Vision and Pattern Recognition (CVPR'05)*, June 2005, pp. 886-893.

[19]. F. Suard, A. Rakotomamonjy, A. Bensrhair, A. Broggi, Pedestrian detection using infrared images and histograms of oriented gradients, in *Proceedings of the Intelligent Vehicles Symposium (IV'06)*, 2006, pp. 206-212.

[20]. T. Ojala, M. Pietikäinen, D. Harwood, A comparative study of texture measures with classification based on feature distributions, *Pattern Recognition*, Vol. 29, Issue 1, 1996, pp. 51-59.

[21]. L. G. Shapiro, G. C. Stockman, Computer Vision, *Prentice Hall*, 2001.

[22]. T. Mäenpää, M. Pietikäinen, Texture analysis with local binary patterns, in Handbook of Pattern Recognition and Computer Vision, *World Scientific*, 2005, pp. 197-216.

[23]. T. Ojala, M. Pietikäinen, T. Maenpaa, Multiresolution gray-scale and rotation invariant texture classification with local binary patterns, *IEEE Transactions on Pattern Analysis and Machine Intelligence,* Vol. 24, Issue 7, 2002, pp. 971-987.

[24]. J. P. Antoine, R. Murenzi, P. Vandergheynst, S. T. Ali, Two-Dimensional Wavelets and Their Relatives, *Cambridge University Press*, 2008.

[25]. S. G. Mallat, A theory for multiresolution signal decomposition: The wavelet representation, *IEEE Transactions on Pattern Analysis and Machine Intelligence*, Vol. 11, Issue 7, 1989, pp. 674-693.

[26]. A. Bruce, H. Y. Gao, Applied Wavelet Analysis with S-plus, *Springer,* New York, 1996.

[27]. A. Haar, Zur theorie der orthogonalen funktionensysteme, *Mathematische Annalen*, Vol. 69, Issue 3, 1910, pp. 331-371.

[28]. I. Daubechies, Ten Lectures on Wavelets, *Society for Industrial and Applied Mathematics*, 1992.

[29]. S. N. Wood, Generalized Additive Models: An Introduction with R, *CRC Press*, 2017.

[30]. C. De Boor, A Practical Guide to Splines, *Cambridge University Press*, 1978.

[31]. J. Friedman, T. Hastie, R. Tibshirani, The Elements of Statistical Learning, *Springer*, New York, 2001.

[32]. G. James, D. Witten, T. Hastie, R. Tibshirani, An Introduction to Statistical Learning, *Springer*, 2013.

[33]. B. D. Ripley, Pattern Recognition and Neural Networks, *Cambridge University Press*, 2007.

[34]. S. Zhang, X. Li, M. Zong, X. Zhu, R. Wang, Efficient KNN classification with different numbers of nearest neighbours, *IEEE Transactions on Neural Networks and Learning Systems*, Vol. 29, Issue 5, 2018, pp. 1774-1785.

[35]. L. Cucala, J-M. Marin, C. P. Robert, D. M. Titterington, A Bayesian reassessment of nearest-neighbour classification, *Journal of the American Statistical Association*, Vol. 104, Issue 485, 2009, pp. 263-273.

[36]. N. Friel, A. N. Pettitt, Classification using distance nearest neighbours, *Statistics and Computing*, Vol. 21, Issue 3, 2011, pp. 431-437.

[37]. C. C. Holmes, N. M. Adams, A probabilistic nearest neighbour method for statistical pattern recognition, *Journal of the Royal Statistical Society: Series B (Statistical Methodology)*, Vol. 64, Issue 2, 2002, pp. 295-306.

[38]. B. Ranneby, J. Yu, Nonparametric and probabilistic classification using NN-balls with environmental and remote sensing applications, in Advances in Directional and Linear Statistics, *Physica-Verlag HD*, 2011, pp. 201-216.

[39]. D. Meyer, E. Dimitriadou, K. Hornik, A. Weingessel, F. Leisch, Misc Functions of the Department of Statistics, Probability Theory Group (Formerly: E1071), R Package, Version 1.7-3, https://CRAN.R-project.org/package = e1071

[40]. A. Beygelzimer, S. Kakadet, J. Langford, S. Arya, D. Mount, S. Li, FNN: Fast Nearest Neighbor Search, Algorithms and Applications, R Package, Version 1.1.3, https://CRAN.R-project.org/package = FNN

Chapter 8
Feature Selection from Time-Frequency Images for Change Detection

Dorel Aiordăchioaie, Theodor D. Popescu and Anisia Culea-Florescu

8.1. Introduction

Change detection is an important activity in many engineering processes from industry – for condition monitoring, medicine – for diagnosis and health monitoring, remote sensing – for land and surface monitoring. A common point of all these processes is the type of data input, i.e. the set of recorded signals organized in mono or multidimensional arrays. The standard choice is the 2D arrays, interpreted as images obtained at different time moments and, possible, from different states of the process and measuring conditions.

Change detection is used in fault detection, process diagnosis and predictive maintenance. There are mainly two approaches, based on parity equations and signal processing paradigm as, e.g. in [1-3], including machine learning [4] and artificial intelligence [5]. Both approaches use models based on equations. The first one is a process model and the second one is a model of the processed signals, directly from the analyzed process or residuals from other processing structures. This work uses the second approach, i.e. based on signal processing paradigm.

Change detection based on signal processing is a dynamic research domain and is based mainly on signal processing statistical techniques, as in [6-8]. New trends and results are obtained with advanced signal processing techniques and combinations of classic methods, i.e. time-frequency transforms, multi-scale transforms, information fusion, as well as artificial intelligence techniques.

Seven categories for change detection are commonly used, as: Direct Comparison (DC); Classification-based Methods (CM); Object-Oriented Method (OOM); Model Methods

Dorel Aiordăchioaie
'Dunarea de Jos' University of Galati, Romania

343

(MM); Time-Series Analysis (TSA); Visual Analysis (VA); and Hybrid Method (HM), as described in [9-10].

Change detection methods based on image processing could be classified into two classes, namely, bi-temporal change detection and temporal trajectory analysis. The former makes a comparison at two-time instants. The latter considers an analysis on a quasi-continuous time scale by defining/computing the trajectories or curves from temporal image data. It is applied for the cases when a high temporal resolution is available. An important application of temporal trajectory analysis is real-time detection, such as video image sequences analysis [11].

An important framework of change detection in image processing is represented by the maximum likelihood method, part of a more general area, named change detection based on statistical signal processing.

A change detection algorithm must identify the changed regions from one image to another one. The regions could be at the level of the pixel or at the level of more wide areas. More theoretical details and considerations could be found in [12-14]. In [15], an example of Change Vector Analysis (CVA) is presented, working on landsat images. In [16] an adaptive change detection technique via a significance test is proposed for remote sensing images. In [17] advanced techniques for change detection is used based on image fusing paradigm and using synthetic images to improve the accuracy of the change detection algorithm. In [18] a review of change detection methods is considered in the field of remote sensing.

Another point of view related to the classification of Change Detection Techniques (CDT) generates two categories: supervised and unsupervised. The former needs a set of images for training, for input, and a set of labels, for output.

The unsupervised CDT uses the set of the images at the input only and has some advantages related to the robustness to the external perturbations. The most used unsupervised methods are CVA (Change Vector Analysis); Principal Component Analysis (PCA); Object-Level Change Detection (OLCD); Multivariate Alteration Detection (MAD). Finally, [19] provides a nice introduction in change detection methods and [20] promotes a survey of change detection methods in Synthetic Aperture Radar (SAR) Images.

Vibration and acoustic emissions are common effects of the incipient and advanced faults in machinery and intensively used as sources of signals, as it is described, e.g., in [21-22]. This work considers the problem of change detection of faults in bearings, based on the vibration measurements and advanced signal processing techniques. A special attention is paid to feature generation and selection from time-frequency images.

Fig. 8.1 presents the main common processing blocks in the processing chain for CD objectives. Raw data coming from measurements are pre-processed by filtering and pre-processing, especially for source separation and stationarity analysis as, e.g. [23-26]. After applying the time-frequency transform, the associated images are pre-processed in

344

terms of scaling and registration. Image registration is necessary for pixel-based methods and not mandatory for feature-based methods. The trend is to develop change detection algorithms which integrate image registration and change detection. The efficiency and effectiveness of these methods must be assessed based on the results obtained by post classification and their accuracy. Finally, the diagnosis is made with the help of classification.

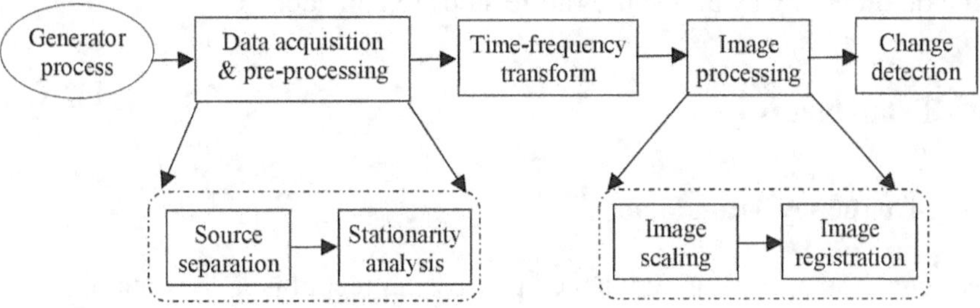

Fig. 8.1. The main processing blocks in change detection and diagnosis, based on TF images.

The chapter promotes two well-known signals transforms, Time-Frequency (TF) and Cosine Transform (CT), in extracting and processing of the main specific features, for change detection problem.

The level of research in this field, i.e. fault detection in bearings, is well covered by [27-29].

The representation of the transform's coefficients is an image, which will be addressed as Time-Frequency Image (TFI). As a data structure is a 2D array. Such image contains a huge information, and it offers solutions to problems from various fields where non-stationary signals are involved. As an example, in the case of bearings, [30-33] use TFIs.

This chapter promotes methods of feature design, under the hypothesis that the TF images are easy to compute, i.e. a high capacity of processing data is available. The next section considers the basic aspects of signal processing with data transforms, i.e. time-frequency and cosine transforms. From variously available time-frequency transforms (TFT), the Choi-Williams (CWT) transform is used, based on its good mathematical properties and fast algorithms of computation. Cosine Transform (CT) is also used based on his power to concentrate the information in its coefficients.

Section 8.3 presents three methods for feature design. The first one uses statistical parameters of the analyzed objects contained in the processed images with physical meaning, e.g. mean and variance. The second method considers virtual parameters, obtained from a transformed time-frequency image, with the help of contours. The third method selects the coefficients of the Cosine Transform (CT), in a similar approach as in image compression.

Selection of one method for feature vector depends on data sources and performance of the change detection criteria. Physical data are described and discussed in Section 8.4, in fact real signals recorded from mechanical vibration of faults in bearings of the rotating machines.

The results of the experiments based on computer simulation using real data records are presented in Section 8.5. Finally, the conclusion section presents a synthesis of the main results in a qualitative way and some future steps to consider.

8.2. Data Transforms

8.2.1. Time-frequency Transform

Time-frequency transforms are advanced processing techniques for data processing, and especially for data coming from non-stationary signals, as audio and speech signals [34], mechanical vibrations or some biological signals [35]. Three main methods are currently used for time-frequency representation and analysis: (i) Short-Time Fourier Transform (STFT); (ii) Wavelet Transform (WT); (iii) Cohen class.

The STFT of a signal $x(t) \in L^2(\mathbf{R})$ considers a window $w(t)$, as

$$STFT_{xw}(t, f) = \int_{-\infty}^{\infty} x(\tau)w(\tau - t)e^{-j2\pi f\tau}d\tau \tag{8.1}$$

The squared modulus is called spectrogram, as

$$S_{xw}(t, f) = |STFT_{xw}(t, f)|^2, \tag{8.2}$$

and constitutes a signal energy distribution in the time-frequency plane. The spectrogram constitutes one of the widely used methods for the analysis of non-stationary signals. In some cases it is unsuitable for the compromise needed for time and frequency resolutions, i.e. it is not possible to simultaneously have good time and frequency resolutions. Consequently, especially considering the proximity and evolution of the signal components in time and in frequency, the user must choose the characteristics of the analysis window depending on the signal structure. The transform is linear and depends on the chosen window, $w(t)$. Details on how to choose the parameters of the observation window, as length and shape, and the discrete-time version, are presented e.g. in [36-38].

The Wavelet Transform (WT) was promoted to solve the time-frequency resolution problem of Fourier-type methods. A concept called "multi-resolution" or "multi-scale" is promoted. In the case of continuous time wavelet transform, a basis of translated and dilated functions called wavelets are used as

$$\psi_{t\prime,a}(t) = \frac{1}{\sqrt{a}}\psi\left(\frac{t-t\prime}{a}\right) \tag{8.3}$$

The wavelet transform is then

$$WT_{xw}(t,a) = \int_{-\infty}^{\infty} x(\tau)\frac{1}{\sqrt{a}}\psi^*\left(\frac{\tau - t\prime}{a}\right) d\tau \tag{8.4}$$

The Cohen class method is the set of all bilinear representations covariant under time and frequency translations, and described by the equation

$$C_x(t,f) = TFT(x(t,f)) = \int_{-\infty}^{\infty}\int_{-\infty}^{\infty} K(t_1,t_2;t,f)\,x(t_1)x^*(t_2)dt_1dt_2 =$$

$$= \int_{-\infty}^{\infty}\int_{-\infty}^{\infty} k(t,f;v,\tau)\,x\left(v+\frac{\tau}{2}\right)x^*\left(v-\frac{\tau}{2}\right)dvd\tau, \tag{8.5}$$

with

$$k(t,f;v,\tau) = K\left(v+\frac{\tau}{2}, v-\frac{\tau}{2}; t,f\right), \tag{8.6}$$

and where the kernel $k(t,f;v,\tau)$ has some special properties, as discussed in [24-25]. By an equivalent parameterization, the equation (8.5) becomes

$$C_x(t,f) = \int_{-\infty}^{\infty}\int_{-\infty}^{\infty} \varphi_{t-f}(t-v, f-v)\cdot W_x(u,v)dvd\tau, \tag{8.7}$$

with

$$\varphi_{t-f}(t,\tau) = K(t,0;0,\tau), \tag{8.8}$$

and

$$W_x(t,f) = \int_{-\infty}^{\infty} x\left(t+\frac{\tau}{2}\right)x \times \left(t-\frac{\tau}{2}\right)e^{-j2\pi f\tau}d\tau \tag{8.9}$$

The function $W_x(t, f)$ is called the Wigner-Ville distribution (WVD), [23], being one of the most important members of the Cohen's class method). It may be the only distribution with real values that satisfies the properties necessary for the classical applications of signal processing. It is also the only distribution to provide perfect localization for impulse signals and signals with a linearly modulated frequency [23].

In this work, the Choi-Williams Distribution (CWD) [39] is used, where the kernel function is

$$\varphi(\theta,\tau) = exp[-(\theta\tau)^2/\sigma^2] \tag{8.10}$$

This distribution function adopts exponential kernel to suppress the cross-terms that result from the components that differ in both time and frequency centers.

The discrete WVD is defined by

$$W(n,m) = \frac{1}{2N}\sum_{k=0}^{N-1} x(kT)\cdot x^*((n-k)T)\,exp\left(-\frac{j\pi\cdot m\cdot(2k-n)}{N}\right) \tag{8.11}$$

It is easy to verify that $W(n, m)$ is a periodic function of period $2N$ in both time and frequency [39]. In the range $0 \leq n < 2N - 1$, $0 \leq m < 2N - 1$, representing one complete period, the WVD needs only be calculated over the range $0 \leq n < N - 1$,

$0 \leq m < N - 1$, having an area of one quarter that of the complete period. To obtain an aliasing-free discrete WVD, the maximum frequency does not exceed $1/2T$ and the time length of the signal must be less than $NT / 2$, which means

$$W_x(t, f) = 0 \quad for \quad f < 0, \quad f > \frac{1}{2T}, \tag{8.12}$$

$$W_x(t, f) = 0 \quad for \quad t < 0, \quad t > \frac{NT}{2} \tag{8.13}$$

These conditions can almost be satisfied simultaneously. For more details specialized literature is available, e.g. [39]. The coefficients of the time-frequency transform define an image, which will be called a time-frequency image. The CWD is used here with a rectangular window and unit variance.

8.2.2. Cosine Transform (CT)

Image compression is very used in many applications, mainly for storage and transmission. A well-known theory, standards and examples as well, are available. Some examples are presented in [40-42]. For transmission applications, a compression ratio or capacity gain are used to estimate the gain after compression in strong correlation with the accepted distortion at the end user site. The most used transforms are Fourier, Karhunen-Loeve [43], Haar-Hadamard and Walsh-Hadamard [44], and Cosine [45].

Discrete Cosine Transform (DCT) is used in this chapter. For change detection context, only the selection and extraction of the main coefficients are the mandatory steps. The reconstructed image with the selected coefficients is not important. Some processing techniques used in image compression could be used also here, i.e. a zigzag trajectory in the reading of the DCT coefficients and extraction of a limited number based on their values and imposed error of detection or classification. The raw image could be also divided into blocks of interest, preferably of the same size.

An important improvement of the change detection performances is to consider the spectrum distribution over the frequency axis, and to define frequency bands for analysis, as low, medium and high frequency bands.

The matrix A of the transformation is defined by

$$A(u, v) = \frac{2}{n} \sum_{i=0}^{n-1} \sum_{j=0}^{n-1} C(i)C(j) \cdot \cos \frac{\pi u(2i+1)}{2n} \cos \frac{\pi v(2j+1)}{2n},$$

$$C(k) = \begin{cases} \frac{1}{\sqrt{2}}, & k = 0 \\ 1, & k \neq 0 \end{cases} ; \quad u, v = 0, \dots, n - 1, \tag{8.14}$$

and it is used as

$$B^* = A \cdot B \cdot A^T \tag{8.15}$$

The size of the output array, B^*, is the same as the input array, B. Fortunately, many of the transform's coefficients, i.e. the elements of B^*, are close to zero and can be ignored.

8.3. Description of the Feature Selection Methods

8.3.1. Preliminary

Change detection is assimilated as a binary hypothesis detection problem. The structure of the method is presented in Fig. 8.2. For each record of N samples coming from the observed process, a sliding non-overlapping window of length $n<<N$ is considered. Data are processed and evaluates a detection criterion and, after comparing with a threshold, γ, a decision is taken, i.e. D_0 for no change and D_1 for a change. The threshold depends on the prior information about data, the monitored process, and the imposed performance of the detection criteria based on custom probabilities.

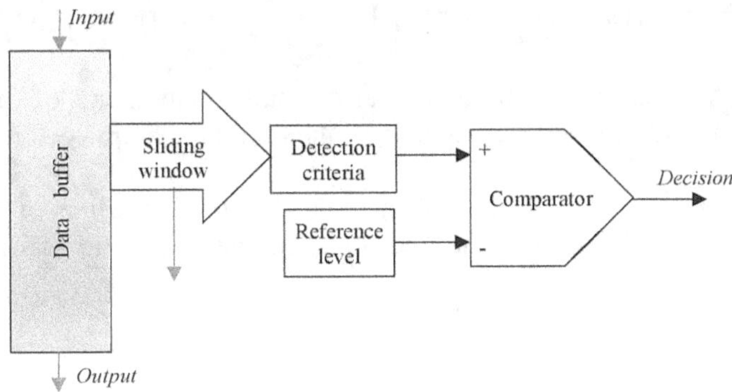

Fig. 8.2. The basic structure of the data processing for change detection.

Each window of observation is considered as a data frame and, by transformation, it generates a time-frequency image (TFI). These operations are made inside of the block for computation of the detection criteria. There are two aspects of the analysis activity: (i) quantitative, based on the change of the energy/information contain in TFI, from one frame to another one. A change in the energy/information is a first sign of difference between current and previous content of the frames. Thus, change detection might be implemented; (ii) qualitative, based on the content of each TFI at the level of patterns/shapes and based on the evolution of these when the frames are changing. This aspect is the main contribution to the base knowledge for diagnosis.

The analysis of the TFIs considers the main components, i.e. the components with the highest amplitudes. In the simplest case, only one component is considered. A more complex method takes a predefined number of peaks and builds a 2D object, in order to describe the behavior or the state of the analyzed process.

A change in the working regime or the appearance of faults in bearings will be reflected by a change in the position of the spectral components in the time-frequency image. In the case of the object, the shape will change. A change in the working regime, e.g. a change in the rotation speed, is reflected by a movement of the spectral components on the frequency axis. Various faults could generate also some extra components.

The parameters of the selected components, e.g. the coordinates on X and Y directions, the amplitude, the frequency bandwidth, the energy, etc. are the elements of the feature vector associated with a frame i as

$$F_i = [f_{i1} \cdots \ f_{ij} \ \cdots f_{ip}], f_{ij} \in \mathbb{R}^{1 \times n}, j = \overline{1,p}, i = 1,2,\dots, \qquad (8.16)$$

where p is the number of the peaks/components and n is the number of the parameters. Scaling to $[0,1]$ is considered for all column vectors, f_j, by dividing it to the maximum value of the considered vector.

To reduce the number of the features, an average of the parameters could be computed and considered as

$$\overline{F}_i = E[F_i] = [\dots \ E[f_{ij}] \ \dots] = [\cdots \ \mu_j \ \cdots] = \mu \qquad (8.17)$$

Some supplementary parameters based on various statistic moments of various orders or extra knowledge could be also used, by adding them to the feature vector, if necessary.

If γ is the threshold vector, computed from the a priori probabilities of hypotheses and desired a posteriori probability of the decisions, the decision criterion is based on the general rule

$$\text{IF} \quad (\mu > \gamma) \quad \text{THEN} \quad (CD = \text{TRUE}), \qquad (8.18)$$

with CD being a binary variable associated to the decision process. In the next sections, when the results of the experiments are presented, the point change detection is estimated by using the CUSUM criterion, a classic sequential technique based on cumulative sum [46-47].

8.3.2. Feature Design by Physical Parameters

In the beginning, only one component is considered, i.e. the component with the highest amplitude, which could be enough for change detection. If diagnosis is imposed, features from more components must be considered and properly processed.

In the simplest case, two parameters are selected, the coordinates x and y, which could be enough for change detection but not for diagnosis purposes, in general. This is the price for simplicity.

A finite number of frames is considered, nw. The base feature vector is

$$F_i = \begin{bmatrix} x_i \\ y_i \end{bmatrix}, i = 1,2,\dots, n_w, \qquad (8.19)$$

with its mean defined as

$$\overline{F}_i = E[F_i] = [(x_i + y_i)/2] = [\mu_i], i = 1,2,\dots, n_w \qquad (8.20)$$

An augmented feature vector could consider the variance (or standard deviation) of the coordinates related to the component of the same frame or on number of previous frames as

$$F_i = [x_i \quad y_i \quad \sigma_{xi} \quad \sigma_{yi}]^T, i = 1,2,\ldots,n_w, \tag{8.21}$$

and

$$\overline{F}_i = [\mu_i \quad \sigma_i]^T, i = 1,2,\ldots,n_w \tag{8.22}$$

8.3.3. Feature Design by Synthetic Parameters

The method considers the content of the image by selecting a pre-defined number of peaks/components, e.g. up to 3...5, depending on the content and complexity of the image. A new image is obtained and defined in terms of contours, defined by the above peaks, which will be called transformed image (TI) or contour-based image (CBI).

In order to extract the right information from the transformed images, the following parameters could be considered in defining the necessary features for change detection and classifications:

i) The number of contours, Nc, as a measure of the complexity;

ii) The area of the polygons, Ac, generated by the above contours, as a measure of the spreading in the plane;

iii) The variance of the above areas, $var(Ac)$, as a measure of the complexity or differences among components of the main image;

iv) The average of the area of the polygons, $E\{A_c\}$;

v) The mean of the squared values of areas, $E\{A^2{}_c\}$;

vi) The entropy of transformed images, $RH\ (TI)$;

and the list could continue. Most of the previous parameters do not have physical meaning, they are synthetic (virtual) parameters.

For each frame i, a vector of features could be defined by using the above variables, as

$$F_i = \begin{bmatrix} N_c & \Sigma A_c & var(Ac) & \overline{A_c} & \overline{A_c^2} & RH \end{bmatrix}, i = 1,2,\ldots \tag{8.23}$$

If M is the number of classes from the pattern space, the effect of the selected features is estimated by a general discriminant or dissimilarity matrix D with elements

$$D(k,j) = \Sigma_{k=1}^{M}\Sigma_{j=1}^{M}(F_k - F_j)^2, k,j = \overline{1,M} \tag{8.24}$$

The number of the selected features depends on the image complexity and on the performance of the detector and classifier used in the process of CD. High similarity images or poor performance classifiers need high quality virtual parameters. More considerations are available in [48-49].

8.3.4. Feature Design by Transformed Parameters

The structure of the method is based on data compression paradigm, and it is presented in Fig. 8.3. Data of the observation window are processed to obtain a time-frequency image and afterward a cosine transform is applied. The matrix of the discrete CT coefficients is considered the basic matrix of the features and it is processed in order to select the relevant features for detection. The new parameters are called transformed parameters, being the output of an auxiliary transform.

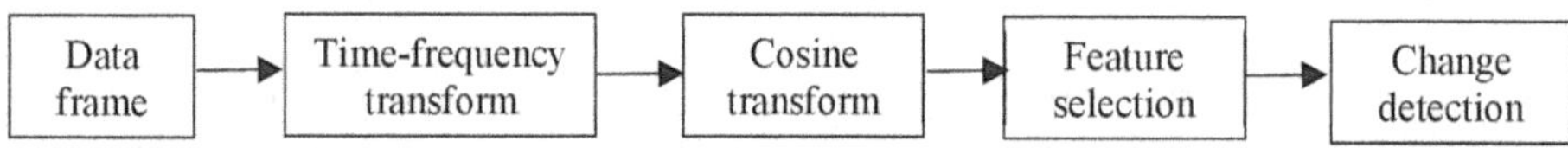

Fig. 8.3. Feature processing by transformed parameters.

Let be the matrix **I**, of size $n \times n$, a time-frequency image of the current window. The main steps for features selection are:

1). If A is the matrix of DCT, compute the compression transform:

$$B = A \cdot I \cdot A^T \qquad (8.25)$$

2). Select and extract the important coefficients (features). The method of zigzag scanning is considered, beginning with the DC coefficient. In the image compression context, the decision is made under a constraint of image reconstruction **I***, which should be like the original image **I**, from a measure represented here by the criterion J as

$$B \rightarrow B^*, \quad card(B^*) << card(B) \quad \backslash \quad J(I - I^*) = min \qquad (8.26)$$

This means that the number of the selected coefficients of B should generate a reconstructed image, I^*, within the constraint of minimum value of the error criterion J. In the context of time-frequency images and signals representing vibrations, the size of the feature's array B^* is selected to maximize the rate of change detection and/or recognition rate of the classifier for diagnosis purposes.

3). Compute the change detection criterion based on the first m coefficients of B, which defines an array $B^* = B(1{:}m)$ with a reduced size as

$$IF \quad (max(B^*)/|min(B^*)| > \gamma) \quad THEN \quad (CD = TRUE), \qquad (8.27)$$

where γ is a threshold depending on data and imposed performances for detection. This rule was elaborated to explore the specific properties of the coefficients associated with normal, free of faults, and abnormal states, i.e. states with faults. More details are available in [50].

8.4. Data Description and Analysis

8.4.1. Data Description

Data were considered for the case of faults in bearings, available from [51], which are also well explained and analyzed in [52], and briefly described in Table 8.1. The number inside of the round parenthesis indicates the name of the file from the original source of data vibrations, i.e. [51].

Table 8.1. Data Test Set.

Fault size	Faults (with files)					
	F0	**F1**	**F2**	**F3**		
				Outer Race		
	Free	**Inner Race**	**Ball**	**06HH**	**03HH**	**12HH**
0.000"	*d0* (f_97) (case #0)	-	-	-	-	-
0.007"	-	*d1* (f_105) (case #1)	*d2* (f_118) (case #5)	*d3* (f_130) (case #9)	*d4* (f_144)	*d5* (f_156)
0.014"	-	*d6* (f_169) (case #2)	*d7* (f_185) (case #6)	*d8* (f_197) (case #10)	-	-
0.021"	-	*d9* (f_209) (case #3)	*d10* (f_222) (case #7)	*d11* (f_234) (case #11)	*d12* (f_246)	*d13* (f_258)
0.028"	-	*d14* (f_3001) (case #4)	*d15* (f_3005) (case #8)	-		

Three types of faults are available: F1 (Inner race fault), F2 (Ball fault), and F3 (Outer race fault). The case F0 means no faults. In the case of the fault F3, there are three sub-cases, depending to the fault position relative to the load zone: 'centered' (fault in the 6.00 o'clock position), 'orthogonal' (3.00 o'clock) and 'opposite' (12.00 o'clock) [52].

Vibration data from four sizes of the faults are available. The data set has the advantage of consistency, by considering faults from the incipient/small size (0.007") to a larger one (0.028"). The sampling rate is 12,000 Hz, the motor load is 0 hp, and all data are from drive end bearing (DE). The type of bearing is 6205.

In the case of the bearings, the spectrum of the vibrations includes low-frequency components, and the time analysis should consider this. The testing data are generated by an electrical machine with a speed of 1700 rot/min, which means 1700/60 s = 28.33 [rot/s] (Hz). The corresponding number of samples for one period is

$$ n = \frac{T_o}{T_s} = \frac{1/28.33}{1/12,000} = \frac{12,000}{28.33} \cong 424, \tag{8.28} $$

which imposes a minimum value of $n = 400$ samples for the sliding window.

For tests based on computer simulation and for an increased readability of the processed variables, new names for some working variables were considered. All names beginning with "*d*" indicate a vector with 5,000 samples from the raw file. Thus, the variable *d0* contains the first 5,000 elements of the raw file named "f_97", and *d1* has 5,000 samples from the record file "f_105" for the fault F1.

Incipient faults are considered, i.e. faults size of 0.007". Data are not scaled. The Choi-Williams distribution (CWD) was used with a weighting window of Kaiser type and length 99. The associated images are scaled.

For a window time length of 0.2 [s], which means $n = 2,400$ [sample] at a frequency sampling of 12 [kHz], the images of Figs. 8.3-8.6 were obtained, within the cases F0 (f_97), F1(f_105), F2(f_118) and F3(f_130). On the left side of the figures, the power spectral density (PSD) is represented. On the bottom of each figure, the evolution over time is shown.

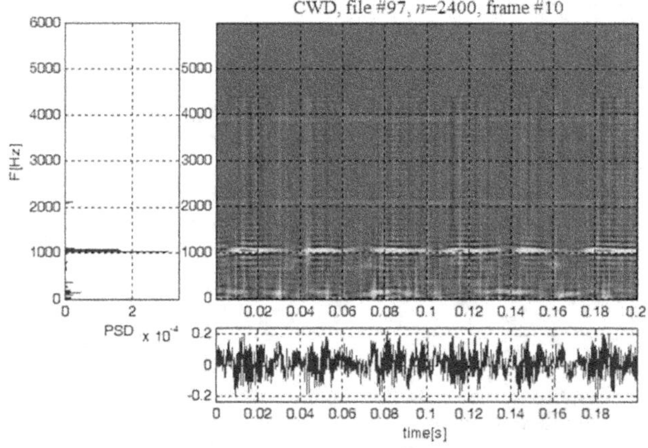

Fig. 8.3. Time-frequency image for the case free of faults, F0 (No faults, file #97).

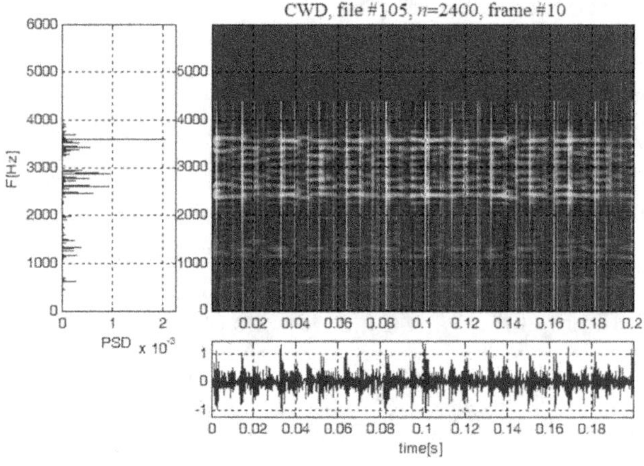

Fig. 8.4. Time-frequency image for the case F1 (Inner race, file #105).

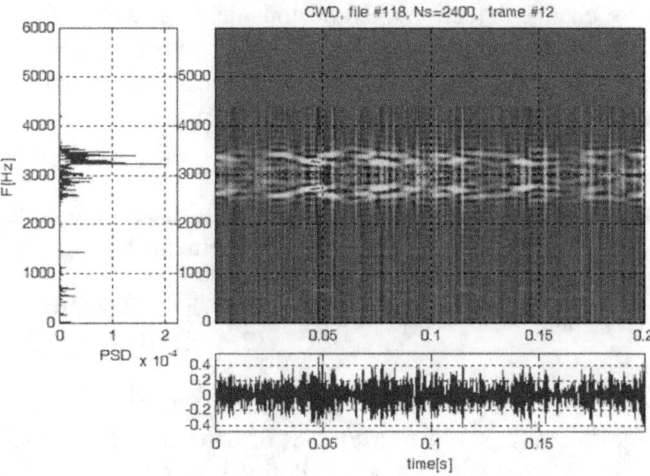

Fig. 8.5. Time-frequency image for the case F2 (Ball fault, file #118).

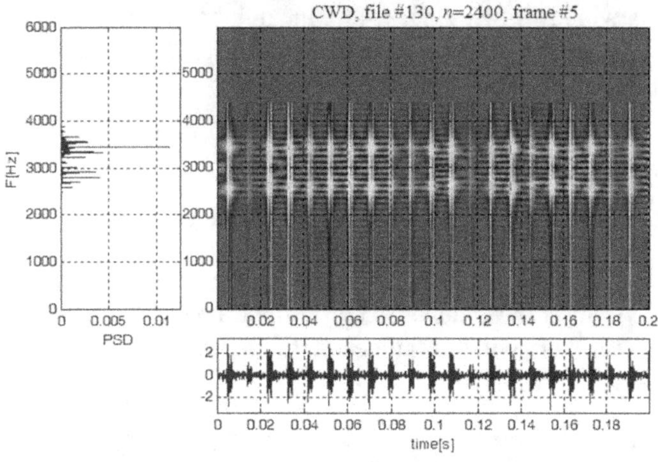

Fig. 8.6. Time-frequency image for the case F3 (Outer race fault, file #130).

8.4.2. Data Analysis

From a primary analysis of these images, the first remark is related to the fact that, in the case without faults, the recorded data have the highest components, around 1 [kHz], and, in the cases with various sizes of faults, the power spectrum is spreading up to 4 [kHz]. Another remark is the possibility to define specific patterns, for each of the considered fault.

In order to simplify the end-user equipment, a small length of the observation window is desired. The length of the data frame is imposed by the time-frequency patterns, in the sense that the TFI must contain enough information for detection and diagnosis purposes.

The requirements of change detection must be considered as well, which needs short data processing windows. In the testing stage, three values for the length of the frames were considered, 400, 1,000 and 2,000 [sample]. A length less than 400 samples does not describe properly the faults in time-frequency (TF) plane. If the length is greater of 2,000 samples the complexity and the cost of the CDD equipment will be raised.

For computer-based simulations and tests, a simple data structure is considered by taking a data matrix D composed of 15 columns of 10,000 elements as

$$D = [D_1 \quad D_2 \quad \ldots \quad D_{15}] = \begin{bmatrix} d0 & d0 & \ldots & d0 & d0 \\ d1 & d2 & \ldots & d14 & d15 \end{bmatrix}_{2x10,000} \tag{8.29}$$

The first 5,000 elements of each column are from the fault free vector, i.e. $d0$. Each column defines a test case. A set of four classes of patterns are considered as: #C0 – no faults, defined by $d0$; #C1 –Inner race fault, defined by $\{d1, d6, d9, d14\}$; #C2 – Ball fault, defined by $\{d2, d7, d10, d15\}$; #C3-outer race fault defined by $\{d3, d8, d11\}$. The change point is 5,000.

Considering a data window of length $n = 400$, the number of the windows is $nw = 25$. The next two figures, Figs. 8.7 and 8.8, present examples of images (frame #8) for each considered class:

- in Fig. 8.7, the classes #C0($d0$), #C1($d9$), #C2($d10$), #C3($d11$) on the left side and #C1($d1, d6, d9, d14$) on the right side;

- in Fig. 8.8, the class #C2($d2, d7, d10, d15$) on the left side and #C3($d3, d8, d11, d12$) on the right side.

At the level of the frames, the changes occur in the frame #13. This is a transition frame. The frames from 1 to 12 are without faults and the frames from 14 to 25 are with changes.

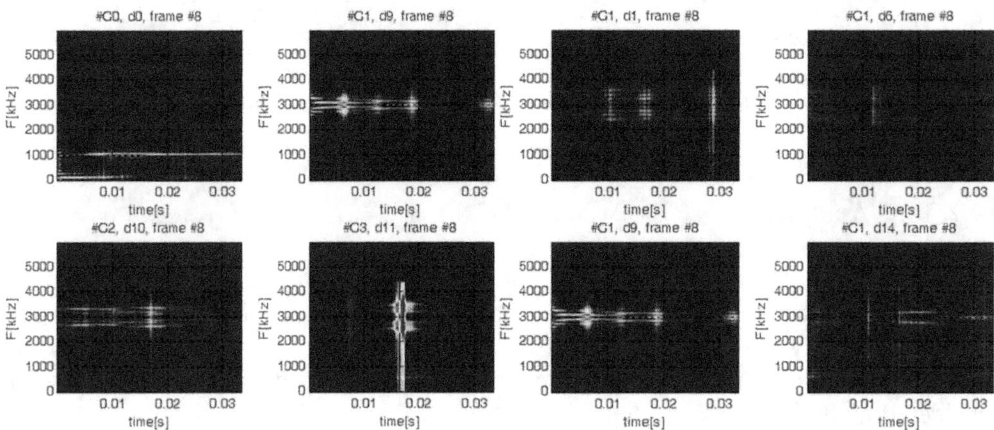

Fig. 8.7. Time-frequency images of the classes #C0($d0$), #C1($d9$), #C2($d10$), and #C3($d11$) (left side); #C1($d1, d6, d9, d14$) (right side).

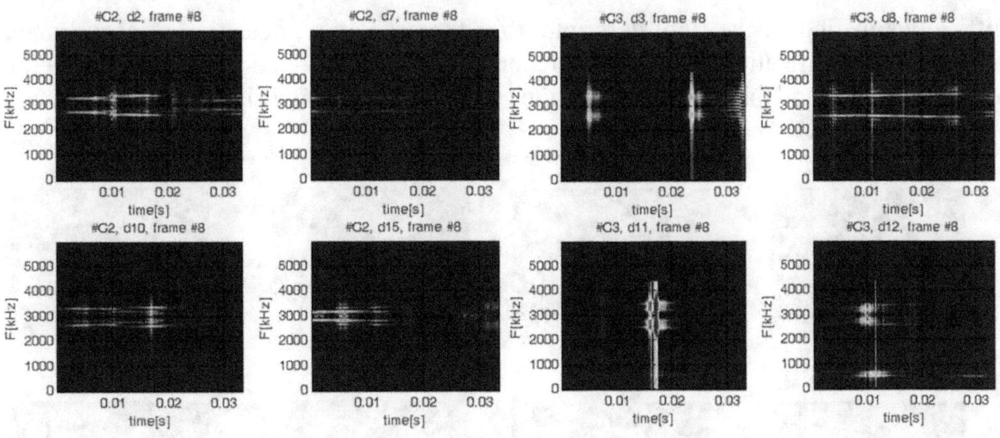

Fig. 8.8. Time-frequency images of the classes #C2(*d2, d7, d10, d15*), and #C3(*d3, d8, d11, d12*).

By considering a data window of length $n = 1,000$, the number of the frames is $nw = 10$. The Figs. 8.9 and 8.10 present examples of such images (frame #8) for each of the considered class:

- in Fig. 8.9, the classes #C0(*d0*), #C1(*d9*), #C2(*d10*), #C3(*d11*) on the left side and #C1(*d1, d6, d9, d14*) on the right side;

- in Fig. 8.10, the class #C2(*d2, d7, d10, d15*) on the left side and #C3(*d3, d8, d11, d12*) on the right side.

At the level of the frames, the changes occur between the frames #5 and #6. The frames from 1 to 5 are without faults and the frames from 6 to 10 are with changes.

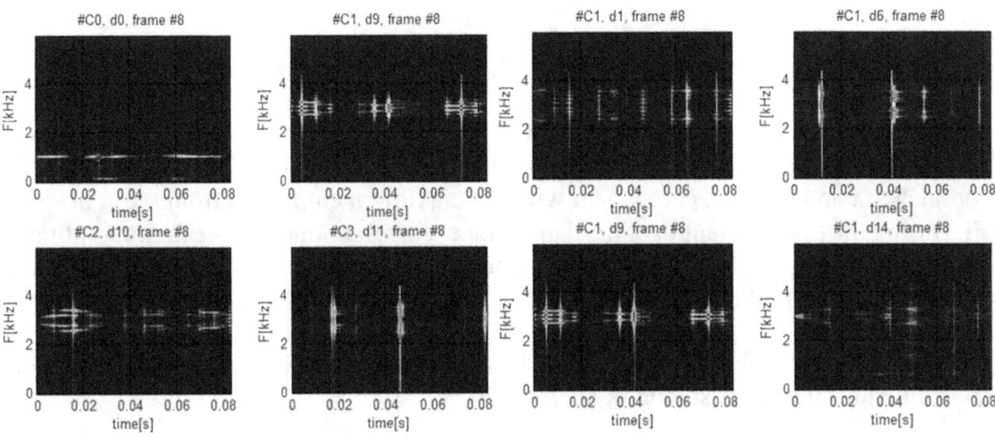

Fig. 8.9. Time-frequency images of the classes #C0(*d0*), #C1(*d9*), #C2(*d10*), and #C3(*d11*) (left side); #C1(*d1, d6, d9, d14*) (right side).

Faults in bearings generate complex patterns in time-frequency images. Short length windows have parts/fragments only of the specific pattern and thus could generate difficulties in finding the right descriptors for representation, detection of changes and further for classification. Long term records reveal some stationary patterns in the analyzed images and thus the simplification and the success of the classification processes.

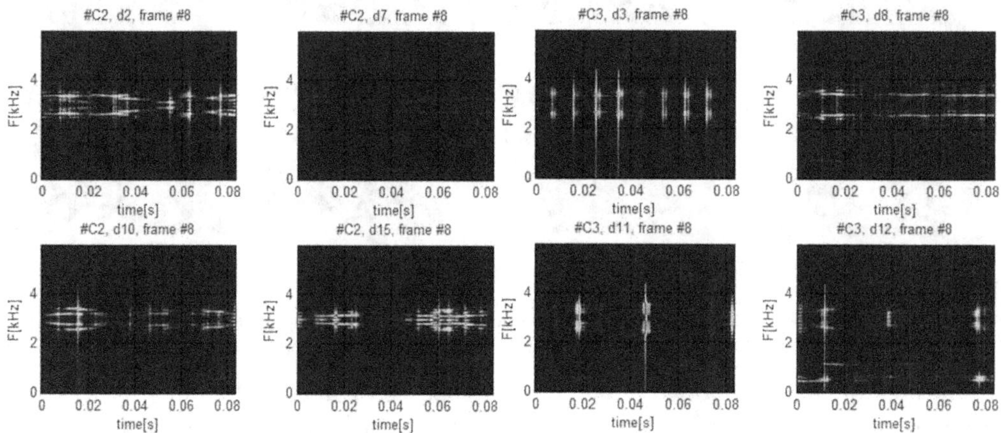

Fig. 8.10. Time-frequency images of the classes #C2 (left) and #C3 (right).

Working with non-overlapping frames could generate systematic errors in the estimation process; however, these are not relevant in the context of high values of the sampling frequency. If it is necessary, partial overlapping windows could be considered for an improvement of the detection accuracy.

8.5. Results of the Experiments

8.5.1. Results from Feature Design with Physical Parameters

This section presents the evolution of the selected features to discriminate between normal and abnormal conditions. The selected features are the mean and the variance of the coordinates, x and y, of the component with the maximum amplitude from the considered TFI. During the first half length of the data test, i.e. corresponding to free faults conditions, both features are constant. A transition is made in the second half of the data test, where the features vary significantly.

The performances of the change detection depend on the window's length. The best results are obtained for the shortest length, i.e. $n = 400$.

In Fig. 8.11, in the red color, the averages of the features on x and on y directions are represented. Pairs of figures are promoted by representing the evolution of the features and, in the pair figure, the decision made. Thus, for case #1 the pair of the Figs. 8.11 and 8.12; for case #2, the pair of the Figs. 8.13 and 8.14; case #3, in Fig. 8.15, the feature evolution only.

358

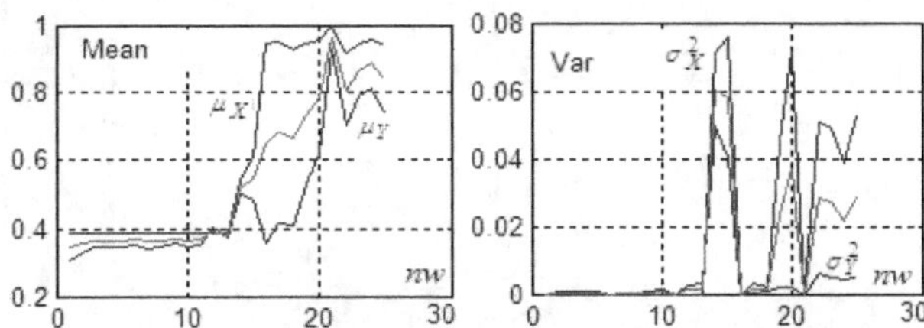

Fig. 8.11. The average of the coordinates of the spectral components, case #1, **D1** = [**d0 d1**].

Fig. 8.12 presents the data test in the test case #1 and the decision, in red. A low level of the decision variable means no change, and a high level indicates a change. The decision is made at the level of frames, in this case, frame #13, where the point of change is between 5,201 and 5,600. The point of the change detection can be considered at the middle of the detected interval, which means 5,401, but also the beginning of the frame, i.e. 5,201. In this work, the estimated point of the change is the beginning of the frame.

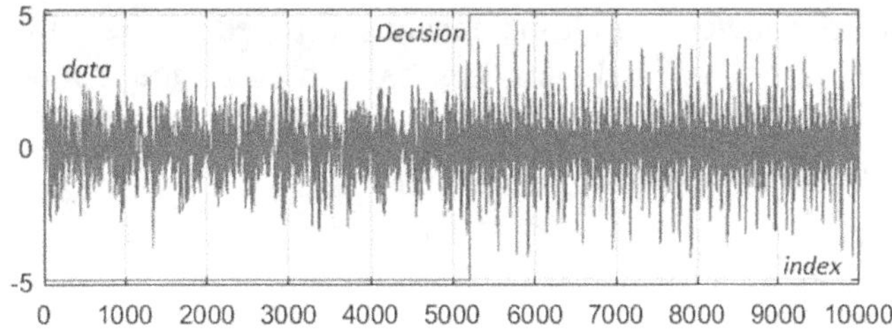

Fig. 8.12. Example of change detection, case #1, **D1** = [**d0 d1**].

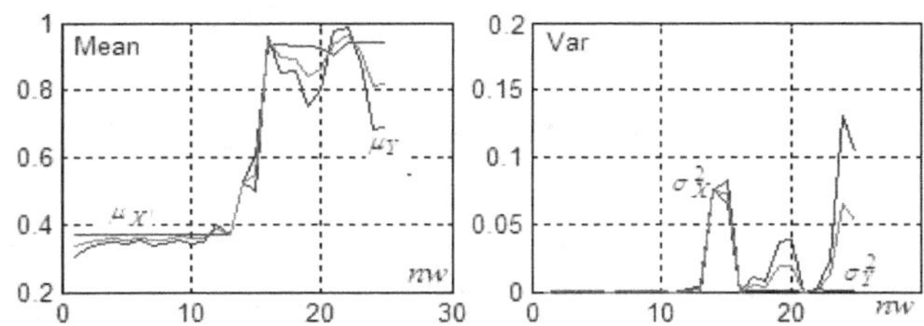

Fig. 8.13. The coordinates of the spectral components, case #2, **D2** = [**d0 d2**].

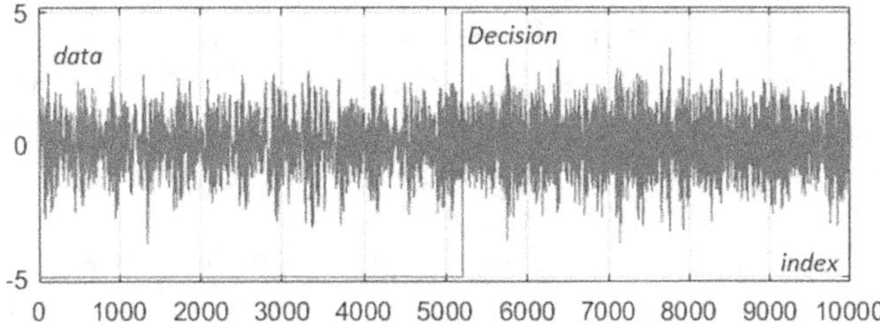

Fig. 8.14. Example of change detection, case #2, **D2** = [**d0 d2**].

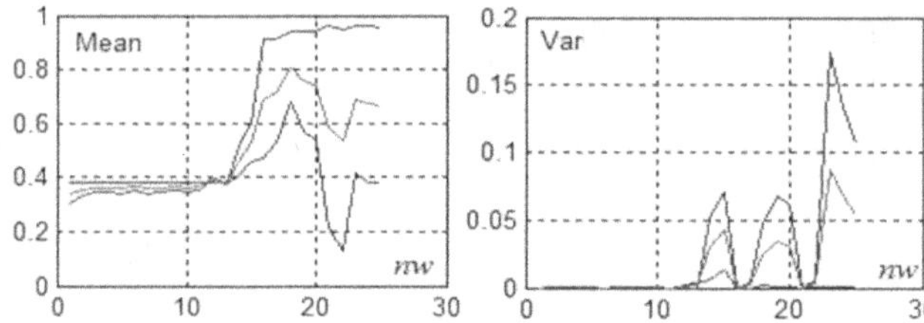

Fig. 8.15. The coordinates of the spectral components, case #3, **D3** = [**d0 d3**].

8.5.2. Results from Feature Design with Virtual Parameters

Fig. 8.16 presents a pair of images, the original (on blue background) and transformed (white background), for class #0. Some details from the transformed image (contour-based) of the class #3 are presented in Fig. 8.17, for two values of the number of contours ($nc = 1$, and 2). As the number of contours is rising, the shape is becoming more complex. In order to extract the right information as features from the transformed images, some elements or virtual parameters should be considered based on the information provided by contours. As in [48], the virtual parameters are based on the number and size of the contours, mixed with the values of some statistical moments and of other entropy-based variable.

8.5.3. Results from Feature Design with Transformed Parameters

The third method for feature design looks on Cosine Transform (CT), and it is inspired by image compression application field. The following steps are considered:

(a). <u>Visualization</u> of the transforming results, concerning the evolution of the DCT coefficients;

(b). <u>Selection</u> of the number of important features, e.g. $m = 10...15$, the non-zero coefficients;

(c). <u>Modelling and parameter</u> estimation, in order to represent the behavior/evolution of the selected features. Practically, by considering some general properties of DCT, the DC and CA coefficients could be represented separately;

(d). <u>Computation</u> of the change criterion, in order to detect the change among running states.

The analysis of the DCT coefficients reveals two sets of values. In one set, of size 100 elements, the values are greater of zero. The second set contains values close to zero. Also, the DCT coefficients are localized around the origin (0,0). These observations allow reducing the size of the DCT matrix, from size of 100×100 to 10×10.

The sample sets of these coefficients are presented in the next figures, in two representations, i.e. 2D and 1D modes. Fig. 8.18 presents the DCT results for class #0 (**d0**), and class #1 (**d9**); Fig. 8.19 presents the DCT results for class #2 (**d10**) and class #3 (**d11**).

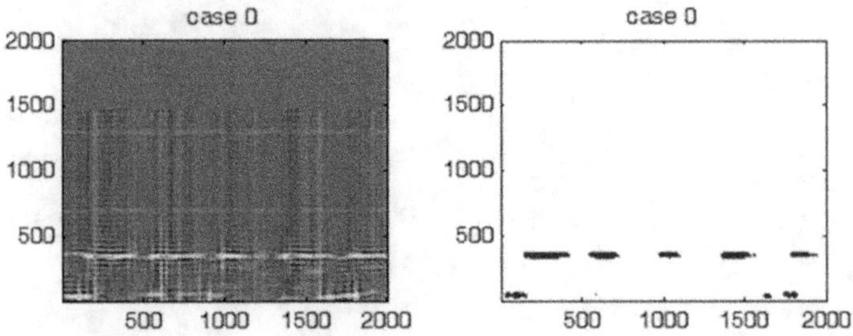

Fig. 8.16. Original and transformed image, class #0.

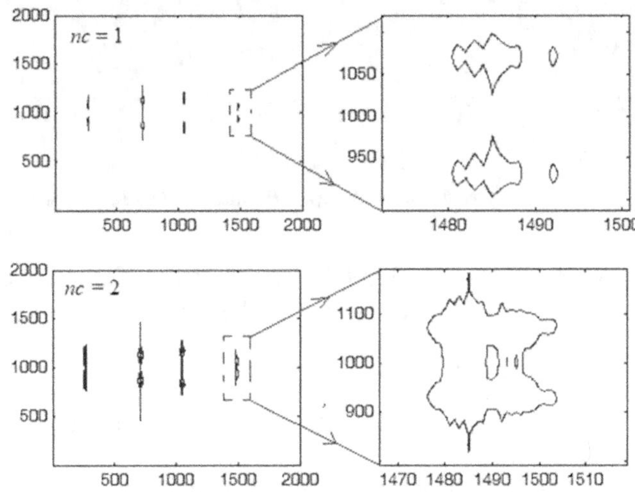

Fig. 8.17. Details of contours based images, class #3.

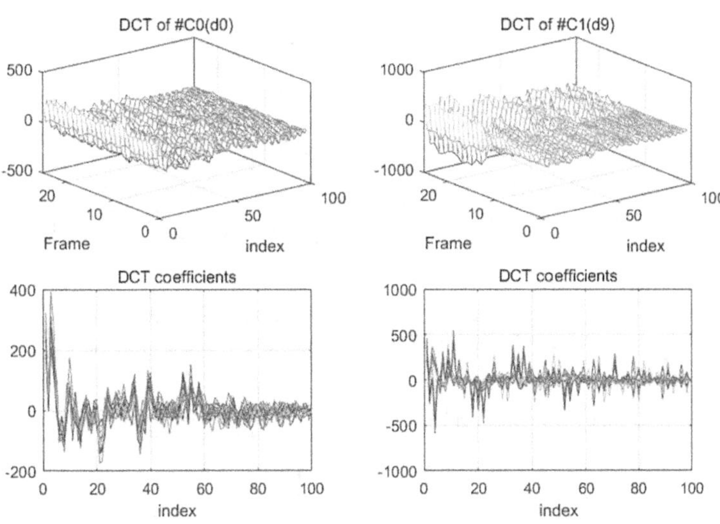

Fig. 8.18. DCT results for #C0(***d0***), and #C1(***d9***); $n = 400$.

361

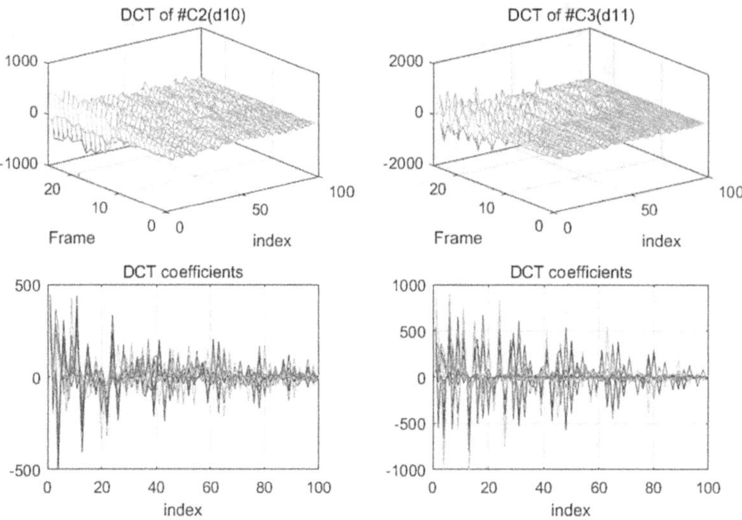

Fig. 8.19. DCT results for #C2(***d10***), and #C3(***d11***); $n = 400$.

If ***px*** represents the vector of the selected DCT coefficients, of size 1×100, the prototype vectors are computed for each class, as average of the vectors which belong to the class

$$P0 = E\{p0\}, P1 = E\{p1, p6, p9, p14\},$$

$$P2 = E\{p2, p7, p10, p15\}, P3 = E\{p3, p8, p11\} \tag{8.30}$$

From Fig. 8.20, where the prototype vectors are offered, a fault means the apparition of the negative coefficients with a higher amplitude than the positive one. Variation of the frame's length does not change significantly the structure of the DCT coefficients [50].

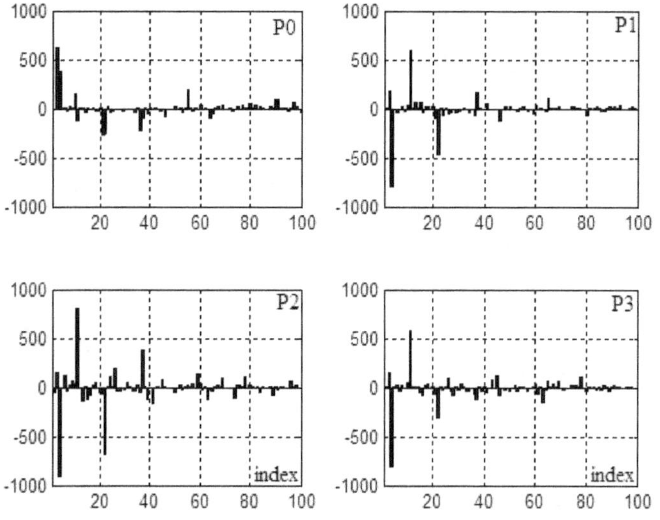

Fig. 8.20. The prototype vectors of classes associated with faults; $n = 400$.

A change detection criterion could be considered by selecting the first 5 to 10 coefficients from the DCT vectors. For the case of normal conditions, the main components are positive. When a change occurs, at least one coefficient is higher and negative, excepting the first one. More results are available in [50].

Better results are obtained with data frames of lengths 1,000 and 2,000 samples.

From the change point detection point of view, the estimation results are biased. For the case of incipient faults bearings, the results are considered more than acceptable.

8.6. Conclusions

The objective of the chapter was to present some solutions to the problem of feature selection for change detection purposes, with application to change detection of (incipient) faults in bearings. The features are extracted from time-frequency images or designed with auxiliary transforms. Static loads are considered, processes under stationary regimes, i.e. the constant speed of rotating machinery. The particularity of the low size faults (incipient) over big size faults (dangerous) is on the available signals, with low amplitude and small signal-to-noise ratios.

The rationale of time-frequency transforms is based on the fact that the associated images are rich in information related to the working regimes of the bearings. It is the task of the operator to select and process the right features to describe the states of the bearings, and thus to solve the change detection problem.

Data coming from vibration signals were considered. Raw data are explored in frames, with a length imposed by a trade-off between the accuracy of the change detection, which requires small lengths, and performance of the diagnosis, which needs longer lengths. Each frame is processed by a suitable time-frequency transform, in order to generate a time-frequency image. This is the input for the change detection method developed in this work.

Three methods of features design were considered.

The first one looks on the main components from the time-frequency image and use its physical parameters. The features design is made by selecting the main component(s), and next by estimating its coordinates on X and Y axes. These are the features used in the detection and diagnosis process. An example is presented for the case of one component, which have the maximum amplitude in the frequency axis. A change is detected if its features change in mean or variance.

The second method uses virtual parameters extracted from a transformed image based on contours. The number of contours depends on the image complexity and on the classification performance. The feature vector is defined by considering several parameters, which properly describes the contour-based images, as the number of the closed curves, the aria and their variance.

The third method uses a data compression paradigm, by selecting and processing the features obtained from the cosine transform. The features are selected by the size and positions of the coefficients in the matrix of DCT. Zigzag scanning is used, and the first 100 elements are extracted. The analysis of the features reveals that change detection is possible by evaluating the signs of the first five to ten coefficients. For normal conditions, all important values are positive. A change is reflected in the different signs for at least one coefficient from the considered set. Based on these features, classifiers based on minimum distance could be considered.

The proposed and implemented solutions have potential on other processes, where measured or estimated signals are available, e.g. in medicine.

An important aspect is to explore the full length of the available data and to observe the changes in time-frequency patterns and on the performance of the detector. The right feature design method depends on data and on the performance of the detector. An optimization procedure is then necessary in order to optimize the processing chain composed of data sources, features, and detector, by using an advanced processing system as described in [53].

Acknowledgements

The authors thank the Executive Agency for Higher Education, Research, Development and Innovation Funding (UEFISCDI) and Ministry of Research and Innovation, for the support under Contract PN-II-PT-PCCA-2013-4-0044 and 2019-2022 Core Program, Project PN 301, RO-SmartAgeing.

References

[1]. R. Isermann, Supervision, fault-detection and fault-diagnosis methods, *Control Engineering Practice*, Vol. 5, Issue 5, 1997, pp. 639-652.

[2]. J. Gertler, Fault Detection and Diagnosis in Engineering Systems, *Marcel Dekker*, 1998.

[3]. J. Chen, R. J. Patton, Robust Model-Based Fault Diagnosis for Dynamic Systems, *Kluwer Academic Publishers*, 1999.

[4]. C. Sobie, C. Freitas, M. Nicolai, Simulation-driven machine learning: Bearing fault classification, *Mechanical Systems and Signal Processing*, Vol. 99, 2018, pp. 403-419.

[5]. R. Liu, B. Yang, E. Zio, X. Chen, Artificial intelligence for fault diagnosis of rotating machinery: A review, *Mechanical Systems and Signal Processing*, Vol. 108, 2018, pp. 33-47.

[6]. A. Rai, S. H. Upadhyay, A review on signal processing techniques utilized in the fault diagnosis of rolling element bearings, *Tribology International*, Vol. 96, 2016, pp. 289-306.

[7]. F. Gustafsson, Adaptive Filtering and Change Detection, *John Wiley & Sons Ltd*, 2000.

[8]. M. Timusk, M. Lipsett, C. K. Mechefske, Fault detection using transient machine signals, *Mechanical Systems and Signal Processing*, Vol. 22, 2008, pp. 1724-1749.

[9]. R. J. Radke, S. Andra, O. Al-Kofahi, B. Roysam, Image change detection algorithms: A systematic survey, *IEEE Transactions on Image Processing*, Vol. 14, Issue 3, 2005, pp. 294-307.

[10]. C. Benedek, X. Descombes, J. Zerubia, Building Extraction and Change Detection in Multitemporal Aerial and Satellite Images in a Joint Stochastic Approach, Raport de Recherche, N° 7143, *INRIA*, 2009.

[11]. S. Minu, A. Shetty, A comparative study of image change detection algorithms in MATLAB, *Elsevier Aquatic Procedia*, Vol. 4, 2015, pp. 1366-1373.

[12]. S. Haigang, L. Deren, G. Jianya, Automatic change detection for road networks from images based on GIS, *Geo-spatial Information Science*, Vol. 6, Issue 4, 2003, pp. 44-50.

[13]. Y. Han, A. Chang, S. Choi, H. Park, J. Choi, An unsupervised algorithm for change detection in hyperspectral remote sensing data using synthetically fused images and derivative spectral profiles, *Journal of Sensors*, Vol. 17, 2017, 9702612.

[14]. B. Boashash, L. Boubchir, G. Azemi, A methodology for time-frequency image processing applied to the classification of nonstationary multichannel signals using instantaneous frequency descriptors with application to new born EEG signals, *EURASIP Journal on Advances in Signal Processing*, 2012, 117.

[15]. H. Yuan, J. Chen, J. G. Dong, Machinery fault diagnosis based on time-frequency images and label consistent K-SVD, *Proceedings of the Institution of Mechanical Engineers, Part C: Journal of Mechanical Engineering Science*, Vol. 232, Issue 7, 2018, pp. 1317-1330.

[16]. D. Lu, P. Mausel, E. Brondízio, E. Moran, Change detection techniques, *International Journal of Remote Sensing*, Vol. 25, Issue 12, 2010, pp. 2365-2401.

[17]. A. Singh, Digital change detection techniques using remotely sensed data, *International Journal of Remote Sensing*, Vol. 10, Issue 6, 1989, pp. 989-1003.

[18]. S. Xiaolu, C. Bo, Change detection using change vector analysis from Landsat TM images in Wuhan, *Procedia Environmental Sciences*, Vol. 11, 2011, pp. 238-244.

[19]. K. Ling, Y. Lin, Z. Zeng, L. Zhang, L. Meng, Adaptive change detection with significance test, *IEEE Access*, Vol. 6, 2018, pp. 27442-27450.

[20]. H. G. Ashok, D. R. Patil, Survey on change detection in SAR images, *International Journal of Computer Applications*, Vol. 2, 2014, pp. 4-7.

[21]. K. Umapathy, B. Ghoraani, S. Krishnan, Audio signal processing using time-frequency approaches: Coding, classification, fingerprinting, and watermarking, *EURASIP Journal on Advances in Signal Processing*, Vol. 28, Issue 1, 2010, 451695.

[22]. B. Boashash, G. Azemi, N. A. Khan, Principles of time-frequency feature extraction for change detection in nonstationary signals: Applications to newborn EEG abnormality detection, *Pattern Recognition*, Vol. 48, Issue 3, March 2015, pp. 616-627.

[23]. P. D. McFadden, W. Wang, Time-Frequency Domain Analysis of Vibration Signals for Machinery Diagnostics. I Introduction to the Wigner-Ville Distribution, Report OUEL 1859/92, *University of Oxford*, 1990.

[24]. L. Cohen, Time-frequency distributions – A review, *Proceedings of the IEEE*, Vol. 77, Issue 7, 1989, pp. 941-980.

[25]. F. Auger, P. Flandrin, P. Gonçalvès, O. Lemoine, Time-frequency Toolbox, *CNRS France – Rice University*, 1996.

[26]. F. Hlawatsch, F. Boudreaux-Bartels, Linear and quadratic time-frequency signal representations, *IEEE Signal Processing Magazine*, Vol. 9, 1992, pp. 21-67.

[27]. M. Cerrada, R. V. Sánchez, C. Li, F. Pacheco, D. Cabrera, J. V. de Oliveira, R. E. Vásquez, A review on data-driven fault severity assessment in rolling bearings, *Mechanical Systems and Signal Processing*, Vol. 99, 2018, pp. 169-196.

[28]. R. B. Randall, J Antoni, Rolling element bearing diagnostics – A tutorial, *Mechanical Systems and Signal Processing*, Vol. 25, 2011, pp. 485-520.

[29]. I. El-Thalji, E. Jantunen, A summary of fault modelling and predictive health monitoring of rolling element bearings, *Mechanical Systems and Signal Processing*, Vols. 60-61, 2015, pp. 252-272.

[30]. D. Aiordăchioaie, Th. D. Popescu, Change detection by feature extraction and processing from time-frequency images, in *Proceedings of the IEEE 10th International Conference on Electronics, Computers and Artificial Intelligence (ECAI'18)*, Iasi, Romania, 2018, pp. 1-7.

[31]. D. Aiordăchioaie, N. Nistor, M. Andrei, Change detection in time-frequency images by feature processing in compressed spaces, in *Proceedings of the IEEE 24th International Symposium for Design and Technology in Electronic Packaging (SIITME'18)*, Iaşi, Romania, 2018, pp. 181-186.

[32]. C. Lu, Y. Wang, M. Ragulskis, Y. Cheng, Fault Diagnosis for Rotating Machinery: A Method based on Image Processing, *PLoS ONE*, Vol. 11, Issue 10, 2016, e0164111.

[33]. H. Yuan, J. Chen, J., G. Dong, Machinery fault diagnosis based on time-frequency images and label consistent K-SVD, *Proceedings of the Institution of Mechanical Engineers, Part C: Journal of Mechanical Engineering Science*, Vol. 232, Issue 7, 2018, pp. 1317-1330.

[34]. K. Umapathy, B. Ghoraani, S. Krishnan, Audio signal processing using time-frequency approaches: coding, classification, fingerprinting, and watermarking, *EURASIP Journal on Advances in Signal Processing*, Vol. 28, Issue 1, 2010, 451695.

[35]. B. Boashash, G. Azemi, N. A. Khan, Principles of time-frequency feature extraction for change detection in nonstationary signals: Applications to newborn EEG abnormality detection, *Pattern Recognition*, Vol. 48, Issue 3, March 2015, pp. 616-627.

[36]. K. S. Shanmugan, A. M. Breipohl, Random Signals: Detection, Estimation and Data Analysis, *John Wiley & Sons,* NY, 1988.

[37]. S. M. Kay, Fundamentals of Statistical Signal Processing, Vol. II, Detection Theory, *Prentice-Hall,* 1998.

[38]. M. Barkat, Signal Detection and Estimation, 2nd Edition, *Artec House*, 2005.

[39]. D. T. Barry, Fast calculation of the Choi-Williams time-frequency distribution, *IEEE Transactions on Signal Processing*, Vol. 40, Issue 2, 1992, pp. 450-455.

[40]. W. Kou, Principles of digital image compression, in Digital Image Compression, *Springer*, Boston, MA, 1995.

[41]. T. Sikora, MPEG digital video-coding standards, *IEEE SP Magazine*, Vol. 14, Issue 5, 1997, pp. 82-100.

[42]. A. B. Watson, Image compression using the discrete cosine transform, *Math. Journal*, Vol. 4, Issue 1, 1994.

[43]. Y. Yamashita, Image compression by weighted Karhunen-Loeve transform, in *Proceedings of the 13th International Conference on Pattern Recognition (ICPR'96)*, Vol. 2, Issue 2, 1996, pp. 636-640.

[44]. K. R. Rao, M. A. Narasimhan, K. Revuluri, Image data processing by Hadamard-Haar transform, *IEEE Transactions on Computers*, Vol. 24, Issue 9, 1975, pp. 888-896.

[45]. K. R. Rao, P. Yip, Discrete Cosine Transform. Algorithms, Advantages and Applications, *Academic Press*, 1990.

[46]. L. Xin, P. H. Yu, K. Lam, An application of CUSUM chart on financial trading, in *Proceedings of the Ninth International Conference on Computational Intelligence and Security (CIS'13)*, 2013, pp.178-181.

[47]. D. Tam, A theoretical analysis of cumulative sum slope statistic for detecting signal onset and offset trends from background noise level, *The Open Statistics and Probability Journal*, Vol. 1, 2009, pp. 43-51.

[48]. D. Aiordachioaie, Th. D. Popescu, S. M. Pavel, A feature selection and extraction method from time-frequency images, *International Journal on Advances in Intelligent Systems*, Vol. 13, Issues 1-2, 2020, pp. 119-128.

[49]. D. Aiordachioaie, S. M. Pavel, Qualitative analysis of the time-frequency images of vibrations in faulty bearings, in *Proceedings of the IEEE 11th Int. Conference on Electronics, Computers and Artificial Intelligence (ECAI'19)*, Romania, 2019, pp. 1-6.

[50]. D. Aiordachioaie, Th. D. Popescu, Aspects of features selection and extraction from time-frequency images of vibration signals, in *Proceedings of the 6th International Symposium on Electrical And Electronics Engineering (ISEEE'19)*, Galati, Romania, 2019, pp. 1-5.

[51]. Case Western Reserve University Bearing Data Center, http://csegroups.case.edu/bearingdatacenter/home

[52]. W. A. Smith, R. B. Randall, Rolling element bearing diagnostics using the Case Western Reserve University data: A benchmark study, *Mechanical Systems and Signal Processing*, Vols. 64-65, 2015, pp. 100-131.

[53]. D. Aiordachioaie, Th. D. Popescu, VIBROCHANGE—a development system for condition monitoring based on advanced techniques of signal processing, *International Journal of Advanced Manufacturing Technology*, Vol. 105, 2019, pp. 919-936.

Chapter 9
Information Entropy for Image Segmentation and Registration

Lei Wang, Qian Chang and Hao Chen

9.1. Information Entropy

In image processing, information entropy [1, 2] is an important measure to assess the statistical characteristics among pixel intensities in global or local regions. It is constructed based on different probability distributions to detect various structural textures depicted on images [3]. These probability distributions are generally derived from the statistical relationship among pixel intensities (*e.g.*, frequency of occurrence or intensity histogram). This makes information entropy independent of a certain imaging modality and suitable for different image segmentation and registration applications. In addition, the entropy is defined leveraging different information estimation schemes, and able to characterize the potential correlation between two images or two regions within a given image. These definitions determine, to some extent, the inherent properties of image entropy and the performances of segmentation and registration methods. The definitions of the entropy and the probability distributions are given in detail in the following.

9.1.1. Definition of Entropy

Entropy is a measure of the uncertainty of a random variable in information theory. It generally refers to the Shannon entropy [4], which is widely used in image processing to assess the amount of information contained in a random variable. Let X be a discrete random variable with possible values $V = \{x_1, x_2, \cdots, x_n\}$, and its probability mass function $p(x) = P\{X = x\}, x \in V$, then the information size of the value x is defined as:

$$h(x) = \log \frac{1}{p(x)} = -\log p(x), \qquad (9.1)$$

Lei Wang
School of Ophthalmology and Optometry, Eye Hospital, Wenzhou Medical University, Wenzhou, China

where $h(x)$ is used to estimate the information containing in the variable value x. $\log(\cdot)$ denotes the logarithm function, which is usually to the base a constant such as 2, natural constant, or 10. Unless stated otherwise, we use the logarithm to base 2, and then the entropy will be measured in bits. Based on the above equation, the whole entropy of the random variable X can be given by

$$H(X) = \sum_{x \in V} p(x) h(x) = -\sum_{x \in V} p(x) \log p(x), \tag{9.2}$$

where the probability $p(x)$ generally satisfies the formulations of (1) $\sum_{x \in V} p(x) = 1$ and (2) $0 \le p(x) \le 1$. The entropy assesses the average uncertainty in the random variable X and is a functional of the probabilities of different variable values. It has several important properties as follows: (1) it is non-negative, *i.e.*, $H(X) \ge 0$, due to the constraint of $0 \le p(x) \le 1$; (2) it takes values in a relatively small range, as illustrated in Fig. 9.1; (3) it gets the maximum value when the variable has the same probability for each possible value, *i.e.*, $p(x_1) = p(x_2) =, \cdots, = p(x_n)$.

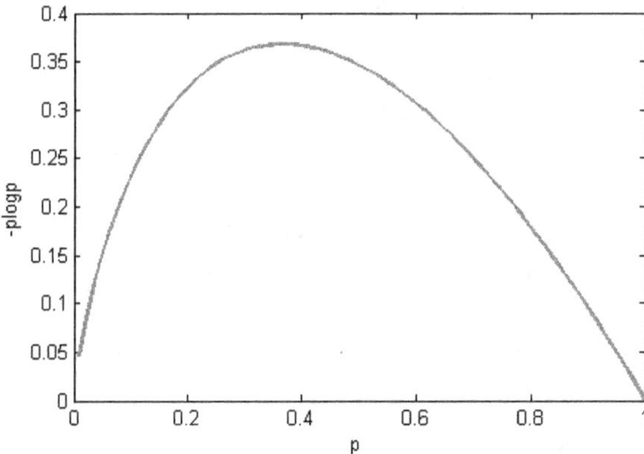

Fig. 9.1. Illustration of the formula of $-p \log p$.

9.1.2. Relative Entropy and Mutual Information

Based on the above definition, we extend the entropy to a pair of random variables (X, Y). Suppose that the variables X and Y have the probability mass functions of $p(x)$ and $p(y)$, respectively, and their joint probability is $p(x, y)$, the statistical relationship between two variables can be measured by the relative entropy [2] and mutual information [1]. They are defined as:

$$R(X,Y) = \sum p(x)\log\left(\frac{p(x)}{p(y)}\right), \tag{9.3}$$

$$I(X,Y) = \sum_{x \in X}\sum_{y \in Y} p(x,y)\log\left(\frac{p(x,y)}{p(x)p(y)}\right), \tag{9.4}$$

The relative entropy (also termed cross entropy or Kullback-Leibler measure) is a measure of the distance between two probability distributions, and generally employed in image segmentation to estimate the differences between the results obtained by a given method and the ground truths (or manual annotations). This entropy is asymmetric and non-negative, and takes the maximum value when $p(x) = p(y)$. Mutual information attempts to assess the amount of information that one random variable contains about another random variable, or the dependence between the two variables. This dependence makes the mutual information very suitable for aligning images with varying modalities, and have the following properties:

(1) $I(X,Y) = I(Y,X)$;

(2) $I(X,X) = H(X)$;

(3) $I(X,Y) \leq H(X)$ and $I(X,Y) \leq H(Y)$;

(4) $I(X,Y) \geq 0$;

(5) $I(X,Y) = 0$ if and only if variables X and Y are independent each other.

9.1.3. Different Entropy

The information entropy defined by C. Shannon [5] was very popular and widely used in a variety of image processing tasks. The entropy, however, has some limitations and may be incompetent to accurately assess the statistical characteristics of pixel intensities in certain cases where an image has very weak contrast or severe motion artefacts. To overcome the limitations, different types of information entropy have been proposed in past few decades leveraging various image estimation strategies. Several commonly used information entropy will be presented in details in the following:

(a) Renyi entropy

Similar to the Shannon entropy, Renyi entropy [6] is also widely used in image segmentation and registration, and defined as:

$$R_a(X) = \frac{1}{1-a}\log\left(\sum_{x \in X} p(x)^a\right), \tag{9.5}$$

where the parameter $0 < a \neq 1$ is used to adapt the sensitivity of the probability $p(x)$. This entropy is additive, *i.e.*, $R_a(X,Y) = R_a(X) + R_a(Y)$, and has different characteristics by assigning different values for the parameter a.

(b) Kapur entropy

Kapur entropy [7] utilizes two parameters to improve the sensitivity of the probability distribution and is defined as:

$$K_{a,b}(X) = \frac{1}{b-a} \log \left(\frac{\sum_{x \in X} p(x)^a}{\sum_{x \in X} p(x)^b} \right), \tag{9.6}$$

where the parameters a and b are used to adapt the sensitivity of $p(x)$, similar to the counterpart in the Renyi entropy. These two parameters satisfy the constraints of $a \neq b$, $a > 0$ and $b > 0$.

(c) Tsallis entropy

Unlike the Renyi and Kapur entropy, Tsallis entropy [8] is not based on the logarithmic function and defined as:

$$T_a(X) = \frac{k}{a-1} \left(1 - \sum_{x \in X} p(x)^a \right), \tag{9.7}$$

where the parameter k is a constant and usually set to 1 in image processing. The entropy has the property of $T_a(X,Y) = T_a(X) + T_a(Y) - (1-a)T_a(X)T_a(Y)$ and the relationship between the Tsallis and Renyi entropy is given by

$$R_a(X) = \frac{1}{1-a} \log \left(1 + (1-a)T_a(X) \right) \tag{9.8}$$

(d) Havrda-Charvat entropy

Similar to the Tsallis entropy, Havrda-Charvat entropy [9, 10] is also not based on the logarithmic function. It uses a single parameter a to adapt the sensitivity of the probability distribution $p(x)$ and can be given by:

$$HC_a(X) = \frac{1}{1-a} \left(\sum_{x \in X} p(x)^a - 1 \right) \tag{9.9}$$

(e) Arimoto entropy

The entropy [11] can be given by

$$A_a(X) = \frac{a}{a-1}\left(1 - \left(\sum_{x \in X} p(x)^a\right)^{\frac{1}{a}}\right) \qquad (9.10)$$

This entropy is symmetric and not less than zero.

(f) Exponential entropy

Exponential entropy [12, 13] uses the exponential function to measure the amount of information and avoid undefined values in the case of $p(x) = 0$ for the entropy based on the logarithmic function. The entropy can be given by

$$E(X) = \sum_{x \in X} p(x)e^{(1-p(x))} \qquad (9.11)$$

The exponential definition can largely speed up the calculation of information entropy, as compared with the logarithmic definitions This definition is able to reduce the computational cost of certain image processing methods in some degrees when they are derived from the exponential entropy. Fig. 9.2 illustrates the differences of information size between the exponential entropy and the Shannon entropy.

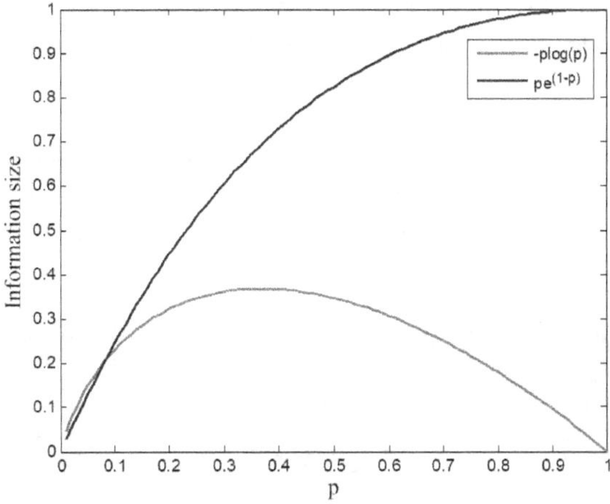

Fig. 9.2. The differences of information size between the exponential entropy and the Shannon entropy.

9.1.4. Different Probability Distributions

In information entropy, the probability distribution $p(x)$ plays an important role and determines the properties of the entropy to some extent. Therefore, it is necessary and advisable to construct a proper probability distribution function to quantify the statistical relationship among different values of a random variable. There are various probability distributions in image processing, which are often used to calculate the information entropy based on the Shannon definition and given as follows.

(a) Probability based on gray level histogram

The probability [14] is widely used in image segmentation and defined based on the gray level histogram (frequencies), as shown in Fig. 9.3. Let $f_1, f_2, \cdots, f_k$ represent the gray level histogram of a given two-dimensional grayscale image with a size of $N \times N$, the probability p_i can be expressed as

$$p_i = \frac{f_i}{N^2}, \tag{9.12}$$

$$\sum_{i=1}^{k} f_i = N^2, \tag{9.13}$$

$$f_i \leftarrow f_i + 1, \tag{9.14}$$

where the parameter k is the number of gray levels in the histogram. This probability can be computed based not only on global image regions, but also on the local neighborhood of a given pixel, leading to two different probability versions, (*i.e.*, the global and local probabilities). With the local probability, the Shannon entropy is large as the number of the gray levels in the local region (*i.e.*, k) increases. The larger entropy means the larger intensity changes in local regions, suggesting that some edge information may be contained in the regions.

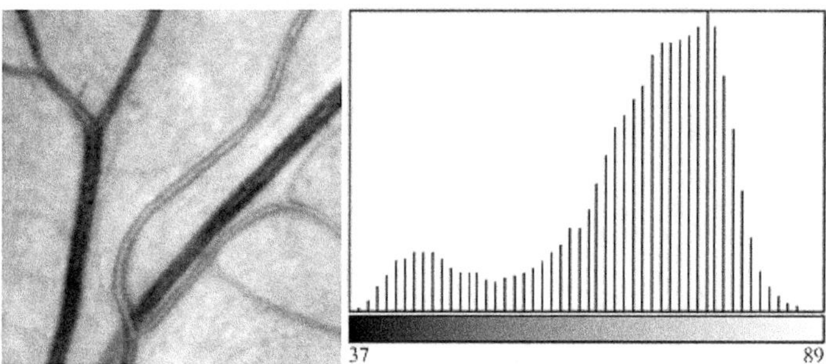

Fig. 9.3. An example of a given image (left) and its intensity histogram (right).

(b) Probability based on spatially-weighted gray level histogram

The above probability based on gray level histogram is simple and useful, but it ignores a wide variety of subtle image textures among adjacent pixels in local regions (*e.g.*, space information). To preserve the texture information, a variant of the gray level histogram based probability [15] is developed and defined as:

$$p_i = \frac{f_i}{\sum_{j=1}^{k} f_j}, \tag{9.15}$$

$$f_i \leftarrow f_i + \omega_i, \tag{9.16}$$

where ω_i is a Gaussian weighting function about the distance between pixel i and the center of a given local region. The spatially-weighted probability takes the intensities and space coordinates of pixels into account, making the Shannon entropy able to capture certain inherent texture information, as shown in Fig. 9.4. The texture information is very helpful for image segmentation and registration.

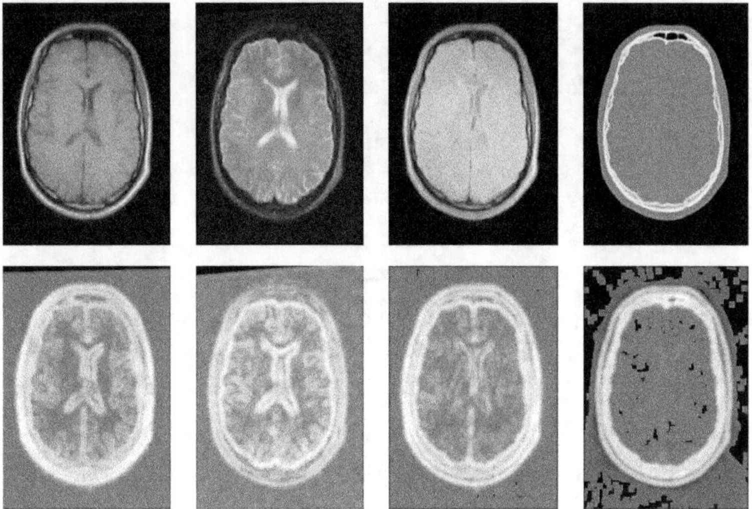

Fig. 9.4. Illustration of four medical images (the first row) and their local Shannon entropy maps (the second row) based on the probability derived from spatially-weighted gray level histogram [15].

(c) Probability based on local gray summation

The probability [16] is defined as

$$p_i = \frac{I_i}{\sum_{j=1}^{n} I_j}, \tag{9.17}$$

where I_i denotes the intensity of the pixel i in a given local region, which has a total of n pixels. With the probability, the Shannon entropy is small when the intensity differences among pixels in a local region is large. In addition, it is independent of pixel positions and thus rotation invariant. To highlight intensity differences, four neighboring pixels or 3×3 neighborhood may be suitable for the entropy. Fig. 9.5 illustrates examples of the feature map obtained by the Shannon entropy based on the local gray summation probability.

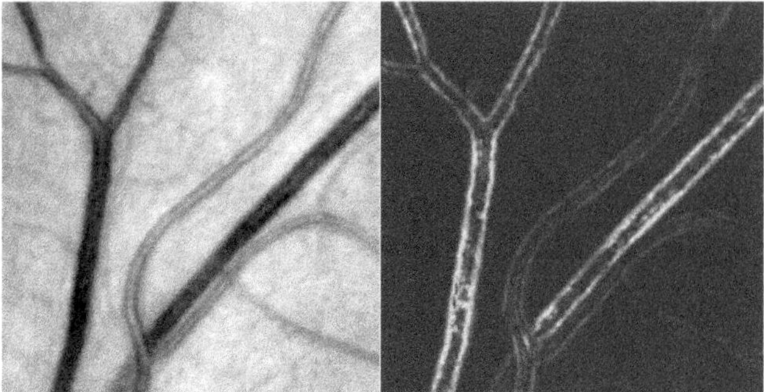

Fig. 9.5. An example of a given grayscale image (left) and its entropy map (right) obtained by the Shannon entropy based on the local gray summation probability.

(d) Probability based on local gray contrast

The local gray contrast based probability [17] is defined as

$$p_i = \frac{c_i}{\sum_{j=1}^{n} c_j}, \tag{9.18}$$

$$c_i = \frac{|I_i - m|}{|I_i + m|}, \tag{9.19}$$

where m is either the average or median of gray levels in a local region, c_i is the intensity contrast of the pixel i, relative to m. Based on the formulation, p_i satisfies the formula of $\sum_i p_i = 1$ and $0 \le p_i \le 1$. The probability causes that the Shannon entropy tends to highlight some feature points, as displayed in Fig. 9.6.

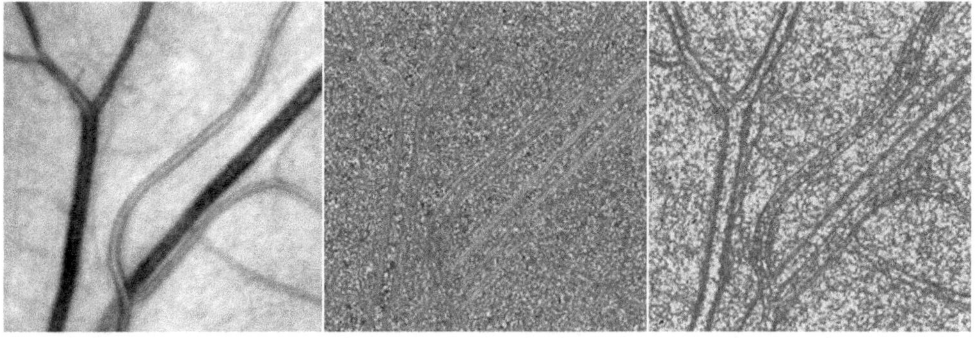

Fig. 9.6. An example of a given image and its Shannon entropy maps based on the local gray average and median contrast probabilities, respectively.

(e) Probability based on local gray residual

There are two different definitions for the local gray residual probability [18, 19], which are given, respectively, by:

$$p_i = \frac{\mathrm{mod}\left(I_i, m\right)}{m},\qquad (9.20)$$

$$p_i = \frac{\mathrm{mod}\left(\left|I_i^2 - m^2\right|, m\right)}{m},\qquad (9.21)$$

where $\mathrm{mod}(\cdot)$ is the modulus operator, which plays a critical role in identifying small intensity variations in local regions. These two probabilities do not satisfy the formula of $\sum_i p_i = 1$ and thus are not true probabilities. They can be regarded as normalized intensity factors in image processing. The Shannon entropy based on these two probabilities is termed as local edge entropy and local inhomogeneity entropy, respectively, both of which are large for local edge regions, as displayed in Fig. 9.7.

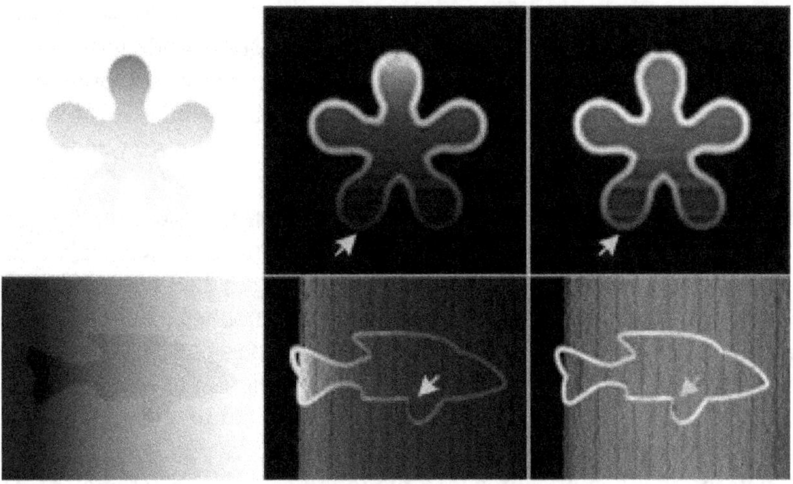

Fig. 9.7. The differences between local edge entropy (the middle column) and local inhomogeneity entropy (the right column) on two given images (the left column) [19].

9.1.5. Summary

Information entropy, especially the Shannon entropy, is widely used in a large number of image processing applications, and its properties are mainly determined by the definitions of the entropy and the probability distribution, as mentioned above. Between the two types of definitions, the probability distribution has been studied extensively in image segmentation and registration because it can detect subtle information differences (*e.g.*, intensities or coordinates) among different pixels in global or local regions. Therefore, the heart of the entropy based image processing is to construct a proper probability distribution to highlight the intensity and position differences within or across images.

9.2. Entropy Based Image Segmentation

Image segmentation [20, 21] is a fundamental process to partition an image into several disjoint regions, each of which corresponds to an individual object or a part of an object. The process can capture the morphological changes of some tissues or lesions depicted on images, and thus plays an important role in aidding the detection and diagnosis of various diseases. To accurately extract regions of interest in images, a large number of segmentation methods have been developed by combining pixel intensities and various image features for different applications. In this section, we focus on a type of segmentation methods, which mainly employ information entropy to discriminate regions of interest from irrelevant image background. These entropy based segmentation methods will be simply summarized as follows.

9.2.1. Segmentation Based on Entropy Features

Information entropy has very different characteristics when it is computed based on different probability distributions (*e.g.*, local gray contrast or residual probability). These probability distributions are constructed using various intensity estimation strategies in a local region. When the intensity differences among adjacent pixels in local regions are highlighted, these probability distributions could make the entropy able to effectively detect / enhance certain structural features such as linear structures, image edges, and object boundaries. This is primarily attributed to the relatively large intensity changes in the neighborhood of these structures. Therefore, this information entropy is widely used as a kind of important features to highlight various image edges associated with regions of interest and provide a valuable complement to pixel intensities in image segmentation. For example, Zhang et al. [22] developed a region based active contour model by introducing a local entropy factor and a linear speed function to reduce the influence of false edge problems. Mylona et al. [23] introduced a unique segmentation framework for the unsupervised parameterization of a region based active contour [24] and utilized the orientation entropy to automatically adjust the involved contour parameters. Zhang et al. [25] constructed an improved geometric active contour model, which used a signed pressure force function based on local Shannon entropy to attract initial contour curves towards object boundaries. He et al. [26] used local image entropy as a weighted function to improve the region-scalable fitting contour model for segmentation purposes. Wang et al. [18] developed an active contour model by defining an edge entropy descriptor to reduce the impact of intensity inhomogeneity, and then further introduced a local inhomogeneity entropy descriptor [19] to overcome the disadvantages of the edge entropy in segmentation. Mirkamali et al. [27] proposed a depth-wise segmentation method to extract target objects from three-dimensional images, where image entropy was used to characterize object boundaries. Sun et al. [28] combined color and entropy information of an image to construct a content adaptive measure, which was used to exclude undesirable feature points. These segmentation methods employed different image estimation schemes to discriminate between regions of interest and their surrounding background, and all of them took advantage of local image entropy to detect a variety of features associated with target objects. Hence, it is of great importance to construct a proper feature factor or weighting function based on the information entropy in image segmentation. The factor or function can effectively highlight the image differences between the object regions and irrelevant background, and reduce the impact of weak contrast and/or severe intensity inhomogeneity. On the other hand, the design of the feature entropy and local probability distribution is also crucial and able to aid the exclusion of certain background surrounding target objects.

9.2.2. Segmentation Based on Maximum Entropy

Unlike the segmentation methods mentioned above, the kind of methods attempt to optimize a cost function constructed by image entropy [29, 30]. The cost function is mainly used to assess the information differences between two images or regions within an image. This leads to two types of segmentation methods: the threshold based methods and cross entropy based methods.

The threshold based methods use a given threshold to discriminate the foreground from the background in an image. The threshold is generally obtained by maximizing the sum of the foreground and background entropy, and given by:

$$S = H_F + H_B = -\sum_{i=1}^{t} p_i^f \log\left(p_i^f\right) - \sum_{i=(t+1)}^{k} p_i^b \log\left(p_i^b\right), \tag{9.22}$$

$$p_i^f = \frac{p_i}{\sum_{j=1}^{t} p_j}, \tag{9.23}$$

$$p_i^b = \frac{p_i}{\sum_{j=(t+1)}^{k} p_j}, \tag{9.24}$$

where the parameters t and k denote a given threshold and the number of intensity histogram, H_F and H_B are the foreground and background entropy, respectively, and their sum is S. Maximizing the entropy S with respect to t can find an optimal threshold to make the foreground and background separate. This is mainly due to the fact that the larger the entropy is, the smoother the intensity distribution within a local region is, and thus the closer pixel intensities are to a constant in this region. Based on the observation, a large number of segmentation methods have been developed in the past few decades. For example, Ajlan et al. [31] introduced a segmentation method by maximizing image entropy based on Gamma distribution. Yin et al. [32] proposed an unsupervised segmentation method to detect blood vessels from retinal fundus images using curvature evaluation and entropy filtering techniques. Abutaleb et al. [33] thresholded the gray level values of an image based on information entropy to identify the different homogeneous components of the image. Qi et al. [34] proposed a multi-level threshold method by maximizing an exponential entropy for segmentation purposes. Oswal et al. [35] introduced an entropy based thresholding approach for automatically extracting cell nuclei from histological images. Khattak et al. [36] performed a comparison of different threshold segmentation methods based on three types of information entropy, including the Shannon, Renyi, and Tsallis entropy, respectively. Lang et al. [37] combined a hybrid whale optimization algorithm and Kapur entropy to perform the segmentation of color images. This kind of segmentation methods is relatively simple and has limited performance for images with varying noise and contrast.

The cross entropy based methods attempt to segment an image by minimizing the cross entropy between the original image and its approximately segmented version, or the segmentation result and its ground truth (or manual annotation). The cross entropy can be given by:

$$R(P,Q) = \sum_{x \in X} p(x) \log \left(\frac{p(x)}{q(x)} \right), \qquad (9.25)$$

where P and Q are the probability distributions for two different images, and they take the values of $p(x)$ and $q(x)$, respectively, for the pixel x. X denotes a kind of pixel characteristics (*e.g.*, pixel intensity or image gradient). The entropy can obtain an optimal value when $p(x)$ is equal to $q(x)$. According to the characteristics, many segmentation methods have been proposed. For example, Song et al. [38] proposed an active contour method by minimizing the cross entropy of the original image and its approximately segmented version, which assumed that the image to be segmented is piece-wise smooth and the intensities of each segmented regions are approximately close to a constant. Wang et al. [39] presented a boundary and entropy driven adversarial learning framework to accurately segment the optic disc and cup from retinal fundus images, where information entropy was calculated based on the boundary prediction map to suppress high uncertainty segmentation on ambiguous boundary regions. Zhao et al. [40] developed a region mutual information loss to model the dependencies among adjacent pixels, which was maximized for semantic segmentation. Zhang et al. [41] presented an entropy based evaluation measure to estimate the uniformity of pixel intensities within a local region, and used the measure to compare the performances of different segmentation methods. Ronneberger et al. [42] used the binary cross entropy loss function with weighting scheme to train the classical U-Net network for segmenting cell border from biomedical images. These methods can effectively assess the information differences between two images and achieve promising segmentation performance, as compared with the threshold based methods.

9.3. Entropy Based Image Registration

The objective of entropy based registration is to align two images under consideration leveraging a predefined similarity measure derived from information entropy. There are a wide variety of entropy based measures [43, 44], which are widely used in image registration and able to determine the final registration accuracy. They are either based on mutual information measures or on feature entropy descriptors, leading to two different types of registration methods, *i.e.*, mutual information based registration and feature entropy based registration.

9.3.1. Mutual Information Based Registration

This type of registration methods typically use the mutual information (MI) as a similarity measure to quantify the amount of the information shared between two images. The MI is often formed by the Shannon entropy and the gray level histogram based probability distribution. This makes it inherently independent of imaging modalities and very popular in various registration tasks to establish the space correspondence between two images. Despite a relatively high popularity, the MI may suffer from limited registration accuracy because it merely takes pixel intensities into account and ignores a large number of other valuable image information (*e.g.*, spatial position or intensity gradient). Additionally, it uses the same weights for both pixel intensities and their probabilities in the calculation of the Shannon entropy, and hence cannot highlight some key object features within regions of interest, as illustrated in Eq. (9.4). To improve the registration accuracy, many methods have been proposed by constructing different variants of the MI. Studholme et al. [45] proposed a normalized mutual information (NMI) metric to tackle the problem caused by varying overlap regions. Tomazevic et al. [46] defined a multi-feature mutual information (MMI) measure, which combined pixel intensities, spatial information, and intensity gradients for deformable image registration. Rivaz et al. [47] presented a contextual conditioned mutual information (CoCoMI) measure to align ultrasound (US) and magnetic resonance (MR) images in the free form deformation (FFD) framework. Liu et al. [48] developed a deformable registration method by incorporating the predefined structural features into the mutual information measure. Gong et al. [49] introduced a nonrigid registration method by defining a correlation ratio based mutual information to align MR images and transrectal ultrasound images. These methods generally achieve reasonable accuracy for both mono- and multi-modal image registration by incorporating different texture feature (especially local features) into mutual information. The feature incorporation is able to provide some local textures for each pixel and ensure that global and local texture information can be considered simultaneously in mutual information. Hence, the design of different local features and their combination strategies for the mutual information plays a crucial role in this kind of registration methods and still remains an active area of research.

9.3.2. Feature Entropy Based Registration

Although mutual information based methods are widely used for various registration applications and have promising performances in terms of accuracy and robustness, they are relatively time-consuming and tend to ignore a large number of texture features associated with regions of interest (*e.g.*, pathological tissues or anatomic structures). The drawback may limit their application in clinical practice. To effectively detect these local textures and reduce registration time, feature entropy based methods have been developed, which directly align feature maps obtained by a predefined entropy descriptor using some existing similarity measures such as the sum of squared differences (SSD) and cross correlation (CC). For example, Wachinger et al. [15] proposed a spatially-weighted patch entropy to obtain the structural representations of multi-modal images and employed the SSD measure to quantify the information differences between these structural representations for registration purposes. Cun et al. [50] developed a stochastic second order entropy descriptor to overcome the shortcomings of structural representation in multi-modal medical registration. Zhang et al. [51] introduced a two-stage nonrigid registration method on the basis of joint structural information and local entropy to align multi-modal medical images. These registration methods may have limited performance since they merely employ a small number of entropy features and a simple similarity measure to align images, and ignore other image information. Additionally, the entropy features are potentially incompetent to effectively preserve small texture characteristics among adjacent pixels in a local region.

9.4. Conclusions

This chapter summarizes different information entropy and the probability distributions, which are widely applied in image segmentation and registration applications. The entropy has different properties based on different definitions of the information size and the probability distribution. This makes it able to detect various image features and assess the information differences between global or local image regions. The detection of these features and its integration with other information (*e.g.*, pixel intensities) is the focus of image segmentation and registration research, and largely determine the final performances of entropy based segmentation and registration methods.

Acknowledgements

This work is supported by Natural Science Foundation of Jiangsu Province (Grant No. BK20170391) and Wenzhou Science and Technology Bureau (Grant No. Y20190180).

References

[1]. J. Pluim, J. Maintz, M. Viergever, Mutual-information-based registration of medical images: A survey, *IEEE Transactions on Medical Imaging*, Vol. 22, Issue 8, 2003, pp. 986-1004.

[2]. T. M. Cover, J. A. Thomas, Elements of Information Theory, *Wiley-Interscience*, 2nd Edition, 2006.

[3]. D. Tsai, Y. Lee, E. Matsuyama, Information entropy measure for evaluation of image quality, *Journal of Digital Imaging*, Vol. 21, Issue 3, 2008, pp. 338-347.

[4]. P. Bromiley, N. Thacker, E. Thacker, Shannon Entropy, Renyi Entropy, and Information, Tina Memo No. 2004-004, *Tina*, 2004.

[5]. C. Shannon, A mathematical theory of communication, *Bell Syst. Tech. J.*, Vol. 27, 1948, pp. 379-423, 623-656.

[6]. A. Renyi, Probability Theory, *North-Holland*, Amsterdam, 1970.

[7]. C. Lang, H. Jia, Kapur`s entropy for color image segmentation based on a hybrid whale optimization algorithm, *Entropy*, Vol. 21, Issue 318, 2019, pp. 334-351.

[8]. C. Tsallis, Possible generalization of Boltzmann-Gibbs statistics, *Journal of Statistical Physics*, Vol. 52, 1988, pp. 479-487.

[9]. S. Khattak, G. Saman, I. Khan, A. Salam, Maximum entropy based image segmentation of human skin lesion, *International Journal of Information, Control and Computer Sciences*, Vol. 9, Issue 5, 2015, 667.

[10]. B. Khehra, A. Pharwaha, Digital mammogram segmentation using Non-Shannon measures of entropy, in *Proceedings of the International Conference of Signal and Image Engineering (ICSIE'11)*, Vol. 2, 2011.

[11]. B. Li, H. Shu, Z. Liu, Z. Shao, C. Li, M. Huang, J. Huang, Nonrigid medical image registration using an information theoretic measure based on Arimoto entropy with gradient distributions, *Entropy*, Vol. 21, Issue 2, 2019, 189.

[12]. N. R. Pal, S. K. Pal, Object-background segmentation using new definitions of entropy, *IEEE Proceedings*, Vol. 136, Issue 4, 1989, pp. 284-295.

[13]. N. R. Pal, S. K. Pal, Entropy: A new definition and its application, *IEEE Transactions on Systems, Man and Cybernetics*, Vol. 21, Issue 5, 1991, pp. 1260-1270.

[14]. L. Wang, X. Gao, Z. Zhou, X. Wang, Evaluation of four similarity measures for 2D/3D registration in image-guided intervention, *Journal of Medical Imaging and Health Informatics*, Vol. 4, 2014, pp. 416-421.

[15]. C. Wachinger, N. Navab, Entropy and Laplacian images: Structural representations for multi-modal registration, *Medical Image Analysis*, Vol. 16, Issue 1, 2012, pp. 1-17.

[16]. A. Shiozaki, Edge extraction using entropy operator, *Computer Vision, Graphics and Image Processing*, Vol. 36, 1986, pp. 1-9.

[17]. A. Khellaf, A. Beghdadi, H. Dupoisot, Entropic Contrast Enhancement, *IEEE Transactions on Medical Imaging*, Vol. 10, Issue 4, 1991, pp. 589-592.

[18]. L. Wang, G. Chen, D. Shi, Y. Chang, S. Chan, J. Pu, Active contours driven by edge entropy fitting energy for image segmentation, *Signal Processing*, Vol. 149, 2018, pp. 27-35.

[19]. L. Wang, L. Zhang, X. Yang, P. Yi, H. Chen, Level set based segmentation using local fitted images and inhomogeneity entropy, *Signal Processing*, Vol. 167, 2020, 107297.

[20]. W. Yang, L. Cai, F. Wu, Image segmentation based on gray level and local relative entropy two dimensional histogram, *PloS ONE*, Vol. 15, Issue 3, 2020, e0229651.

[21]. L. Wang, Y. Chang, H. Wang, Z. Wu, J. Pu, X. Yang, An active contour model based on local fitted images for image segmentation, *Information Sciences*, Vols. 418-419, 2017, pp. 61-73.

[22]. Q. Zheng, E. Dong, Z. Cao, W. Sun, Z. Li, Active contour model driven by linear speed function for local segmentation with robust initialization and applications in MR brain images, *Signal Processing*, Vol. 97, 2014, pp. 117-133.

[23]. E. A. Mylona, M. A. Savelonas, D. Maroulis, Entropy-based spatially-varying adjustment of active contour parameters, in *Proceedings of the IEEE International Conference on Image Processing (ICIP'12)*, 2012, pp. 2565-2568.

[24]. E. A. Mylona, M. A. Savelonas, D. Maroulis, Self-parameterized active contours based on regional edge structure for medical image segmentation, *Springer Plus*, Vol. 3, 2014, 424.

[25]. J. Zhang, X. Ma, Y. Chen, L. Fang, J. Wang, MR image segmentation based on local information entropy geodesic active contour model, *International Journal of Multimedia and Ubiquitous Engineering*, Vol. 9, Issue 4, 2014, pp. 127-136.

[26]. C. He, Y. Wang, Q. Chen, Active contours driven by weighted region-scalable fitting energy based on local entropy, *Signal Processing*, Vol. 92, 2012, pp. 587-600.

[27]. S. Mirkamali, P. Nagabhushan, Depth-wise segmentation of 3D images using entropy, *International Journal of Computer Engineering Science*, Vol. 2, Issue 5, 2012, pp. 44-52.

[28]. Y. Sun, J. Xin, Content Adaptive Image Matching by Color-Entropy Segmentation and Inpainting, Computer Analysis of Images and Patterns, *Springer*, Berlin, Heidelberg, 2011.

[29]. M. Smieja, J. Tabor, Image segmentation with use of cross-entropy clustering, in *Proceedings of the 8th International Conference on Computer Recognition Systems (CORES'13)*, 2013, pp. 403-409.

[30]. L. Wang, H. Liu, J. Zhang, H. Chen, J. Pu, Computerized assessment of glaucoma severity based on color fundus images, *Proceedings of SPIE*, Vol. 10953, 2019, 1095322.

[31]. A. Ajlan, A. E. Zaart, Image segmentation using minimum cross-entropy thresholding, in *Proceedings of the IEEE International Conference on Systems, Man, and Cybernetics (ICSMC'09)*, San Antonio, TX, USA, 2009, pp. 1776-1781.

[32]. X. Yin, B. Ng, J. He, Y. Zhang, D. Abbott, Unsupervised segmentation of blood vessels from colour retinal fundus images, in *Proceedings of the International Conference on Health Information Science (HIS'14)*, 2014, pp. 194-203.

[33]. A. Abutaleb, Automatic Thresholding of gray-level pictures using two-dimensional entropy, *Computer Vision, Graphics, and Image Processing*, Vol. 47, 1989, pp. 22-32.

[34]. C. Qi, Maximum Entropy for image segmentation based on an adaptive particle swarm optimization, *Applied Mathematics & Information Sciences*, Vol. 8, Issue 6, 2014, pp. 3129-3135.

[35]. V. Oswal, A. Belle, R. Diegelmann, K. Najarian, An entropy-based automated cell nuclei segmentation and quantification: application in analysis of wound healing process, *Computational and Mathematical Methods in Medicine*, Vol. 2013, 2013, 592790.

[36]. S. Khattak, G. Saman, I. Khan, A. Salam, Maximum entropy based image segmentation of human skin lesion, *International Journal of Information, Control and Computer Sciences*, Vol. 9, Issue 5, 2015, pp. 1094-1098.

[37]. C. Lang, H. Jia, Kapur`s entropy for color image segmentation based on a hybrid whale optimization algorithm, *Entropy*, Vol. 21, Issue 318, 2019, pp. 334-351.

[38]. Y. Song, Y. Wu, Y. Dai, A new active contour remote sensing river image segmentation algorithm inspired from the cross entropy, *Digital Signal Processing*, Vol. 48, 2016, pp. 322-332.

[39]. S. Wang, L. Yu, K. Li, X. Yang, C. Fu, P. Heng, Boundary and entropy-driven adversarial learning for fundus image segmentation, *arXiv*, 2019, arXiv:1906.11143v2.

[40]. S. Zhao, Y. Wang, Z. Yang, D. Cai, Region mutual information loss for semantic segmentation, *arXiv*, 2019, arXiv:1910.12037v1.

[41]. H. Zhang, J. Fritts, S. Goldman, An Entropy-based Objective Evaluation Method for Image Segmentation, *The International Society for Optical Engineering on Storage and Retrieval Methods and Applications for Multimedia*, 2004.

[42]. O. Ronneberger, P. Fischer, T. Brox, Unet: Convolutional networks for biomedical image segmentation, in *Proceedings of the Conference on Medical Image Computing and Computer-assisted Intervention (MICCAI'15)*, 2015, pp. 234-241.

[43]. L. Wang, X. Gao, Q. Fang, A novel mutual information-based similarity measure for 2D/3D registration in image guided intervention, in *Proceedings of the 1st International Conference on Orange Technologies (ICOT'13)*, 2013, pp. 135-138.

[44]. L. Wang, X. Gao, R. Zhang, W. Xia, A comparison of two novel similarity measures based on mutual information in 2D/3D image registration, in *Proceedings of the IEEE International Conference on Medical Imaging Physics and Engineering (ICMIPE'13)*, 2013, pp. 215-218.

[45]. C. Studholme, D. Hill, D. Hawkes, An overlap invariant entropy measure of 3D medical image alignment, *Pattern Recogn.*, Vol. 32, 1999, pp. 71-86.

[46]. D. Tomazevic, B. Likar, F. Pernus, Multi-feature mutual information image registration, *Image Anal. Stereol.*, Vol. 31, 2012, pp. 43-53.

[47]. H. Rivaz, Z. Karimaghaloo, V. Fonov, D. Collins, Nonrigid registration of ultrasound and MRI using contextual conditioned mutual information, *IEEE Transactions on Medical Imaging*, Vol. 33, Issue 3, 2014, pp. 708-725.

[48]. X. Liu, Z. Tang, M. Wang, Z. Song, Deformable multi-modal registration using 3D-FAST conditioned mutual information, *Computer Assisted Surgery*, Vol. 22, 2017, pp. 295-304.

[49]. L. Gong, H. Wang, C. Peng, Y. Dai, M. Ding, Y. Sun, X. Yang, J. Zheng, Non-rigid MR-TRUS image registration for image-guided prostate biopsy using correlation ratio-based mutual information, *Biomed. Eng. Online,* Vol. 16, Issue 1, 2017, 8.

[50]. X. Cun, C. Pun, H. Gao, Applying stochastic second-order entropy images to multi-modal image registration, *Signal Processing: Image Communication*, Vol. 65, 2018, pp. 201-209.

[51]. J. Zhang, J. Wang, X. Wang, D. Feng, Multimodal image registration with joint structure tensor and local entropy, *International Journal of Computer Assisted Radiology and Surgery*, Vol. 10, 2015, pp. 1765-1775.

Chapter 10
EEG Signal Analysis with a Statistical Entropy-based Measure for Alzheimer's Disease Detection

Nesma Houmani, Majd Abazid, Kylliann De Santiago, Jerome Boudy, Bernadette Dorizzi, Jean Mariani and Kiyoka Kinugawa-Bourron

10.1. Context

With the unprecedented aging of the population in the world, dementia and neurodegenerative disorders have become a major societal concern, imposing an important burden on modern societies. Alzheimer's disease (AD) is the most common form of dementia [1]; it affects 11 % of the world population aged over 65 [2] and its incidence increases exponentially with age, and doubles every 5 years after the age of 65 [3, 4]. The number of individuals with AD is expected to reach 115 million in 2050 [5].

AD is a chronic neurodegenerative disorder, characterized by irreversible brain damages, associated with memory impairments and a wide range of cognitive dysfunctions. The evolution of AD frequently follows five stages. AD dementia is proceeded by the asymptomatic "preclinical" stage, characterized by the absence of overt symptoms [6], but the brain lesions of AD exist. At this stage, the concept of Subjective Cognitive Impairment (SCI) has been proposed recently, defined by a self-experienced persistent decline in cognitive capacity in comparison with a previously normal status [7]. These subjective complaints are considered as elderly at-risk for AD [8, 9]. Then, in "prodromal" (Mild Cognitive Impairment, MCI) stage, patients exhibit measurable memory impairments, but maintain their functional capacities [9, 10]; 6 to 25 % of MCI patients later develop AD. In the "Mild AD" stage, cognitive deficits are more marked, such as memory and learning impairments. These deficits become more severe in the "moderate" stage, and in the final "severe" stage of the disease, almost all cognitive and motor functions are significantly deteriorated and patients lose autonomy in daily life activities,

Nesma Houmani
Electronics and Physics Department, Telecom SudParis, Institut Polytechnique de Paris, Evry, France

becoming completely dependent on caregivers. The average duration of survival of AD patients is 5-8 years after clinical diagnosis [11].

Growing evidence shows that AD is clinically characterized by a pathological accumulation of amyloid-beta (Aβ) and hyperphosphorylated tau peptides that alter excitatory and inhibitory synaptic transmission [12-14]. In particular, at the early stage, AD affects limbic regions related to episodic memory, which leads to a relative inability to retain new information. In addition, a disruption of fronto-hippocampal connections has been observed from the early stage of AD, in parallel with hippocampal atrophy, and it has been reported that it may contribute to the initial memory impairment in AD patients [15]. Over time, AD spreads to other brain regions [16] and affects other cognitive functions, such as executive functions, attention, visuospatial and language abilities.

No medication exists for curing this pathology, and many therapeutics trials failed. Therefore, the early detection of AD becomes an important issue for the scientific and medical community. Even if the available treatments cannot stop or reverse the disease progression, early therapeutic interventions may delay its evolution. Moreover, AD diagnosis at the early stage can help the patient and his caregivers to anticipate the future.

According to the most recent guidelines [17, 18], AD can be diagnosed in preclinical and prodromal stages, before the manifestation of any cognitive and behavioral symptoms. This could be possible based on pathophysiological markers, revealed by cerebrospinal fluid (CSF) and positron emission tomography (PET) biomarkers of Aβ42 and tau in the brain. Although these clinical methodologies are relevant for AD assessment, these guidelines underline the need to extend research to non-invasive and inexpensive instrumental techniques that can be deployed in clinical environment and used for large scale assessment over time of a great number of individuals.

10.2. Electroencephalography (EEG)

Recent advances in functional neuroimaging techniques have greatly enhanced clinical research and practice to assess brain neural networks involved in normal brain functions as well as neurological disorders. Functional Magnetic Resonance Image (fMRI) has considerably developed during past decades and is now commonly used for brain connectivity analysis. In the meantime, numerous studies demonstrated that electroencephalography (EEG) associated with appropriate signal processing methods can also bring valuable information on normal and impaired brain networks [19].

EEG is a non-invasive, relatively inexpensive, and potentially mobile technology. It is characterized by a high temporal resolution (about milliseconds), which is crucial for the analysis of fast dynamics in the cortex over very short duration and at different frequency ranges (1-4 Hz, delta; 4-8 Hz, theta; 8-12, alpha; 12-30 Hz, beta; and >30 Hz, gamma). Each frequency band conveys a specific physiological information on brain functional activity.

EEG has been exploited successfully to investigate AD-related alterations in the brain dynamics using spontaneous resting-state EEG (rsEEG) [20, 21] and/or event-related responses [22, 23]. The rsEEG with eyes-closed represents a simple acquisition procedure and has three main interests. First, it may be carried out rapidly in clinical setting. Second, the recording at rest does not require auditory or visual stimuli that could induce fatigue commonly observed during task performance. Third, EEG signals can be recorded in relatively comparable experimental conditions on healthy subjects and patients suffering from neuropathological disorders.

Nevertheless, diagnosing AD with EEG signals at the early stage remains a challenge. This is mainly due to the complex nature of EEG signals, which must be modeled as nonstationary, nonlinear and multidimensional time series. Moreover, EEG signal is known to have a low Signal-to-Noise Ratio (SNR), since it is strongly affected by different sources of noise, be they biological (e.g. ocular and neuromuscular activities) or electronic.

10.3. State-of-the-art

EEG was largely investigated as a tool for AD diagnosis, by comparing EEG recordings of AD patients to those of control subjects (healthy subjects) [24-28]. Several studies have highlighted that one of the major effects of AD is the reduction in complexity of the EEG signal compared to that of healthy subjects. However, it is not always easy to detect such effects because of the large inter-variability between AD patients and symptoms of AD are often dismissed as normal consequence of aging.

Several methods have been used to assess the complexity of EEG signals [29-31]. The fractal dimension was exploited as a potential discriminative feature for AD diagnosis [31-34]. The correlation dimension and the first positive Lyapunov exponent were frequently used [30, 35-39]. The correlation dimension ($D2$) quantifies the number of independent variables that are necessary to describe the dynamics of the system. The Lyapunov exponent ($L1$) reflects the divergence of trajectories starting at nearby initial states. It has been found that EEG signals from AD patients display lower values of such measures (lower complexity) than signals from age-matched normal subjects in almost all EEG channels. However, it has been reported in the literature that these two measures are computationally expensive because of the reconstruction of a phase space trajectory.

Alternative methods more appropriate for sparse data were also suggested to quantify signal complexity, inherited from information theory. In this setting, the complexity of a signal relies to its unpredictability: irregular signals are more complex than regular ones since they are more unpredictable. Several measures were thus proposed; most of them exploit the concept of entropy: sample entropy [40], Tsallis entropy [41], approximate entropy [42, 43], multi-scale entropy [44], and Lempel-Ziv complexity [45].

In parallel to the computation of EEG complexity, other studies focused on the analysis of the abnormalities of the functional connectivity across brain regions, since AD can be viewed as a disconnection syndrome. A large variety of measures has been proposed to quantify linear and nonlinear relationship between EEG channels: correlation coefficient [46], coherence [46-48], Granger causality [46-49], phase synchrony [46, 50, 51], state space based synchrony [46, 50, 52], stochastic event synchrony [46, 51, 48], and mutual information [53].

The majority of these studies reported decreased rsEEG synchrony in MCI and AD patients compared to healthy subjects. However, conflicting results exist in the literature mainly due to methodological differences and the use of different sparse databases. Some studies reported decreased spectral coherence in posterior alpha and beta in AD patients comparatively to age-matched control subjects. This phenomenon was also observed in temporo-parieto-occipital brain region [54-57] and in fronto-central region in other works [58-60]. Conversely, other studies provided less straightforward findings at low frequency bands [56, 57, 67, 68]. Some studies reported a decrease of coherence in AD at central electrodes in the theta band [57, 67]. A previous study [53] showed that mutual information is lower for AD than in healthy subjects, especially in frontal and antero-temporal regions. Moreover, there was a decrease in information transmission between corresponding inter-hemispheric electrodes and between distant electrodes in the right hemisphere.

Other studies reported a reduction of the synchronization likelihood across all electrode pairs at beta in AD patients compared with MCI and control subjects [61-64], and a reduction of alpha synchronization between frontal and parietal regions in AD and MCI patients compared with control subjects [65, 66].

All the above-mentioned measures share in particular two main drawbacks. First, these measures did not consider the EEG signal as a multidimensional time series. Indeed, the prevailing paradigms extract information from EEG signals by averaging them over channels. The EEG being a multidimensional signal provided by a number of electrodes, it is of high potential interest to exploit its spatio-temporal nature, through signal processing techniques, which can take into account inter-channel relations.

Secondly, such measures were computed on the whole EEG time series without addressing the problem of their non-stationarity. Nevertheless, it is a known fact that most physiological signals, such as EEG are non-stationary. Some studies reported that EEG time series are quasi-stationary and described as a piecewise stationary process, i.e. EEG data can be segmented into stationary segments (epochs), with different probabilistic characteristics, separated by abrupt transitions [69, 70]. Other studies [71-73] identified quasi-stationary states in EEG, called "microstates". These states are supposed to reflect coherent neural activities. Also, in [74], the author suggested that perception is based on sequences of stationary patterns demarcated by discontinuities. In some studies, when the non-stationarity of EEG signals has been considered to extract EEG features, the authors proposed to segment the signal into a given number of epochs of the same size fixed empirically.

The purpose of this chapter book is to review an entropy measure, termed "epoch-based entropy" (EpEn), already introduced and published in [75-77]. This entropy measure relies on a refined characterization of the local statistical properties of the EEG signal using a

Hidden Markov Model (HMM), which takes into account the non-stationarity and multi-dimensionality of the EEG time series.

The present study addresses the problem of AD detection based on the analysis of a database containing EEG time series acquired in real clinical conditions at Charles-Foix Hospital in France. This database contains EEG data from patients with subjective cognitive impairment (SCI), AD patients and mild cognitive impairment patients (MCI).

The objective of the study is to extend our previous results and demonstrate the effectiveness of the epoch-based entropy measure for AD detection. Four alternative functional connectivity measures are also used as ground truth to assess the effectiveness of our approach: phase synchrony, granger causality, coherence and mutual information. We will show that the proposed statistical measure allows a better characterization of the underlying neuronal dynamics in the context of AD detection, since it relies on the refined statistical modeling of EEG signal considering its spatio-temporal nature.

10.4. EEG Database Description

The present study was conducted on a database containing EEG signals of 72 subjects recorded in real clinical conditions between 2009 and 2013 at Charles-Foix Hospital (Ivry-sur-Seine, France). Subjects who complained of memory impairment were referred to the outpatient memory clinic of the hospital to undergo a battery of clinical and neuropsychological tests for brain disorders.

Each subject was given a diagnosis at the memory clinic based on the clinical assessment, brain imaging, psychometric findings, interviews and neuropsychological tests, conducted by a multidisciplinary medical staff, according to the standard diagnostic criteria: DSM-IV, NINDS, Jessen criteria for SCI, Mc Keith criteria for Lewy body dementia [7, 8, 78]. Patients with epilepsy were excluded from the cohort, and it is worth noticing that EEG was not exploited to establish the diagnosis of patients in this cohort.

The local ethical committee of the Sorbonne Université approved this retrospective study. The database reflects what medical practitioners are facing in reality, as opposed to databases used in the literature [38, 46-51, 76, 79] that are prone to experimental constraints that do not match the reality on the ground.

The study population includes EEG recordings of 22 SCI subjects, 28 probable AD patients and 22 MCI patients. Table 10.1 reports information about demographic and clinical characteristics of the patients.

EEG data were recorded with a Deltamed digital EEG acquisition system from 30 scalp derivations (electrodes or channels) placed over the whole head according to the 10-20 international system. Thirteen electrodes from Fp1, Fp2, F7, F3, Fz, F4, F8, FT7, FC3, FCz, FC4, FT8, T3, C3, Cz, C4, T4, TP7, CP3, CPz, CP4, TP8, T5, P3, Pz, P4, T6, O1, Oz, and O2 were considered as displayed in Fig. 10.1.

Table 10.1. Clinical characteristics of the cohort. AD: Alzheimer's disease; MCI: mild cognitive impairment; SCI: subjective cognitive impairment; MMSE: mini mental state examination; BZD: benzodiazepine.

	SCI (n = 22)	MCI (n = 22)	AD (n = 28)
Age (mean ± SD)	68.9 ± 10.3	76.6 ± 10.2	80.8 ± 10.5
Female (%)	81,8 %	77,2 %	67,8 %
MMSE (mean ± SD)	28.3 ± 1.6	23.3 ± 6.4	18.3 ± 6.1
BZD use (%)	4 (18.2 %)	2 (9 %)	8 (28.6 %)
Antidepressant use (%)	2 (9 %)	3 (13.6 %)	12 (42.8 %)
Neuroleptic use (%)	0	2 (3.8 %)	5 (17.8 %)
Hypnotic use (%)	5 (22.7 %)	4 (18.1 %)	7 (25 %)

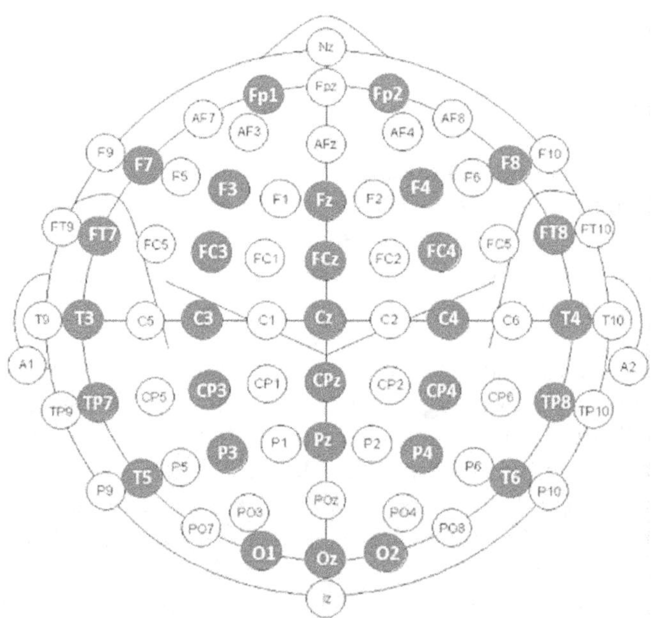

Fig. 10.1. Placement of the 30 electrodes used for EEG signal recordings (marked in red).

EEG data were obtained in resting-state conditions (rsEEG), while subjects were sitting relaxed, with their eyes closed. All data were digitalized in a continuous recording mode for a minimum of 20 minutes using a sampling rate of 256 Hz.

The rsEEG recordings were pre-processed off-line on Matlab. For each subject, continuous epochs of 20 seconds, free from artefacts (ocular, muscular, instrumental), were preliminary selected manually. Then, EEG signals were band-pass filtered with a third-order digital Butterworth filter in four conventional frequency bands: delta (1-4 Hz), theta (4-8 Hz), alpha (8-12 Hz), and beta (12-30 Hz).

10.5. EEG Functional Connectivity Measures

Since AD is hypothesized to induce disconnections between brain regions, several studies of the literature investigated functional connectivity for AD diagnosis through different measures [46-53]. We include in this study four frequently used measures, which rely on different theoretical concepts: coherence, phase synchrony, mutual information and Granger causality. All these measures are presented in details in the sequel. We will see that the three first measures quantify the spatial interdependence between the electrodes while the last one measures the temporal causality between pairs of electrodes.

Additionally to these commonly used measures, we include the recently proposed epoch-based entropy measure (EpEn) [75-77], which estimates the functional connectivity in terms of the information content by means of statistical modeling of EEG signals. This will allow a better estimation of the spatio-temporal characteristics of EEG time series. Contrary to the other measures, EpEn is computed both locally at the time scale of a quasi-stationary epoch and spatially by estimating the inter-channel connectivity.

10.5.1. Coherence

The coherence measure captures the linear component of the functional coupling of the paired EEG oscillations x and y as a function of a frequency f [46-48]. In order to compute the magnitude square coherence, the signals x and y are subdivided in M segments of equal length L, then the coherence function is computed by averaging over those segments. The magnitude square coherence $c(f)$ is calculated as:

$$c(f) = \frac{|\langle X(f)Y^*(f)\rangle|^2}{|\langle X(f)\rangle||\langle Y(f)\rangle|}, \tag{10.1}$$

where $X(f)$ and $Y(f)$ are the Fourier transforms of x and y respectively; Y^* is the complex conjugate of Y; $|Y|$ is the magnitude of Y, and $\langle X(f)\rangle$ stands for the average of $X(f)$ computed over the M segments, likewise $\langle Y(f)\rangle$ and $\langle X(f)Y^*(f)\rangle$.

10.5.2. Phase Synchrony

The general principle of phase synchrony is to detect the existence of phase locking between two oscillatory signals. Phase synchrony refers to the interdependence between instantaneous phases ϕ_x and ϕ_y of two signals x and y [46, 50, 51]. The instantaneous phase ϕ_x of a signal x is defined as:

$$\phi_x(t) = arg[x(t) + i\tilde{x}(t)], \tag{10.2}$$

where $\tilde{x}$ is the Hilbert transform of x.

The phase synchrony index γ for two instantaneous phases ϕ_x and ϕ_y is computed as:

$$\gamma = |\langle e^{i(\phi_x - \phi_y)}\rangle|, \tag{10.3}$$

where $\langle . \rangle$ denotes average over time.

Phase synchrony range is between 0 and 1: for uncorrelated signals, γ is close to 0, whereas it tends to 1 for strong phase synchronization.

10.5.3. Mutual Information

Mutual information is derived from Shannon's information theory to estimate the information gained from observations of one random variable X on another Y:

$$I(X,Y) \; = \; H(X) + H(Y) - H(X,Y), \tag{10.4}$$

where $H(X)$ and $H(Y)$ is the Shannon entropy of X and Y respectively, and $H(X,Y)$ is the joint entropy of X and Y. It is always positive, and it vanishes when X and Y are statistically independent.

Applied to EEG signals, mutual information quantifies dynamical coupling or information transmission between electrode pairs [53]. As reported in [46], computing mutual information by quantizing the signals from the resulting histograms generally leads to unreliable estimation of the measure. Therefore, we compute it in time–frequency domain using the normalized spectrograms as follows:

$$C_x(k,f) \; = \; \frac{|X(k,f)|^2}{\sum_{k,f}|X(k,f)|^2}, \tag{10.5}$$

where the summation in the denominator is carried out over the time window k and frequency range f. The normalized spectrograms can be managed as probability distributions. Accordingly, the mutual information of the normalized spectrograms is defined as:

$$I_w\big(C_x, C_y, C_{xy}\big) \; = \; \sum_{k,f} C_{xy}(k,f)\, log\, \frac{C_{xy}(k,f)}{C_x(k,f)C_y(k,f)}, \tag{10.6}$$

where the normalized cross time–frequency distribution of x and y is computed as follows:

$$C_{xy}(k,f) \; = \; \frac{|X(k,f)Y^*(k,f)|}{\sum_{k,f}|X(k,f)Y^*(k,f)|} \tag{10.7}$$

10.5.4. Granger Causality

Granger causality is based on the general concept that the prediction of a given time series could be improved by considering the information of past values of another time series. In this setting, the latter time series is said to have a *causal* influence on the former one [46-49]. Ganger causality suggests that a variable X causes another variable Y, if the past of X contains information that help predict the future of Y, over and above the information already in the past of Y itself.

This measure requires the estimation of vector autoregregressive (VAR) models, in which the value of a variable $X(t)$ in time domain is modeled as a linear weighted sum of its own past and of the past of another variable $Y(t)$:

$$Y(t) = \sum_{n=1}^{p} a_n Y_{t-n} + \varepsilon_1(t), \tag{10.8}$$

$$Y(t) = \sum_{n=1}^{p} a_n Y_{t-n} + \sum_{n=1}^{p} b_n X_{t-n} + \varepsilon_2(t), \tag{10.9}$$

where $\varepsilon_1(t)$ and $\varepsilon_2(t)$ are the prediction errors, a_n and b_n are the coefficients (gain factors) of the model, and p is the maximum number of lagged observations included in the autoregressive model ($p \ll T$).

The linear influence from $X(t)$ to $Y(t)$ can be calculated as the log ratio between the variance of the residual errors:

$$GC_{X \to Y} = log \frac{var(\varepsilon_1)}{var(\varepsilon_2)} \tag{10.10}$$

Finally, Granger causality is given by the ratio of the variance of the prediction-error terms for the reduced (when omitting the signal of the potential cause) and full regressions (when including the signal of the potential cause).

10.5.5. Epoch-based Entropy

The entropy of a random variable depends on its probability density value [80]. Epoch-based entropy measure (EpEn) relies on the fundamental assumption that the EEG signal is piecewise stationary, i.e. can be viewed as being stationary at the time scale of an epoch.

In this context, Hidden Markov Models (HMM) are good techniques for estimating the information content in piecewise stationary signals: they can segment the EEG signals into stationary epochs, and at the same time perform a local estimation of the probability density on each epoch. The use of HMM is also motivated by the fact that HMM's structure is adapted for modeling "hidden" neural dynamics underlying the observed EEG signals.

As in our previous works [75-77], EEG signals are modeled by a continuous left-to-right HMM (see Fig. 10.2) that allows transitions from each state to itself and to its immediate right-hand neighbors only. The states of the HMM correspond to the stationary parts of the EEG signal, and the transitions of the HMM correspond to the abrupt variations in the signal [84, 85]. EEG signal of one subject is thus considered as a succession of epochs, obtained by segmenting such a signal with the Viterbi algorithm using the associated subject's HMM [84, 85]. Each epoch S_i corresponds to a state of the HMM and contains a given number of observations (sample points). For each epoch, the probability density function is modeled by a mixture of M Gaussian functions, considering a diagonal covariance matrix for each multivariate Gaussian.

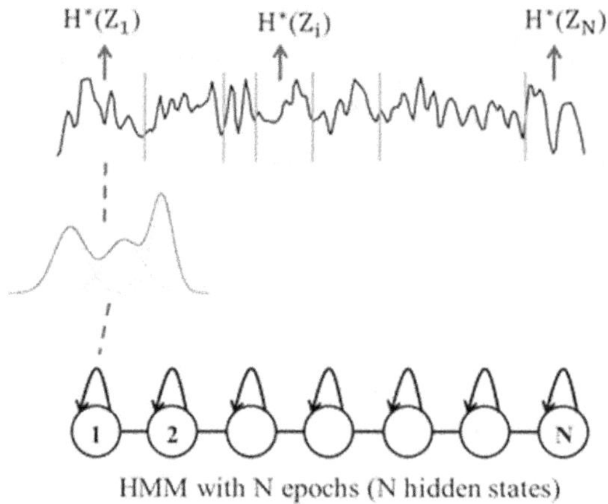

Fig. 10.2. Epoch-based entropy computation of a univariate EEG signal [76].

Then, each observation z in a given epoch S_i is considered as a realization Z_i of a random variable Z that follows a given observation probability distribution $P_i(z)$ modeled by the Gaussian mixture. Consequently, each stationary epoch of the signal is associated to a random variable, and the entropy $H^*(Z_i)$ of the epoch S_i is that of an ensemble of realizations of Z_i:

$$H^*(Z_i) = -\sum_{z \in S_i} P_i(z). log_2 P_i(z) \qquad (10.11)$$

By averaging the entropy over the N epochs of the EEG signal of the subject, we obtain an entropy-based value $EpEn(Z)$ of the signal:

$$EpEn(Z) = \frac{1}{N}\sum_{i=1}^{N} H^*(Z_i) \qquad (10.12)$$

Note that the EEG sampling period (typically 4 ms in the case of the current study) is small with respect to the epoch length (typically 57 ms). Thus, although Z_i is a discrete variable, we can take advantage of the continuous emission probability law estimated on each epoch by the HMM.

The use of HMM is further motivated by the multi-channel EEG analysis, since EEG data are often correlated time series from multiple electrodes on the scalp: HMM can manage multidimensional signals by applying multivariate probability density functions on such signals. Hence, they are appropriate for modeling the inter-relations between EEG time series recorded from multiple electrodes. In this case, for each subject, an HMM is trained on a set of D EEG signals captured from D electrodes (see Fig. 10.3).

At time t, a hidden state emits a D-dimensional observation vector. By applying the Viterbi algorithm, N epochs are generated for each EEG signal and the entropy $H^*(Z_i)$ of each epoch S_i is computed considering the probability density estimated by the HMM on the observations of the D epochs S_i.

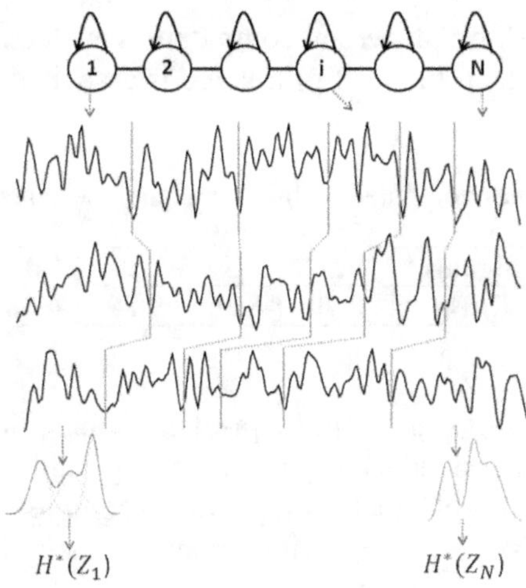

Fig. 10.3. Illustration of multi-channel ($D = 3$) EEG signal modeling with HMM [76].

Although all N epochs are matched between EEG channels, it is worth noticing that the model does not constrain these epochs to be of equal length for all channels. Finally, by averaging the entropy over all the N epochs, an epoch-based entropy value (EpEn) associated to the multi-channel EEG of the considered subject is computed.

10.5.6. Illustration of EpEn Functioning

In this section, we illustrate the functioning of EpEn for measuring the information content of multivariate piecewise stationary EEG signals. To this end, we compute EpEn value of four signals displayed in Fig. 10.4, considering them first separately in a univariate analysis, then as pairs of signals for a multivariate analysis.

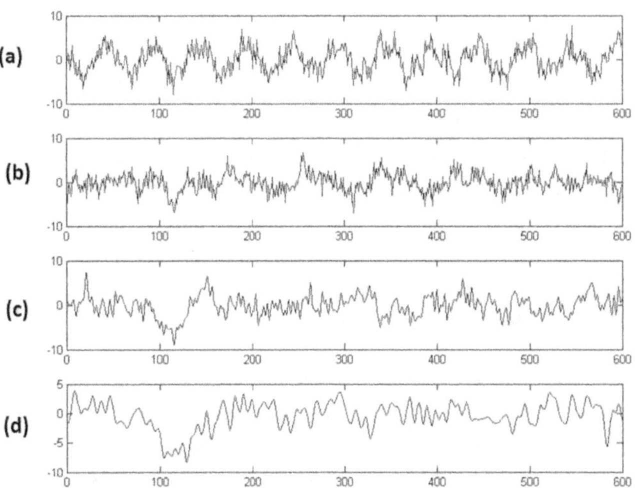

Fig. 10.4. Examples of four signals of different complexities.

Visually, these signals exhibit different complexities reflected by their EpEn values reported in Table 10.2. In fact, higher EpEn value is associated with high "irregular" or "complex" signals.

Table 10.2. EpEn values computed on the four signals when considered separately.

Signals	a	b	c	d
EpEn	8.79	8.56	7.89	6.38

Then, considering four pairs of signals ([a-a], [a-b], [a-c] and [a-d]), EpEn measure detects the inter-channel statistical dependencies, as reported in Table 10.3. The proposed EpEn measure reflects both intra-channel complexity (complexity over time) and the inter-channel complexity (spatial complexity, or heterogeneity between all the signals).

When computing the EpEn value on identical signals [a-a], there is no inter-channel difference and thus the combined distribution becomes more regular. This leads to a reduction of the EpEn value to when the signal is considered alone (from 8.79 in Table 10.2 to 7.51 in Table 10.3). Nevertheless, the combined entropy is still nonzero since it considers intra-channel disorder.

Table 10.3. Epoch-based entropy computed on pairs of signals.

Signals	a-a	a-b	a-c	a-d
EpEn	7.51	7.38	7.81	8.70

When computing entropy on two signals of different complexities, for example on the most complex signal (a) with a signal of lower complexity (c or d), the entropy increases as well as the difference increases between signals (inter-channels) and also over time for each signal (intra-channel).

The statistical estimation of entropy with HMM allows to quantify the information content of multivariate EEG signals at two levels simultaneously: at the time level, EpEn quantifies the information content or the disorder on piecewise stationary epochs of EEG signals over time; and at the spatial level, EpEn quantifies the functional connectivity in terms of the heterogeneity of piecewise stationary epochs between multi-channel EEG signals.

10.6. Experimental Study and Results

In this work, we investigate the reliability of EpEn measure for the discrimination of AD patients from SCI and MCI patients. We confront this statistical measure to coherence, phase synchrony, mutual information and Granger causality in terms of classification accuracy using SVM classifier. This comparison of the five measures will allow a better understanding on the mechanisms of EpEn in the framework of AD detection.

10.6.1. Study Design

This study is carried out on a real life database containing EEG recordings of 22 SCI subjects, 28 AD patients and 22 MCI patients. For each person, we compute all the EEG measures on the four frequency bands (delta, theta, alpha and beta) considering different brain regions defined arbitrarily, using sets of channels located in regions susceptible to be sensitive to changes due to AD.

We defined seven regions of interest: prefrontal (Fp1, Fp2), occipital (O1, O2), frontal (F7, F3, Fz, F4, F8), temporal (T6, T4, F8, T5, T3, F7), central (FCz, C3, CPz, C4), occipito-prefrontal (Fp1, Fp2, O1, O2) and parieto-occipital (T6, P4, Pz, P3, T5, O1, O2).

To distinguish automatically between each pair of classes, i.e. AD vs. SCI and AD vs. MCI, a linear single-feature SVM classifier was first used with a leave-one-out procedure, and the threshold that gave the best correct classification rate was selected. The performance was assessed for each brain region and each frequency band. Then, in order to improve the performance, we estimated the classification performance per frequency band using a linear SVM combining two brain regions.

In order to compute for each person the four classical EEG features (coherence, phase synchrony, mutual information and Granger causality), we first calculated such measures between all pairs of electrodes belonging to the considered brain region. Then, we averaged over all those signal pairs to obtain a functional connectivity measure per region.

By contrast, EpEn computes in one single step the information content conveyed by coupling different EEG signals of a given region, by means of the statistical modeling of the multidimensional EEG signal. Note that the optimal values of the hyper parameters needed for a reliable estimation of the EpEn measure, such as number of Gaussians and epochs (states of the HMM), were fixed based on our experimental findings in previous works [75-77].

10.6.2. Discriminating AD Patients from SCI Subjects

Table 10.4 presents the correct classification rate for the five measures per brain region and frequency band. We report the configurations that led to the best classification performance in terms of accuracy.

To obtain an EpEn value for each subject, per brain region and frequency band, the HMM was trained on a set of EEG signals captured by the electrodes of the considered brain region, as explained in Section 10.5.5. The other measures were computed for each person, per brain region and frequency band, by averaging all the connectivity values computed between pairs of electrodes of the considered region. Note that Granger causality measure is computed on the complete EEG time series in the time domain.

Results show that the five EEG features are not reliable to discriminate AD patients from SCI subjects, when computed per brain region and per frequency band. Coherence and EpEn are those giving the best accuracy value of 70 % but with an unbalanced performance in terms of sensitivity (percentage of AD well classified) and specificity (percentage of SCI well classified).

Besides, Table 10.4 highlights different brain regions of interest and frequency bands when discriminating AD from SCI, dependent on the functional connectivity measure under consideration. This finding reflects in part the disparity of the conclusions in the literature on the brain regions and frequency bands that could be considered in the framework of AD.

Table 10.4. Best classification performance (in %) when discriminating AD from SCI with each EEG feature.

AD vs. SCI	Coherence	Phase synchrony	Granger causality	Mutual information	EpEn
Brain region	Parieto-occipital	Prefrontal	Frontal	Temporal	Occipital
Frequency band	Delta	Alpha	/	Alpha	Theta
Accuracy	70 %	66 %	66 %	68 %	70 %
Sensitivity (AD)	59.1 %	68.2 %	41 %	89.3 %	59.1 %
Specificity (SCI)	78.6 %	64.3 %	85.7 %	40.9 %	78.6 %

For a refined comparative analysis, Fig. 10.5 shows the boxplots of the five features considering the brain regions and the frequency band that gave the best accuracy value to distinguish AD patients from SCI. Fig. 10.5a and Fig. 10.5e show that AD patients have lower values of coherence and EpEn than control subjects, respectively, in low frequency bands. AD induces a decreased coherence on delta band in parieto-occipital regions (Mann-Whitney $p = 1.42 \times 10^{-2}$), and a decreased information content conveyed by the multidimensional EEG time series on theta band in the occipital region (Mann-Whitney $p = 5.22 \times 10^{-4}$).

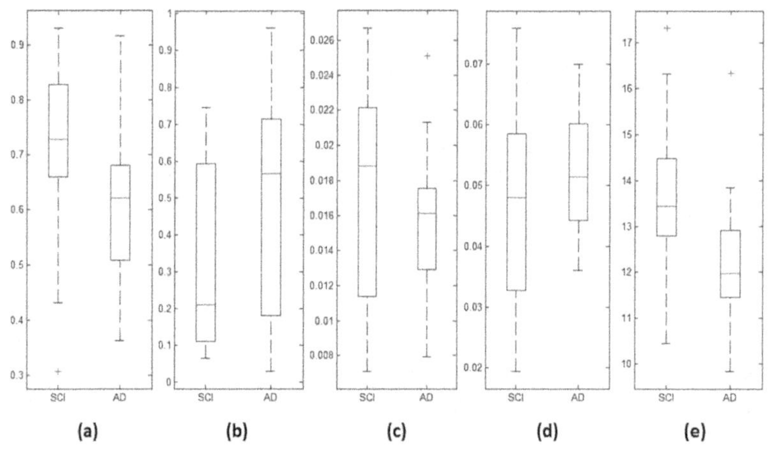

Fig. 10.5. Box plots of the five EEG features when discriminating AD patients from SCI subjects with: (a) Coherence; (b) Phase synchrony; (c) Granger causality; (d) Mutual information; and (e) EpEn measure, computed on the region and the frequency band reported in Table 10.4.

Table 10.5 shows the classification performance when discriminating AD from SCI with each measure, considering the functional connectivity values of two brain regions as input to the linear SVM classifier. We performed experiments on different combinations of brain regions and reported in Table 10.5 only those leading to the best accuracy value.

Results clearly show that, except for coherence measure, the performance are improved for all measures when combining different brain regions. However, we observe that for the four widely used measures, an unbalanced specificity and sensitivity values remains.

The improvement of performance is significant with EpEn measure. A correct classification rate of 98 % is reached with a specificity of 100 % and a sensitivity of 95.5 %. This result shows the reliability of the used feature to detect AD, even the control subjects of this database are not healthy subjects since they have some memory complaints.

Also, we observe that this good discrimination of AD from SCI is obtained with different combinations of brain regions (temporal & parieto-occipital, frontal & occipital, central & occipital, central & prefrontal). This finding suggests that EpEn is less sensitive to brain regions changes, and thus is more stable and reliable comparatively to the other measures.

In addition, the selected frequency band and brain regions with EpEn are in accordance with some results reported in the literature [54-57, 67] and clinical knowledge [81-83]: on the one hand, AD detection has been shown on theta band; on the other hand, temporo-parieto-occipital regions are the first affected regions in the early stage of AD.

Table 10.5. Best classification performance (in %) when discriminating AD from SCI with each EEG feature, considering a combination of two brain regions.

AD vs. SCI	Coherence	Phase synchrony	Granger causality	Mutual information	EpEn
Brain regions	Parieto-occipital	Central & Occipito-prefrontal	Frontal & Temporal Frontal & Occipital	Temporal & Occipito-prefrontal	Temporal & Parieto-occipital Frontal & Occipital Central & Occipital Central & Prefrontal
Frequency band	Delta	Theta	/	Alpha	Theta
Accuracy	70 %	72 %	74 %	70 %	98 %
Sensitivity	59.1 %	63.6 %	54.5 %	89.3 %	95.5 %
Specificity	78.6 %	78.6 %	89.3 %	45.5 %	100 %

Another interesting finding emerge with EpEn: the central region is informative for AD detection considering long-range dynamics with distant regions, i.e. the extreme posterior (occipital) or the extreme anterior (prefrontal) regions. Actually, we notice that the discrimination between AD and SCI with EpEn relies on the quantification of the long-range information transmission among regions of the brain, as displayed in Fig. 10.6.

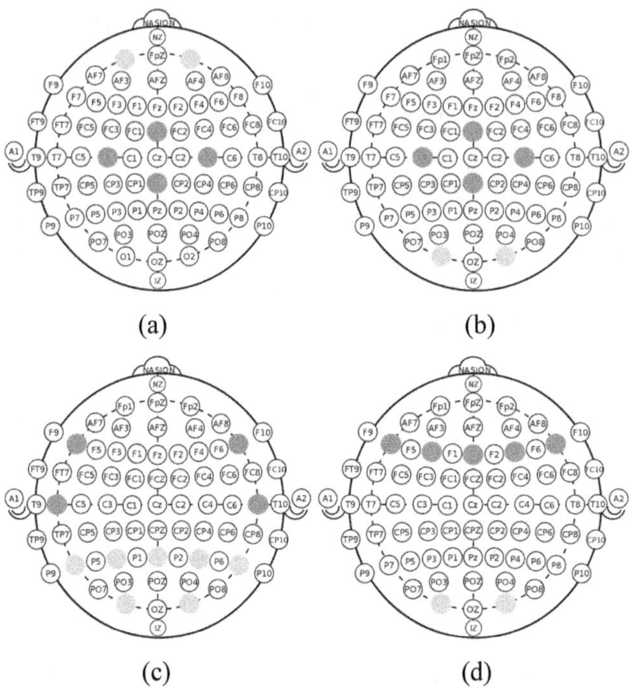

Fig. 10.6. The electrodes that belong to the regions leading to the best discrimination between AD and SCI with EpEn in theta band: (a) Central & prefrontal; (b) Central & occipital; (c) Temporal & parieto-occipital; (d) Frontal & occipital regions.

In order to go deeper in our understanding, Fig. 10.7 shows the distribution of EpEn values in theta band for the two populations (AD and SCI) considering the two combinations of two regions among those reported in Table 10.5 (frontal & occipital regions, central & occipital regions). A linear separation between the two populations appears clearly in both cases; this result highlights the discrimination efficiency of EpEn measure.

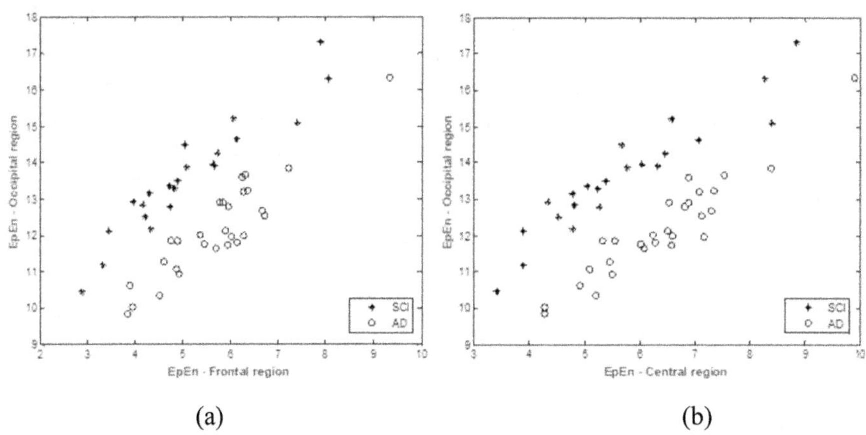

Fig. 10.7. The distribution of EpEn values of AD patients and SCI subjects computed on theta band considering: (a) Frontal & occipital regions, and (b) Central & occipital regions.

10.6.3. Discriminating AD Patients from MCI Patients

Table 10.6 shows the performance of the linear SVM classifier when discriminating AD patients from MCI patients, with a leave-one-out procedure. We report the best classification performance obtained for each EEG feature.

Table 10.6. Best classification performance (in %) when discriminating AD from MCI with each EEG feature.

AD vs. MCI	Coherence	Phase synchrony	Granger causality	Mutual Information	EpEn
Brain region	Central	Parieto-occipital	Occipital	Temporal	Occipital
Frequency band	Delta	Delta	/	Beta	Theta
Accuracy	66 %	60 %	54 %	70 %	70 %
Sensitivity (AD)	60.7 %	75 %	100 %	57.1 %	64.3 %
Specificity (MCI)	72.7 %	40.9 %	0 %	86.4 %	77.3 %

Mutual information and EpEn measures are those giving the best accuracy of 70 % when discriminating AD patients from MCI patients. However, EpEn offers a better balance between sensitivity (percentage of AD well classified) and specificity (percentage of MCI well classified).

When comparing the obtained results with EpEn in Table 10.6 (AD vs. MCI) to those reported in Table 10.4 (AD vs. SCI), we notice that the occipital region and theta band are selected in both cases. This means that computing EpEn in the occipital region on theta band allows discriminating AD from the two first stages of the disease, i.e. SCI and MCI. This behavior is not observed for the other measures.

We display in Fig. 10.8 the boxplots for each EEG features considering the brain regions and the frequency band that gave the best accuracy value to distinguish AD patients from MCI. Fig. 10.8b and Fig. 10.8e show that AD patients have lower values of phase synchrony and EpEn than MCI subjects, respectively. Moreover, it appears clearly that the EpEn values of AD patients are significantly lower than those of MCI patients (Mann Whitney $p = 1.22 \times 10^{-4}$). This result is coherent with those reported in the literature [61-64]: AD induces a reduction of synchronization and information transmission in AD compared with MCI.

Table 10.7 reports the classification performance for each measure, when combining two brain regions. We carried out experiments on different combinations of brain regions per frequency band and reported only those leading to the best accuracy value.

EpEn measure outperforms significantly all the other functional connectivity measures: an accuracy of 100 % is reached in the theta band when considering occipital and frontal regions.

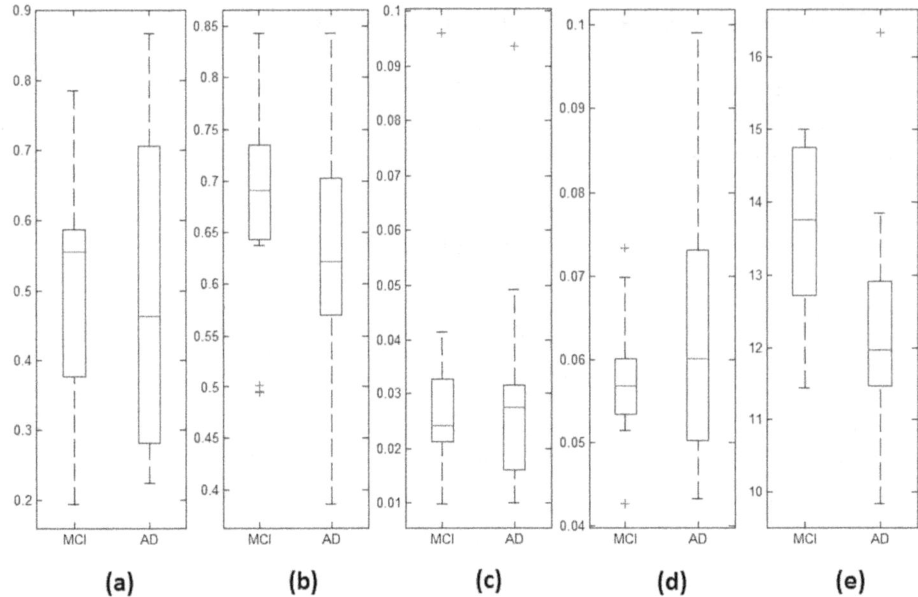

Fig. 10.8. Box plots of the five EEG features when discriminating AD patients from MCI patients with: (a) Coherence; (b) Phase synchrony; (c) Granger causality; (d) Mutual information; and (e) EpEn measure, computed on the region and the frequency band reported in Table 10.6.

Table 10.7. Best classification performance (in %) when discriminating AD from MCI with each EEG feature, considering a combination of two brain regions.

AD vs. MCI	Coherence	Phase synchrony	Granger causality	Mutual information	EpEn
Brain regions	Occipital Parieto-occipital	Frontal & Temporal	Temporal & Parieto-occipital	Occipito-prefrontal &Temporal	Occipital & Frontal
Frequency band	Alpha	Delta	/	Alpha	Theta
Accuracy	68 %	64 %	58 %	74 %	100 %
Sensitivity (AD)	60.7 %	85.7 %	60.7 %	96.4 %	100 %
Specificity (MCI)	77.3 %	36.4 %	54.6 %	45.5 %	100 %

Compared to the previous results obtained in Table 10.5 (AD vs. SCI), we notice that EpEn is the best feature for AD detection (versus SCI or MCI) when computed on theta band in the Occipital and Frontal regions.

Fig. 10.9 displays the distribution of EpEn values in theta band for AD and MCI patients considering the combinations of the occipital and frontal regions. We clearly observe a linear separation between the two populations, which indicates the discrimination efficiency of EpEn measure.

404

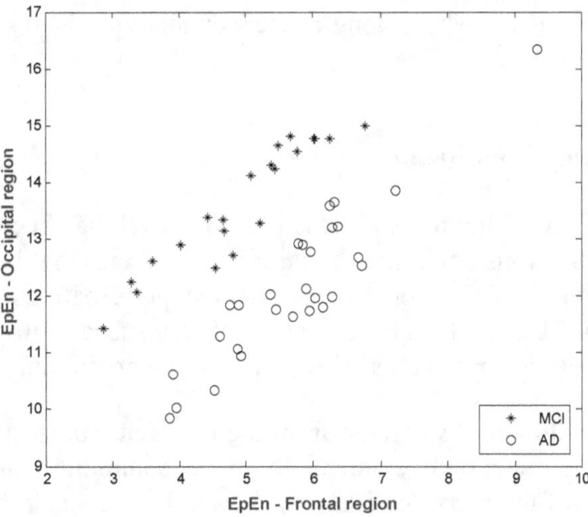

Fig. 10.9. The distribution of EpEn values of AD and MCI patients computed on theta band considering the frontal and occipital regions.

10.6.4. Discriminating SCI Subjects from MCI Patients

To go deeper in our analysis, we carried out additional experiments to discriminate SCI subjects from MCI patients considering only EpEn measure, since it outperforms the other measures for AD detection.

As shown in Table 10.8, a classification accuracy of 77.3 % is obtained with a specificity of 77.3 % and a sensitivity of 77.3 %, considering all frequency bands and the combination of EpEn values computed on the occipital region and the occipito-parietal region. A polynomial SVM of order two was used for this two-class classification using eight features as input to the SVM. This result proves the difficulty to differentiate these two populations, which are both at the first stage of the disease.

Table 10.8. Best classification performance (in %) when discriminating SCI from MCI with EpEn measure.

SCI vs. MCI	EpEn
Brain regions	Occipital & Parieto-occipital
Frequency band	Delta - Theta - Alpha - Beta
Accuracy	**77.3 %**
Sensitivity (MCI)	77.3 %
Specificity (SCI)	77.3 %

Interestingly, we observe that the selected brain region to distinguish MCI from SCI subjects with EpEn is localized in the posterior brain area, i.e. in the parietal and the occipital regions. This result is in contrast with the obtained results on AD patients (AD vs. SCI or AD vs. MCI), where long-range connectivity is taken into account to detect the disease.

10.7. Discussion and Conclusion

The utility of rsEEG in Alzheimer's disease (AD) research has been demonstrated over several decades in numerous studies. The major EEG changes that have been reported in AD are the reduction of EEG complexity and the perturbations of EEG functional connectivity. Several EEG markers have been employed successfully to investigate these two AD-related alterations, based on sophisticated signal processing techniques.

However, three main drawbacks emerge from the research works in the state-of-the-art. First, complexity and functional connectivity were commonly quantified separately. Second, the majority of the extracted EEG markers did not consider the EEG signal as a multidimensional time series. Third, such measures were computed on the whole EEG time series without addressing the problem of their non-stationarity, even it is well known that most physiological signals, such as EEG are non-stationary.

The purpose of the present study was to review a recently proposed entropy-based functional connectivity measure, called epoch-based entropy, and to investigate its potential application to the detection of AD based on multi-channel EEG signals. This entropy measure is computed on piecewise stationary epochs using a HMM, which performs local density estimation at the epoch level. The use of HMM is motivated by the fact that its structure is suitable for modeling neural dynamics underlying the observed EEG signals. In addition, HMM can manage multidimensional signals by applying multivariate probability density functions on the signals.

The originality of this statistical measure lies on the fact that it estimates the information content or the disorder in EEG signals on piecewise stationary epochs over time; and at the spatial level, by quantifying the functional connectivity in terms of the heterogeneity of piecewise stationary epochs between multi-channel EEG signals. This will allow a better estimation of the spatio-temporal characteristics of EEG signals merged into a single figure.

By comparing epoch-based entropy to four alternative functional connectivity measures, namely coherence, phase synchrony, Granger causality and mutual information, we showed that the statistical measure is by far a more reliable feature for AD detection, on our experimental data. We obtained a high accuracy for the classification of AD vs. SCI (98 % accuracy, 100 % specificity, 95.5 % sensitivity). Then, by comparing AD and MCI, we reached an accuracy of 100 %. In both cases, a common finding was observed: computing the epoch-based entropy on theta in frontal and occipital regions allows a good discrimination between AD and its early stages (SCI and MCI).

In conclusion, our study demonstrates the effectiveness of the statistical modeling of EEG with HMM for analyzing the dynamics of neural activity in patients with AD. However, by comparing SCI and MCI patients, we obtained an accuracy of 77.3 % due to the similarities between SCI patients and MCI patients. Indeed, a proportion of SCI patients are actually at an early stage of MCI [86].

Further experiments should be conducted on additional patients to assess the generalization of our method to clinical usage. Furthermore, in this work, entropy was computed on EEG time series locally at the epoch level, and then averaged over all the epochs. It could be interesting to keep the entropy values per epoch in order to characterize more finely how the EEG signal fluctuates over time.

References

[1]. C. Reitz, C. Brayne, R. Mayeux, Epidemiology of Alzheimer disease, *Nat. Rev. Neurol.*, Vol. 7, 2011, pp. 137-152.

[2]. C. P. Ferri, M. Prince, C. Brayne, H. Brodaty, et al., Alzheimer's disease international, Global prevalence of dementia: A Delphi consensus study, *Lancet*, Vol. 366, Issue 17, 2005, pp. 2012-2017.

[3]. C. Qiu, M. Kivipelto, E. von Strauss, Epidemiology of Alzheimer's disease: Occurrence, determinants, and strategies toward intervention, *Dialogues in Clinical Neuroscience*, Vol. 11, Issue 2, 2009, pp. 111-128.

[4]. M. M. Corrada, R. Brookmeyer, A. Paganini-Hill, D. Berlau, C. H. Kawas, Dementia incidence continues to increase with age in the oldest old: The 90+ study, *Ann. Neurol. Jan.*, Vol. 67, Issue 1, 2010, pp. 114-121.

[5]. L. Lili, D. Ruau, R. Chen, S. Weber, A. J. Butte, Systematic identification of risk factors for Alzheimer's disease through shared genetic architecture and electronic medical records, *Pac. Symp. Biocomput.*, 2013, pp. 224-235.

[6]. J. L. Price, D. W. McKeel, V. D. Buckles, C. M. Roe, et al., Neuropathology of nondemented aging: Presumptive evidence for preclinical Alzheimer disease, *Neurobiol. Aging*, Vol. 30, 2009, pp. 1026-1036.

[7]. F. Jessen, R. E. Amariglio, M. Van Boxtel, M. Breteler, et al., A conceptual framework for research on subjective cognitive decline in preclinical Alzheimer's disease, *Alzh. Dementia*, Vol. 10, Issue 6, 2014, pp. 844-852.

[8]. A. J. Mitchell, H. Beaumont, D. Ferguson, M. Yadegarfar, et al., Risk of dementia and mild cognitive impairment in older people with subjective memory complaints: Meta-analysis, *Acta Psychiatrica Scandinavica*, Vol. 130, Issue 6, 2014, pp. 349-351.

[9]. B. Dubois, H. Hampel, H. H. Feldman, P. Scheltens, et al., Preclinical Alzheimer's disease: Definition, natural history, and diagnostic criteria, *Alzh. Dementia*, Vol. 12, Issue 3, 2016, pp. 292-323.

[10]. R. C. Petersen, R. O. Roberts, D. S. Knopman, B. F. Boeve, et al., Mild cognitive impairment: Ten years later, *Arch. Neurology*, Vol. 66, Issue 12, 2009, pp. 1447-1455.

[11]. E. P. Helzner, N. Scarmeas, S. Cosentino, M. X. Tang, et al., Survival in Alzheimer disease: A multiethnic, population-based study of incident cases, *Neurology*, Vol. 71, Issue 19, 2008, pp. 1489-1495.

[12]. C. W. Wong, V. Quaranta, G. G. Glenner, Neuritic plaques and cerebrovascular amyloid in Alzheimer disease are antigenically related, *Proceedings of Natl. Acad. Sci.*, Vol. 82, Issue 24, 1985, pp. 8729-3872.

[13]. M. A. Daulatzai, Early stages of pathogenesis in memory impairment during normal senescence and Alzheimer's disease, *J. Alzheimers Dis.*, Vol. 20, Issue 2, 2010, pp. 355-367.

[14]. Z. X. Shen, Brain cholinesterases: III. Future perspectives of AD research and clinical practice, *Med. Hypotheses*, Vol. 63, Issue 2, 2004, pp. 298-307.

[15]. F. Remy, N. Vayssière, L. Saint-Aubert, E. Barbeau, J. Pariente, White matter disruption at the prodromal stage of Alzheimer's disease: Relationships with hippocampal atrophy and episodic memory performance, *Neuroimage Clin.*, Vol. 27, Issue 7, 2015, pp. 482-492.

[16]. E. Braak, K. Griffing, K. Arai, J. Bohl, et al., Neuropathology of Alzheimer's disease: What is new since A. Alzheimer ?, *Eur. Arch. Psychiatry Clin. Neurosci.*, Vol. 249, Suppl. 3, 1999, pp. 14-22.

[17]. G. M. McKhann, D. S. Knopman, H. Chertkow, B. T. Hyman, et al., The diagnosis of dementia due to Alzheimer's disease: Recommendations from the National Institute on Aging-Alzheimer's Association workgroups on diagnostic guidelines for Alzheimer's disease, *Alzheimers Dement*, Vol. 7, 2011, pp. 263-269.

[18]. B. Dubois, H. H. Feldman, C. Jacova, H. Hampel, et al., Advancing research diagnostic criteria for Alzheimer's disease: The IWG-2 criteria, *Lancet Neurol.*, Vol. 13, 2014, pp. 614-629.

[19]. H. Hampel, N. Toschi, C. Babiloni, F. Baldacci, et al., Alzheimer Precision Medicine Initiative (APMI). Revolution of Alzheimer precision neurology. Passageway of systems biology and neurophysiology, *J. Alzheimers Dis.*, Vol. 64, Issue s1, 2018, pp. S47-S105.

[20]. S. Giaquinto, G. Nolfe, The EEG in the normal elderly: a contribution to the interpretation of aging and dementia, *Electroencephalogr. Clin. Neurophysiol.*, Vol. 63, 1986, pp. 540-546.

[21]. R. C. Briel, I. G.McKeith, W. A. Barker, Y. Hewitt, et al., EEG findings in dementia with Lewy bodies and Alzheimer's disease, *J. Neurol. Neurosurg. Psychiatry*, Vol. 66, 1999, pp. 401-403.

[22]. P. M. Rossini, S. Rossi, C. Babiloni, J. Polich, Clinical neurophysiology of aging brain: From normal aging to neurodegeneration, *Prog. Neurobiol.*, Vol. 83, Issue 6, 2007, pp. 375-400.

[23]. G. Yener, B. Güntekin, E. Başar, Event-related delta oscillatory responses of Alzheimer patients, *Eur. J. Neurol.*, Vol. 15, Issue 6, 2008, pp. 540-547.

[24]. J. Dauwels, F. B. Vialatte, A. Cichocki, Diagnosis of Alzheimer's disease from EEG signals: Where are we standing ?, *Curr. Alzheimer Res.*, Vol. 7, Issue 6, 2010, pp. 487-505.

[25]. J. Dauwels, K. Srinivasan, M. R. Reddy, T. Musha, et al., Slowing and Loss of Complexity in Alzheimer's EEG: Two Sides of the Same Coin?, *Int. J. Alzheimers Dis.*, Vol. 2011, 2011, 539621.

[26]. F. Vecchio, C. Babiloni, R. Lizio, V. FallaniFde, et al., Resting state cortical EEG rhythms in Alzheimer's disease: Toward EEG markers for clinical applications: A review, *Suppl. Clin. Neurophysiol.*, Vol. 62, 2013, pp. 223-236.

[27]. A. Albredi, A. Aztiria, A. Basarab, On the early diagnosis of Alzheimer's disease from multimodal signals: A survey, *Artificial Intelligence in Medicine*, Vol. 71, 2016, pp. 1-29.

[28]. N. Malek, M. R. Baker, C. Mann, J. Greene, Electroencephalographic markers in dementia, *Acta Neurol. Scand.*, Vol. 135, Issue 4, 2016, pp. 388-393.

[29]. J. Jeong, EEG dynamics in patients with Alzheimer's disease, *Clin. Neurophysiol.*, Vol. 115, Issue 7, 2004, pp. 1490-1505.

[30]. H. Adeli, S. Ghosh-Dastidar, N. Dadmehr, Alzheimer's disease: Models of computation and analysis of EEGs, *Clin. EEG Neurosci.*, Vol. 36, Issue 3, 2005, pp. 131-140.

[31]. H. Adeli, S. Ghosh-Dastidar, N. Dadmehr, A spatio-temporal wavelet-chaos methodology for EEG-based diagnosis of Alzheimer's disease, *Neurosci. Lett.*, Vol. 444, Issue 2, 2008, pp. 190-194.

[32]. A. Ahmadlou, H. Adeli, A. Adeli, Fractality and a wavelet-Chao Methodology for EEG-based diagnosis of Alzheimer's disease, *Alzheimer Dis. Assoc. Disorders*, Vol. 25, Issue 1, 2011, pp. 85-92.

[33]. P. Grassberger, Generalized dimensions of strange attractors, *Phys. Lett. A*, Vol. 97, Issue 6, 1983, pp. 227-230.

[34]. P. Grassberger, I. Procaccia, Measuring the strengeness of strange attractors, *Physica D*, Vol. 9, 1983, pp. 189-208.

[35]. B. Jelles, J. H. Van Birgelen, J. P. Slaets, R. E. Hekster, et al., Decrease of nonlinear structure in the EEG of Alzheimer patients compared to healthy controls, *Clin. Neurophysiol.*, Vol. 110, Issue 7, 1999, pp. 1159-1167.

[36]. J. Jeong, S. Y. Kim, S. H. Han, Non-linear dynamical analysis of the EEG in Alzheimer's disease with optimal embedding dimension, *Electroencephalogr. Clin. Neurophysiol.*, Vol. 106, Issue 3, 1998, pp. 220-228.

[37]. J. Jeong, J. H. Chae, S. Y. Kim, S. H. Han, Nonlinear dynamic analysis of the EEG in patients with Alzheimer's disease and vascular dementia, *J. Clin. Neurophysiol.*, Vol. 18, Issue 1, 2001, pp. 58-67.

[38]. T. Takahashi, Complexity of spontaneous brain activity in mental disorders, *Prog. Neuropsychopharmacol. Biol. Psychiatry*, Vol. 45, 2013, pp. 258-266.

[39]. T. Yagyu, J. Wackermann, M. Shigeta, V. Jelic, et al., Global dimensional complexity of multichannel EEG in mild Alzheimer's disease and age-matched cohorts, *Dement. Geriatr. Cogn.*, Vol. 8, Issue 6, 1997, pp. 343-347.

[40]. D. Abásolo, R. Hornero, P. Espino, D. Alvarez, J. Poza, Entropy analysis of the EEG background activity in Alzheimer's disease patients, *Physiol. Meas.*, Vol. 27, Issue 3, 2006, pp. 241-253.

[41]. T. De Bock, S. Das, M. Mohsin, N. B. Munro, et al., Early detection of Alzheimer's disease using nonlinear analysis of EEG via Tsallis entropy, in *Proceedings of the Biomedical Sciences and Engineering Conference (BSEC'10)*, 2010, pp. 1-4.

[42]. D. Abásolo, R. Hornero, P. Espino, J. Poza, et al., Analysis of regularity in the EEG background activity of Alzheimer's disease patients with approximate entropy, *Clin. Neurophysiol.*, Vol. 116, Issue 8, 2005, pp. 1826-1834.

[43]. S. Pincus, Approximate entropy as a measure of irregularity for psychiatric serial metrics, *Bipolar Disord.*, Vol. 8, Issue 5, 2006, pp. 430-440.

[44]. J. Escudero, D. Abásolo, R. Hornero, P. Espino, M. López, Analysis of electroencephalograms in Alzheimer's disease patients with multiscale entropy, *Physiol. Meas.*, Vol. 27, 2006, pp. 1091-1106.

[45]. D. Abásolo, R. Hornero, C. Gomez, M. Garcia, M. López, Analysis of EEG background activity in Alzheimer's disease patients with Lempel-Ziv complexity and central tendency measure, *Med. Eng. Phys.*, Vol. 28, 2006, pp. 315-322.

[46]. J. Dauwels, F. B. Vialatte, T. Musha, A. Cichocki, A comparative study of synchrony measures for the early diagnosis of Alzheimer's disease based on EEG, *Neuroimage*, Vol. 49, Issue 1, 2010, pp. 668-693.

[47]. J. Escudero, S. Sanei, D. Abásolo, R. Hornero, Regional coherence evaluation in mild cognitive impairment and Alzheimer's disease based on adaptively extracted magnetoencephalogram rhythms, *Physiol. Meas.*, Vol. 32, Issue 8, 2011, pp. 1163-1180.

[48]. Z. Sankari, H. Adeli, A. Adeli, Wavelet coherence model for diagnosis of Alzheimer's disease, *Clin. EEG Neurosci.*, Vol. 43, Issue 3, 2012, pp. 268-278.

[49]. C. Babiloni, M. Pievani, F. Vecchio, C. Geroldi, et al., White-matter lesions along the cholinergic tracts are related to cortical sources of EEG rhythms in amnesic mild cognitive impairment, *Human Brain Mapping*, Vol. 30, Issue 5, 2009, pp. 1431-1443.

[50]. B. Czigler, D. Csikós, Z. Hidasi, Z. Anna Gaál, et al., Quantitative EEG in early Alzheimer's disease patients – Power spectrum and complexity features, *Int. J. Psychophysiol.*, Vol. 68, Issue 1, 2008, pp. 75-80.

[51]. Y. M. Park, H. J. Che, C. H. Im, H. T. Jung, et al., Decreased EEG synchronization and its correlation with symptom severity in Alzheimer's disease, *Neurosci. Res.*, Vol. 62, Issue 2, 2008, pp. 112-117.

[52]. M. A. Kramer, F. L. Chang, M. E. Cohen, D. Hudson, A. J. Szeri, Synchronization measures of the scalp EEG can discriminate healthy from Alzheimer's subjects, *Int. Journal of Neural Systems*, Vol. 17, Issue 2, 2007, pp. 61-69.

[53]. J. Jeong, J. Gore, B. Peterson, Mutual information analysis of the EEG in patients with Alzheimer's disease, *Clinical Neurophysiology*, Vol. 112, Issue 5, 2001, pp. 827-835.

[54]. V. Jelic, P. Julin, M. Shigeta, A. Nordberg, et al., Apolipoprotein E epsilon 4 allele decreases functional connectivity in Alzheimer's disease as measured by EEG coherence, *J. Neurol. Neurosurg Psychiatry*, Vol. 63, 1997, pp. 59-65.

[55]. V. Jelic, S. E. Johansson, O. Almkvist, M. Shigeta, et al., Quantitative electroencephalography in mild cognitive impairment: longitudinal changes and possible prediction of Alzheimer's disease, *Neurobiol. Aging*, Vol. 21, 2000, pp. 533-540.

[56]. T. Locatelli, M. Cursi, D. Liberati, M. Franceschi, G. Comi, EEG coherence in Alzheimer's disease, *Electroencephalogr. Clin. Neurophysiol.*, Vol. 106, 1998, pp. 229-237.

[57]. G. Adler, S. Brassen, A. Jajcevic, EEG coherence in Alzheimer's dementia, *J. Neural. Transm.*, Vol. 110, 2003, pp. 1051-1058.

[58]. A. F. Leuchter, J. J. Dunkin, R. B. Lufkin, Y. Anzai, et al., Effect of white matter disease on functional connections in the aging brain, *J. Neurol. Neurosurg. Psychiatry*, Vol. 57, 1994, pp. 1347-1354.

[59]. C. Besthorn, H. Förstl, C. Geiger-Kabisch, H. Sattel, et al., EEG coherence in Alzheimer disease, *Electroencephalogr. Clin. Neurophysiol.*, Vol. 90, 1994, pp. 242-245.

[60]. L. C. Fonseca, G. M. A. S. Tedrus, P. N. Carvas, E. C. F. A. Machado, Comparison of quantitative EEG between patients with Alzheimer's disease and those with Parkinson's disease dementia, *Clin. Neurophysiol.*, Vol. 124, 2013, pp. 1970-1974.

[61]. C. J. Stam, B. W. van Dijk, Synchronization likelihood: an un-biased measure of generalized synchronization in multivariate data sets, *Phys. D*, Vol. 163, 2002, pp. 236-251.

[62]. C. J. Stam, Y. van der Made, Y. A. L. Pijnenburg, P. H. Scheltens, Synchronization of brain activity in mild cognitive impairment and early Alzheimer's disease, *Acta Neurol. Scand.*, Vol. 108, 2003, pp. 90-96.

[63]. C. Babiloni, G. Binetti, E. Cassetta, D. Cerboneschi, et al., Mapping distributed sources of cortical rhythms in mild Alzheimer's disease. A multi-centric EEG study, *NeuroImage*, Vol. 22, 2004, pp. 57-67.

[64]. Y. A. Pijnenburg, Y. v d Made, A. M. van Cappellen van Walsum, D. L. Knol, P. Scheltens, C. J. Stam, EEG synchronization likelihood in mild cognitive impairment and Alzheimer's disease during a working memory task, *Clin. Neurophysiol.*, Vol. 115, Issue 6, 2004, pp. 1332-1339.

[65]. C. Babiloni, R. Ferri, D. V. Moretti, A. Strambi, et al., Abnormal fronto-parietal coupling of brain rhythms in mild Alzheimer's disease: A multicentric EEG study, *Eur. J. Neurosci.*, Vol. 19, 2004, pp. 2583-2590.

[66]. C. Babiloni, R. Ferri, G. Binetti, A. Cassarino, et al., Fronto-parietal coupling of brain rhythms in mild cognitive impairment: A multicentric EEG study, *Brain Res. Bull.*, Vol. 69, 2006, pp. 63-73.

[67]. V. Knott, E. Mohr, C. Mahoney, V. Ilivitsky, Electroencephalographic coherence in Alzheimer's disease: Comparisons with a control group and population norms, *J. Geriatr. Psychiatry Neurol.*, Vol. 13, 2000, pp. 1-8.

[68]. C. Babiloni, G. B. Frisoni, F. Vecchio, M. Pievani M, et al., Global functional coupling of resting EEG rhythms is related to white-matter lesions along the cholinergic tracts in subjects with amnesic mild cognitive impairment, *J. Alzheimers Dis.*, Vol. 19, 2010, pp. 859-871.

[69]. G. Praetorius, H. M. Bodenstein, Feature extraction from the electroencephalogram by adaptive segmentation, *Proceedings of IEEE*, Vol. 65, Issue 5, 1977, pp. 642-652.

[70]. A. Y. Kaplan, A. A. Fingelkurts, A. A. Fingelkurts, S. V. Borisov, B. S. Darkhovsky, Nonstationary nature of the brain activity as revealed by EEG/MEG: Methodological, practical and conceptual challenges, *Signal Process.*, Vol. 85, Issue 11, 2005, pp. 2190-2212.

[71]. D. Brandeis, D. Lehmann, M. Michel, W. Mingrone, Mapping event-related brain potential microstates to sentence endings, *Brain Topogr.*, Vol. 8, Issue 2, 1995, pp. 145-159.

[72]. D. Lehmann, W. Skrandies, Reference-free identification of components of checkerboard-evoked multichannel potential fields, *Electroenceph. Clin. Neurophysiol.*, Vol. 48, 1980, pp. 609-621.

[73]. H. Bauer, A. Flexer, Discovery of common subsequences in cognitive evoked potentials, in *Proceedings of the European Symposium on Principles of Data Mining and Knowledge Discovery (PKDD'98)*, 1998, pp. 309-317.

[74]. W. Freeman, A cinematographic hypothesis of cortical dynamics in perception, *Int. J. Psychophysiol.*, Vol. 60, 2006, pp. 149-161.

[75]. N. Houmani, F. B. Vialatte, C. Latchoumane, J. Jeong, et al., Stationary epoch-based entropy estimation for early diagnosis of Alzheimer's disease, in *Proceedings of the IEEE Faible Tension Faible Consommation (FTFC'13)*, Paris, France. 2013, pp. 1-4.

[76]. N. Houmani, F. B. Vialatte, G. Dreyfus, Epoch-based entropy for early screening of Alzheimer's disease, *Int. J. of Neural Systems*, Vol. 25, Issue 8, 2015, 1550032.

[77]. N. Houmani, F. B. Vialatte, E. Gallego-Jutglà, G. Dreyfus, et al., Diagnosis of Alzheimer's disease with Electroencephalography in a differential framework, *PloS ONE*, Vol. 13, Issue 3, 2018, e0193607.

[78]. I. G. McKeith, D. W. Dickson, J. Lowe, M. Emre, et al., Diagnosis and management of dementia with Lewy bodies, Third report of the DLB consortium, *Neurology*, Vol. 65, Issue 12, 2005, pp. 1863-1872.

[79]. T. De Bock, S. Das, M. Mohsin, N. B. Munro, et al., Early detection of Alzheimer's disease using nonlinear analysis of EEG via Tsallis entropy, in *Proceedings of the Biomedical Sciences and Engineering Conference (BSEC'10)*, 2010, pp. 1-4.

[80]. T. M. Cover, J. A. Thomas, Elements of Information Theory, 2nd Edition, *John Wiley & Sons*, 2006.

[81]. H. Braak, E. Braak, Neuropathological staging of Alzheimer-related changes, *Acta Neuropathol.*, Vol. 82, Issue 4, 1991, pp. 239-259.

[82]. J. Pantel, P. Schönknecht, M. Essig, J. Schröder, Distribution of cerebral atrophy assessed by MRI reflects pattern of neuropsychological deficits in Alzheimer's dementia, *Neurosci. Lett.*, Vol. 361, 2004, pp. 17-20.

[83]. A. Brun, E. Englund, Regional pattern of degeneration in Alzheimer's disease: Neuronal loss and histopathological grading, *Histopathology*, Vol. 5, 1981, pp. 459-564.

[84]. L. R. Rabiner, B. H. Juang, An introduction to hidden Markov models, *IEEE ASSP Magazine*, Vol. 3, Issue 1, 1986, pp. 4-16.

[85]. L. R. Rabiner, B. H. Juang, Fundamentals of Speech Recognition, *Prentice Hall*, 1993.

[86]. R. Wurtman, Biomarkers in the diagnosis and management of Alzheimer's disease, *Metabolism*, Vol. 64, Issue 3, Suppl. 1, 2015, pp. S47-S50.

Chapter 11
Artefacts Detection in EEG Signals

Antonio Quintero-Rincón, Carlos D'Giano and Hadj Batatia

11.1. Introduction

Electroencephalography (EEG) is a non-invasive and widely available biomedical modality that is used to measure brain activity in order to diagnose different neurological pathologies and plan treatment. Neurologists trained in EEG are able to determine the correct medical diagnostics by identifying visually different waveforms, known as spikes, sharp waves, or the mix of both.

The standardized international 10-20 system is generally used to record EEG activity. This system has 21 electrodes located symmetrically on the surface of the scalp. These positions are computed as percentages of standard distances, the resulting records are comparable between different patients. EEG electrode positions are determined as follows: the reference points are the nasion, which is the delve at the top of the nose, at the level of the eyes; and the inion, which is the bony lump at the base of the skull on the midline at the back of the head. From these points and once the central point (Cz) is localized, the skull perimeters are measured in the transverse and median planes. Electrode locations are determined by dividing these perimeters into 10 % and 20 % intervals, see Fig. 11.1. Additionally, the EEG measurement provides temporal and spatial information about the synchronous firing of many neurons inside the brain with a dominant frequency according to the brain rhythms [1]. The EEG measurement can use a unipolar montage configuration, where the potential of each electrode is compared either to a neutral electrode or to the average of all electrodes; or bipolar montage configuration, where the potential difference between a pair of electrodes spatially close is measured.

An artefact in EEG signals is defined as an electrical potential that is not originated in the brain. The two basic artefact types are physiological and non-physiological. A physiological artefact is generated from the electrical activity associated with the patient's body normal functioning, e.g., eye movement and blinking, normal and fast breathing,

Antonio Quintero-Rincón
Department of Electronics, Catholic University of Argentina (UCA), and Epilepsy and Telemetry Integral Center, Foundation for the Fight against Pediatric Neurological Disease (FLENI), Buenos Aires, Argentina.

chewing, bruxism, swallowing, tongue movement, skin potentials, body tremor, cardiac activity, muscle activity, sweat glands, pulse in the tissues, and artificial cardiac pacemaker. A non-physiological artefact is generated by electromagnetic fields outside the body, such as bad signal recording, line frequency (50/60 Hz), misplaced or malfunctioning electrodes, medical equipment, cell phones, lights, and the environmental movement.

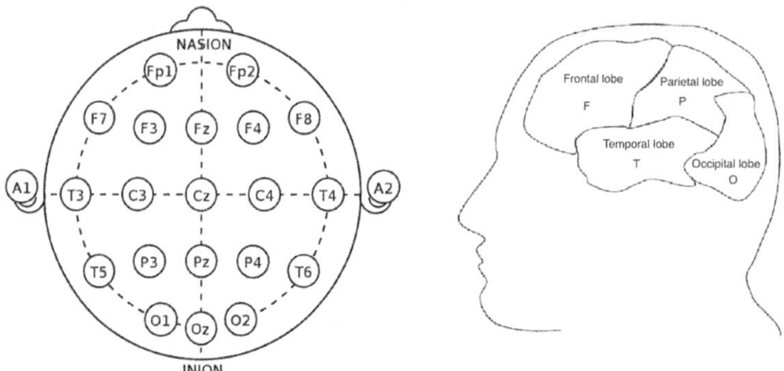

Fig. 11.1. Standard 10-20 system montage. Location and nomenclature of the electrodes: temporal lobe (T), parietal lobe (P), occipital lobe (O), and frontal lobe (F). The odd number corresponds to the left hemisphere and the even number to the right hemisphere.

Noises are as important as artefacts. Acquiring EEG signal properly means mainly measuring biosignals with safety, high signal to noise ratio (SNR) and no data loss. The system electronics include circuitry and printed circuit board design, filtering stages, electronic amplifier's noise control, correct signal conversion, data storing, contact resistance skin-electrodes, and background noise. See [2-4] for more details.

Significant sources of fluctuating electric potential, proper to the human body, cannot be simply ruled out. They must be filtered out during the EEG measurement. Three examples are the face muscle contractions caused by blinking, the chest motion due to respiration, and the electrocardiogram [5].

Due to the large variability and dynamics of EEG signals, diverse artefacts are common during the acquisition such as sampling errors, noises, unusually small or large values, individual cases that violate the nominal relationship between specific variables, and missing data values. It is very important to know the origins, characteristics, sources, and influence of these anomalies on the data, in order to improve analysis. The goal of this chapter is to review and evaluate some classical methods used to detect different types of EEG artefacts in clinical applications. We review supervised detection methods with a variety of features, including mainly temporal-domain curve fitting, multiple signal frequency-domain, empirical wavelet transforms, t-location-scale statistical modeling, and multivariate analysis with independent vector analysis. Multivariate analysis is widely used to remove artefacts. We show the potential of this approach, using the Hampel filter to correct different types of artefacts. In addition, the chapter provides a complete state-of-the-art along with a recommended bibliography.

11.2. Artefacts

The artefacts caused by involuntary body movements are frequent. They are generated by electrical muscular activity, which can be measured through electromyographic signals (EMG). Muscle contraction, bruxism chewing, swallowing, and tongue movement that include the face, the jaw, and the neck, are examples of EMG. Usually, these types of artefacts occur when the patient is stressed, anxious with difficulties to relax or stay still. To control involuntary movements and correct them, electrodes are placed so that simultaneous reference signals are acquired and correlated with EEG recordings. Involuntary movement can also be controlled using neuromuscular-blocking drugs [6]. For illustration see Figs. 11.2 and 11.3.

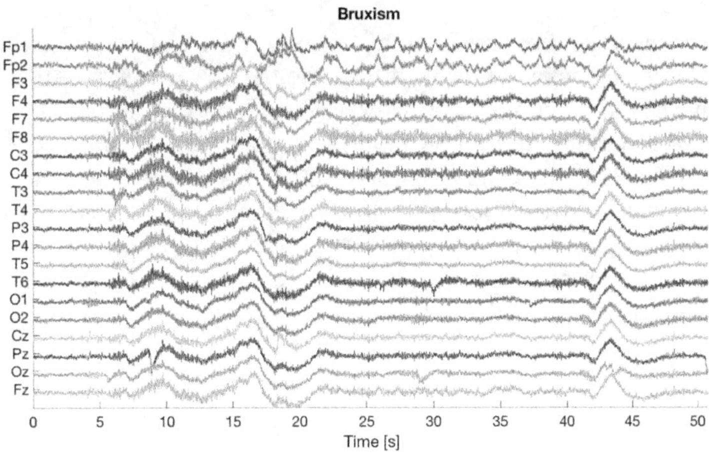

Fig. 11.2. Bruxism artefact example. The figure shows that all the channels are affected starting at 5 seconds abruptly, especially visible between 10 and 20 seconds, and 40 and 45 seconds.

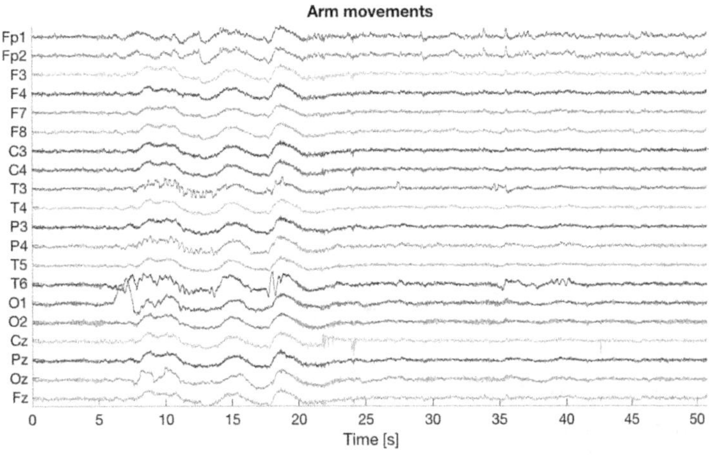

Fig. 11.3. Arm movements artefact example. The figure shows a train of low-amplitude spikes in all channels starting at 5 seconds, and occasionally high amplitude spikes, as between 5 and 25 sec.

415

However, depending on the EEG application, muscular artefacts might be highly desired. This is the case in brain-computer interface (BCI) approaches where motor EEG signal imagery is analyzed [7].

Eye movement artefacts, called electrooculogram (EOG), are generated when the eye potential changes between the cornea and the retina. While electroretinogram (ERG) artefacts occur as a response of the eyes-retinal cell to photic stimulation. In general, these artefacts are present in the frontal electrodes such as Fp1, Fp2, F3, F4, and F7, implying many high and low frequencies, depending on their duration and amplitude [6, 8]. ERG artefacts can be corrected by blocking the light source to the eye. Ocular movement can be cancelled by placing a reference electrode over the nose and using a common-mode rejection ratio from a differential amplifier [5].

The artefacts produced by the movement of the tongue are called Glossokinetic potential (GKP). These artefacts make the electric field around the mouth and the jaw change as the tongue behaves like a dipole. GKP is best detected in the ramifications located near the mouth, such as the lips, the infraorbital leads, and the Fp1 and Fp2 electrodes. To understand this phenomenon, some tests can be made in order to detect possible artefacts. For example, comparing the *not-talk* with repeating words that cause significant movement of the tongue, such as saying the expressions "lalala" or "tom thumb". In children or infants, this artefact is manifested when chewing, sucking, crying, swallowing, or hiccup [6]. See Figs. 11.4 and 11.5 for some examples.

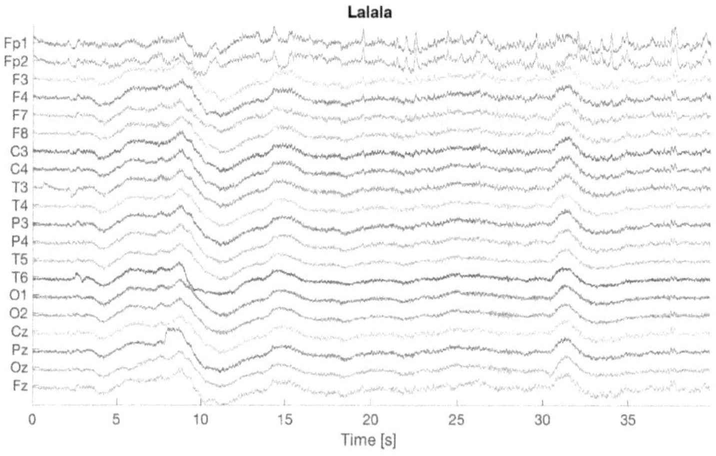

Fig. 11.4. "lalala" artefact example. The figure shows that all channels are affected starting at 5 seconds, especially from 6 to 15 sec and 30 to 35 sec.

Electrocardiogram (ECG or EKG) artefacts are produced by heart electrical activity. Usually, they have a spike or sharp waveform and can be confused by non-expert with epileptiform activity. The heart rate of the ECG can be recorded by placing two EEG electrodes in any non-cephalic part of the body. The ECG heart rate variability (HRV) can be recorded by placing two EEG electrodes in any non-cephalic part of the body. Due to their high waveform amplitude, it is possible to distinguish ECG artefacts from natural EEG signals. The voltage of such artefacts is also reduced by using a bipolar instead of a unipolar montage with OA1 and OA2 ear reference electrodes.

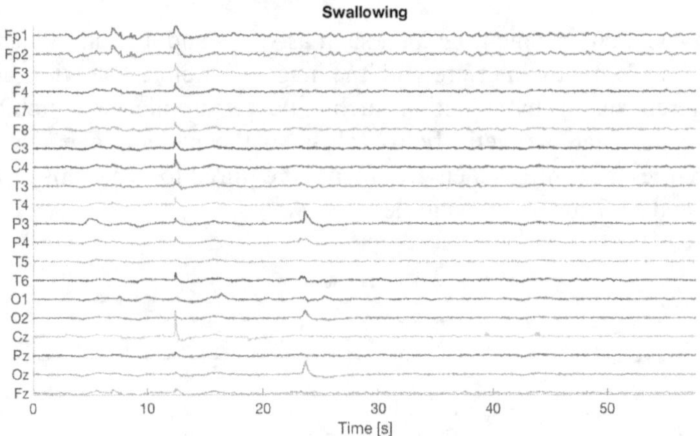

Fig. 11.5. Swallowing artefact example. The figure shows high spikes present between 10-sec and 25-sec on almost all channels.

Power supply (50 or 60 Hz) artefacts are caused by electrical and/or mechanical devices such as the stretcher, ventilation, intravenous infusion devices, sequential compression devices, ECG monitors, dialysis machines, fluorescent lights, heating/cooling, lighting, and rugs. Note that, the 50/60 Hz interference may produce an imprecise reference that can be confused with muscular artefacts or fast brain activity.

This same artefact may be caused by EEG wires in contact with the ground or with electrical wires from other devices, including the power cord from the EEG instrument itself. It can also be caused by poor electrode contact, inadequate preparation of the skin, faulty wiring, or faulty ground making. This artefact leads to high impedance, which can be amplified and detected in one or more channels [9]. In the presence of 50/60 Hz artefacts, and where the impedance of the electrodes is less than 5 KΩ, the plug of each electrical device should be disconnected one at a time while checking the EEG signal improvement. This simple cyclical action permits identifying the source to be removed. At last, a notch filter with a cut-off frequency equal to 50/60 Hz can be used. It must be taken into account that the cerebral signals within this range would also be attenuated. This hinders the detection of epileptiform spikes that have a low amplitude frequency similar to the cut-off frequency [10].

Interruptions between electrodes and the skin can be source of artefacts due to inadequate contact, lack of conductive gel, broken or damaged wires. These are easily distinguished due to the abrupt change in the electrode impedance that creates unexpected potential. Such electrode "pop" may appear in EEG signals as focal spikes or sharp waves with characteristic morphology waveforms having a very steep ascent and a deep fall resembling the shape of the direct current (DC) calibration signal [6]. The movement of the electrodes can also break their balance that can be identified through the movement magnitude. However, the restoration of the equilibrium point may require a long time, with the possibility of compensation using a low-pass filter with a small cut-off frequency that does not affect the EEG signal [5].

In addition, physical environment factors, such as wires and human movement, can also generate artefacts when they interfere with the magnetic fields. The wires act as antennas and collect different signals by induction, such as power supplies or switching equipment. Human movements near magnetic fields (close to the patient) generate capacitive or electrostatic charges. A controlled environment is required to avoid this type of artefacts. See Figs. 11.6 and 11.7 for some examples.

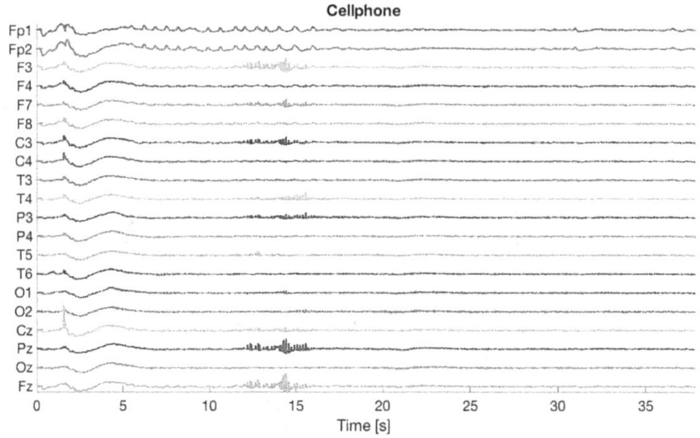

Fig. 11.6. Example of cell phone artefact. The figure shows an effect on all channels from 0 to 20 sec.

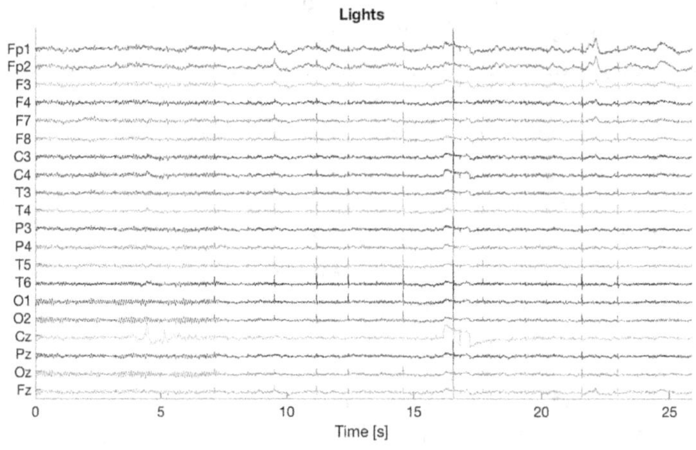

Fig. 11.7. Example of light artefacts. The figure shows that spike waveforms are exhibited in almost all channels, especially between 5 and 25 sec.

11.3. Database

For this study, we created a database of EEG signals from two healthy subjects at the Foundation for the Fight against Pediatric Neurological Disease (FLENI) hospital, Buenos Aires. Each signal has 20 unipolar channels with a sampling frequency of 200 Hz, and duration of 1 minute. The location and nomenclature of the electrodes were placed according to the International 10-20 system (Fig. 11.1). Physiological artefacts such as arm and legs movements, swallowing, bruxism, and movement of the tongue with the expression "lalala", were created. While cell phone and light interferences were introduced as non-physiological artefacts. Artefacts were measured successively, with a resting epoch with closed eyes as a control signal after each measurement. Signals measured from the two subjects with 20 channels were partitioned into 1-second segments for subsequent analysis. This resulted in a large dataset with a variety of situations suitable for analysis.

11.4. Artefacts Removal Methods

It is very important that the actions taken to cancel artefacts do not cause new artefacts or loss of neurological valuable information. Therefore, adequate algorithms depend on the intended application. Visual inspection is often required in order to guarantee good signals. This section presents briefly the most common methods used for artefacts cancellation.

Let $X \in \mathbb{R}^{NxM}$ the EEG matrix, where M is the number of channels measured simultaneously. X columns are naturally correlated as electric brain sources are *projected* on different channels. Assuming, reasonably, that all brain signals arrive at the electrodes instantaneously at the same time, N represents the time-instants of observations

$$x(n) = [x_1, x_2, \ldots, x_m, \ldots, x_M]^T, \tag{11.1}$$

We define W^K as a moving window so that

$$X^K = \mathrm{W}^K X, \tag{11.2}$$

where K is the positive integer representing the width of the window. Selecting the width is as crucial as the time interval of epochs and segments to capture the phenomenon under study. Typically, in EEG signal processing, the width of 1 or 2 seconds is recommended to guarantee stability, especially with low sampling frequencies [5].

11.4.1. Time-domain Analysis

Empirically, artefacts are determined by using an amplitude threshold. Thus, any signal-segment that crosses the threshold is considered an artefact. Since this is an insufficient condition, some criteria related to the data are usually added. A typical criterion is to combine the threshold with the boundary conditions of the algorithm [11]. For example, a signal-segment is considered an artefact when the instantaneous amplitude exceeds six times the average amplitude of the recording over the preceding 10 seconds [5].

Usually, the electrical signals generated by the muscular activity have steeper slopes than the average EEG signal. Therefore, the first and second-order derivatives can be used to measure EEG signal mobility [8]. The methodology of differential registration is used in order to improve spatial resolution. The idea is to analyze the signal amplification and digitization systems independently for each electrode with CMOS technology [12].

The time-resolution can be improved by the feed-forward displacements of an order of 10 ms, on the selected time segment. This is used for applications where an analyzing fast signal change is required, such as detecting epileptic seizures [5].

In previous work [13], we found that EEG signals can be analytically represented by using a quadratic linear-parabolic model:

$$y = a\sin(x - \pi) + b(x - 10)^2 + c. \tag{11.3}$$

In this work, we make use of this model in order to differentiate EEG resting states signals from artefacts. The underlying hypothesis is that parameters a, b and c are characteristics of these two classes. Here we estimated a, b and c using the least-squares method [14]. Table 11.1 shows the resulting coefficients for different types of signals. One notices that a threshold approach can be used in order to detect artefacts. Figure11.8 illustrates the detection based on this model.

Table 11.1. Estimated quadratic linear-parabolic coefficients and their associated 95% confidence bounds (CB). These results show that it is possible to use a threshold to distinguish between a resting signal and an artefact.

Artefact	a coefficient		b coefficient		c coefficient	
	Value	95 % CB	Value	95 % CB	Value	95 % CB
Resting	-7.536	[-7.933, -7.140]	0.778	[0.777, 0.778]	-91.910	[-92.24, -91.58]
Bruxism	-9.787	[-11.91, -7.659]	0.936	[0.935, 0.936]	-99.64	[-101.3, -97.99]
Arms movements	-9.193	[-11.27, -7.118]	0.939	[0.939, 0.940]	-112.4	[-114, -110.80]
Legs movements	-9.052	[-10.73, -7.377]	0.929	[0.928, 0.929]	-116.3	[-117.50, -115]
lalala	-10.630	[-13.89, -7.373]	0.920	[0.919, 0.920]	-13.61	[-16.08, -11.14]
Swallowing	-7.616	[-9.898, -5.333]	1.010	[1.010, 1.010]	-168.6	[-170.30, -167]
Cell phone	-11.130	[-14.42, -7.837]	0.994	[0.994, 0.995]	-151.1	[-153.50, -148.70]
Lights	-13.400	[-15.18, -11.63]	1.047	[1.046, 1.047]	-183.5	[-184.70, -182.20]

This time-domain experimentation with the quadratic linear-parabolic model brings a different approach to artefact detection in EEG signals, as compared to existing techniques. The confidence intervals estimated for the coefficients (eq.11.3) provide a good discrimination between resting and artefact signals. This simple model can be implemented easily in real-time.

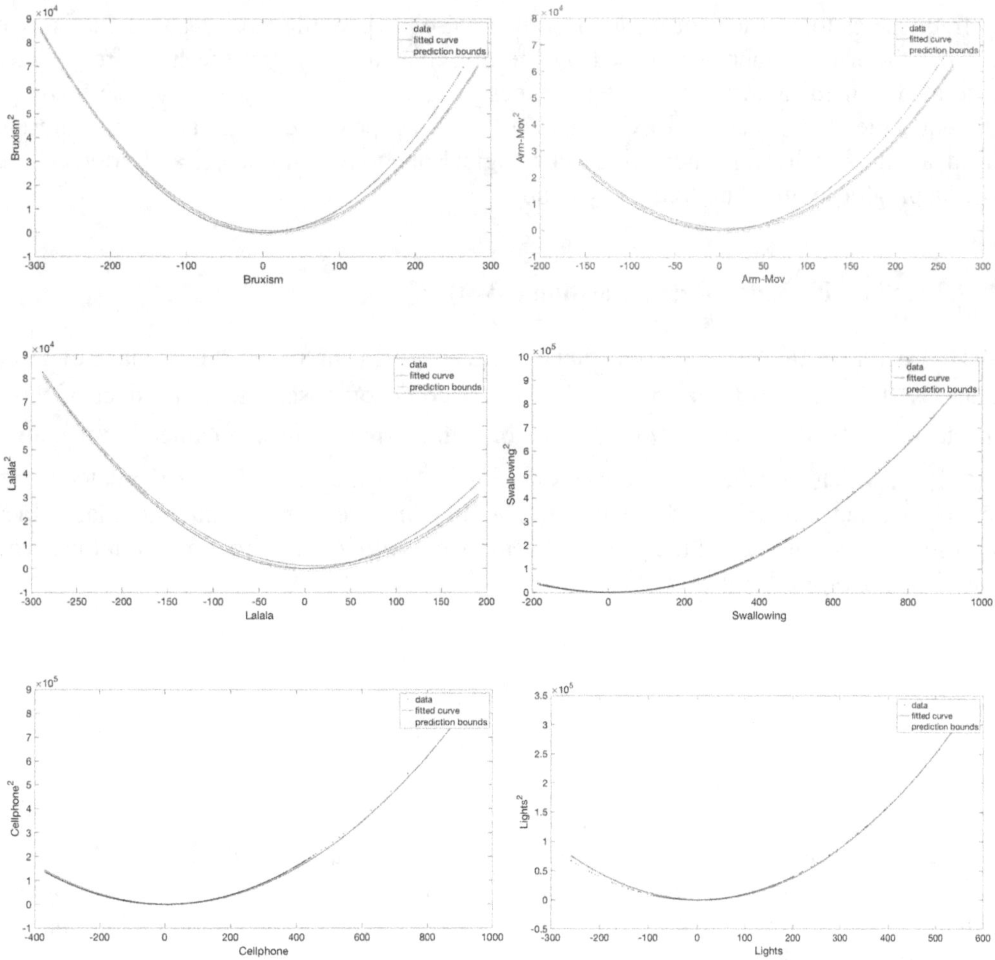

Fig. 11.8. Examples of fitting the quadratic linear-parabolic
model to different artefact signals.

11.4.2. Frequency-domain analysis

Fast Fourier transform (FFT) is the principal linear time-invariant (LTI) method to use with the signal power spectrum density (PSD), $PSD = |x(f)|^2$. This is easily achieved by calculating the discrete Fourier transform (DFT) from the correlation function [15]. For example, PSD shows irregular patterns on the higher harmonics of the frequency spectrum for artefacts [8]. PSD can also be used to decompose the EEG signal into different brain rhythms, following current medical practices. It is well known that a large time-interval for the frequency transformation gives a better frequency-resolution [11].

In general, LTI filters are used to reduce EMG and line interference artefacts [16, 17]. Notch and band-pass filters are two classical filters. A notch adaptive digital filter allows all frequencies to pass except for interference noise [18]. While a band-pass filter is used to filter the low frequencies caused by the muscular activity and the high frequencies generated by medical instrument interferences [5]. This allows attenuating a specific range of frequencies to extract the EEG signal in the purest possible form. The main filtering disadvantage is when an interesting neurological phenomenon occurs, and artefacts are overlapping or in the same frequency band.

11.4.2.1. Singular Value Decomposition (SVD)

SVD is a linear algebra operation that reduces a given matrix to the product of three matrices. Thus, the EEG matrix $X \in \mathbb{R}^{NxM}$ can be decomposed as the product of three matrices $X^T = USV^T$, where $U \in \mathbb{R}^{MxN}$ is an orthonormal matrix obtained from matrix $V \in \mathbb{R}^{NxN}$, which contains the eigenvectors. $S \in \mathbb{R}^{MxN}$ is the diagonal matrix related to the actual covariance matrix of the data set. The diagonal elements are the singular values representing the variance of the principal components ordered by order of magnitude. The classical expression is given by

$$X\vec{v} = \lambda\vec{v}$$
$$X\vec{v} = (\lambda I)\vec{v} \qquad (11.3)$$
$$(X - \lambda I)\vec{v} = \vec{0}$$

where λ are the square roots of the eigenvalue associated with the variance eigenvector $\vec{v}$. The eigenvectors have the same order as the eigenvalues and should be normalized and scaled by their respective singular values given by the square root of the eigenvalues, to become the principal components with decreasing order of magnitude. Note that, S and V depend on the source.

11.4.2.2. Multiple Signal Classification (MUSIC)

MUSIC is based on the Pisarenko harmonic decomposition method used to estimate the spectrum of the noise subspace, not the signal subspace. The idea is based on the assumption that after eigenvalue decomposition using SVD for example, the signal and noise subspaces are orthogonal, with a small noise spectrum at frequencies where a signal is present [15]. MUSIC narrowband estimator from the PSD is given by

$$PSD(\omega) = \frac{1}{\sum_{k=p+1}^{M} \left| FFT(V_k) \right|^2 / \lambda_k}, \qquad (11.4)$$

where M is related to the dimension of the eigenvector, p is the dimension of the signal subspace. Note that, $1 < \text{signal subspace} < p$, and $p+1 < \text{noise subspace} < M$. λ_k is the eigenvalue of the *k*th eigenvector, V_k is the *k*th eigenvector estimated from the correlation matrix of the input signal, which is ordered from the highest to the lowest power, see SVD for more details. Note that the denominator term λ_k is the summation of the scaled power spectra of noise subspace components. If the number of narrowband processes is known, then the dimensions of the signal subspace can be estimated as twice the number of present sinusoids or narrowband processes. If it is unknown, the signal subspace can be determined using the size of the eigenvalues. The idea is based on the fact that the noise components have about the same energy with a flat slope, and the signal components decrease in energy as they are ordered by the energy level. Using a scree plot, it is possible to find the number of components where the eigenvalue plot switches from a downslope to a relatively flat slope.

Fig. 11.9 shows the MUSIC performance to detect the different artefacts in EEG signals through the narrowband estimator from its PSD. The resting signal (black color) and legs movements artefacts (gray color) have the same PSD, therefore it is difficult to distinguish them. To the contrary, comparing the resting signal (black color) with the other artefacts, one can clearly see the possibility of distinguishing them using a threshold. Note that, the arm movements artefact (yellow color) and cell phone artefact (yellow color) have similar PSD. However, they are very different from the resting signal. Another interesting threshold is between the resting signal (black color) and the light artefact (magenta color), which has a larger amplitude with respect to the control signal. In physiological artefacts, it is also interesting to note that the resting state has the greatest amplitude with respect to the artefacts. This interesting result is useful in brain-computer interfaces (BCI) and neuron mirror studies with mu rhythm suppression [19]. The mean value and the confidence bounds show in Table 11.2 corroborate these results.

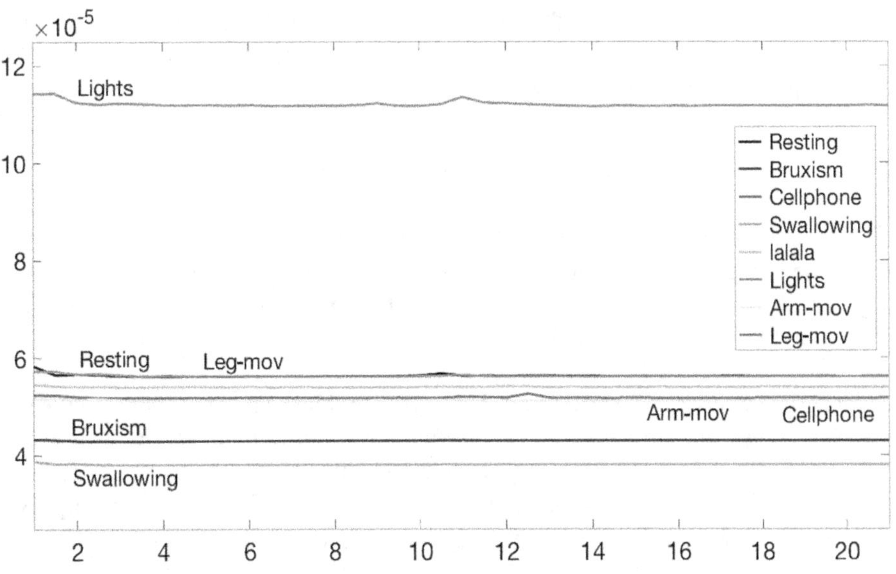

Fig. 11.9. Examples of artefacts using MUSIC narrowband estimator from the PSD. A threshold can be used to differentiate between resting signal and artefact signals.

Table 11.2. Values associated with 95 % confidence bounds (CB) from the MUSIC narrowband estimator from Fig. 11.9. All values are multiplied by e-05.

Artefact	MUSIC PSD	
	Value	95 % CB
Resting	5.619	[5.436, 5.803]
Bruxism	4.292	[4.245, 4.339]
Arms movements	5.121	[5.048, 5.195]
Legs movements	5.617	[5.472, 5.763]
lalala	5.401	[5.343, 5.459]
Swallowing	3.804	[3.745, 3.862]
Cell phone	5.178	[5.062, 5.29]
Lights	11.167	[10.804, 11.530]

11.4.3. Wavelet-domain Analysis

Many applications rely on the detection and recognition of waveform patterns from the background noise signal. Such waveforms, with spike or wave shapes, are typical in EEG signals. They manifest randomly in a short period of time, and due to their fast transient change, the Fourier analysis does not allow their detection. Therefore, mother wavelet with specific waveform can be used to help detecting such patterns, while determining the exact time at which they occur.

Wavelets used as filter banks analysis are very appropriate for denoising and filtering EEG signals. Typically, the low scales that are related to the high frequencies identify the additive noise. Thus, the signal can be filtered by removing the low-scale components. The next section introduces the 2D empirical Littlewood-Paley wavelet transform as a tool for artefacts detection in EEG signals. The method is based on detecting the Fourier boundaries with an empirical number of filters.

11.4.3.1. 2D Empirical Littlewood-Paley Wavelet Transform

The idea of the empirical wavelet transform (EWT) is to build a family of wavelets adapted to the signal of interest across two steps: 1) Detecting the Fourier supports and building the corresponding wavelet, 2) Filtering the input signal with the obtained filter bank to extract the different components [20].

The 2D empirical Littlewood-Paley transform (β) is widely used to filter images with 2D wavelets defined in the Fourier domain on annuli supports, centered around the origin. The inner and outer radius of these supports are fixed upon a dyadic decomposition of the Fourier plane [20]. The wavelets come from the iteration of filters with scaling, with the scale being the inverse of the frequency. Thus, wavelets are obtained from a single prototype mother wavelet by rescaling and shifting. The idea is to use the EEG signal with its rows and columns like a 2D signal as if it was an image [21]. In order to detect the radius of each annuli, the Fourier plane is considered as a polar representation $F_P(\bullet)$ since finding such boundaries is equivalent to working with the frequency modulus $|\omega|$. To avoid discontinuities in the output components, the spectrum $\overline{F}_P(\cdot)$ is averaged with respect to each angle θ, see equation (11.5) where N_θ is the number of discrete angles. The set of the spectral radius $\Omega = \left\{\omega^n\right\}_{n=0,\ldots,N}$, with $\omega^0 = 0$ and $\omega^N = \pi$, is obtained by estimating the Fourier boundaries on $\overline{F}_P(|\omega|)$.

The following 2D empirical Littlewood-Paley wavelet transform algorithm [20] is used:

1. *Input*: EEG signal $f(x) = x(n)$, number of filters N.

2. Compute the pseudo-polar FFT, $F_P(f)(\theta,|\omega|)$ and take the average with respect to the angle θ:

$$\overline{F}_P(|\omega|) = \frac{1}{N_\theta} \sum_{i=0}^{N_\theta-1} \left|F_P(f)(\theta_i,|\omega|)\right|, \tag{11.5}$$

3. Perform the detection of the Fourier boundaries on $\overline{F}_P(|\omega|)$ to get Ω and build the corresponding filter bank $\beta = \left\{\phi_1(X), \left\{\psi_n(X)\right\}_{n=1}^{N-1}\right\}$ according to

$$F_2(\phi_1)(\omega) = \begin{cases} 1 & \text{if } |\omega| \leq (1+\gamma)\omega^1 \\ \cos\left[\frac{\pi}{2}\beta\left(\frac{1}{2\gamma\omega^1}\left(|\omega|-(1-\gamma)\omega^1\right)\right)\right] & \text{if } (1-\gamma)\omega^1 \leq |\omega| \leq (1+\gamma)\omega^1, \\ 0 & \text{otherwise} \end{cases} \tag{11.6}$$

if $n \neq N-1$:

$$F_2(\Psi_{N-1})(\omega) = \begin{cases} 1 & \text{if } (1+\gamma)\omega^n \leq |\omega| \leq (1-\gamma)\omega^{n+1} \\ \cos\left[\frac{\pi}{2}\beta\left(\frac{1}{2\gamma\omega^{n+1}}\left(|\omega|-(1-\gamma)\omega^{n+1}\right)\right)\right] & \text{if } (1-\gamma)\omega^{n+1} \leq |\omega| \leq (1+\gamma)\omega^{n+1} \\ \sin\left[\frac{\pi}{2}\beta\left(\frac{1}{2\gamma\omega^n}\left(|\omega|-(1-\gamma)\omega^n\right)\right)\right] & \text{if } (1-\gamma)\omega^n \leq |\omega| \leq (1+\gamma)\omega^n \\ 0 & \text{otherwise} \end{cases} \tag{11.7}$$

if $n = N-1$:

$$F_2\left(\phi_{N-1}\right)\left(\omega\right) = \begin{cases} 1 & \text{if } \left(1+\gamma\right)\omega^{N-1} \leq \left|\omega\right| \\ \sin\left[\dfrac{\pi}{2}\beta\left(\dfrac{1}{2\gamma\omega^{N-1}}\left(\left|\omega\right|-\left(1-\gamma\right)\omega^{N-1}\right)\right)\right] & \text{if } \left(1-\gamma\right)\omega^{N-1} \leq \left|\omega\right| \leq \left(1+\gamma\right)\omega^{N-1}, \\ 0 & \text{otherwise} \end{cases}$$

$$(11.8)$$

4. Filter f by using equations (11.9) and (11.10)

$$W_f\left(n,x\right) = F_2^*\left(F_2\left(f\right)\left(\omega\right)\overline{F_2\left(\Psi_n\right)\left(\omega\right)}\right),\qquad(11.9)$$

$$W_f\left(0,x\right) = F_2^*\left(F_2\left(f\right)\left(\omega\right)\overline{F_2\left(\phi_1\right)\left(\omega\right)}\right),\qquad(11.10)$$

where F_2 is the 2D Fourier transform and F_2^* its inverse.

5. *Output*: β and $W_f\left(x,n\right)$.

Figure 11.10 illustrates the 2D empirical Littlewood-Paley wavelet transform reconstruction of a signal by using this algorithm. Through visual inspection, we note that the resting signal has all the components while the other signals are outliers or artefacts.

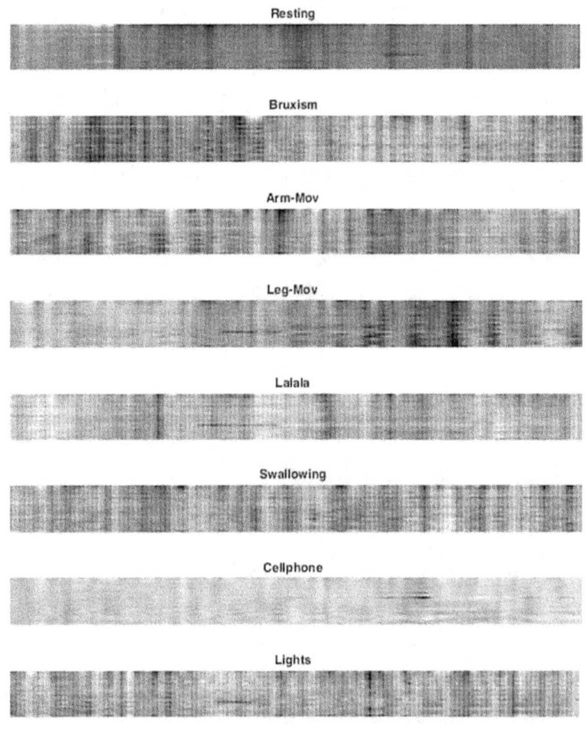

Fig. 11.10. Signal reconstruction using 2D empirical Littlewood-Paley wavelet transform. The error-values are defined as the difference between resting signal and each artefact signal.

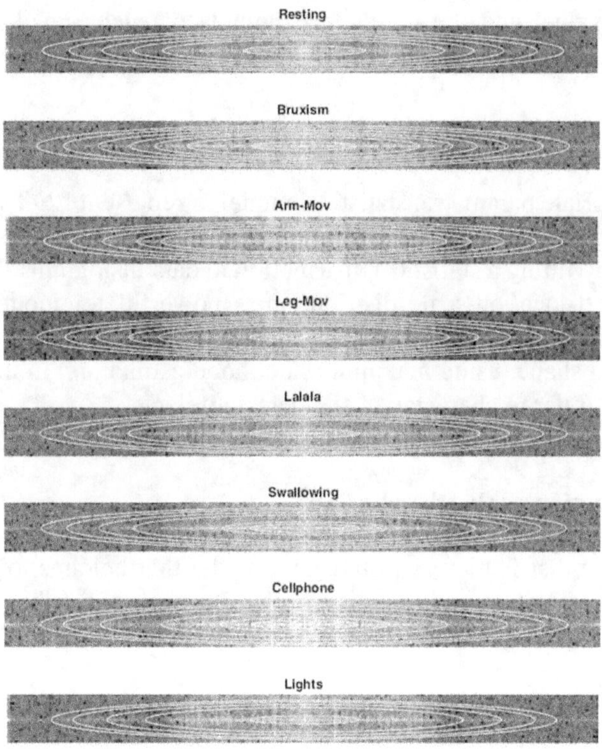

Fig. 11.11. Concentric boundaries from the 2D empirical Littlewood-Paley wavelet transform.

Table 11.3. Mean and bounds of boundaries from 2D empirical Littlewood-Paley wavelet transform.

Artefact	Boundaries	
	Value	**95 % BC**
Resting	1.924	[0.662, 2.822]
Bruxism	1.810	[0.552, 2.810]
Arms movements	2.169	[0.154, 2.810]
Legs movements	1.679	[0.294, 2.810]
lalala	1.983	[0.576, 2.785]
Swallowing	1.781	[0.576, 2.773]
Cell phone	1.740	[0.564, 2.785]
Lights	1.936	[0.589, 2.736]

We propose to use the boundaries of this 2D empirical Littlewood-Paley wavelet transform in order to detect the artefacts in EEG signals. The boundaries estimated using the algorithm introduced above allows us to define a threshold that detects artefacts. Figure 11.11 illustrates the polar representation for the 2D empirical Littlewood-Paley wavelet transform. We can observe the concentric boundary variations. But the visual inspection is complex due to the dynamics and variability of the EEG signals. Table 11.3

shows the mean value, and the confidence bounds of each signal. We can derive a threshold to detect artefacts.

11.4.4. Statistical Modeling

Creating an appropriate parametric statistical model to represent EEG signals can be very useful to capture the characteristics of the underlying physical process. When feasible, this is often done by fitting a statistical distribution to data histograms. In previous works, we found that the t-location-scale distribution is powerful for modeling EEG signals [22-24]. Fitting this distribution was performed by estimating its three parameters, namely location, scale, and shape, using maximum likelihood estimators. In this section, we use this distribution in order to characterize and detect artefacts.

11.4.4.1. The t-location-scale Distribution

The t-location-scale distribution is a statistical model that belongs to the location-scale family formed by translation and rescaling of the Student's t-distribution. Its probability density function (PDF) is given by

$$f\left(x\mid\mu,\sigma,\upsilon\right)=\frac{\Gamma\left(\dfrac{\upsilon+1}{2}\right)}{\sigma\sqrt{\upsilon\pi}\Gamma\left(\dfrac{\upsilon}{2}\right)}\left[\frac{\upsilon+\left(\dfrac{x-\mu}{\sigma}\right)^2}{\upsilon}\right]^{-\left(\frac{\upsilon+1}{2}\right)}, \tag{11.11}$$

where $-\infty<\mu<\infty$ is the location parameter, $\sigma>0$ is the scale parameter, $\upsilon>0$ is the shape parameter, and $\Gamma(.)$ is the Gamma function. Figure 11.12 shows the fit of the distribution to our EEG data histograms. This shows a good fit and makes it possible to correctly characterize different artefacts the distribution parameters.

From Table 11.4, one notices that it is possible to discriminate artefacts from a resting signal. The scale and shape parameters can be used in a straightforward manner to discriminate between artefacts and non-artefacts. While values of the location parameter associated with a muscular artefact and resting signal are close. Nevertheless, discrimination is still possible.

11.4.5. Multivariate Analysis

The principle of multivariate analysis is to consider the relationship between multiple variables. Precisely, this analysis operates on all data or measurements but treats them as a single entity.

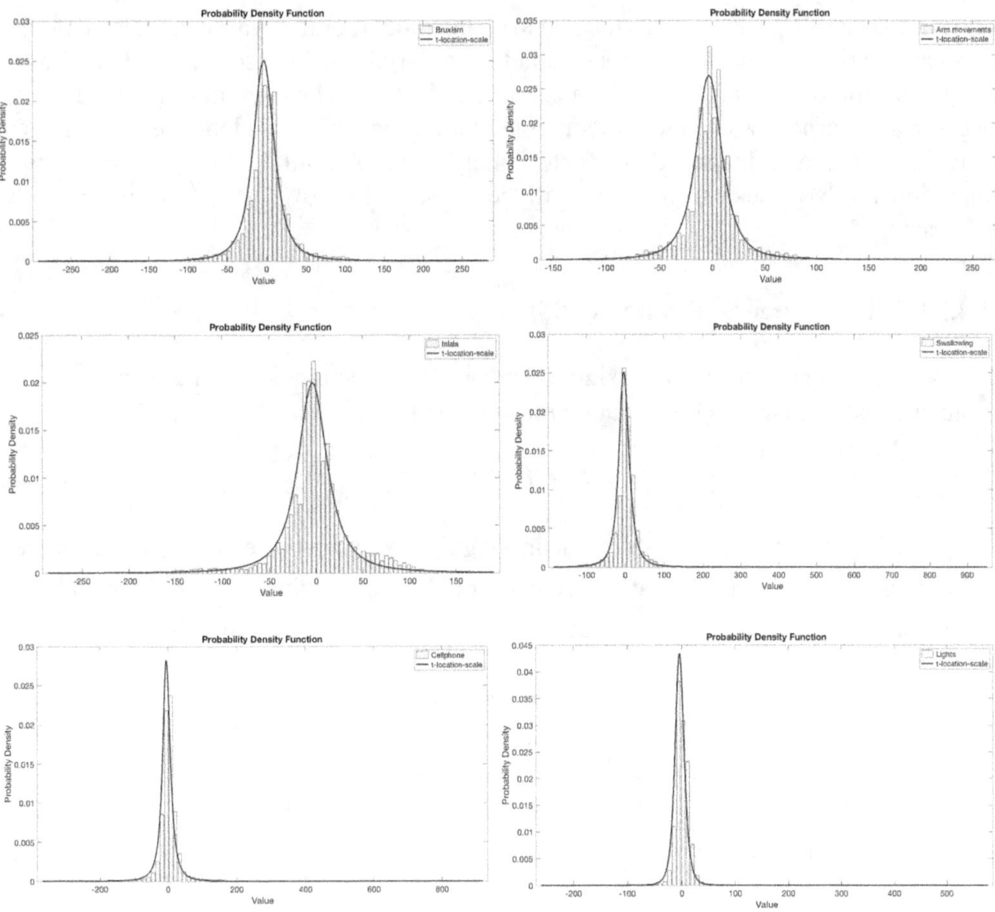

Fig. 11.12. Fits of the t-location-scale distribution to histograms of data for different artefacts.

Table 11.4. Mean values associated with 95 % confidence bounds (CB) from the t-location-scale parameters. These results show that it is possible to use intervals to differentiate between normal (resting) signals and artefacts.

Artefact	Location parameter		Scale parameter		Shape parameter	
	Value	95 % CB	Value	95 % CB	Value	95 % CB
Resting	-3.243	[-3.258, -3.229]	9.763	[9.749, 9.777]	6.598	[6.553, 6.644]
Bruxism	-3.580	[-3.631, -3.529]	14.257	[14.199, 14.316]	2.197	[2.179, 2.215]
Arm movements	-3.205	[-3.257, -3.153]	13.351	[13.293,13.410]	2.349	[2.328, 2.370]
Legs movements	-3.220	[-3.266, -3.174]	12.444	[12.396, 12.493]	2.958	[2.929, 2.987]
lalala	-3.747	[-3.819,-3.676]	17.309	[17.219, 17.400]	1.653	[1.639, 1.666]
Swallowing	-2.789	[-2.837, -2.740]	14.123	[14.069, 14.177]	2.069	[2.054, 2.084]
Cell phone	-4.513	[-4.565, -4.462]	1.677	[1.666, 1.689]	1.677	[1.666, 1.689]
Lights	-4.203	[-4.250, -4.155]	3.733	[3.674, 3.793]	3.733	[3.674, 3.793]

A classic example is the covariance matrix that incorporates variance related to the individual variables, and the covariance between variables. In essence, the multivariate data transformations search for a dimensional reduction. Three multivariate techniques used for this purpose are principal component analysis (PCA), independent component analysis (ICA), and independent vector analysis (IVA), all based on blind source separation (BSS). These techniques will be presented below in the context of EEG processing.

11.4.5.1. Blind Source Separation (BSS)

BSS assumes that the source signals reach the electrodes at the same time t instantaneously, equation (11.1) can be reformulated as

$$x(n) = H\,s(n) + v(n), \qquad (11.12)$$

where $x(n)$ is the EEG matrix, H is the mixing matrix, $s(n)$ is the matrix of the sources, and $v(n)$ is the noise. The separation is carried out by means of the matrix W, which uses only the information about $x(n)$ to reconstruct the original source signals as:

$$y(n) = W\,x(n) \qquad (11.13)$$

In BBS, the goal is to estimate the unmixing matrix W by using the singular value decomposition method (SVD) such that $Y = WX$ best approximates the independent sources S, where $y(n) = \left[y_1(n), \ldots, y_M(n) \right]^T$ and $x(n) = \left[x_1(n), \ldots, x_M(n) \right]^T$, therefore $S \rightarrow Y = WX$, see Fig. 11.13.

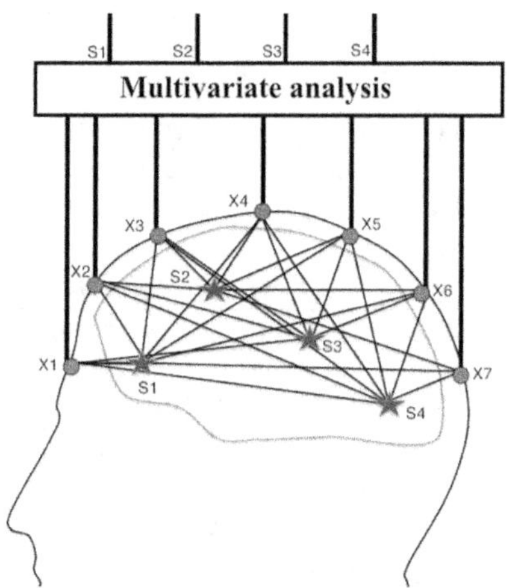

Fig. 11.13. Illustration of mixing information from brain sources $s(n)$ to create an EEG channel $x(n)$. Multivariate analysis methods aim at finding the brain sources that generated the EEG signals (source [29]).

During the BSS process, the method separates EEG signals into components of the EEG signals. Components attributed to artefacts are identified and discarded. The signal is then reconstructed based only on the remaining components. This method does not require having the reference of artefacts. BBS is typically used to correct EOG, EMG, and ECG signals [25-27].

11.4.5.2. Principal Component Analysis (PCA)

In PCA the idea is to transform the original data into a new smaller dimension dataset without loss of information. This transformation between a dataset of correlated variables into a new dataset of decorrelate variables is called principal components because it includes the most significant information from the original data. PCA uses the eigenvectors of the covariance matrix from the original data set in order to transform the data into another coordinate system, estimating the input projection data with their higher variability. Once this is done, the data set components are extracted by the eigenvectors data projection, see the previous two sections for more details. In the PCA algorithm, three recommendations have to be considered: i) data must be centered by removing the mean values $(x - \bar{x})$., ii) data must be rotated in order to decorrelate it, iii) the data must be normalized according to its variance before the transformation. The normalization process removes variables with very small values. For example, if $\lambda_1 >> \lambda_2$, the information energy will be mostly concentrated in the dominant matrix $X_1 = \sqrt{\lambda_1} u_1 v_1$ associated with the first singular values. This will be higher with respect to the other values; therefore, many values could be removed. According to [28], PCA cannot completely separate eye-movement artefacts, EMG, and ECG artefacts from the EEG signal, especially when they have comparable amplitudes. PCA also does not necessarily decompose similar EEG features into the same components when applied to different segments. Additionally, PCA is not consistent with respect to the interference suppression from overlapping brain activity [30]. One solution is to use the artefact subspace reconstruction (ASR) approach. The idea is to learn from a clean calibration statistical data model in order to attenuate the EEG artefacts by decomposing it in small segments and comparing them to calibration data in the PCA component subspaces [31]. A recent interesting study was done in simultaneous EEG-fMRI data, where the helium-pump artefacts given by the mechanical vibrations in the MRI system were removed using the EEG-segment-based principal component analysis in the gamma band [32].

11.4.5.3. Independent Component Analysis (ICA)

In PCA, the data decorrelation is not sufficient to produce independence between the variables at least when the variables have non-Gaussian distributions [15]. Thus, the ICA goal is to transform the original data into statistically independent variables, which will be the most significant variables. The assumption is that the original variables are independent and non-Gaussian. The signals are assumed to be linear combinations of these original variables, with negligible propagation delay.

As such, the EEG signals are considered as combinations of the underlying neural sources with added artefacts. Signals from multiple sources S are assumed to arrive simultaneously at electrodes X see Fig. 11.8. The number of sources is usually greater than the number of electrodes. Each electrode consists of a mixture of all sources, which are assumed independent. Therefore, applying ICA to EEG signals is a way for estimating the sources. The goal is not to reduce the number of signals, but to produce signals that are more meaningful. The typical model that relates the sources with the measurements is given by

$$x = As, \qquad (11.14)$$

where x vector of EEG signals, s are the sources, and A is the mixing matrix. This assumes that the measured signal vector x, is related to the underlying source signals vector s by a linear transformation. The objective is to determine the unknown mixing matrix. Assuming that the mixing matrix is square, the independent components can be obtained through matrix inversion

$$s = A^{-1}x, \qquad (11.15)$$

Since the size of x is supposed to be higher than s, inversion cannot be done immediately. PCA is usually applied first to reduce the number of components to be equal to the considered number of sources. Then the ICA algorithm is applied. While assuming sources to be independent and non-Gaussian, this method does not require their distribution to be known [15]. The following standard ICA algorithm is used:

ICA algorithm

1. Center the original data by removing the mean values $(x - \overline{x})$. Note that, the signals are assumed to be correlated.

2. Apply a whitening process by rotating the data in order to decorrelate them until the non-Gaussianity of the transformed data set is maximized. This process helps to trace variance information. After whitening, the components must be scaled in order to have unit variance $\sigma^2 = 1$.

3. Estimate the data Gaussianity. The estimation builds on the result of the central limit theorem that states that the distribution of a set of k random variables that are independent and identically distributed with mean μ and variance $0 < \sigma^2 < 1$, converges to a Gaussian distribution as k grows, regardless of the distribution of the independent variables. The kurtosis measurement is used to quantify the lack of data Gaussianity. Kurtosis is the fourth order cumulant defined as $kurt(x) = E\{x^4\} - 3\left[E\{x^2\}\right]^2$, where E is the expectation operator. Kurtosis may compress the data using a nonlinear function to restrain outliers before taking the fourth power [15]. Note that, for real data that have zero mean, the variance $E\{x^2\} = \sigma^2$, therefore $kurt(x) = \text{mean}(x^4) - 3$. Depending on the shape of the distribution, there are three types of denominations. If $kurt = 0$ one speaks about non-Gaussian distribution,

432

if $kurt < 0$ that data has sub-Gaussian distribution with shape broad and flat, and if $kurt > 0$ data has super-Gaussian distribution with shape spiky.

4. Estimate eigenvalues of x.

5. Apply JADE algorithm for a real-valued signal [33] to find the unmixing matrix A^{-1} with the number of independent components.

EEG signals have the assumption that the volume conduction is linear and instantaneous and the underlying brain signals are mixed together with their artefacts. Thus, the eye and muscular activity sources, line noise, and cardiac signal usually are not separated from the EEG sources activity. Therefore, ICA assumes that there exists an effective number of statistically independent signals that contribute to the scalp activity and that are not known. But it is possible to find them and discriminate between the artefacts when the signal is reconstructed using the unmixing signal, see equation (11.14) and equation (11.15). ICA was adopted as a strong algorithm used within the EEGLAB toolbox for eliminating multiple artefacts [34] and continues to evolve with the help of the scientific community as will be exposed by using the independent vector analysis approach in the next subsection.

11.4.5.4. Independent vector analysis (IVA)

IVA is an extension from the univariate source signals formulation to multivariate source signals formulation [35]. We use the algorithm proposed in [36], where the idea is to minimize the BSS mutual information through the sources to achieve independent vector analysis with second-order statistical information across datasets by assuming that the source component vector distributions are multivariate Gaussians. Precisely, this type of assumption has been considered in EEG signals [21, 37]. In [36], the proposed approach has the particularity of relying on a cost function equivalent to a particular multiset canonical correlation analysis (MCCA). Thus, the IVA cost function minimization by using a decoupled gradient-based optimization scheme permits us to find the set of nonorthogonal unmixing matrices.

Let us consider the multivariate formulation based on the classical BBS approach problem, see equation (11.12), for data observations from K datasets each formed from linear mixtures of N independent sources

$$X^{[K]} = A^{[K]} s^{[K]}, \qquad 1 \leq k \leq K, \tag{11.16}$$

where $s^{[K]} = \left[s_1^{[k]}, \ldots, s_N^{[k]} \right]^T$ is the zero-mean source vector, and $A^{[K]}$ is the invertible mixing matrix $A^{[K]}$, see equation (11.14). They are unknown real-valued quantities to be estimated. The n-th source component vector $s_n = \left[s_n^{[1]}, \ldots, s_n^{[K]} \right]^T$ is independent of all other source component vectors within the dataset and exactly dependent on at most one source in each of the other datasets. Therefore, their joint distribution for all the independent sources is given by $p\left(s_1,\ldots,s_N\right)=\prod_{n=1}^{N}p(s_n)$. The BBS approach solution is given by $y^{[K]}=W^{[K]}x^{[K]}$, where the goal is to find K unmixing matrices $W^{[K]}$ and source vector estimates for each dataset $y^{[K]}$, see equation (11.13).

The identification of the independent source component vectors is achieved by minimizing the mutual information among the source component vectors

$$
\begin{aligned}
\mathcal{I}_{IVA} &= \sum_{n=1}^{N}H\left[y_N\right]-\sum_{k=1}^{K}\log\left|\det\left(W^{[k]}\right)\right|-C_1 = \\
&= \sum_{n=1}^{N}\left(\sum_{k=1}^{K}H\left[y_N^{[k]}\right]-I\left[y_N\right]\right)-\sum_{k=1}^{K}\log\left|\det\left(W^{[k]}\right)\right|-C_1,
\end{aligned}
\tag{11.17}
$$

where $y_N=\left[y_n^{[1]},\ldots,y_n^{[K]}\right]^{T}$ is the n-th source component vector estimation s_n, and C_1 denotes the constant term $H\left[x^{[1]},\ldots,x^{[K]}\right]$. The equation (11.17) shows that minimizing the cost function simultaneously minimizes the entropy of all components and maximizes the mutual information within each source component vector.

The zero-mean and real-valued multivariate Gaussian distribution is useful to solve the joint BSS approach problem using second-order statistics and is given by

$$
p\left(y_n\mid\Sigma_n\right)=\frac{1}{\left(2\pi\right)^{\frac{K}{2}}\det\left(\Sigma_n\right)^{\frac{1}{2}}}\exp\left(-\frac{1}{2}y_n^{T}\Sigma_n^{-1}y_n\right),
\tag{11.18}
$$

where K is the source component vector dimension and Σ is the covariance matrix. Substituting in the equation (11.17) the expression $H(y)=\frac{1}{2}\log\left[\left(2\pi e\right)^{K}\prod_{k=1}^{K}\lambda_k\right]$, where λ_k is the k-th eigenvalue of the covariance matrix Σ. Then, the cost function for the multivariate Gaussian distribution is given by

$$
\mathcal{I}_{IVA-MG}=\frac{NK\log\left(2\pi e\right)}{2}+\frac{1}{2}\log\left(\prod_{n=1}^{N}\prod_{k=1}^{K}\lambda_{k,n}\right)-\sum_{k=1}^{K}\log\left|\det\left(W^{[k]}\right)\right|-C_1,
\tag{11.18}
$$

where $\lambda_{k,n}$ is the k-th eigenvalue of the covariance matrix associated with the n-th source component vector. The cost function from equation (11.18) indicates that the product of the source component vector covariance eigenvalues should be minimized. Under the constraint that the sum of the eigenvalues is fixed, then the minimal cost is met when each covariance matrix is as ill-conditioned as possible. Thus, the eigenvalues are maximally spread apart. See [36] for a general description of this multivariate Gaussian formulation.

In order to check the potentiality of the independent vector analysis (IVA), we use the bruxism signal that has artefacts in all channels, see Fig. 11.2. For illustration, the resulting demixing matrices $W^{[K]}$ are illustrated in Fig. 11.14, and the signal reconstruction is shown in Fig. 11.15. Note that, this process can help to remove artefacts from the signal, as it can be seen in the difference between Fig. 11.2 and Fig. 11.15. However, it is important to make a visual inspection with the expert physician.

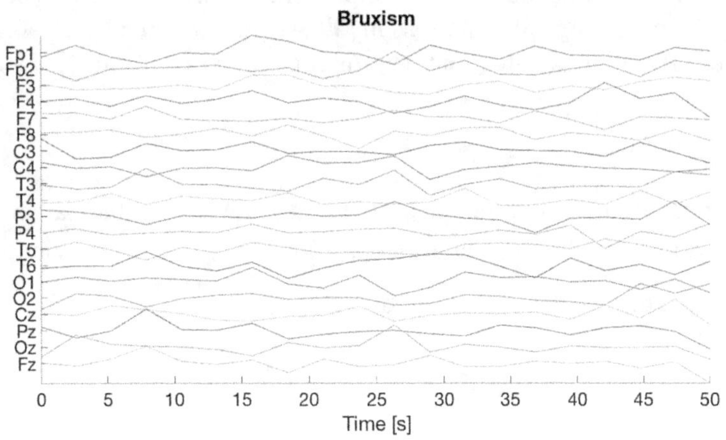

Fig. 11.14. Example of unmixing matrices obtained using IVA for Bruxism signals.

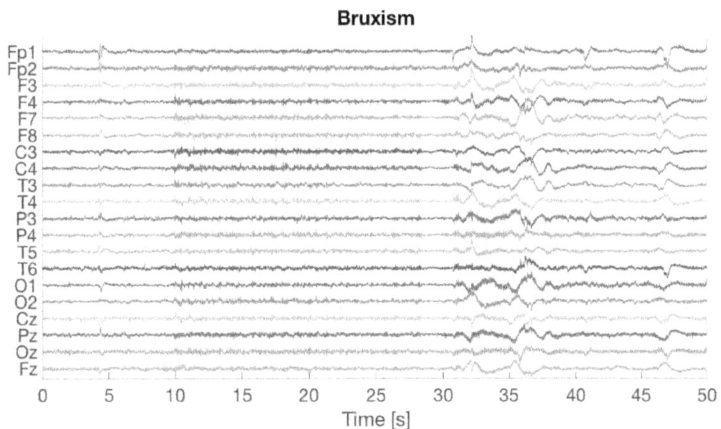

Fig. 11.15. Example of reconstructed Bruxism signals using IVA.

11.4.6. Data-mining Domain

Data mining is usually defined as the use of automated procedures to extract useful information and insight from large datasets such as the EEG signal [38]. In practice, the EEG data contains various types of anomalous measurements, which are typical during the signal acquisition. These anomalous measurements can complicate the analysis because of the prevalence of artefacts, outliers, missing or incomplete data, and other more subtle phenomena such as misalignments. Results that are essential to the medical diagnosis might be invalidated due to these phenomena. We propose the Hampel filter to correct artefacts in EEG signals. This is a non-linear filter for data cleaning that searches outliers or abnormal local data in a temporal sequence, making it interesting for looking for abnormalities in EEG signals. The motivation of using this filter is related to a previous no published work [39] where a descriptive statistical analysis and visual inspection were used to study the performance of the Hampel filter. This filter has been used in the literature to study physiological signals, as in [40, 41] for removing artefacts in deep brain stimulation (DBS), and in [42] for compensating motion artefacts in ECG signals.

11.4.6.1. Hampel Filter

Let K in the equation (11.1) be the positive tuning integer flexible parameter from the window half-width W^K. The standard median filter m_k using the moving data window for each channel sequence $\{X_k\}$ is given by

$$m_k = \text{median}\{x_{k-K}, \ldots, x_k, \ldots, x_{k+K}\}, \qquad (11.19)$$

The main advantage of the median is its extreme resistance to local outliers or impulsive noise. While its main disadvantage is the possible introduction of significant distortions.

The Hampel filter H_K belongs to the class of decision-based filters [38] that detect outliers based on the median and the MAD scale estimator. It is based on the moving-window implementation of the Hampel identifier [43]. The filter response is given by

$$y_k = \begin{cases} x_k & |x_k - m_k| \leq tS_k \\ m_k & |x_k - m_k| > tS_k \end{cases}, \qquad (11.20)$$

where m_k is the median value from the moving data window and S_k is the MAD scale estimate, defined by

$$S_k = 1.4826 \quad \text{median}_{j \in [-K, K]}\{|x_{k-j} - m_k|\}. \qquad (11.21)$$

The factor S allows the MAD scale estimator of the standard deviation of Gaussian data to be unbiased. Precisely, EEG signals have been shown to follow a generalized Gaussian distribution [37]. Consequently, the Hampel filter is a good candidate to cancel outliers and artefacts in EEG. Note that if the threshold parameter $t = 0$, the standard median filter corresponds to $y_{K|t=0} = m_k$. This allows using the Hampel filter as a generalization of the median filter, with t as an additional tuning parameter. Note that the MAD scale estimator is sensitive to implosion. If more than 50 % of the data values are the same, the MAD scale estimate is zero, independently of the other values and the parameter t. Thus, data must implosion-free in order to obtain a correct characterization of the signal. Then, for an implosion-free sequence $\{x_k\}$, the Hampel filter reduces to an identity filter for some sufficiently large but finite value of t restricted to the range $0 \leq t \leq T$, where T is the threshold value. Thus,

$$t \geq \frac{\max_k \left| x_k - m_k \right|}{\min_k S_k} \tag{11.22}$$

We refer the reader to [43, 38, 44] for a comprehensive treatment of the mathematical properties of the Hampel filter.

In order to evaluate this filter, we run experiments on short epochs from each artefact signals. These epochs were annotated by a neurologist to indicate the location of the artefact in the signal. Figure 11.16 shows the results. We can notice that for physiological artefacts, when the variability of the EEG signal to that of the artefact, the Hampel filter corrections are close to the original signal. However, with high peaks, the filter brings over-corrections. We, therefore, recommend choosing a correct moving window in order to not lose relevant information. Inversely, for non-physiological artefacts, the high peaks are the principal outliers to correct making the Hampel filter useful for this type of outliers. However, visual inspection remains essential to avoid losing relevant information.

11.5. Conclusions

In this chapter, we reviewed some classical signal processing methods for detecting and removing artefacts in EEG signals. The goal was to use a supervised analysis to estimate a threshold capable of discriminating between normal EEG and artefact signals. Experiments were conducted using a signal database created specifically for this purpose. A resting-state signal was used as a control signal for comparing with artefact signals of different types. First, a time-domain analysis was used to detect artefacts with a linear-parabolic model-fitting method. Results showed that the parabolic waveform is interesting in analyzing EEG signals with a convex waveform. Second, the MUSIC algorithm was used to conduct frequency-domain analysis using a narrowband estimator from the power spectrum density of the signal. Results showed interesting discrimination between normal signals and artefacts, but with a high computational cost. The third review was with wavelet-domain analysis, where we evaluated the 2D empirical Littlewood-Paley wavelet transform across its boundaries. Using boundaries as features turned out to be useful to differentiate between normal and artefact signals. Finally, the study of the statistical parameters of the t-location-scale distribution showed a good fit with EEG data. Using these parameters as features were shown to be useful for detecting artefacts in EEG signals. In conclusion, the study of all these methods showed the feasibility of detecting artefacts in EEG signals, with good performance. This makes possible the design a multi-class or multi-label detection-classification [45] with unbalanced data [46], which is would be very useful during the acquisition of EEG signals.

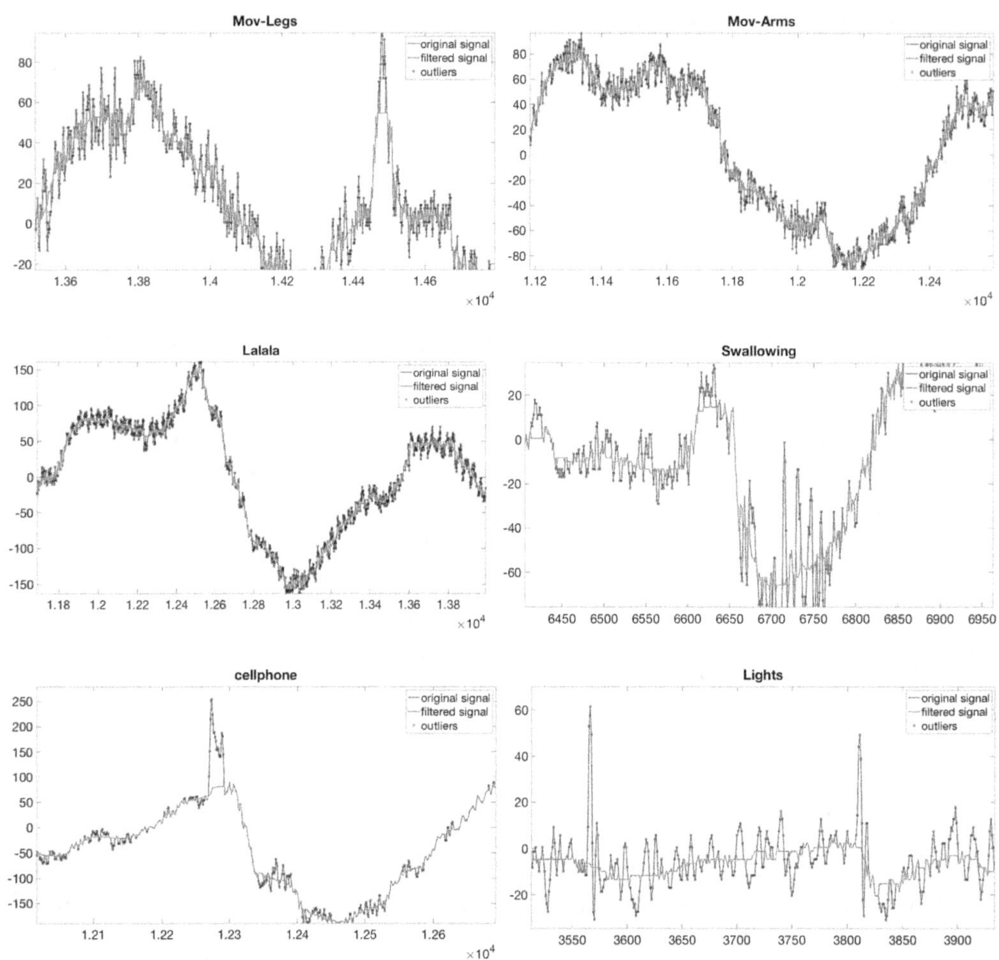

Fig. 11.16. Examples of artefact epochs extracted from one EEG channel. The blue line is the original signal, and the red line is the signal with the outliers removed using the Hampel filter.

For artefacts removal, we evaluated the IVA algorithm from the multivariate analysis domain and the Hampel filter as a data-mining technique. These have proven to be effective techniques to remove outliers and artefacts. Such methods can be implemented in combination with the detection methods described above to achieve automatic detection and filtering processes. However, it is important to conduct such an analysis under the supervision of expert neurologists in order to not lose essential medical information.

References

[1]. J. Malmivuo, R. Plonsey, Bioelectromagnetism, *Oxford University Press*, 1995.

[2]. C. M. Sinclair, M. C. Gasper, A. S. Blum, Basic electronics in clinical neurophysiology, *The Clinical Neurophysiology Primer, Humana Press*, 2007, pp. 3-18.

[3]. S. V. Vaseghi, Advanced Digital Signal Processing and Noise Reduction, *Wiley*, 2006.

[4]. A. B. Usakli, Improvement of EEG signal acquisition: An electrical aspect for state of the art of front end, *Computational Intelligence and Neuroscience*, Vol. 2010, 2010, 630649.

[5]. K. Najarian, R. Splinter, Biomedical Signal and Image Processing, *CRC Press*, 2016.

[6]. D. M. White, C. A. Van Cott, EEG artefacts in the intensive care unit setting, *American Journal of Electroneurodiagnostic Technology*, Vol. 50, Issue 1, 2010, pp. 8-25.

[8]. L. Sörnmo, P. Laguna, Bioelectrical Signal Processing in Cardiac and Neurological Applications, *Academic Press*, 2005.

[7]. L. Frølich, I. Winkler, K.-R. Müller, Investigating effects of different artefact types on motor imagery BCI, in *Proceedings of the 37th Annual International Conference of the IEEE Engineering in Medicine and Biology Society (EMBC'15)*, 2015, pp. 1942-1945.

[9]. A. Peper, C. A. Grimbergen, EEG measurement during electrical stimulation, *IEEE Transactions on Biomedical Engineering*, Vol. 30, Issue 4, 1983, pp. 231-233.

[10]. S. Sanei, J. A. Chambers, EEG Signal Processing, *Wiley-Interscience*, 2013.

[11]. W. van Drongelen, Signal Processing for Neuroscientists, *Academic Press*, 2018.

[12]. S.-F. Yen, D. Schoonover, J. C. Sanchez, J. C. Principe, J. G. Harris, Differential EEG, in *Proceedings of the 4th International IEEE/EMBS Conference on Neural Engineering (NER'09)*, 2009, pp. 18-21.

[13]. A. Quintero-Rincón, C. D'Giano, H. Batatia, A quadratic linear-parabolic model-based EEG classification to detect epileptic seizures, *The Journal of Biomedical Research*, Vol. 34, Issue 3, pp. 205-212, 2020.

[14]. P. C. Hansen, V. Pereyra, G. Scherer, Least Squares Data Fitting with Applications, *Johns Hopkins University Press*, 2013.

[15]. J. L. Semmlow, Benjamin Griffel, Biosignal and Medical Image Processing, *CRC Press*, 2014.

[16]. L. P. Panych, J. A. Wada, M. P. Beddoes, Practical digital filters for reducing EMG artefact in EEG seizure recording, *Electroencephalography and Clinical Neurophysiology*, Vol. 72, Issue 3, 1989, pp. 268-276.

[17]. M. van de Velde, G. van Erp, P. J. Cluitmans, Detection of muscle artefact in the normal human awake EEG, *Electroencephalography and Clinical Neurophysiology*, Vol. 107, Issue 2, 1998, pp. 149-158.

[18]. S. P. Cho, M. H. Song, Y. C. P. Choi, H. Seon, K. J. Lee, Adaptive noise canceling of electrocardiogram artefacts in single channel electroencephalogram, in *Proceedings of the 29th Annual International Conference of the IEEE Engineering in Medicine and Biology Society (EMBS'07)*, 2007, pp. 3278-3281.

[19]. A. Quintero-Rincón, C. D'Giano, H. Batatia, Mu-suppression detection in motor imagery electroencephalographic signals using the generalized extreme value distribution, in *Proceedings of the International Joint Conference on Neural Networks (IJCNN' 2020)*, 2020, pp. 1-5.

[20]. J. Gilles, G. Tran, S. Osher, 2D Empirical transforms. Wavelets, ridgelets and curvelets revisited, *SIAM Journal on Imaging Sciences*, Vol. 7, Issue 1, 2014, pp. 157-186.

[21]. A. Quintero-Rincón, J. Prendes, M. A. Pereyra, H. Batatia, M. Risk, Multivariate Bayesian classification of epilepsy EEG signals, in *Proceedings of the 12th IEEE Workshop on Image, Video, and Multidimensional Signal Processing (IVMSP'16)*, 2016, pp. 1-5.

[22]. A. Quintero-Rincón, J. Prendes, V. Muro, C. D'Giano, Study on spike-and-wave detection in epileptic signals using t-location-scale distribution and the k-nearest neighbors classifier, in *Proceedings of the IEEE URUCON Congress on Electronics, Electrical Engineering and Computing*, 2017, pp. 1-4.

[23]. I. Zorgno, et al., Epilepsy seizure onset detection applying 1-NN classifier based on statistical parameters, in *Proceedings of the Biennial Congress of Argentina (ARGENCON'18)*, 2018, pp. 1-5.

[24]. A. Quintero-Rincón, V. Muro, C. D'Giano, J. Prendes, H. Batatia, Statistical Model-Based Classification to Detect Patient-Specific Spike-and-Wave, *Computers,* Vol. 9, Issue 4, 2020, pp. 1-14.

[25]. L. Shoker, S. Sanei, M. A. Latif, Removal of eye blinking artefacts from EEG incorporating a new constrained BSS algorithm, in *Proceedings of the Annual International Conference of the IEEE Engineering in Medicine and Biology Society (EMBS'04)*, 2004, pp. 909-912.

[26]. J. Gao, P. Lin, P. Wang, Y. Yang, Online EMG artefacts removal from EEG based on blind source separation, in *Proceedings of the 2nd International Asia Conference on Informatics in Control, Automation and Robotics (ICINCO'10)*, 2010, pp. 28-31.

[27]. H. Hallez, A. Vergult, R. Phlypo, I. Lemahieu, Muscle and eye movement artefact removal prior to EEG source localization, in *Proceedings of the Annual International Conference of the IEEE Engineering in Medicine and Biology Society (EMBS'06)*, 2006, pp. 1002-1005.

[28]. G. Geetha, S. N. Geethalakshmi, Scrutinizing different techniques for artefact removal from EEG signals, *International Journal of Engineering Science and Technology (IJEST)*, Vol. 3, Issue 2, 2011, pp. 1167-1172.

[29]. A. Quintero-Rincón, M. Flugelman, J. Prendes, C. D'Giano, Study on epileptic seizure detection in EEG signals using largest Lyapunov exponents and logistic regression, *Revista Argentina de Bioingeniería*, Vol. 23, Issue 2, 2019, pp. 17-24.

[30]. N. T. Haumann, M. Huotilainen, P. Vuust, E. Brattico, Applying stochastic spike train theory for high-accuracy human MEG/EEG, *Journal of Neuroscience Methods*, Vol. 340, 2020, 108743.

[31]. T. R. Mullen, et al., Real-time neuroimaging and cognitive monitoring using wearable dry EEG, *IEEE Transactions on Biomedical Engineering*, Vol. 62, Issue 11, 2015, pp. 2553-2567.

[32]. H.C. Kim, S.-S. Yoo, J.-H. Lee, Recursive approach of EEG-segment-based principal component analysis substantially reduces cryogenic pump artefacts in simultaneous EEG-fMRI data, *Neuroimage*, Vol. 104, 2015, pp. 437-451.

[33]. J. F. Cardoso, A. Souloumiac, Blind beamforming for non-Gaussian signals, *IEE Proceedings F – Radar and Signal Processing*, Vol. 140, Issue 6, 1993, pp. 362-370.

[34]. D. Arnaud, S. Makeig, EEGLAB: An open source toolbox for analysis of single-trial EEG dynamics including independent component analysis, *Journal of Neuroscience Methods*, Vol. 134, 2004, pp. 9-21.

[35]. T. Kim, I. Lee, T.-W. Lee, Independent vector analysis: Definition and algorithms, in *Proceedings of the Fortieth Asilomar Conference on Signals, Systems and Computers (ACSSC'06)*, 2006, pp. 1393-1396.

[36]. M. Anderson, X.L. Li, T. Adali, Nonorthogonal independent vector analysis using multivariate Gaussian model, *Lecture Notes in Computer Science*, Vol. 6365, 2010, pp. 354-361.

[37]. A. Quintero-Rincón, M. Pereyra, C. Risk, M. D'Giano, H. Batatia, Fast statistical model-based classification of epileptic EEG signals, *Biocybernetics and Biomedical Engineering*, Vol. 38, 2018, pp. 877-889.

[38]. R.K. Pearson, Mining Imperfect Data Dealing with Contamination and Incomplete Records, SIAM: Society for Industrial and Applied Mathematics, 2005.

[39]. A. Quintero-Rincón, M. Risk, S. Liberczuk, EEG preprocessing with Hampel filters, in *Proceedings of the Biennial Congress of Argentina (ARGENCON'12)*, Vol. 89, 2012, pp. 1-6.

[40]. M. Dagar, N. Mishra, A. Rani, S. Agarwal, J. Yadav, Performance comparison of Hampel and median filters in removing deep brain stimulation artefact, *Innovations in Computational Intelligence, Studies in Computational Intelligence*, Vol. 713, 2018, pp. 17-28.

[41]. D. P. Allen, E. L. Stegemöller, C. Zadikoff, J. M. Rosenow, C. D. Mackinnon, Suppression of deep brain stimulation artefacts from the electroencephalogram by frequency-domain Hampel filtering, *Clinical Neurophysiology*, Vol. 121, Issue 8, pp. 1227-1232, 2010.

[42]. F. A. Ghaleb, M. B. Kamat, M. Salleh, M. F. Rohani, S. A. Razak, Two-stage motion artefact reduction algorithm for electrocardiogram using weighted adaptive noise cancelling and recursive Hampel filter, *PLoS ONE*, Vol. 13, Issue 11, 2018, e0207176.

[43]. A. Davies, U. Gather, The identification of multiple outliers, *Journal of the American Statistical Association*, Vol. 88, Issue 423, 1993, pp. 782-792.

[44]. R. K. Pearson, Y. Neuvo, J. Astola, M. Gabbouj, Generalized Hampel filters, *EURASIP Journal on Advances in Signal*, 2016, 87.

[45]. F. Herrera, F. Charte, A. J. Rivera, M. J. del Jesus, Multilabel Classification, *Springer*, 2016.

[46]. A. Fernández, et al., Learning from Imbalanced Data Sets, *Springer*, 2018.

Chapter 12

BeE-nose – An In-hive Multi-gas-sensor Extension to the IndusBee 4.0 System for Hive Air Quality Monitoring and Varroa Infestation Level Estimation

Andreas König

12.1. Introduction

The impact of insects, wild bees, and honey bees in particular, on our ecosystem and our agricultural production system is a well-known fact and found its place in public awareness. Documentaries like 'More than Honey' of M. Imhoof et al. or the book of Tautz [5] etc. contributed to communicate lucid to the public. The dependence of food production on pollination is so significant, that in the recent pandemic the systemic role of hive keepers was honored by assuring their roaming privilege without the restrictions applicable to most of the population.

However, the happy and easy days of bee keeping unfortunately seem to be past, as numerous threats from pesticides to parasites, threaten these essential contributors to our ecosystem. For instance, to name just a few, chalk brood, tracheal and varroa mites, viruses, colony collapse disorder (CCD), nosema, the small hive beetle (SHB), or phorid flies, have arrived or are currently approaching and spreading. The *varroa destructor* (VD) in the case of honey bees, is a major issue for bee keeping and the observed increasing number of colony collapses are largely attributed to this parasite. Without numerous effective initiatives and counter measures, one consequence is the need for manual pollination, as already exercised in Asia, or the intentions to automate the natural pollination service by bee-like drones. A most recent patenting activity to be observed was by Walmart (cf. [20]) on a bee-like pollination robot. Numerous similar activities can be found in the web on different realization scales. In case of pesticides, neonicotinoides and others cause heavy losses to bee populations trying to exercise pollination on the so treated plant cultivation. The weakened bee populations are easy prey for illnesses and

Andreas König
Institute of Integrated Sensor Systems, TU Kaiserslautern, Kaiserslautern, Germany

parasites then, including also wax moths, like the *Galleria Mellonella*, increasingly aggravating the necessary attention and time of hive keepers as well as requiring more and more elaborate skills.

Traditionally, hive tending works on a scheduled and *post mortem* basis with numerous opening and inspection of the hives, which is invasive, disturbing for the bees and weakening their capabilities to resist the above named increasing challenges. The surging advance in sensor & actuator integration technology, MEMS [7], in particular, ,Machine Learning, and general automation, e.g., Industry 4.0, which also opens the door to the conception of bee-like robots or drones (cf. [20]), allows to introduce capable yet affordable hive instrumentation solutions to closely monitor and manage an increasing number of hives and conduct less invasive *event-driven* tending of bees with minimized hive openings, helping the hive keeper to do the right thing at the right time. Thus, ***digital*** bee keeping moves from scheduled and *post mortem* to *predictive* and *event-driven* operation, thus reducing effort, cost, and risk. Major bee research institutions, e.g., in Germany Hohenheim, Veitshöchheim etc. [28-30] have dealt with related issues and goals employing in numerous cases rather costly and lab-size equipment.

The ongoing vivid advance in sensor, actuator, electronics, (RF) communications, and packaging/integration technology (cf. [20]) provides a remarkable richness in sensory diversity and capability as well as an unprecedented leverage for well performing, yet cheap and disappearing application systems, that seem to make Berkeley's Smart-Dust vision getting reality. The driving force of both in-line as well as mobile sensory and analytic devices, step by step rivaling the performance of bulky and expensive desktop analytic equipment, provides also for bee keeping numerous promising opportunities. For instance, the concurrently developing field of handheld food and drug scanners as well as Point-of-Care (PoC) devices is an excellent example and bears numerous relationships to be exploited for bee hive condition monitoring and hive keeper assistance systems, e.g., honey quality assessment tool as an extension of established hygrometers. Miniaturized visual, IR, or hyperspectral sensors and cameras, MEMS microphones and vibration sensors, ultrasonic transducers, e.g., capacitive-micromachined-ultrasonic-transducers (CMUTS), and last not least integrated gas sensors and gas sensor systems offer powerful monitoring options at low cost. From major research institutes to a vivid maker scene, an increasing amount of projects and initiatives of hive instrumentation and bee monitoring can be found., e.g., in [13-15] and in the reference list of [20] Even in the maker scene and those institutions, that base on off-the-shelf electronic components and systems, an on-going miniaturization of the measurement systems for hive monitoring and integration of those inside of the hives, e.g., in the comb, can be observed. Though these solutions still differ substantially in size and occupation of the comb, the trend to have a small and potentially cheap instrumentation, leaving the majority of the comb for the bees, can be observed. The motivation to carry this on to dedicated monolithical sensor and measurement systems is also one motivation of the **IndusBee 4.0** project.

A most remarkable research work in that direction was announced by Intel in cooperation with the Australian CSIRO (Commonwealth Scientific and Industrial Research Organization) already in 2015 [7] and carried on for several years. An RFID-chip with sensing and additional capabilities, e.g., power generation by harvesting from vibration, was employed in a system based on Intel Edison to study details of bee behavior. For that purpose, the chips or tags, which amount to a third of the average worker bees weight, were mounted as *backpack* on bees. A large number of bees, in the order of 10.000 and more have been tagged and data, e.g., on flight pattern and foraging behavior, collected [8, 9]. Data is read from bees returning to the hives by base stations at the flight hole or in the hive. This, and related work [10] is intriguing, shows the potential of moving from off-the-shelf packaging solutions to dedicated monolithic and MEMS solutions, and offers interesting baseline for further hive monitoring and hive keeper assistance system. This will be briefly revisited in Section 12.3.

Additionally, there is a striking similarity to the field of Industry 4.0, where the condition of technical systems, mainly for production, is monitored to predict maintenance requirement and optimize up-time and reliability of these systems. Concepts of so called Self-X-Systems and Machine Learning/AI are combined to achieve the aspired level of (self)-awareness and adaptation capability. This surging development of Technical Cognition Systems, comprised of integrated sensors, electronics, RF, energy harvesting, and Machine Learning/AI, offers a leverage and an answer for many issues met in bee keeping. This kindled the **IndusBee 4.0** project, employing experience from prior Industry 4.0 related work and experience in sensory system micro technological integration. Two major lines of activities can be observed and are pursued in **IndusBee 4.0**, i.e., *Bee Hive Monitoring* by sensors, embedded systems and custom software, and *Hive Keeper Assistance Systems* by discrete appliances for a group of hives in an apiary.

In this contribution, we will in general report on the status of the **IndusBee 4.0** hive monitoring system and put a focus of its extension on introducing contemporary integrated gas sensing to the monitoring palette. This can naturally serve for hive air quality monitoring, but more interesting is the potential of extracting hints on colony health condition. In particular, the estimation and prediction of *varroa* infestation level [2, 36] based on direct or indirect indicators is one possible objective of the gas analysis. In contrast to existing approaches, in **IndusBee 4.0** hive integration of the measurement systems are pursued, as this avoid a couple of real-world problems, e.g., due to dew point, in more discrete and hive external devices. Several conventional as well as (semi)-automated methods for varroa screening [20] have been pursued in prior work and will be revisited to provide and estimated *Ground Truth* for the actual infestation level assumed to be found in gas sensing measurement.

The main part of the chapter is structured in a surveying second section on integrated gas sensors, chips, and application systems. Here, also the issue of general hive air analysis will be addressed and a selection for the first **IndusBee 4.0** prototypes will be motivated. The third section will summarize the status of the **IndusBee 4.0** hive monitoring system with the current gas sensing unit. The fourth section will describe the in-field application and data acquisition, including the estimation of the *varroa* infestation level *Ground Truth* by conventional methods. The fifth section will describe the evaluation methods employed, present the experiments, and discuss the first results. The concluding section will summarize the approach, present and discuss the intermediate results from the running bee season, as well as give an outlook on the next steps in gas sensing or **BeE-Nose** advance and general aspired extensions to the **IndusBee 4.0** activity.

12.2. Integrated Gas Sensors and Sensing Systems

In addition to the extremely common acquisition of temperature and moisture at various locations in the hives, and the hive weight by various versions of scales, the hive air has also been in the focus of monitoring activities for a long time, e.g., [15, 14, 41]. In particular, the monitoring of CO_2 concentration has been part of numerous hive condition monitoring systems for many years. The advance of sensor technology allows both to get more apt and more versatile gas sensors of diminishing size ready for application to bee colony monitoring. The monitoring of CO_2 concentration has been a topic of interest by itself, trying to understand, in how far bees are also able to regulate not only temperature and moisture but also CO_2 concentration in the hive.

The gradual loss of that ability, or deviation from normal regulation ability and concentration values, of course in the context of other sensory registrations, basically also provides a principal indicator to the hive health condition and emerging issues. This very basic idea kindles the interest in more profound analysis of hive air and composites for possible alerting indicators, direct or indirect in nature. However, this is a rather intricate and demanding problem requiring advanced gas analysis laboratory equipment for the examination of hive air and possible pathological changes in this gas composition, that are possibly correlated with problematic hive conditions.

In this context, another field of research becomes increasingly interesting. For a long time, bees have not only been highly esteemed for their pollination services and delivery of honey, wax, pollen, or royal jelly, but also the hive air has achieved the reputation to be healthy for persons with respiration system issues. Consequently, inhalation systems attached to a hive have been conceived, that allow human patients to inhale hive air. Naturally, this heuristic approach triggered scientific curiosity and review including thorough analysis of hive air with regard to its composition and ingredients that could indeed have the attributed positive health aspect. Such investigations use state-of-the-art analysis equipment that extracts hive air for external measurement. One such activity can be found at TU Dresden, faculty of chemistry, Prof. Speer, who is active in hive air analysis, which is described in a very intriguing way in an on-line available document [11]. As this work is very promising from the point of view of the **IndusBee 4.0** activity to clarify possible direct indicators of hive issues with parasites or illnesses, a contact was established in the first quarter of 2020 This personal communication delivered, that until that time instant, only the absolute composition of the hive air was in the focus of interest, not the relative variations or deviations due to possible hive infestation by a parasite, e.g., *varroa destructor,* or other illness, e.g., *foul brood*. A bit sobering was to learn of the opinion, that even with full-fledged macroscopic analytic equipment, the possible finding of such abnormality indicators could potentially be hard without further cues or context.

This implies, that the aspired use of affordable and small scale gas sensors, with reduced analytical capability compared to lab equipment, would make it even less probable to find the useful and stable (direct) indicators. The creation of a database from hive air analysis complemented by concurrent varroa infestation level monitoring of the analyzed hive was considered as a point to be pursued in future activities. While this definitely most interesting and superior database is not yet available, in the **IndusBee 4.0** project, the more moderate, but rather easy to implement approach of using integrated gas sensors or gas sensor systems for hive air monitoring, while proceeding with concurrent conventional varroa screening for the regarded hive(s), is pursued. The approach bases on the somewhat heuristic hope, that the sensitivity and selectivity of today's available integrated gas sensors will be able to detect a direct or indirect indication of the pest of interest, e.g., the varroa infestation level. Possible cues could origin from understanding the parasites metabolism, e.g., the varroa excretions and their effect on hive air for appropriate selection of suiting sensor sensitivity. Indirect indications could also be unusual or abnormal patterns in CO_2 concentrations, as well as unusual or abnormal volatile organic compound (**VOC**) concentrations and regulation deficiencies. This is also a general air quality monitoring issue and very able and convenient sensor have become available, e.g., the Sensirion SGP30 [21], that been selected first in 2017 for the **IndusBee 4.0** project.

Gas sensing and the achievement of affordable small scale gas sensors and possibly hand-held devices has been an issue of interest and amazing progress for more than three decades by now. Applications deal with the detection of fires and related smoke, hazardous gases detection like carbon monoxide, hydrogen, or methane, as well as ethanol, oxygen or carbon dioxide detection and quantity or concentration measurement. From fire detectors, to alcohol level estimators to food scanners, food sniffers [32] in particular, applications have been pursued and E-nose-systems based on available sensors been achieved. Two concepts have found particular attention and interest, these are so called metal-oxide-gas-sensors (MOX) and gas sensitive field effect transistors, GASFETs, which bear similarity to the ISFETs employed in chemical analysis of liquids. One very interesting gas sensor MEMS/microelectronics implementation and system was achieved by former MICRONAS, Freiburg, e.g., [22], which was of major interest for the **IndusBee 4.0** realization and activities, but unfortunately has stopped being available any more. More activities seem to be pursued, but not immediately applicable to the bee monitoring task.

Due to their simplicity, potential low-cost realization, and MEMS/microelectronics integration potential, MOX sensors have found widespread use and in this work [35] and related work [2], variants of those have been regarded. Oxide layers, like, e.g., SnO_2, ZnO_2 or WO_2, serve as sensory layers. MOX sensors require active heating for their operation, so a heater and a temperature sensor are commonly integrated in MEMS solution. In particular, it is of great interest, that the selectivity and sensitivity of this class of sensors can be effectively influenced by temperature modulation [12], i.e., by imposing transient or cyclic thermal modulation on the sensor via the heater by the power electronics. Though this gives interesting leverage and has raised also the term of *virtual gas sensor array*, it also makes the conception of required sensor electronics more challenging and more space consuming in discrete solutions.

A major provider of MOX gas sensors is Figaro [23] and a collection of those discrete sensors with various sensitivities have been used in [2] in an intriguing work on varroa infestation detection by a hive-external gas sensor unit based on a selection of Figaro sensors. From the rich portfolio of Figaro, the CDM7160 CO_2 sensor and the TGS8100 MEMS air quality sensor are in particular interesting for the work pursued in **IndusBee 4.0** work.

Renesas ZMOD4410 for indoor air quality and ZMOD4510 for outdoor air quality monitoring gas sensor and AQI generation [24] provide eCO_2 and TVOC detection and are also interesting candidates for the work aspired in our research. The Sensirion SGP30 [21] offers similar feature of eCO_2 and TVOC detection and, thus, has been selected as first candidate for inclusion in the modular **BeE-Nose** module of our **IndusBee 4.0** activity. Another very interesting sensor is the Bosch Sensortec BME680 [25] that includes also temperature, moisture, and pressure sensing along with the gas sensor unit. In contrast to SGP30, the software for AQI generation is not open source, which is not a mandatory requirement for the sought after indicators of varroa infestation level. However, one big advantage is the possible heater control, e.g., for cyclic modulation, that allows with some restrictions to realize concepts, e.g., described in [12] and other work on virtual gas sensor arrays, in a highly integrated embodiment, contrasting realizations with Figaro sensors. The SGP30 multi-pixel gas sensor unfortunately does not allow this influence on the sensor for the standard user/customer. The BME680 has also been included in our **BeE-Nose** module.

Another interesting MEMS gas sensor in common use in related projects is the SGX Sensortech MiCS-6814 [26] for carbon monoxide, ammonia and other gases related to agricultural/industrial odors, the MiCS 4515 [27] for carbon monoxide, nitrogen dioxide, ammonia and others, or the MiCS 5914 for ammonia and other gases detection. However, these MEMS sensor contain sensing and heater elements only in their packages and require substantial external electronics. Nevertheless, the MiCS-6814 with its agricultural odors background seems to be an interesting complement to the **BeE-Nose** module. Also, there are MEMS gas sensors CCS801 [29] and the CCS811 [28] with integrated electronics and heater control for thermal cycling provided by AMS for eCO_2 and TVOC detection, the latter not unsimilar to the already mentioned SGP30 of Sensirion or the BME680 of Bosch Sensortec. Further, there is a very interesting German sensor manufacturer, the Umweltsensortechnik GmbH (UST) [30], in Geschwenda, Thuringia. They provide a range of gas sensor and the two most interesting ones are the Triplesensor, which features three different gas-sensitive MOX-layers for reducible, easily oxidizable and heavily oxidizable gases. The packaged sensor for VOC detection comprises sensory and heater elements, external electronics are required for operation, e.g., a USB Kit and proprietary software, which makes integration into the **BeE-Nose** module a trifle harder. In addition to this very interesting sensor, UST provides a VOC/CO_2 sensor module [31], that employs a photo acoustical principle for the detection of CO_2, i.e., provides a real CO_2 sensor and not a eCO_2 one as in many other air quality sensors. Further, the photo acoustical evaluation requires a MEMS microphone, which could also serve for acoustical hive monitoring. Ignoring the substantial price of that unit, it was considered for inclusion in the **BeE-Nose** research and promising negotiations with UST had been initiated, but unfortunately did not come to a conclusion before this year's lock-down.

Though there are other integrated gas sensors worth to be included in this brief survey, for the purpose and constraint of the first BeE-Nose the presented ones comprise the most interesting selection. There have been numerous other recent activities creating compact and application-specific e-nose systems. This section will conclude with a brief rehearsal of representative implementations with closest relation to the aspired bee hive air monitoring.

Activities of integrated gas sensor systems with specific electronics and evaluation methods/ software can be found in tasks from food inspection to medical applications. A particular food scanner is provided by the commercial Food-Sniffer [32], a handheld device to probe the freshness of food, e.g., meat based on the gaseous emissions. Unfortunately, the underlying sensor technology has not been disclosed, even on request. So, linking the functionality to one or several of the sensors summarized above is unfortunately not possible. Another intriguing food application is given in [34]. In intelligent food packaging, a monitoring system is included in the package that monitors food quality parameters, like temperature or emitted gases. For the latter purpose, the BME680 and MQ series gas sensors [42]. The latter ones come with rich selection of different gas sensitivities, not dissimilar to the Figaro palette, but they also deliver only the sensor and heater arrangement in the package and require substantial external electronics.

For air quality analysis, a system with four sensors, the BME680, the SGP30, the CCS811, and the iAQ-Core also from AMS, has been developed and most recently been presented [1]. A neural network has been specifically conceived for the information processing in this activity, which from the point of sensing and neglecting the iAQ-Core incorporates a subset of the one in [4, 5]. Another air quality system is presented in [3] based on the BME680 and the MG-811 CO2 sensor [33].

A very interesting work, pursuing a modular approach as seems to be necessary for our BeE-Nose due to possible unknown indicators, has been conceived for breath analysis in the embodiment of a modular handheld device [4, 5], employing most of the sensors summarized above but the ones from UST. Purely analog sensors, requiring additional sensing and actuation electronics, and digital complete sensors or sensor systems are combined in this intriguing approach in a dedicated and very compact embodiment. Several of these sensors allow temperature modulation, which gives an even richer sensitivity and selectivity offering of that system.

Clearly, our **BeE-Nose** comes closest to this last application system, but substantially differs in the application, recognition goal, and long term measurement requirements. An open issue, not so feverishly discussed in related papers is the application-specific finding of thermal modulation or cycling parameter. Similar to the finding and computation of meaningful features, also the finding of these parameters is an issue of design automation, which we have pursued in prior work to **IndusBee 4.0** for quite some time now [16-19].

12.3. IndusBee 4.0 In-hive Monitoring System

The **IndusBee 4.0** in-hive monitoring system resembles in many details the organization and structure of other smart monitoring systems, e.g., from smart food or medical packages to Industry 4.0 applications. Fig. 12.1 shows the block diagram of the current **IndusBee 4.0** apiary monitoring system and its relation with emerging hive keeper assistance systems.

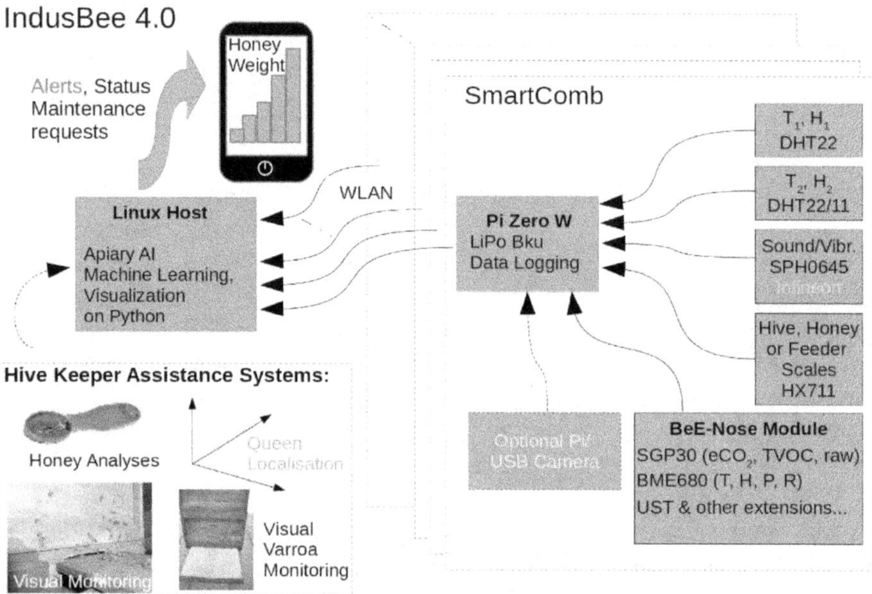

Fig. 12.1. Block diagram of **IndusBee 4.0** apiary monitoring system and interfacing to emerging hive keeper assistance systems.

The **BeE-Nose** described in this chapter, is a module of the **SmartComb** [20], that goes the same way as, e.g., the work reported in [4, 5]. In the first embodiment, only the Sensirion SGP30 and the Bosch Sensortec BME680 are incorporated. Further extensions, e.g., from the UST palette, such as Triplesensor and the CO_2/VOC-module, are pursued. In the first in-hive realization, standard or off-the-shelf components and break-out boards have been used for simplicity. The work reported in [4, 5] clearly shows the next step of integration, but there are even more efficient yet affordable System-in-Package (SiP) integration options, which will be pursued in the wake of this work, when the indicators and corresponding sensors for the varroa infestation level estimation have been determined from the first-cut prototypes. These can lead to extremely small and cost effective solutions, so that more than one comb per hive could be instrumented and give spatially distributed information on the hive status. In any case, even current **BeE-Nose** is small enough not to obstruct the bees use of the comb, which can be seen in some related approaches, and it is keeping the gas sensor system, in contrast to, e.g., [2], close to the origin of the aspired measurand and in very stable environmental conditions regulated by the bees themselves.

For long term measurement campaigns, e.g. for several weeks or for the whole bee season, the current node power budget will not allow to rely only on local energy storage. Thus, the **SmartComb** has been extended with a local LiPo-accu and charger that supports autonomous operation for several hours. Each hive in addition is connected in a distributional hierarchy to a wired 5 V supply of the apiary, currently fed from the power grid. In the frequent required hive inspections, **SmartCombs** can be detached and removed from the hive without interrupting and reengaging operation, only measurements collected in that time could be surprising and have to be eliminated from analysis as outliers by taking the context of the inspection time into account. The chosen Raspberry Pi Zero W is not efficient with regard to its power consumption, but it allows a very flexible approach, easy standard communication, transfer of existing solutions from other Linux computers, and its size allows the aspired first in-hive implementation. More power efficient cores and overall electronics could be considered after a stable system design has been explored, e.g., the suitable indicators and sensors for varroa infestation level and related tasks have been identified.

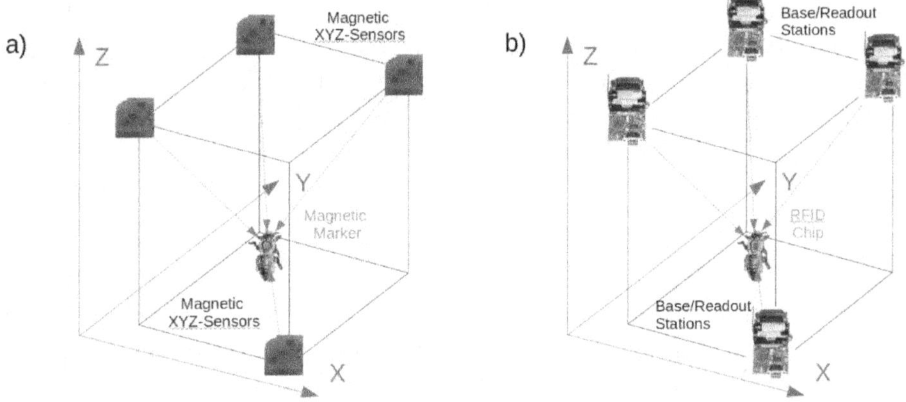

Fig. 12.2. Concept of **IndusBee 4.0** queen localization assistance system: a) Based on magnetic marker tracking, b) Use of RFID-chip as tag or marker.

From Fig. 12.1, it becomes obvious, that multi-sensor evaluation, e.g., fusing information from acoustical readings of the MEMS microphone and gas sensor readings can also be considered and pursued to get more reliable cues on possible issues of the bee colony.

In Fig. 12.2, another frequent issue for hive keepers is picked up. Assuring the presence of the queen and, in particular, being aware of the queen's location are common requirements, e.g., to avoid accidental bruising or dislocation from the hive of the queen. The desire for such a localization functionality in the community is confirmed in many newsgroups or logs. The current practice is, to give the queen a color tag, both as age identification as well as spotting aid. A similar localization issue is met in numerous technical and medical problems, e.g., in wireless sensor networks (WSN), general indoor localization, or smart pill tracking. Magnetic methods have been investigated and employed in large number, in particular in environments, where wireless communication works not robust enough for localization [37]. Several magnetic localization and tracking systems, e.g., from Matesy (cf. [37]), use spatially arranged sensors and a magnetic markers, e.g., as part of a smart pill, which is localized and pursued. Adapting this approach to localizing and tracking the queen in a hive seems to be attractive as magnetic markers with significant materials improvement become very small, while still observably strong. Magnetic XYZ-sensors in hive lid and bottom and a tiny magnetic marker, possible also colored, attached to the queen, as suggested in Fig. 12.2 (a), could allow the adoption of this concept. In addition to possible marker weight issues, several studies, e.g., [38, 39], have shown, that bees display a sensitivity to electro-magnetic fields and seem to be disturbed or degraded in performance by immanent magnetic influence or electro-magnetic fields. Loss of orientation or regularity in comb building are reported. Both the named marker tracking as well as the method employed in [37] with magnetic beacons have to be revisited with regard to this issue. Another alternative for localization links to the intriguing work Intel and CSIRO [7] are doing.

Tagging the queen with one of the employed RFID-chips, as suggested in Fig. 12.2(b), could be employed similar to localization in WSN by several base or readout stations in or at the hive, where the radio signal itself and its attenuation (Received-Signal-Strength-Indicator (RSSI)) and following triangulation or comparable methods can serve as baseline for localization. The longevity of the queen of several years compared to just 4-6 weeks of a worker bee, will also require the investigation of possible side effects and compatibility with queen and hive health [39]. But nevertheless, localization seems to be a promising, so far not visited, and practically relevant additional research topic for **IndusBee 4.0**.

Four **SmartComb** measurement stations given in Figs. 12.3 and 12.4 have been setup from 2019 until present and insertion in selected fully developed hives has started in May 2020 before the expected onset of significant varroa infestation. The first prototypes from 2019 have only been equipped with SGP30 gas sensor and applied for basic tests at hive [35]. By now four prototypes, extended to both SGP30 and BME680 inclusion, are available and insertion in bee hives for long term measurement and evaluation has started. Details of the **SmartComb** application, planned experiments, and intermediate results will be described in the following section.

The **IndusBee 4.0** activity also tries to exploit synergy between the ***digital*** bee keeping activities and **Industry 4.0** activities, e.g., intelligent condition monitoring based on advanced machine learning (ML) and artificial intelligence (AI) methods, including artificial neural networks (ANN) of various modeling and complexity levels. The creation of a varroa infestation level detector or estimator practically is a process of a virtual sensor creation, similar to e.g., [37], based on ANN and ML methods. Additionally, the robustness of sensors and sensor systems, as requested in current white papers for **Industry 4.0** and related initiatives, requires self-x-properties, i.e., self-monitoring, – trimming, or -repairing/-healing. Both reuse of Industry 4.0 techniques for bee keeping as well as options for advancing self-x-sensor systems in an application context free of industrial confidentiality requirements are offered by the pursued **IndusBee 4.0** activity.

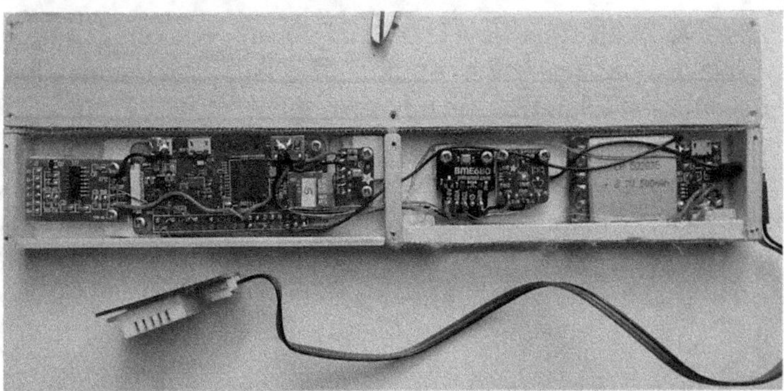

Fig. 12.3. IndusBee 4.0 autonomous data logging system of **SmartComb** with current gas sensor or **BeE-Nose** unit and power extension before completion and insertion into comb frame with opened lid. One DHT22 sensor is temporarily attached, as the SGP30 needs a humidity value, which at first-hand has been provided by the DHT22. Connection to second DHT22, both scales and the external power have to be complemented for this otherwise fully functional module.

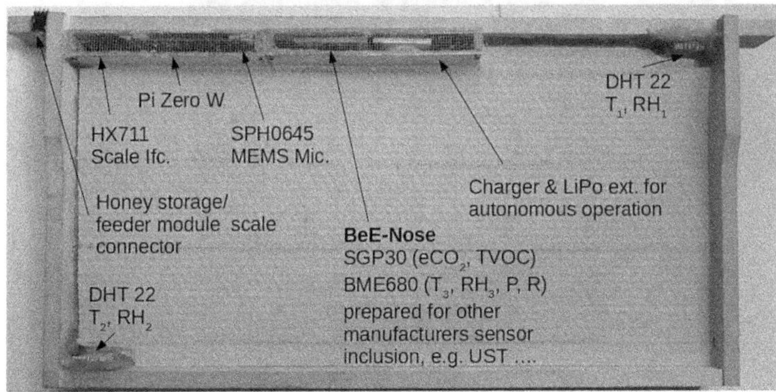

Fig. 12.4. Completed autonomous **SmartComb** prototype no. 3 with current **BeE-Nose** module and fresh support ready for hive insertion.

12.4. In-field Application and Data Acquisition

For the **IndusBee 4.0** activities an apiary of eight hives has been set-up in a rural location close to a house and the power grid. Currently, four of the hives are subject to the digital approach, while the other four are kept analog. In the group for the digital approach, two hives have been allowed to swarm and for the two remaining ones swarming has been prevented. The first hive instrumented with the current **BeE-Nose** is shown in Fig 12.5.

The **SmartComb** is placed in the second super close to the hive center, to be allocated as close as possible to the geometrical hive center, commonly also the center of activity and breeding. No other instrumentation, e.g., scales, has been applied to this particular hive, which has been prevented from swarming. Fig. 12.6 shows the partially extracted **SmartComb** prototype no. 4 with current **BeE-Nose** module, which is active in the hive for more than one month by now. Obviously, the instrumentation part consumes only a negligible part of the comb and, along with the choice of materials, is as little intrusive for the bees as possible with this level of simple integration technology. Possible incompatibility due to the operation of the processor, communication, and sensor electronics and need for shielding or spatial separation should be checked in concurrent investigation before a more widespread use of such systems. On the hive left, a cable can be seen entering the super, which serves for supplying the apiary 5 V power to all hives and **SmartCombs** to keep them going for weeks or months. This power comes straight from the grid via a power supply in the case of the lab apiary, but it could also come from solar harvesting and backup storage. The same cable also connects to the hive scale in the foot, if installed.

Fig. 12.5. First hive instrumented with **SmartComb** prototype no.4 with current **BeE-Nose** module. No other instrumentation, e.g., scales, is currently attached here. For this hive, swarming was prevented.

The **SmartComb** is delivering an extended data set compared to prior versions [20] and currently extracts data at very short, possibly redundant intervals of about two seconds. Fig. 12.7 shows an excerpt of a campaign running this June. The time stamp of the registration along with temperature and relative humidity values of the upper and lower DHT 22 are complemented by the eCO_2 and VOC readings of the SGP30, followed by the two scale entries, void, as no scales are currently connected to this hive, followed by the hive sound classification result and the two raw data values also offered from SGP30 [21]. Then, the output is complemented by the BME680 data on temperature, relative humidity, pressure, and the gas sensor resistance at a fixed heater-induced temperature of $T = 400°$ C. The BME680 nearly offers the data of a weather station, if complemented by a simple external light sensor. As the Bosch software for the BME680, including AiQ-index calculation and calibration issues, is unfortunately kept proprietary, currently only the straight resistance value is recorded for the investigations.

Fig. 12.6. SmartComb prototype no.4 with current **BeE-Nose** module temporarily extracted from its place in the hive, as practiced for the weekly check-ups in the swarm season. The instrumentation part consumes only a negligible part of the comb and is as little intrusive for the bees as possible.

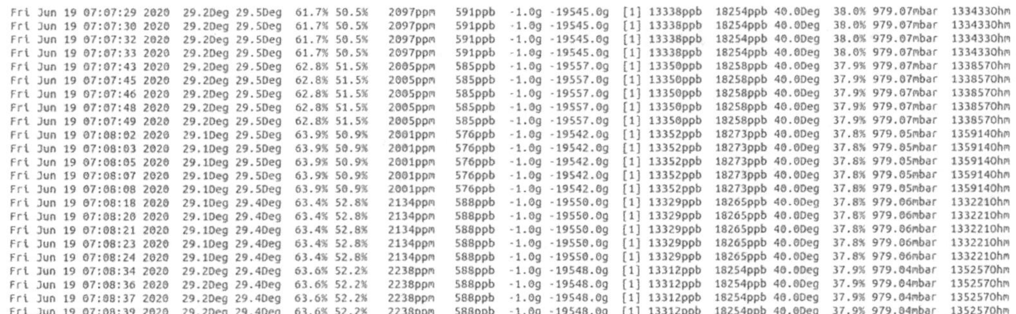

```
Fri Jun 19 07:07:29 2020  29.2Deg 29.5Deg  61.7% 50.5%   2097ppm  591ppb  -1.0g -19545.0g  [1] 13338ppb  18254ppb 40.0Deg  38.0% 979.07mbar  1334330hm
Fri Jun 19 07:07:30 2020  29.2Deg 29.5Deg  61.7% 50.5%   2097ppm  591ppb  -1.0g -19545.0g  [1] 13338ppb  18254ppb 40.0Deg  38.0% 979.07mbar  1334330hm
Fri Jun 19 07:07:32 2020  29.2Deg 29.5Deg  61.7% 50.5%   2097ppm  591ppb  -1.0g -19545.0g  [1] 13338ppb  18254ppb 40.0Deg  38.0% 979.07mbar  1334330hm
Fri Jun 19 07:07:33 2020  29.2Deg 29.5Deg  61.7% 50.5%   2097ppm  591ppb  -1.0g -19545.0g  [1] 13338ppb  18254ppb 40.0Deg  38.0% 979.07mbar  1334330hm
Fri Jun 19 07:07:43 2020  29.2Deg 29.5Deg  62.8% 51.5%   2005ppm  585ppb  -1.0g -19557.0g  [1] 13350ppb  18258ppb 40.0Deg  37.9% 979.07mbar  1338570hm
Fri Jun 19 07:07:45 2020  29.2Deg 29.5Deg  62.8% 51.5%   2005ppm  585ppb  -1.0g -19557.0g  [1] 13350ppb  18258ppb 40.0Deg  37.9% 979.07mbar  1338570hm
Fri Jun 19 07:07:46 2020  29.2Deg 29.5Deg  62.8% 51.5%   2005ppm  585ppb  -1.0g -19557.0g  [1] 13350ppb  18258ppb 40.0Deg  37.9% 979.07mbar  1338570hm
Fri Jun 19 07:07:48 2020  29.2Deg 29.5Deg  62.8% 51.5%   2005ppm  585ppb  -1.0g -19557.0g  [1] 13350ppb  18258ppb 40.0Deg  37.9% 979.07mbar  1338570hm
Fri Jun 19 07:07:49 2020  29.2Deg 29.5Deg  62.8% 51.5%   2005ppm  585ppb  -1.0g -19557.0g  [1] 13350ppb  18258ppb 40.0Deg  37.9% 979.07mbar  1338570hm
Fri Jun 19 07:08:02 2020  29.1Deg 29.5Deg  63.9% 50.9%   2001ppm  576ppb  -1.0g -19542.0g  [1] 13352ppb  18273ppb 40.0Deg  37.8% 979.05mbar  1359140hm
Fri Jun 19 07:08:03 2020  29.1Deg 29.5Deg  63.9% 50.9%   2001ppm  576ppb  -1.0g -19542.0g  [1] 13352ppb  18273ppb 40.0Deg  37.8% 979.05mbar  1359140hm
Fri Jun 19 07:08:05 2020  29.1Deg 29.5Deg  63.9% 50.9%   2001ppm  576ppb  -1.0g -19542.0g  [1] 13352ppb  18273ppb 40.0Deg  37.8% 979.05mbar  1359140hm
Fri Jun 19 07:08:07 2020  29.1Deg 29.5Deg  63.9% 50.9%   2001ppm  576ppb  -1.0g -19542.0g  [1] 13352ppb  18273ppb 40.0Deg  37.8% 979.05mbar  1359140hm
Fri Jun 19 07:08:08 2020  29.1Deg 29.5Deg  63.9% 50.9%   2001ppm  576ppb  -1.0g -19542.0g  [1] 13352ppb  18273ppb 40.0Deg  37.8% 979.05mbar  1359140hm
Fri Jun 19 07:08:18 2020  29.1Deg 29.4Deg  63.4% 52.8%   2134ppm  588ppb  -1.0g -19550.0g  [1] 13329ppb  18265ppb 40.0Deg  37.8% 979.06mbar  1332210hm
Fri Jun 19 07:08:20 2020  29.1Deg 29.4Deg  63.4% 52.8%   2134ppm  588ppb  -1.0g -19550.0g  [1] 13329ppb  18265ppb 40.0Deg  37.8% 979.06mbar  1332210hm
Fri Jun 19 07:08:21 2020  29.1Deg 29.4Deg  63.4% 52.8%   2134ppm  588ppb  -1.0g -19550.0g  [1] 13329ppb  18265ppb 40.0Deg  37.8% 979.06mbar  1332210hm
Fri Jun 19 07:08:23 2020  29.1Deg 29.4Deg  63.4% 52.8%   2134ppm  588ppb  -1.0g -19550.0g  [1] 13329ppb  18265ppb 40.0Deg  37.8% 979.06mbar  1332210hm
Fri Jun 19 07:08:24 2020  29.1Deg 29.4Deg  63.4% 52.8%   2134ppm  588ppb  -1.0g -19550.0g  [1] 13329ppb  18265ppb 40.0Deg  37.8% 979.06mbar  1332210hm
Fri Jun 19 07:08:34 2020  29.2Deg 29.4Deg  63.6% 52.2%   2238ppm  588ppb  -1.0g -19548.0g  [1] 13312ppb  18254ppb 40.0Deg  37.9% 979.04mbar  1352570hm
Fri Jun 19 07:08:36 2020  29.2Deg 29.4Deg  63.6% 52.2%   2238ppm  588ppb  -1.0g -19548.0g  [1] 13312ppb  18254ppb 40.0Deg  37.9% 979.04mbar  1352570hm
Fri Jun 19 07:08:37 2020  29.2Deg 29.4Deg  63.6% 52.2%   2238ppm  588ppb  -1.0g -19548.0g  [1] 13312ppb  18254ppb 40.0Deg  37.9% 979.04mbar  1352570hm
Fri Jun 19 07:08:39 2020  29.2Deg 29.4Deg  63.6% 52.2%   2238ppm  588ppb  -1.0g -19548.0g  [1] 13312ppb  18254ppb 40.0Deg  37.9% 979.04mbar  1352570hm
```

Fig. 12.7. Excerpt of the **SmartComb** prototype no. 4 with current **BeE-Nose** module data from a campaign running in June 2020. As no scales are connected, these fields are void, here.

Additionally, the BME680 allows temperature modulated or cycled operation, i.e., the sensor can be heated to a sequence of temperatures and measurements can be conducted on reaching these temperature levels. This effectively gives a set of values like a sensor array, cf. e.g., [12], with fortunate information for pattern recognition in general. The degrees of freedom, e.g., number and levels of temperature steps, require knowledge of the measurand, which can be made available for known target gas, but might be hard to define for the yet unknown indicators of varroa infestation level or other bee diseases. Nevertheless provisions of extending the footprint by temperature modulation has been provided and data is acquired for a first-cut setting of number and levels of temperature steps (8 levels, from 50° C to 400° C in equidistant steps of 50° C). Further, the acoustic data, which might also contain cues for the varroa infestation level estimation, is both classified to assess the hive state [20], and optionally recorded for both future analysis on cues after knowing the varroa infestation level ground truth as well as refining acoustic hive state classification in general.

To provide the required ground truth with regard to current varroa infestation level conventional or advanced methods of analysis are required. A brief summary can be found in [20]. Complementing the automation options described in [20], a non-destructive method similar to flight hole bee inspection for '*on-board*' varroa was recently suggested based on an instrumented bee escape in [35]. The flotation method, for instance, was used in [2, 36] to provide required ground truth for the following evaluation and estimation steps.

The destructive nature of the flotation method, however, let us prefer in the work presented here the traditional slider or varroa board human-based analysis. The hive of Fig. 12.5 under investigation, was made subject to frequent varroa inspections for about two days duration each. The expectation of this approach to be detailed in the following section is, that the varroa countings and the corresponding varroa population in the hive increase in June and July and in proportion to this increase changes in the gas measurements, possibly also acoustic patterns, can be found to be employed for the design of a varroa infestation level estimator (VILE).

12.5. Experiments and Results

The **SmartComb** with the **BeE-Nose** extension (see Figs. 12.4 and 12.6) has been inserted in the hive shown in Fig. 12.5 in mid of May 2020. An earlier insertion would have been favorable, but circumstances due to unexpected lock-down did unfortunately not allow this. Continuous measurements with the system started in the last week of May and were accompanied by frequent conventional varroa counting or inspections of the varroa board.

Table 12.1 shows the results of the first sequence of inspection of the varroa board. The varroa board has been inserted in the hive currently in intervals of one week for the recommended two days. To prevent result falsification due to varroa removal by ants or other insects, openings between varroa board and hive were sealed or at least minimized and the recommended oiled tissue was used on the varroa board. The third column takes the achieved count from the second column, divided by the days of hive insertion, i.e., 2, and factorized by the rule-of-thumb value 300 to estimate varroa population in the hive.

As becomes clear from Table 12.1, the varroa infestation level for the regarded hive is on a level that is fortunate for the colony, however, not so fortunate for the objectives of this work. It also implies, that the late insertion has not been harmful for this hive. The monitoring will be extended into July and August until treatment becomes mandatory from the varroa count and the corresponding estimated VILE. With an onset of varroa number increase, which seems to announce itself with the last value from June 29th, the pollings will be increased, e.g., up to three inspections per week, which practically implies the insertion of a fresh varroa board, when the current one is extracted for counting.

Table 12.1. Varroa infestation level ground truth determined by traditional counting of varroa on slider after hive insertion for <u>two days</u>. To prevent result falsification due to varroa removal by ants or other insects, oiled tissue was used on the varroa board.

Date of extraction from hive	Counting result	Estimate of varroa infestation level
28.05.2020	0	0
08.06.2020	2	300
14.06.2020	0	0
21.06.2020	1	150
29.06.2020	4	600

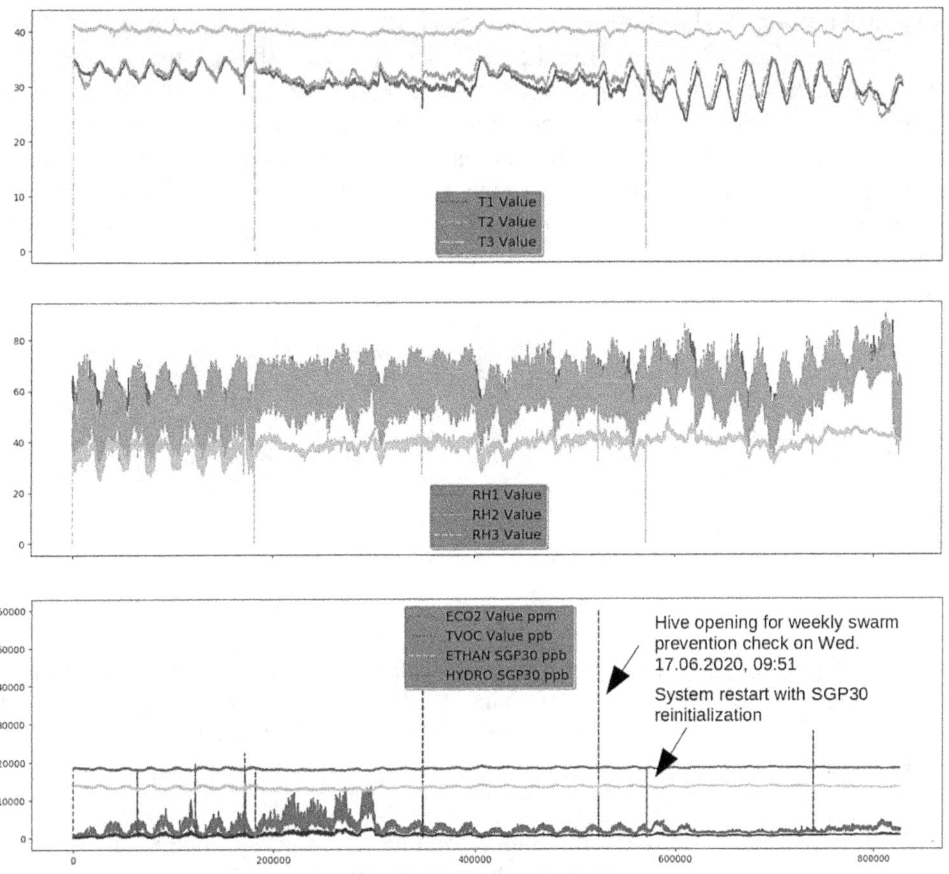

Fig. 12.8. Display of sensory registrations of the **SmartComb** prototype no.4 with current **BeE-Nose** module data from May 27 to June 29 2020, concurrent to the counting values of Table 12.1.

Reasons for the lag in the varroa development could be, that this hive was weakened due to a case of massive bee mortality on all the apiary, killing about 30 % of all populations, which judging from the observable symptoms, was due to poisoning, probably pesticide application at the end of May to early June. Breeding activity recently increased, which will also boost the varroa population.

In Fig. 12.8, the collected and concatenated measurements roughly overlapping in acquisition time from May 27 to June 29 with the discrete values from varroa board monitoring are given for the **SmartComb** sensors. A couple of extreme peaks or outliers can be observed. They are due to for instance to three (re)starts of the system due to technical issues, which implies a reinitialization of the SGP30 and the underlying algorithm for eCO_2 and VOC calculation from the raw data. The last restart is indicated in Fig. 12.8. Related data could be removed from the database without distortion of meaning. Other peaks, going to very high values in contrast to the previous ones going to zero, are due to hive opening with smoker application for the mandatory weekly swarm prevention check-ups. The peak related to the check of Wednesday, Jun 17, 09:51, has been indicated in Fig. 12.8. Left of it, the peak of Wednesday, Jun 10, 09:56, can be seen. There is one more less pronounced from Friday, June 26, 08:22, where swarm checks at the end of the swarm season have been done more superficially without taking the hive completely apart and checking all the combs for concealed swarm cells and with less smoke. Clearly, the check-up/opening times per hive should be noted and due to their expected influence on the measurement data be blocked or removed from the data analysis and ensuing VILE activity.

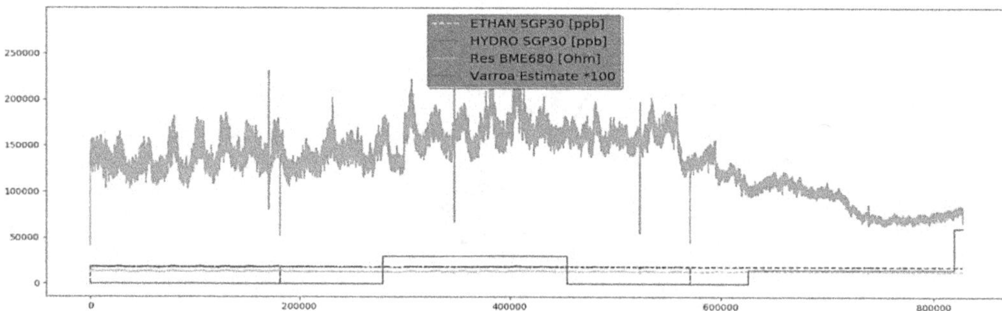

Fig. 12.9. Enhancement of sensory registrations of the **SmartComb** prototype no.4 with current **BeE-Nose** module data from May 27 to June 29 2020 with the BME680 resistance output complemented by the repetition of the SGP30 raw data and in red ink the varroa counting and estimates values of Table 12.1.

Fig. 12.9 gives the BME680 resistance output for the campaign in Fig. 12.8, complemented by the repetition of the SGP30 raw data and in red ink the varroa counting and corresponding model-based estimates of hive infestation level values of Table 12.1. There seems to be an effect on the BME680 with a trend to reduce its resistance towards the end of the campaign. However, whether this correlates with the announcing increase of the varroa population in Table 12.1, cannot be taken from the currently existing data without indulging in speculations. Data of the next two months is expected to provide the needed information, in particular, by extending the BME680 use to thermal modulation and virtual gas sensor array operation [12]. Further, after finding the aspired dependency,

458

to create a generic VILE and apply multiple instances, issues of gas sensor calibration have to be revisited.

In Fig. 12.10, the pursued information processing is outlined, that follows the standard idea of creating a virtual sensor for the aspired VILE as, e.g., employed in [37] for the localization task or in [40] for driver tiredness detection in vehicles. Radial-Basis-Function (RBF) and Support-Vector-Regression (SVM-R) networks as well as popular deep-learning--neural networks according to previous experience are the methods in the focus. In particular, for the RBF and SVM-R approach, the explicit computation of features will be helpful and we intend to reuse experience from prior work in gas sensing and automated design of intelligent sensor systems for that purpose [16-19]. In the first steps, measurement data from the regarded hive and **BeE-Nose** with SGP30 and BME680 will be employed together with the ground truth from varroa board based counting to achieve a mapping. This will serve to create estimates from live hive data. It is intended to continue varroa board based counting, so both the live VILE performance can be assessed and more extensive data analysis and improved VILE training can be carried out.

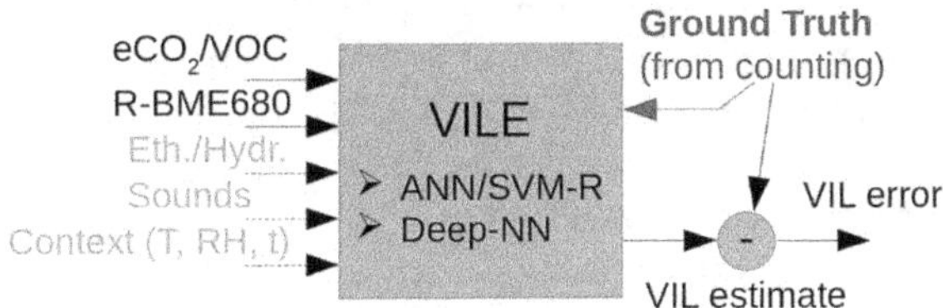

Fig. 12.10. Concept of creating a virtual sensor for the aspired **VILE** from **SmartComb** measurement data and varroa board counting.

Additionally, the raw data of the SGP30 as well as acoustic information from MEMS microphone readings and derived features are candidates to be included in the aspired VILE unit training. Further, available context information of the acquired data, e.g., in-hive temperature, outdoor temperature, time of day, or humidity readings might be useful in the learning process. The need for treatment, e.g., by formic acid, could at first-hand be derived from the achieved VILE index and the standards for thresholds on the varroa board counting procedure results.

12.6. Conclusions and Future Work

This chapter briefly summarized the important role of bees in our ecosystem and food chain, and the increasing amount of issues and detrimental influences that have to be mastered to keep bee colonies in a state to fulfil this role. The increasing burden for hive keepers is counteracted by numerous initiatives and projects, e.g., employing sensors, electronics, integration technologies and smart systems for bee hive condition monitoring. The **IndusBee 4.0** project is one of those activities, with a focus to minimize hive opening and resulting colony disturbance by both in-hive sensor systems and intelligent hive monitoring complemented by the introduction of hive keeper assistance systems, e.g., for automated varroa counting or queen localization, as outlined in Fig. 12.1. The key motivation comes from exploiting synergy from Industry 4.0 related activities and leading edge integration technologies for MEMS sensor and electronics and the bee keeping challenges and activities. The focus of this chapter was on the current extension and first application of the in-hive measuring system **SmartComb** to hive air analysis and related gas sensing, denoted as **BeE-Nose**. Results on hive air quality, e.g., in hive CO_2 levels, conform well to existing investigations, e.g., in [41].

In the first effort for VILE, a heuristic bottom-up approach was chosen, employing low-cost or at least affordable gas sensors, already in use in related indoor/outdoor air quality and gas sensing applications, with the hypothesis, that sufficiently robust indirect or direct indicators on the actual varroa infestation level can be found. A similar approach [2, 36] has been pursued with encouraging results with a set of Figaro 'analog' gas sensors, so that the chosen heuristic approach, possibly by extension with additional sensors from UST or other provider seems not to be too unrealistic. Nevertheless, a top-down approach based on the hive air analysis research, e.g., of Speer et al. [11] and cues from parasites metabolism, e.g., varroa excretions, could lead to more general and robust results, that in an ensuing step could then be melted down into an affordable integrated sensor system. For both bottom-up or top-down approach, major effort invested in the infrastructure and apiary will be needed and is reusable.

Unfortunately, the outlined work beyond air quality and regulation on VILE is still in the flux and only intermediate results for a single hive can be reported here, due to shortage and delayed or failed delivery of electronics and sensors as well as the unexpected lock-down in 2020. More conclusive results on the viability of the chosen bottom-up approach and the current sensor palette can be expected not before August or September. In immediate future work, the data acquisition will be continued and additional three hives will be instrumented to investigate, e.g., generalization of results between hives. Result dependent, the **BeE-Nose** will be extended on availability with additional gas sensors for the next bee season.

References

[1]. P. Arroyo, F. Meléndez, J. I. Suárez, J. L. Herrero, S. Rodríguez, J. Lozano, Electronic nose with digital gas sensors connected via Bluetooth to a smartphone for air quality measurements, *MDPI Sensors,* Vol. 20, Issue 3, 2020, 786.

[2]. A, Szczurek, M. Maciejewska, B. Bąk, J. Wilk, J. Wilde, M. Siuda, Detecting varroosis using a gas sensor system as a way to face the environmental threat, *Science of The Total Environment,* Vol. 722, 2020, 137866.

[3]. G. Chiesa, S. Cesari, M. Garcia, M. Issa, S. Li, Shuyang, Multisensor IoT platform for optimizing IAQ levels in buildings through a smart ventilation system, *MDPI Sustainability,* Vol. 11, Issue 20, 2019, 5777.

[4]. C. Jaeschke, O. Gonzalez, M. Padilla, K. Richardson, J. Glöckler, J. Mitrovics, B. Mizaikoff, A novel modular system for breath analysis using temperature modulated MOX sensors, *MDPI Proceedings,* Vol. 14, Issue 1, 2019, 49.

[5]. C. Jaeschke, O. Gonzalez, J. Glöckler, L. T. Hagemann, K. E. Richardson, F. Adrover, M. Padilla, J. Mitrovics, B. Mizaikoff, A novel modular Enose system based on commercial

MOX sensors to detect low concentrations of VOCs for breath gas analysis, *MDPI Proceedings,* Vol. 2, Issue 13, 2018, 993.

[6]. P. Souza, P. Marendy, K. Barbosa, S. Budi, P. Hirsch, N. Nikolic, T. Gunthorpe, G. Pessin, A. Davie, Low-cost electronic tagging system for bee monitoring, *Sensors (Basel),* Vol. 18, Issue 7, 2018, 2124.

[7]. Engadget, https://www.engadget.com/2015-08-25-australia-honey-bee-research-intel.html

[8]. ZDNet, https://www.zdnet.com/article/intel-and-csiro-create-rfid-bee-backpacks-with-edison/

[9]. H. Nguyen, F. Wang, R. Williams, U. Engelke, A. Kruger, P. Souza, Immersive Visual Analysis of Insect Flight Behaviour, https://pdfs.semanticscholar.org/68ed/1cffd34597ccf0587b574d1d0178f8dac0ed.pdf

[10]. P. Gomes, Y. Suhara, P. Nunes-Silva, L. Costa, H. Arruda, G. Venturieri, V. L. Imperatriz-Fonseca, A. Pentland, P. Souza, G. Pessin, An Amazon stingless bee foraging activity predicted using recurrent artificial neural networks and attribute selection, *Scientific Reports*, Vol. 10, Issue 1, 2020, 9.

[11]. K. Speer, TUD, Stockluftanalyse, https://www.chm.tu-dresden.de/lc3/dateien/stockluft.pdf

[12]. A. P. Lee, B. J. Reedy, Temperature modulation in semiconductor gas sensing, *Sensor. Actuat. B: Chem.,* Vol. 60, 1999, pp. 35-42.

[13]. S. Gil-Lebrero, F. Quiles Latorre, M. Ortiz, V. Sánche Ruiz, V. Gamiz, J. J. Luna-Rodriguez, Honey bee colonies remote monitoring system, *MDPI Sensors,* Vol. 17, Issue 1, 2017, 55.

[14]. S. Cecchi, S. Spinsante, A. Terenzi, S. Orcioni, A smart sensor-based measurement system for advanced bee hive monitoring, *MDPI Sensors,* Vol. 20, Issue 9, 2020, 2726.

[15]. M. Giammarini, E. Concettoni, C. C. Zazzarini, N. Orlandini, M. Albanesi, C. Cristalli, BeeHive Lab project – Sensorized hive for bee colonies life study,, in *Proceedings of the 12th International Workshop on Intelligent Solutions in Embedded Systems (WISES'15),* Ancona, Italy, 2015, pp. 121-126.

[16]. K. Iswandy, A. König, T. Fricke, M. Baumbach, A. Schütze, Towards automated configuration of multi-sensor systems using evolutionary computation – A method and a case study, *Journal of Computational and Theoretical Nanoscience*, Vol. 2, Issue 4, 2005, pp. 574-582.

[17]. K. Iswandy, A. König, A novel fully evolved kernel method for feature computation from multisensor signal using evolutionary algorithms, in *Proceedings of 7th International Conference on Hybrid Intelligent Systems (HIS'07),* Kaiserslautern, Germany, September 2007, pp. 123-127.

[18]. K. Iswandy, A. König, Fully evolved kernel method employing SVM assessment for feature computation from multisensor signals, *International Journal of Computational Intelligence and Applications*, Vol. 8, Issue 1, 2009, pp. 1-15.

[19]. K. Iswandy, A. König, Methodology, Algorithms, and emerging tool for automated design of intelligent integrated multi-sensor systems, *Algorithms (Open Access)*, Vol. 2, 2009, pp. 1368-1409.

[20]. A. König, IndusBee4.0 – Integrated intelligent sensory systems for advanced bee hive instrumentation and hive keeper's assistance systems, *Sensors & Transducers,* Vol. 237, Issues 9-10, 2019, pp. 109-121.

[21]. Sensirion SGP30, https://www.sensirion.com/fileadmin/user_upload/customers/sensirion/Dokumente/9_Gas_Sensors/Datasheets/Sensirion_Gas_Sensors_SGP30_Datasheet.pdf

[22]. M. Fleischer, M. Meixner, E. Simon, U. Lampe, R. Pohle, FET-Based Gas Sensor, U. S. Patent 7,707,869 B2, *USA*, 2010.

[23]. Figaro Gas Sensors, http://www.figarosensor.com/

[24]. Renesas Gas Sensors, https://www.idt.com/eu/en/document/dst/zmod4510-datasheet

[25]. Bosch Sensortec Gas Sensors, https://www.bosch-sensortec.com/media/boschsensortec/downloads/datasheets/bst-bme680-ds001.pdf

[26]. SGX Sensortech, https://www.sgxsensortech.com/content/uploads/2015/02/1143_Datasheet-MiCS-6814-rev-8.pdf

[27]. The MiCS-4514, Data Sheet, https://www.mouser.de/datasheet/2/18/0278_Datasheet-MiCS-4514-rev-16-1144833.pdf

[28]. AMS CCS811 Gas Sensor, Data Sheet, https://cdn.sparkfun.com/assets/2/4/2/9/6/CCS811_Datasheet.pdf

[29]. CCS801, Ultra-Low Power Analog VOC Sensor for Indoor Air Quality Monitoring, https://www.mouser.de/datasheet/2/588/CCS801_DS000457_2-00-1140362.pdf

[30]. UST Gas Sensors, http://www.umweltsensortechnik.de/en/gas-sensors/mox-gas-sensors-overview.html

[31]. UST VOC/CO$_2$ Sensor, http://www.umweltsensortechnik.de/fileadmin/assets/downloads/gassensoren/module/Datasheet_VOC-CO2-Sensor_Rev1700.pdf

[32]. Food Sniffer, http://www.myfoodsniffer.com/

[33]. MG811 CO$_2$ Sensor, https://datasheetspdf.com/pdf-file/576123/ETC/MG811/1

[34]. A. Popa, M. Hnatiuc, M. Paun, O. Geman, D. J. Hemanth, D. Dorcea, L. H. Son, S. Ghita, An intelligent IoT-based food quality monitoring approach using low-cost sensors, *Symmetry*, Vol. 11, Issue 3, 2019, 374.

[35]. A. König, Integrated intelligent sensor systems for in-hive Varroa infestation control in digital bee keeping, in *Proceedings of the Conference Sensors and Measurement Science International (SMSI'20)*, Nuremberg, Germany, 22-25 June 2020, pp. 111-112.

[36]. A. Szczurek, M. Maciejewska, B. Bąk, J. Wilk, J. Wilde, M. Siuda, Detection level of honeybee disease: Varroosis using a gas sensor array, in *Proceedings of the 5th International Conference on Sensors and Electronic Instrumentation Advances (SEIA'19)*, Tenerife, Spain, 25-27 September 2019, pp. 255-256.

[37]. D. Groben, K. Thongpull, A. C. Kammara, A. König, Neural virtual sensors for adaptive magnetic localization of autonomous dataloggers, *Advances in Artificial Neural Systems*, Vol. 2014, 2014, 394038.

[38]. H. Stever, J. Kuhn, C. Otten, B. Wunder, W. Harst, Verhaltensänderung unter elektromagnetischer Exposition, Pilotstudie 2005, *Universität Koblenz-Landau*, AGBI, https://www.buergerwelle.de/assets/files/bienenstudie.pdf?cultureKey=&q=pdf/bienenstudie.pdf, http://www.bienenarchiv.de/forschung/2006/elmagexp_bienen_06.pdf

[39]. S. Kimmel, J. Kuhn, W. Harst, H. Stever, Electromagnetic radiation: Influences on honeybees (Apis mellifera), in *Proceedings of the 19th International Conference on Systems Research, Informatics and Cybernetics (InterSymp'07)*, Baden-Baden, Deutschland, 2007, pp. 1-6.

[40]. L. Li, K. Thongpull, A. König, Optimizing the design of a multi-sensor system for on-line driver state and drowsiness detection, in *Proceedings of the Tagungsband des XXVIII Messtechnisches Symposium des Arbeitskreises der Hochschullehrer für Messtechnik e.V. (AHMT'14)*, Saarbrücken, Sept. 18-20, 2014, pp. 123-127.

[41]. M. Ohashi, R. Okada, T. Kimura, H. Ikeno, Observation system for the control of the hive environment by the honeybee (Apis mellifera), *Behavior Research Methods*, Vol. 41, Issue 3, 2009, pp. 782-786.

[42]. MQ2 and Other MQ Gas Sensors, https://components101.com/mq2-gas-sensor

Index

generalized receiver, 63, 71
Genetic algorithm, 214
geometric active contour model, 376
Gibbs sampling technique, 187
global
    maximum, 180
    minimum, 177
Glossokinetic potential, 414
GLRT-based approach, *235*
Gold sequences, 78
Granger causality, 388, 391-393, 396-399,
    401, 402, 404
Gray coded, 68
gray level histogram, 372
    frequencies, 372

H

Hampel Filter, 434
Harmonic
    chirp model, 156
Havrda-Charvat entropy, 370
Hidden Markov Model, 389
Hilbert operator, 25, 26
    1st–order partial, 26
    2nd–order partial, 26
    3rd – order partial, 26
Hilbert transformation, 152
Hilbert-Huang Transform, 278, 279, 285
hive air analysis, 443, 444, 458
Hjorth Parameter, 275, 276, 285
HMM, 389, 393-397, 404
honey bees, 441
Hurst Exponent, 281, 286
Hybrid Selection, 287, 289
Hypercomplex (Cayley-Dickson) Fourier
    spectrum. *see n-D hypercomplex Fourier
    transform*
Hypercomplex signal, 35
hypothesis, 64

I

i.i.d., 163
identity matrix, 88
Image
    acquisition, 311
    annotation, 312
    edges, 376
    processing, 157, 342, 367

Segmentation, 376
Incipient faults, 352
independent, 379
Independent
    component analysis, 429
    vector analysis, 431
information criterion
    Akaike, 172
    Bayesian, 173
    Kundu, 173
information entropy, 367
    probability distributions, 367
    random variable, 367
information theoretic criteria, 172
intensity
    changes, 376
    differences, 376
    inhomogeneity, 376
interval
    confidence, 167, 174
IQML, 166
iterative
    algorithm, 168
    Ikarm, Abed-Meraim and Hua, 186
    procedure, 166
iterative quadratic maximum likelihood,
    166

J

Jeffrey's prior, 174
joint probability, 368

K

Kapur entropy, 370, 378
KNN, 290, 293-296, 298
Kronecker product, 57
Kurtosis, 275, 285, 294

L

Laplace
    approximation, 174
    density function, 189
    distribution, 189
linear
    mixture model, *233*
    unmixing, 233